Benthic foraminifera from the classic southern Caribbean region are presented in this book, to provide useful information on ranges for biostratigraphers working in the region and beyond. Around 1000 of the more important species are assembled, from the Barremian (Early Cretaceous) to the Middle Miocene, approximately 120 to 10 million years before present. The deeper-water benthic species are tied in to the zonal scheme used in the earlier book, *Plankton Stratigraphy*, published by Cambridge University Press in 1985.

The taxa have been brought up to date generically, and in many cases new comparisons between species have been made; the Late Cretaceous and Early Paleogene are particularly detailed. This information, together with illustrations, will enable the taxa to be used stratigraphically.

**Benthic foraminiferal biostratigraphy
of the south Caribbean region**

Benthic foraminiferal biostratigraphy of the south Caribbean region

HANS M. BOLLI

Paleontological Institute
University of Zurich, Zurich, Switzerland

JEAN-PIERRE BECKMANN

Geological Institute
Federal School of Technology, Zurich, Switzerland

JOHN B. SAUNDERS

Natural History Museum
Basel, Switzerland

CAMBRIDGE
UNIVERSITY PRESS

CAMBRIDGE UNIVERSITY PRESS
Cambridge, New York, Melbourne, Madrid, Cape Town, Singapore, São Paulo

Cambridge University Press
The Edinburgh Building, Cambridge CB2 2RU, UK

Published in the United States of America by Cambridge University Press, New York

www.cambridge.org
Information on this title: www.cambridge.org/9780521415217

First published 1994
This digitally printed first paperback version 2005

A catalogue record for this publication is available from the British Library

Library of Congress Cataloguing in Publication data
Bolli, Hans M.
Benthic foraminiferal biostratigraphy of the South Caribbean
Region / Hans M. Bolli, Jean-Pierre Beckmann, John B.
Saunders.
        p.   cm.
Includes bibliographical references (p.   ) and index.
ISBN 0 521 41521 7 (hc)
1. Foraminifera, Fossil – Caribbean Area.   2. Paleontology,
Stratigraphic.   3. Paleontology – Caribbean Area.   I. Beckmann,
Jean-Pierre.   II. Saunders, John B.   III. Title.
QE772.B577   1994
563′.12′09729 – dc 20   93-29909   CIP

ISBN-13 978-0-521-41521-7 hardback
ISBN-10 0-521-41521-7 hardback

ISBN-13 978-0-521-02253-8 paperback
ISBN-10 0-521-02253-3 paperback

Contents

Illustrations

Introduction

Benthic foraminifera have been used for stratigraphic purposes almost since they began to be studied systematically. Already around the middle of the nineteenth Century, A. E. Reuss realised the stratigraphic value of benthic foraminifera. However, the real breakthrough in the application of foraminifera to biostratigraphy – at the time almost exclusively benthic foraminifera – began with the requirements of the oil industry for dating well sections and comparing them with the strata seen on surface. One of the first areas to use the tool was the Gulf Coast region of the United States and Mexico, above all by the great pioneer in promoting the stratigraphic application of foraminifera, Joseph A. Cushman. His textbook and that by J. J. Galloway were the first modern comprehensive references on the subject. Other significant contributors in this area were Helen J. Plummer, M. P. White, W. L. F. Nuttall and R. W. Barker. The use of benthic foraminifera in stratigraphy spread rapidly, mainly through Cushman and his co-authors, to the Caribbean (mainly Trinidad), then to California (R. M. Kleinpell, V. S. Mallory) and Venezuela (A. Senn). By the middle nineteen-thirties, foraminifera – still almost exclusively benthic forms – were the major microfossil group used worldwide in the micropaleontological laboratories of the oil industry.

Cretaceous and Tertiary planktic foraminifera, that had previously been largely ignored, gradually began to attract more attention from about the mid-nineteen-thirties but this was mainly in sediments where benthic forms were poorly represented. Early studies included those on the Globotruncanids of the Alpine-Mediterranean area followed by investigations on the Tertiary planktics, first in the North Caucasus region but then mainly in Trinidad.

These studies led to the recognition of a great number of stratigraphically restricted taxa which, in turn, formed the base for a detailed planktic zonal subdivision. Due to their floating and drifting lifestyle, planktic foraminifera are widespread in oceanic waters, and consequently zones based on them are applicable over wide distances. Planktic foraminiferal zonal schemes thus became more and more widely used in the dating of open marine rock sequences, replacing earlier attempts using benthic foraminifera. Cretaceous and Cenozoic zonal schemes based on marine planktic micro-organisms are today well established on a worldwide basis. Many of them are summarized in *Plankton Stratigraphy* (1985, edited by Bolli, Saunders and Perch-Nielsen).

We know that the stratigraphic resolution based on benthic forms is less detailed than that based on planktics, and that their distribution, both geographical and bathymetric, is environmentally controlled, although less so in deeper than in shallower waters. In fact, both planktics and benthics suffer from some aspects of provincialism and the former show considerable latitudinal control as well as a connection to major water masses in the oceans. On the other hand, more recent published data on the benthic taxa have shown that, under suitable conditions, some Cretaceous and Tertiary species have a wide geographic distribution. This is especially so for those that inhabited the deep sea floor.

Following the boom in the application of planktic foraminifera for stratigraphy in the nineteen-fifties which led to today's well-established zonal schemes, interest in benthics has grown again. This is expressed

in the 'Benthos' series of conferences of Halifax (1975), Pau (1983), Geneva (1986) and Sendai (1990) that followed the 'Plankton' conferences in Geneva (1967), Rome (1970) and Kiel (1974).

Studies correlating the stratigraphic distribution of deep-water benthic species with planktic zonal schemes have been growing particularly in the last two decades due to the availability of deep-sea cores obtained by ocean drilling (Deep Sea Drilling Project, DSDP and Ocean Drilling Program, ODP). The range of this material covers the complete Cretaceous and Cenozoic. To begin with, the studies were largely taxonomic but notable exceptions include papers by Douglas (1973), Douglas and Woodruff (1981) and Morkhoven *et al.* (1986).

The reason for preparing this volume is to follow this trend and to review and expand our knowledge of stratigraphically significant Cretaceous and Cenozoic deep-water benthic foraminifera from outer shelf and bathyal sediments in an area where their ranges have been tied to planktic zonal schemes. The region chosen, which is suitable for such a purpose, is the southern Caribbean comprising Trinidad, Venezuela and Barbados (Fig. 1). Here we find stratigraphically nearly continuous deep-water marine sequences rang-

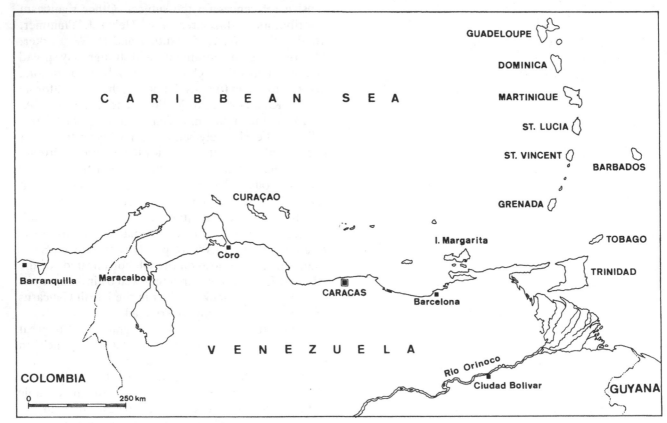

Fig. 1. Map of the southern Caribbean region.

ing from Barremian to Miocene, mostly rich in both planktic and benthic foraminifera. It is also the area where a substantial part of the low-latitude planktic foraminiferal zonal scheme in use today has been developed.

The early studies undertaken by micropaleontologists in the region concerned benthic foraminifera. Attention was first focussed on sediments and faunas of Tertiary (Paleocene to Middle Miocene) age, later also on the Early Cretaceous. The record published so far is incomplete, with a gap particularly in the Late Cretaceous, and the parts issued until about 1950 are in need of taxonomic and stratigraphic revision. Beside all this published work, a great wealth of additional information has been accumulated and used in commercial laboratories supported by oil companies. These unpublished company reports contain enormous quantities of taxonomic and stratigraphic data, some of which were repeatedly revised and re-evaluated. An extensive treatment of all important facts stored in these reports would have far exceeded the scope of a single book. Nevertheless, an effort has been made to incorporate from this source a significant amount of important new data and conclusions. This concerns not only the hitherto neglected Late Cretaceous time interval in Trinidad but also parts of the Cretaceous and Tertiary of Venezuela and the Tertiary of Trinidad and Barbados. Altogether, the faunal record in the present volume consists of about 1000 of a total of some 2500 benthic species known from the area. The approach to the faunas of the southern Caribbean area has always been mainly taxonomic and stratigraphic. The paleoecological aspect has been discussed in a few papers, but thorough evaluations of complete populations using up-to-date data and methods are not available. However, some interpretations of the Late Cretaceous and basal Tertiary faunas of Trinidad are included in the present volume.

The main geographic regions of Trinidad, Venezuela and Barbados provide a frame for the threefold subdivision of this book. Each of the three parts consists of a general discussion with maps and charts, followed by one or more annotated taxonomic lists with illustrations. The taxonomic lists are usually attached to the respective chapters; an exception is made for the late Early Eocene to Middle Miocene, where the Trinidad, Venezuela and Barbados faunas are combined in a single list. By this procedure, frequent repetitions or complicated cross-references could be reduced to a minimum. Normally, the annotated taxonomic lists include a selection of morphologically distinct and/or stratigraphically significant

foraminiferal taxa. In the Late Albian to earliest Cenozoic interval of Trinidad, where the faunas were up to now practically unknown, they are treated in greater detail. The taxonomic lists include, where necessary, revisions of species definitions and, as a rule, are accompanied by illustrations of foraminifera, either original or reproduced from other publications. Everywhere, an effort has been made to indicate the stratigraphic distribution of each species, either in the text or, in many cases, on stratigraphic distribution charts. The stratigraphic ranges are all calibrated with the current planktic foraminiferal and nannofossil zonal schemes.

Consequently, this study of a large number of Early Cretaceous to Middle Miocene deep-water benthic foraminifera correlated with planktic zonal schemes will, we hope, be a useful contribution towards a more accurate application of this group to stratigraphy, particularly in cases where planktic microfossils may be rare or absent. One difficulty associated with the preparation of the present book has been that many groups of benthic foraminifera badly need taxonomic revision. Such time-consuming work goes increasingly slowly as more and more micropaleontologists are concentrating on different aspects of the subject. For the three authors of this book, this has been an opportunity to prevent the loss of stratigraphic and taxonomic information accumulated over the years in the region. Variations in style throughout the book are unavoidable, particularly as the work of one author (J.P.B.) reflects a monographic study of the Late Cretaceous to Early Paleogene that he has been engaged upon for a number of years. We realise that this presentation is by no means exhaustive and calls for additional work to be done. Apart from the southern Caribbean, there are other areas where similar compilations could be carried out, such as the Gulf Coast (United States and Mexico), Northwest Europe, the Alpine–Mediterranean region, Indonesia and oceanic regions from where abundant Deep Sea Drilling/Ocean Drilling data are available.

Acknowledgments, authorship, depository

The authors are indebted to the Paleontological Institute of the University of Zürich, the Geological Institute of the Federal School of Technology in Zürich (ETH) and the Natural History Museum, Basel, for making available the facilities needed for the preparation of text and illustrations. Christoph Kopps gave great assistance in the design and computer production of many charts. Photographs of taxa were most professionally prepared by Urs Gerber. The authors would also like to thank Maria Lourdes Diaz de Gamero and Max Furrer (both Caracas) for kindly reading the parts of the manuscript dealing with Venezuela and for making valuable suggestions for their improvement. They also wish to thank Amoco Production Co., Houston, for their support towards the preparation of illustrations.

The individual authors are largely responsible for the major chapters as follows:

H. M. Bolli for the Barremian to Early Albian of Trinidad

J. P. Beckmann for the Late Albian to Early Eocene of Trinidad

H. M. Bolli and J. B. Saunders for the late Early Eocene to Middle Miocene of Trinidad

H. M. Bolli for the chapter on Venezuela

J. B. Saunders and J. P. Beckmann for the chapter on Barbados

H. M. Bolli and J. P. Beckmann for the annotated taxonomic lists of late Early Eocene to Middle Miocene benthic foraminifera of Trinidad, Venezuela (Falcon) and Barbados.

Specimens of the species from Trinidad, Venezuela and Barbados discussed in the book are deposited at the Natural History Museum, Basel, Switzerland in the following collections:

Barremian to Early Albian of Trinidad: Collection Bartenstein, Bettenstaedt and Bolli

Late Albian to Early Eocene of Trinidad: Collection Beckmann

Late Early Eocene to Middle Miocene of Trinidad: Collections Bolli, Saunders

Early Cretaceous of Eastern Venezuela: Collection Guillaume, Bolli and Beckmann

Oligocene-Miocene of Falcon (Venezuela): Collections H.H. Renz, Bolli, Diaz de Gamero

Eocene-Oligocene of Barbados: Collections Senn, Beckmann, Saunders.

Publications in which the distribution of Cenozoic benthic foraminifera is correlated with planktic foraminiferal zonal schemes

Years before the renewed interest in benthic foraminifera, Blow (1959), in his publication 'Age, correlation, and biostratigraphy of the Upper Tucuyo (San Lorenzo) and Pozon formations, Eastern Falcon, Venezuela' attempted a correlation in the Miocene Agua Salada Formation of the previously established benthic zones (Renz 1948) with the planktic foraminiferal zones proposed from Trinidad. The presence of a mixed benthic and planktic foraminiferal fauna made this possible.

Diaz de Gamero (1977*a*) in a paper entitled 'Estratigrafia y micropaleontologia del Oligoceno y Mioceno Inferior del centro de la Cuenca de Falcon, Venezuela' studied the occurrence and stratigraphic distribution of Oligocene and Early Miocene planktic and benthic foraminifera in several sections and tied the ranges of the benthic forms into the existing planktic foraminiferal zonal scheme.

Based on these papers we possess from the Falcon area a good documentation of a large number of Oligocene and Miocene benthic species and their ranges compared with the planktic foraminiferal zonal scheme. Based on a re-evaluation of the distribution of the benthic foraminifera as presented in Renz (1948) a selection of stratigraphically and morphologically significant taxa has been made for the present volume.

A first attempt to tie the distribution of Paleocene-Eocene benthic foraminifera within the Alpine–Mediterranean area into the planktic foraminiferal zonal scheme was made in the Possagno section in northern Italy (Monografia micropaleontologica sul Paleocene e l'Eocene di Possagno, Provincia di Treviso, Italia; H. M. Bolli editor, 1975). F. Proto Decima and R. de Biase determined the ranges of 69 benthic species between the Paleocene *Morozovella pseudobulloides* Zone and the Middle Eocene *Truncorotaloides rohri* Zone. In the same way G. Braga and A. Gruenig covered 116 benthic species between the Late Eocene *Globorotalia cerroazulensis cerroazulensis* and the *Globorotalia cerroazulensis cunialensis* zones. Here the composition of benthic associations becomes influenced by progressive shallowing, with larger foraminifera like nummulites and discocyclinids appearing in the uppermost Eocene *Globorotalia cerroazulensis cunialensis* Zone.

Also in the southern Alps of Italy, West of the Possagno section, J. P. Beckmann *et al.* (1982) published the monograph 'Micropaleontology and biostratigraphy of the Campanian to Paleocene of the Monte Giglio, Bergamo Province, Italy'. The work covers the occurrences and stratigraphic distributions of calcareous nannofossils and planktic and deep-water benthic foraminifera and in redeposited approximately age-equivalent layers, also shallow-water organisms such as calcareous algae and larger foraminifera. The benthic deep-water foraminifera and the re-deposited organisms are illustrated. In a chapter 'Campanian–Maastrichtian benthic foraminifera' J. P. Beckmann describes 52 benthic taxa in an annotated species list which also shows the occurrence of each species within the *Globotruncana calcarata* and the *Abathomphalus mayaroensis* zones. Similarly, within the same monograph, F. Proto Decima & H. M. Bolli in 'Paleocene smaller benthic foraminifera' recognize 78 species half of which are agglutinated . Their occurrences between the *Morozovella angulata* and the *Planorotalites pseudomenardii* zones are listed.

In 1978, Proto Decima & Bolli investigated the benthic foraminifera of Paleocene to Oligocene age of DSDP Leg 40 sites in the South Atlantic Cape Basin, Walvis Ridge and Angola Basin, and correlated their ranges with the planktic foraminiferal zonal scheme. Their range charts clearly show the restricted stratigraphic distribution of many of the 125 taxa studied.

Beckmann *et al.* (1981) published a paper entitled 'Major calcareous nannofossil and foraminiferal events between the Middle Eocene and the Early Miocene'. Here J. B. Saunders demonstrates the stratigraphic significance of selected Middle Eocene to Middle Miocene benthic smaller foraminifera from Trinidad in respect to the planktic foraminiferal zonal scheme erected from the same formations. The benthic taxa were selected based on their first or last occurrence in relation to the planktic zonal scheme of this stratigraphic interval. It is pointed out that environmental conditions vary from Middle Eocene to Middle Miocene. The fauna of the Middle Eocene is lower bathyal with a shallowing trend setting in towards the Late Eocene causing the deposition of reefal limestones and glauconitic sands. Deepening took place again in the Oligocene reaching upper to middle bathyal depths. In the same paper F. Proto Decima and H. M. Bolli compare stratigraphic ranges of DSDP Leg 40 South Atlantic Paleocene to Oligocene benthic species between the Cape Basin and the Walvis Ridge areas in relation to the planktic foraminiferal zonal scheme. Differences in ranges of numerous species are attributed to different environmental conditions in the two geographically close areas. The same authors also discuss stratigraphic and environmental comparisons of South Atlantic Leg 40 benthic taxa with those of corresponding ages in Trinidad and the Possagno section in northern Italy.

From the central Pacific area Douglas (1973) correlates ranges of a considerable number of Paleocene to Pleistocene benthic forms of DSDP sites 167 and 171 with the P and N planktic foraminiferal zonal scheme. The taxa are illustrated on 25 plates, partly by scanning electron micrographs, partly by line drawings.

A more recent contribution that also focusses on the stratigraphic significance of benthic foraminifera is that by Morkhoven, Berggren & Edwards (1986). In this work 126 predominantly Cenozoic benthic taxa are studied for their taxonomy, biostratigraphy and paleobathymetry. Of particular interest is the listing of observed occurrences which in many of the species dealt with indicate wide geographic distribution. As an example may serve the Late Cretaceous to Eocene *Nuttallides truempyi* which has been reported from the Gulf Coast, Caribbean, Atlantic, Europe, North Africa, former USSR, Indian Ocean, New Zealand, Pacific and California, making it a truly cosmopolitan species. Ranges of the 126 taxa discussed are correlated with the P and N planktic foraminiferal zonal scheme.The authors do not state by what criteria they were guided in the selection of the taxa. The 126 species and forms in open nomenclature (7) belong to 41 genera. About half (64) of the taxa dealt with belong to the four genera *Cibicidoides, Bulimina, Planulina* and *Uvigerina*. The selected taxa and their stratigraphic distributions are very similar to those previously published by Morkhoven (1981) where he included 120 species.

Views of foraminiferal workers concerning the use of deep-water benthic foraminifera are divided even today. Some of them are expressed in the preface of Morkhoven, Berggren & Edwards (1986). Included are the views of Boltovskoy (1965, 1978) which are sceptical because of the huge number of taxa, many of them invalid (synonyms), and slow evolution resulting in long ranges. Douglas (1973) on the other hand is more positive in saying: 'Many of these bathyal species have wide geographic distribution, particularly in the Cretaceous and Paleogene, and limited stratigraphic ranges. For these reasons they offer an excellent basis for stratigraphic correlation, especially to supplement planktonic biostratigraphy'.

In another paper Douglas & Woodruff (1981) discuss the possibilities but also the limitations in applying benthic foraminifera to stratigraphy. The following passages are taken from their paper:

> Many deep-sea benthic foraminifera possess the characteristics of good stratigraphic index fossils: a distinct and easily recognized morphology, a wide distribution in the ocean, and fairly short geologic ranges. Thus deep-sea benthic foraminifera should be useful in stratigraphic correlation and age determination, especially when pelagic microfossils, the usual basis for correlation in oceanic sediments, are poorly preserved or absent.
> The biostratigraphy of benthic foraminifera from

AGE	STAGES	PLANKTIC FORAMINIFERAL ZONES (from Caron 1985)	CALCAREOUS NANNOFOSSIL ZONES (from Sissingh 1977)	TRINIDAD PLANKTIC FORAM. ZONES (Saunders & Bolli 1985)	TRINIDAD FORMATIONS	EASTERN VENEZUELA PLANKTIC / BENTHIC FORAM. ZONES (Guillaume et al. 1972)	EASTERN VENEZUELA FORMATIONS / MEMBERS**	MARACABO BASIN FORMATIONS
LATE CRETACEOUS	MAASTRICHTIAN	Abathomphalus mayaroensis	Nephrolithus frequens	Abathomphalus mayaroensis	Guayaguayare		Vidoño	Mito Juan
		Gansserina gansseri	Arkhangelskiella cymbiformis	Globotruncana gansseri	Guayaguayare		San Juan	Colon
		Globotruncana aegyptiaca	Reinhardtites levis	Globotruncana lapparenti tricarinata				
		Globotruncanella havanensis	Tranolithus phacelosus		not recorded			
	CAMPANIAN	Globotruncana calcarata	Quadrum trifidum				San Antonio	Socuy
			Quadrum sissinghii					
		Globotruncana ventricosa	Ceratolithoides aculeus	Globotruncana stuarti				
			Calculites ovalis					
		Globotruncanita elevata	Apsidolithus parcus					
			Calculites obscurus					
	SANTONIAN	Dicarinella asymetrica	Lucianorhabdus cayeuxii	Globotruncana fornicata	Naparima Hill			
			Reinhardtites anthophorus					
		Dicarinella concavata	Micula decussata	Globotruncana concavata				
	CONIACIAN		Marthasterites furcatus					La Luna
		Dicarinella primitiva	Lucianorhabdus maleformis	Globotruncana renzi				
	TURONIAN	Marginotruncana sigali	Quadrum gartneri	Globotruncana inornata			Querecual	
		Helvetoglobotruncana helvetica						
		Whiteinella archeocretacea			not recorded			
	CENOMANIAN	Rotalipora cushmani	Microrhabdulus decoratus					
		Rotalipora reicheli						
		Rotalipora brotzeni	Eiffellithus turriseiffelii	Rotalipora app. appenninica	Gautier	Rotalipora appenninica / Rotalipora ticinensis	Chimana / El Cantil	Maraca
		Rotalipora appenninica		Favusella washitensis				Lisure
		Rotalipora ticinensis		Rotalipora tic. ticinensis				
EARLY CRETACEOUS	ALBIAN	Rotalipora subticinensis	Prediscosphaera columnata		not recorded	Neobulimina primitiva		
		Biticinella breggiensis				Praeglobotruncana primitiva		
		Ticinella primula				Neobulimina subcretacea	Borracha	
		Ticinella bejaouensis	Chiastozygus litterarius	Praeglobotruncana rohri	Maridale	Praeglobotruncana rohri	Valle Grande	
		Hedbergella gorbachikae		Planomalina maridalensis		Praeglobotruncana infracretacea	Garcia	
		Globigerinelloides algeriana				Biglobigerinella barri		Apon / Cogollo
	APTIAN	Schackoina cabri		Leupoldina protuberans	Cuche	Biglobigerinella cf. barri	Taguarumo	
		Globigerinelloides blowi		Lenticulina ouach. ouachensis		Choffatella decipiens	Picuda / Morro Blanco / Barranquin	
	BARREMIAN	Hedbergella sigali	Micrantholithus hoschulzii	Lenticulina barri	Toco		Venados	Rio Negro
			Lithraphidites bollii					

Fig. 2. Correlation of Cretaceous formations of Trinidad and Venezuela with planktic foraminiferal and calcareous nannofossil zones (zonal correlation after Bolli *et al.* 1985). * Ages adjusted; ** Early Cretaceous: from Guillaume, Bolli & Beckmann (1972), Late Cretaceous from Gonzales de Juana *et al.* (1980).

the deep sea at the present time is underdeveloped, and there are differing opinions regarding the potential usefulness of these microfossils. Boltovskoy (1978) argues that deep-sea species are of limited stratigraphic value compared to planktic foraminifera, because of a variety of taxonomic problems and the long geologic ranges of species. His views are based on a study of Neogene species from the Indian Ocean. As we have discussed above, the Neogene, especially since the Middle Miocene, has been a time of little evolutionary change in deep-sea faunas. For this time period, benthic foraminifera, compared to planktic foraminifera, offer only broad stratigraphic resolution. However, earlier Tertiary faunas are different, and in the Paleocene, Eocene, and Oligocene the first and last appearance and ranges of selected species offer considerable potential for biostratigraphic zonation (Berggren 1977). Besides the preservational advantage offered by many benthic species, their stratigraphic ranges can be directly calibrated in terms of planktic microfossil zones, magnetic stratigraphy, and, indirectly, absolute time.

We agree with Boltovskoy (1978) that several important problems limit the stratigraphic usefulness of deep-sea benthic foraminifera at least at the present. Foremost of these problems is the early stage of development of taxonomic descriptions of benthic species. As noted before this problem is compounded by the range of morphologic variation expressed by many species, the large number of species present in the deep ocean, and the few monographic descriptions of pre-Recent faunas. Dissolution and diagenesis eliminate benthic species, especially agglutinated taxa, and affect apparent stratigraphic occurrences of benthic foraminifera. The relative preservability of different benthic species is generally unknown but must be determined in order to assess the stratigraphic implications of species absence. Benthic foraminifera tend to be distributed by depth and their distributions change with fluctuations in the deep-sea environment. The horizontal and vertical migrations of species modify their occurrence and affect their use for correlation and age determination. Most of these problems are related to our ignorance and the current state of knowledge and will diminish with further investigations. Possibly the greatest drawback to deep-sea benthic foraminifera as stratigraphic indicators is the long geologic range, on the average, of benthic species compared to planktic species (Berggren, 1977). However, this may be offset by the much greater diversity of benthic faunas.

The study of benthic Cenozoic foraminifera from Ecuador by Whittaker (1988) is another example of the correlation of ranges of benthic taxa with planktic zonal schemes. One hundred and thirty two Late Oligocene to Late Pliocene taxa are described and excellently illustrated. Special attention is given to the Buliminida and Rotalida, two groups that also get preferential treatment in the Middle Eocene to Middle Miocene part of the present volume. Whittaker studied the benthic foraminifera in several basins along the North–South running coastal area and tied their distribution for each basin into planktic foraminiferal and nannoplankton zonal schemes. He found that the fauna in the Borbon and Manabi basins to the north is rich and diversified in contrast to the impoverished one in the Progreso and Manglaralto basins immediately to the south. The difference reflects much less favourable oceanic conditions for benthic foraminifera towards the south. The Chongon-Colonche Hills dividing these two areas are regarded as having acted as a high already at the time of deposition of the investigated faunas. This shows how within a short distance of tens of kilometres benthic foraminiferal associations can change drastically.

Correlation of zones and formations in the areas under discussion

Figure 2 shows the correlation of the Cretaceous Barremian to Maastrichtian formations of Trinidad, eastern Venezuela and the Maracaibo Basin with the standard low latitude planktic foraminiferal and nannofossil zonal schemes. Added on Fig. 2 is the correlation with the local, predominantly planktic, zonal schemes as applied in Trinidad (see also Fig. 5, 12) and eastern Venezuela (see also Fig. 67). For the local foraminiferal zonations used to subdivide the Cretaceous of the Maracaibo Basin and its correlation with the standard planktic foraminiferal zonal scheme reference is made to Fig. 83 and 84.

Figure 3 shows the correlation for the Paleocene to Miocene formations recognized in Trinidad, Venezuela (Falcon Basin and Maracaibo Basin) and Barbados with the standard low latitude planktic foraminiferal and nannofossil zonal schemes. Locally developed benthic zonations for the Miocene of the eastern Falcon Sub-basin (Venezuela) are correlated with the standard zonal schemes on Fig. 71. Figure 72 contains the information on the correlation of the Late Eocene to Late Miocene formations recognized in southeastern, northeastern, central and north-central Falcon with the standard planktic foraminiferal scheme. The local benthic zonation applied to the Tertiary of the Maracaibo Basin and its correlation with the formations and the standard planktic foraminiferal zonal scheme is shown on Fig. 83.

AGE	PLANKTIC FORAMINIFERAL ZONES (from Bolli et al 1985)		CALCAREOUS NANNOFOSSIL ZONES (from Bolli et al 1985)	
MIOCENE LATE	N17	Globorotalia humerosa	NN11	Discoaster quinqueramus
	N16	Globorotalia acostaensis	NN10	Discoaster calcaris
MIOCENE MIDDLE	N15	Globorotalia menardii	NN9	Discoaster hamatus
	N14	Globorotalia mayeri	NN8	Catinaster coalitus
	N13	Globigerinoides ruber	NN7	Discoaster kugleri
	N12	Globorotalia fohsi robusta		
	N11	Globorotalia fohsi lobata	NN6	Discoaster exilis
	N10	Globorotalia fohsi fohsi		
	N9	Globorotalia fohsi peripheroronda		
MIOCENE EARLY	N8	Praeorbulina glomerosa	NN5	Sphenolithus heteromorphus
	N7	Globigerinatella insueta	NN4	Helicosphaera ampliaperta
	N6	Catapsydrax stainforthi	NN3	Sphenolithus belemnos
	N5	Catapsydrax dissimilis	NN2	Discoaster druggii
	N4	Globigerinoides primordius	NN1	Triquetrorhabdulus carinatus
OLIGOCENE LATE	P22	Globorotalia kugleri	NP25	Sphenolithus ciperoensis
	P21	Globigerina ciperoensis ciperoensis	NP24	Sphenolithus distentus
OLIGOCENE MIDDLE	P19/20	Globorotalia opima opima	NP23	Sphenolithus predistentus
OLIGOCENE EARLY	P18	Globigerina ampliapertura	NP22	Helicosphaera reticulata
	P17	Cassig. chipolensis / Pseud. micra	NP21	Ericsonia subdisticha
EOCENE LATE	P16	Turborotalia cerroazulensis s.l.	NP19/20	Isthmolithus recurvus
	P15	Globigerinatheka semiinvoluta	NP18	Chiasmolithus oamaruensis
	P14	Truncorotaloides rohri	NP17	Discoaster saipanensis
EOCENE MIDDLE	P13	Orbulinoides beckmanni	NP16	Discoaster nodifer
	P12	Morozovella lehneri	NP15	Nannotetrina fulgens
	P11	Globigerinatheka s. subconglobata	NP14	Discoaster sublodoensis
	P10	Hantkenina nuttalli	NP13	Discoaster lodoensis
EOCENE EARLY	P9	Acarinina pentacamerata	NP12	Tribrachiatus orthostylus
	P8	Morozovella aragonensis	NP11	Discoaster binodosus
	P7	Morozovella formosa formosa	NP10	Tribrachiatus contortus
	P6	Morozovella subbotinae		
PALEOCENE LATE	P5	Morozovella edgari	NP9	Discoaster multiradiatus
	P4	Morozovella velascoensis	NP8	Heliolithus riedeli
		Planorotalites pseudomenardii	NP7	Discoaster mohleri
			NP6	Heliolithus kleinpelli
PALEOCENE MIDDLE	P3	Planorotalites pusilla pusilla	NP5	Fasciculithus tympaniformis
	P2	Morozovella angulata	NP4	Ellipsolithus macellus
		Morozovella uncinata	NP3	Chiasmolithus danicus
PALEOCENE EARLY	P1 c	Morozovella trinidadensis	NP2	Cruciplacolithus tenuis
	P1 b	Morozovella pseudobulloides		
	P1 a	Globigerina eugubina	NP1	Markalius inversus

Regional formations (columns across the figure): S- AND SE-TRINIDAD (rich in planktic Foraminifera): Lengua, Hiatus, Cipero, Hiatus, San Fernando, Navet, Upper Lizard Springs, Lower Lizard Springs, not recorded or Hiatus. — S- AND SE-TRINIDAD (predom. benthic Foraminifera): Cruse, Karamat, Herrera, Brasso, Ste. Croix, Nariva, Pointe-à-Pierre, Chaudière, not recorded or Hiatus.

VENEZUELA — EASTERN: La Pica, Carapita, Chapapotal, Carapita, Naricual, Areo, Los Jabillos, Tinajitas, Caratas, Vidoño. — SE-FALCON: Ojo de Agua, Pozon, San Lorenzo, Guacharaca, Cerro Mision, Agua Salada. — CENTRAL-FALCON: La Vela, Caujarao, Socorro, Cerro Pelado, Agua Clara, Pedregoso, Pecaya, El Paraiso, Jarillal, Santa Rita. — MARACAIBO BASIN: Betijoque, Isnotú, Lagunillas, La Rosa, Icotea, Mene Grande, Pauji, Misoa, Trujillo, Guasare.

BARBADOS: Bissex Hill, Conset, Oceanic.

Fig. 3. Correlation of Tertiary formations of Trinidad, Venezuela and Barbados with planktic foraminiferal and calcareous nannofossil zones (zonal correlation after Bolli *et al.* 1985).

Trinidad

Introduction

Trinidad has a long history of investigation of fossil foraminifera. Following the classical 'Report on the geology of Trinidad' by Wall & Sawkins (1860) who drew attention to the presence of fossils 'which might render important services to geological science', it was Guppy (1863, 1873) who first pointed out the presence of Tertiary foraminifera in Trinidad. He was also the first to illustrate some of them in 1894.

Applications of benthic foraminifera to stratigraphy were initiated about 1917 by W. F. Penny and P. W. Jarvis at the demand of Trinidad's expanding oil industry. In 1929 Cushman & Jarvis published a number of new benthic and planktic species from the Tertiary of Trinidad while, shortly before, Nuttall (1928) published a more comprehensive study on Oligo-Miocene benthic foraminifera from the Naparima region in southern Trinidad. This was also the first study on Trinidad foraminifera where the distribution of the species comprising a total of 144 taxa was plotted on a chart.

Following these pioneer studies which demonstrated the usefulness of the benthic species (the very frequent planktic forms were at the time almost completely ignored) to solve stratigraphic problems, oil companies began to establish their own micropaleontological laboratories. This increased the level of activity, and the initiative of some dedicated company paleontologists resulted in a number of detailed predominantly systematic studies on the Tertiary foraminifera, again still concentrating on benthic forms. Publications include those by Cushman & Stainforth (1945) and Cushman & Renz (1946, 1947a,b, 1948) which span the interval Paleocene to Middle Miocene.

With the exception of the papers by Cushman & Renz (1946, 1947a) little information exists today on the Late Cretaceous, Paleocene and Early Eocene benthic foraminifera. This is in contrast to the planktic foraminifera which have been published in detail. The Early Cretaceous Barremian to earliest Albian benthic foraminifera on the other hand were studied and published by Bartenstein, Bettenstaedt & Bolli (1957, 1966) and by Bartenstein & Bolli (1973, 1977, 1986).

Saunders & Bolli (presented 1979, published 1985) gave an account of 'Trinidad's contribution to world stratigraphy'. Reference is made to this publication for more details on both the general history of biostratigraphic studies in Trinidad, including benthic foraminifera, and the development in Trinidad of a zonal scheme based on planktic foraminifera.

In the present volume, the Early Cretaceous to Middle Miocene benthic foraminifera of Trinidad are discussed under three sections: Barremian to earliest Albian, Albian to Early Eocene, late Early Eocene to Middle Miocene.

The base for the first section are the five papers by Bartenstein, Bettenstaedt and Bolli listed above. Of the 185 taxa originally described and illustrated in these papers, 77 characteristic and stratigraphically significant species were selected for this chapter. They are re-illustrated on Figs. 6–11 and their ranges shown on Fig. 5. In the systematic part are given type and Trinidad references for each species, critical annotations and ages. Illustrations of holotypes are added for comparison of species whose assignments to Trinidad forms are questioned. The localities from where the faunas originate are shown on Fig. 4.

In contrast to the planktic foraminifera, little published data (Cushman & Renz, 1946, 1947a) exist on the Albian to Early Eocene benthic foraminifera. The fauna of this interval as presented in the second sec-

tion are therefore the result of a comprehensive new investigation in which 553 species are dealt with. All are illustrated on Figs. 18–47 by one or more specimens. Ranges of 180 selected taxa are shown on Figs. 15–17. Fig. 12 is a correlation with the planktic foraminiferal zonal scheme for the formations from which the treated species come. Fig. 13 shows the locations from where the faunas originate and Fig. 14 displays the distribution and frequency of major faunal groups. The systematic part contains for each taxon its type and additional references, annotations where deemed useful, and stratigraphic ranges.

The third section deals with the late Early Eocene to Middle Miocene fauna. The illustrations of the 175 selected benthic species are largely taken from the publications by Cushman & Renz (1947b,1948) and Cushman & Stainforth (1945) and to a lesser degree from Nuttall (1928). Where deemed advantageous for comparison purposes holotypes of certain species are also illustrated along with the corresponding Trinidad specimens.

The systematics of section 3 with taxonomic annotations are incorporated with the Tertiary benthics from Falcon and Barbados at the end of the volume. The ranges of the selected taxa are shown in this section on Figs. 50–52, their illustrations on Figs. 53–64. A location map is found on Fig. 48, a correlation of the Trinidad Middle Eocene to Middle Miocene formations with the planktic foraminiferal zonal scheme on Fig. 49. In contrast to section 2, only a selection of taxa is dealt with in sections 1 and 3.

Barremian to Early Albian

Introduction

Bartenstein, Bettenstaedt & Bolli and Bartenstein & Bolli described and illustrated in five papers (1957–1986) the benthic and some planktic foraminifera from the Early Cretaceous Barremian to Early Albian of Trinidad. They distinguished 78 benthic genera, 185 species and subspecies of which 13 were described as new. The idea of these joint publications with Bartenstein and Bettenstaedt dates back to a suggestion by Bolli (1950) which was guided by the following considerations:

Compared with those from western Europe, where Early Cretaceous foraminifera are well developed and can be studied in successive sedimentary sequences including boreholes, those from the Americas were at the time less well known. The study of the European forms began in the nineteenth century with authors such as Roemer, Reuss, Berthelin and Chapman. These early studies were supplemented in the Northwest German Early Cretaceous from 1932 to 1942 by the more stratigraphically orientated investigations of Eichenberg, Hecht and Wicher. This period was followed by modern revisions by authors such as Albers, Bartenstein, Bettenstaedt and Brand.

In their publication 'Stratigraphic correlation of Upper Cretaceous and Lower Cretaceous in the Tethys and Boreal by the aid of microfossils (Germany)' Bettenstaedt & Wicher (1955) for the first time pointed out the worldwide occurrence of numerous Cretaceous benthic foraminifera with apparently the same stratigraphic distribution as in the boreal and tethyan realms. This was based on a comparison of the North German faunas with those of other European countries, Israel, Egypt, Morocco, USA, Mexico and Trinidad.

Most of the material studied in northwestern Germany since 1932 has come from petroleum boreholes. This has led to a well proven stratigraphic-micropaleontologic subdivision of the strata concerned. It seemed natural that Bartenstein and Bettenstaedt with their expertise would be the ideal co-investigators for a detailed study of Trinidad's Early Cretaceous benthic foraminifera. Already in 1949 Bettenstaedt had had an opportunity to identify faunas from Trinidad's Cuche and Maridale Formation which were considered by him, in a private communication, as identical with or very closely related to those of northwestern Germany. These findings pointed for the first time to a much wider regional distribution of Early Cretaceous benthic foraminiferal faunas than was previously assumed, with the likelihood that they could be used for world-wide correlations. This had already been reported for some Late Cretaceous benthics by e.g. Wicher (1949, 1956) and Bettenstaedt & Wicher (1955).

The description and illustration of Trinidad's Early Cretaceous benthic foraminifera appeared in five parts between 1957 and 1986, all published in the *Eclogae geologicae Helvetiae*. The first two publications were authored by Bartenstein, Bettenstaedt & Bolli (1957, 1966), the others by Bartenstein & Bolli (1973, 1977, 1986). Part I (1957) deals with the Toco- and the lower part of the Cuche Formation, Part II (1966) with the type locality of the Maridale Formation, Part III (1973) with the co-type locality of the Maridale Formation, Part IV (1977) with the *Leupoldina protuberans* Zone, upper part of the Cuche Formation, and Part V with the *Hedbergella rohri* Zone, upper part of the Maridale Formation. Selected planktic foraminifera were included in Parts II–IV while Part V contained all planktic taxa recorded at that level. The authors distinguished 78 benthic genera with 185 species and subspecies of which 13 were described as new. An index listing all benthic taxa treated in Parts I to V was attached to part V.

Additional Early Cretaceous planktic foraminifera were described and illustrated in Bolli, Loeblich & Tappan (1957) and in Bolli (1957a, 1959). In his 1959 paper Bolli proposed a zonation of Trinidad's Early Cretaceous based in its lower part on benthic index forms and in its upper part on planktics as these had become more evolved and diversified.

In addtion to the above listed publications on the Trinidad Early Cretaceous foraminiferal faunas and their biostratigraphic significance, papers by the following authors also deal with these subjects:

Publications by Bartenstein (1976, 1985, 1987) contain discussions, range charts, illustrations and correlations by means of stages and zones of Early Cretaceous species from Northwest Germany and Trinidad. In his 1976 paper 'Foraminiferal zonation of the Lower Cretaceous in North West Germany and Trinidad, West Indies. An attempt' Bartenstein presents a chart correlating some benthic index species

of Northwest Germany and Trinidad that are typical for both areas. In the same publication are also pointed out correlations with Early Cretaceous faunas from eastern Canada, the Caucasus, and Iran.

In 'Stratigraphic pattern of index foraminifera in the Lower Cretaceous of Trinidad' Bartenstein (1985) illustrates 31 benthic and planktic species considered the most important stratigraphic index forms for the area under discussion. Their ranges are fitted into the stage subdivision and the zonal scheme as applied in Trinidad. The following species are listed as index forms for a correlation between Europe and Trinidad: *Gaudryina dividens, G. compacta, G. reicheli, Lenticulina (S.) spinosa, Conorotalites aptiensis, C. intercedens, Marssonella kummi, Epistomina hechti* and *Gavelinella barremiana*.

In 'Micropaleontological synopsis of the Lower Cretaceous in Trinidad, West Indies. Remarks on the Aptian/Albian boundary' Bartenstein (1987) illustrates and shows the distribution of 53 benthic and 11 planktic species characteristic for the Trinidad Middle Barremian to Early Cenomanian. Indicated are also the ranges of these taxa as known from outside Trinidad. Bartenstein interprets the correlation of the stratigraphic position of the Trinidad Early Cretaceous zones in relation to the Northwest German microfaunal stage terminology and the somewhat differing standard ammonite zones in relation to the stage and substage boundaries. Bartenstein's table 4 shows the discrepancies between the correlation on one hand of the standard ammonite zones (tethyan and boreal) with the stage subdivisions and on the other hand with the different placing of the stage/substage boundaries related to the Northwest German microfaunal terminology.

Koutsoukos and Merrick (1986) discuss the paleoenvironmental evolution of the Barremian to Maastrichtian of Trinidad based on selected planktic and benthic foraminifera from a number of well sections. The study is based mainly on the taxonomic composition, diversity, abundance and size of the foraminiferal assemblages taken into account. They show the distribution of selected planktic and benthic foraminifera from Barremian to Albian and give lists of faunas characteristic for the environmental conditions of individual stratigaphic intervals from Early Barremian to latest Aptian. The paleoenvironmental evolution from Barremian to Late Aptian is summarized. Regrettably, no documentation on the location of the studied sections is given by Koutsoukos and Merrick.

An unpublished doctoral thesis submitted to the University of Michigan by Stacy (1966) deals with the Early Cretaceous microfauna of Trinidad and adjacent areas. The study is based on a large number of surface and subsurface samples from Trinidad, the Gulf of Paria and eastern Venezuela. Two hundred and six foraminiferal species of which 87 are new and 77 ostracod species with 37 new forms are described and illustrated. The new forms remain invalid. A zonal subdivision based on the shaley Trinidad formations is correlated with the limestone equivalents of Eastern Venezuela and also with the Gulf Coast of the USA.

Because the Trinidad Early Cretaceous zones are based on individual isolated outcrops and not on continuous sections, it is not possible to determine the actual ranges of the zonal markers and other species encountered in the respective samples defining the individual zones. It is therefore possible, or likely, that the ranges of the zonal markers of neighbouring zones may be either (a) separated by intervals in which they would not be present or (b) would to some degree overlap within these intervals.

In the case of (a) it would become necessary to define the additional stratigraphic intervals, in the case of (b) zonal intervals would at least in part have to be defined differently. Considering the general planktic and benthic foraminiferal sequences in Trinidad, it is obvious that our knowledge of their stratigraphic ranges and evolutionary development over some 14 Ma (Barremian to earliest Albian) can only be fragmentary, being restricted to a few observation points. However, it allows us to distinguish through time five distinct faunal associations reflected in the five zones erected.

The Early Cretaceous formations of Trinidad

The Early Cretaceous of Trinidad consists of the following formations: in approximate order of decreasing age (see also Fig. 4, locality map):
1. Tompire Formation first named by Barr (1963), described by Imlay (1954) and containing ammonites. It occurs in the northeast corner of Trinidad at the easternmost end of the Northern Range. It was not dealt with in the publications by Bartenstein, Bettenstaedt & Bolli and Bartenstein & Bolli.
2. Toco Formation, considered slightly younger than the Tompire Formation, also occurs at the eastern end of the Northern Range, with its type locality at Toco Bay (Bartenstein, Bettenstaedt & Bolli, 1957). It was originally described by Trechmann (1935) and by Liddle (1946) as Toco Bay Beds. The

formation consists of black-grey shales which, in contrast to the Cuche shales are generally of a phyllitic appearance, with layers of sandstone and irregular lenses of limestones.

One of the richest fossil localities of the Toco Formation is found at its type locality at Toco Bay, from where the foraminiferal fauna is described in Bartenstein, Bettenstaedt & Bolli (1957). Molluscs and corals were also reported from the same locality by Thomas (1935), Trechmann (1935), Wells (1948) and Kugler (1950). Near the type locality, the Toco shales have a rich fauna of *Trocholina infragranulata* Noth. From some kilometres south of the type locality Maync (1956) mentioned *Choffatella decipiens* from the Toco Formation whose age he gave as Hauterivian to Barremian. Bolli (1959) proposed the *Lenticulina barri* Zone for the benthic foraminiferal association at the type locality of the Toco Formation.

For further information on the Tompire and Toco formations reference is made to Bartenstein, Bettenstaedt & Bolli (1957). Field relationships can be found in Barr (1963).
3. The term Cuche was first introduced by Hutchison (1938) as Cuche River Beds which outcrop in the Central Range. The lithology of the Cuche Formation is close to that of the Toco Formation of the eastern Northern Range: black-grey shales, soft but not phyllitic, often silty, with layers of coarse sandstone and with occasional lenses of marls, conglomerates and limestone components. The foraminiferal fauna described in Bartenstein, Bettenstaedt & Bolli (1957) comes from five different isolated outcrops. Of these the Bon Accord Marl at Station Road in Pointe-à-Pierre is the most prolific. The five localities from where the faunas were described are considered approximately time equivalent and regarded as typical for the lower part of the formation in which – as in the Toco Formation – benthic taxa strongly dominate over a few species of usually rare planktic foraminifera. Bolli (1959) divided the Cuche Formation into an older *(Lenticulina ouachensis ouachensis)* Zone and a younger *Leupoldina protuberans* Zone. The two faunas are quite different. As can be seen from the range chart, a large number of the lower Cuche species do not continue into the upper part, where a considerable number of taxa first appear, many of them continuing into the overlying Maridale Formation.

In contrast to the lower part of the Cuche Formation, its upper part contains, together with the benthic forms, also a rich and diversified planktic

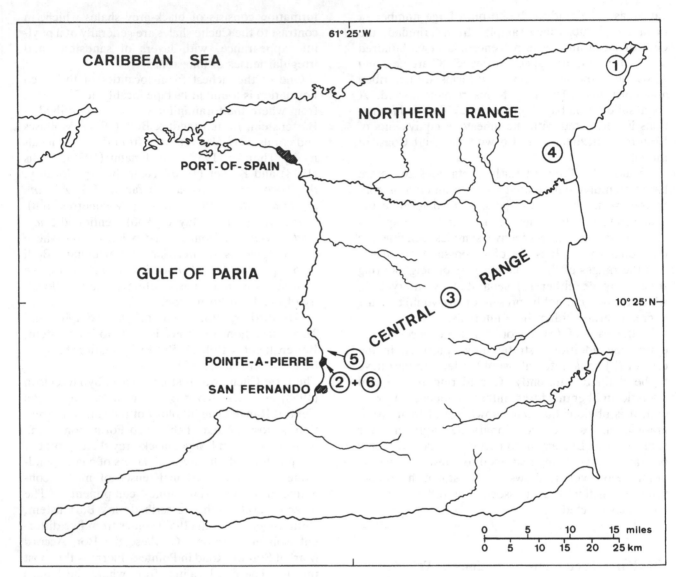

Fig. 4. Map of Trinidad showing type localities 1–6 of Early Cretaceous foraminiferal zones. For locality descriptions see Bartenstein, Bettenstaedt & Bolli (1957, 1966), Bartenstein & Bolli (1977, 1986), Bolli (1957*e*). Type localities 3 to 6 are no longer accessible. 1 *Lenticulina barri* Zone, 2 *Lenticulina ouachensis ouachensis* Zone, 3 *Leupoldina protuberans* Zone, 4 *Planomalina maridalensis* Zone, 5 *Planomalina maridalensis* Zone (co-type locality), 6 *Hedbergella rohri* Zone.

fauna, including the characteristic zonal marker *Leupoldina protuberans*. This planktic fauna is partly also illustrated in Bartenstein & Bolli (1977). The *Schackoina* and *Leupoldina* species were already previously recorded by Bolli (1957*e*) and the remaining planktic taxa also by the same author in 1959. The type locality of the *Leupoldina protuberans* Zone in the Piparo River, Central Range, is shown on fig. 1 in Bolli (1957*e*).

4. The term Maridale was introduced by Hutchison (1938) for dark-grey to bluish-grey, yellowish weathering marls rich in benthic and planktic fora-

minifera and often, as is the case in the co-type locality, also rich in radiolaria. The benthic foraminiferal fauna of the Maridale Formation was dealt with in Bartenstein, Bettenstaedt & Bolli (1966) and Bartenstein & Bolli (1973, 1986). Other fossils present include ostracods, fragments of macrofossils such as echinid spines, fish teeth, sponge spicules and belemnites which are of particular significance.

The formation is only known in the Central Range where it occurs in the form of blocks and slipmasses in younger formations. It could therefore be argued that – like the Bon Accord Marl in the lower part of the Cuche Formation – the Maridale Marl could also occur as lenses in the upper part of the Cuche Formation. A review on observations on the Maridale Formation is given by Bartenstein, Bettenstaedt & Bolli (1966). Bolli (1959) subdivided the Maridale Formation into a lower *Biglobigerinella barri* Zone (later re-named *Planomalina maridalensis* Zone) and an upper *Praeglobotruncana rohri* Zone (now *Hedbergella rohri* Zone).

The type locality of the formation is situated on Maridale Estate at the eastern end of the Central Range. Its benthic foraminiferal fauna was described by Bartenstein, Bettenstaedt & Bolli (1966) where also a map showing the type locality is included. The very similar fauna of the more easily accessible co-type locality at the western end of the island, north of Pointe-à-Pierre (Bartenstein, Bettenstaedt & Bolli 1957, locality 6 on their fig.1), was dealt with by Bartenstein & Bolli (1973). The faunas of both the type locality and the co-type locality are characteristic for the *Planomalina maridalensis* Zone. Today both these localities have become virtually inaccessible.

The fauna placed in the *Hedbergella rohri* Zone described by Bartenstein & Bolli (1986) originates from a boulder of the Maridale Formation that became exposed in a trench dug at Pointe-à-Pierre about 100 metres to the east of the Bon Accord Marl type locality, shown as (3) on the section of fig. 2 in Bartenstein, Bettenstaedt & Bolli (1957). It is no longer accessible.

In contrast to the *Planomalina maridalensis* Zone which is rich in benthic and planktic foraminifera and radiolaria, the fauna of the *Hedbergella rohri* Zone contains amongst the benthic forms comparatively few planktics and no radiolaria.

Based on the strongly differing compositions of the planktic foraminiferal associations of the two successive zones it can be concluded that an appreciable stratigraphic interval divides the samples from which the two faunas originate, possibly equivalent to the *Hedbergella gorbachikae* Zone of the planktic foraminiferal zonal scheme of Caron (1985).

Remarks

The 77 taxa described and illustrated here from the Early Cretaceous of Trinidad have been chosen from the 185 species and subspecies and 52 forms in open nomenclature published in Trinidad I–V. They were selected based on their stratigraphic and morphologic significance. The following information is given for each taxon:

1. Reference to original type.
2. References to species and subspecies described and illustrated in Trinidad I–V.
3. Annotations on each taxon. In some cases including comparisons with the respective primary type.
4. Stratigraphic range of taxon in Trinidad and world-wide.
5. For additional information concerning description of the taxa and their stratigraphic significance reference is made to the species descriptions in Trinidad I–V.

Figures

Figures 6–11 contain illustrations of the 77 selected taxa, all taken from the Trinidad I–V publications. For comparison with the figured Trinidad forms, illustrations of the original types are added in some cases.

Range chart

Ranges of the 77 benthic taxa are plotted on Fig. 5 in stratigraphic order, correlated with stages, formations and the foraminiferal zonal scheme applied in Trinidad.

References to the five publications on the Early Cretaceous foraminifera of Trinidad by Bartenstein, Bettenstaedt & Bolli (1957, 1966) and Bartenstein & Bolli (1973, 1977, 1986) are abbreviated as Td. I, Td. II, Td. III, Td. IV and Td. V, respectively.

Example of a reference, with publication number, page, figure number and illustration number: Td.I: 15, 3/2–4, reads Trinidad I, page 15, figure 3, Illustration 2–4.

Annotated taxonomic list

Family Ammodiscidae

GLOMOSPIRELLA GAULTINA
(Berthelin)
Figure **6.1–3**; 5

Type reference. *Ammodiscus gaultinus* Berthelin
    1880, p. 19, pl. 1, fig. 3.
Trinidad references. II: 140, 1/29; III: 394, 2/1–13
    (in both papers described as *Ammodiscus
    gaultinus*).

The weakly streptospiral early stage that characterizes the genus is usually poorly developed in the Trinidad specimens which vary in size from about 0.5 to 1.2 mm in diameter.

The species is quoted to occur world-wide in the Cretaceous. In the Trinidad Early Cretaceous the species is, however, restricted to the Late Aptian/Early Albian Maridale Formation where it is fairly frequent. This may reflect local environmental conditions.

Family Hormosinidae

REOPHAX GUTTIFER Brady
Figure **6.9–12**; 5

Type reference. *Reophax guttifera* Brady 1884,
    p. 295, pl. 31, figs. 10–15.
Trinidad reference. I: 16, 1/12–15; non II: 137, 1/
    23–28; III: 393, 1/27–33, V: 949, 1/24.

The Trinidad specimens usually consist of only one flattened chamber (Fig. 6.10), rarely of two (Fig. 6.9) attached to each other, and very seldom three. The form of the individual chamber is somewhat heart-shaped with a distinct neck/siphon-like extended apertural area, and a distinctly concave basal line of the chamber. With their characteristic shape the Trinidad specimens do not compare with those described and figured by Brady as *Reophax guttifera*. Under this name he illustrated six specimens that in fact should better be assigned to two different species (his figs. 10–12 and 13–15 respectively). One – an example is shown here on Fig. 6.11 – displays strong constrictions between chambers, formed

by the narrow tubular extensions of the apertural structures connecting the chambers.

The other type illustrated by Brady – a specimen is here shown on Fig. 6.12 – shows little resemblance to the Trinidad forms. This is also the case with the specimens illustrated as *Reophax guttifera* in the Trinidad parts II, III and V, all from the Maridale Formation. There is a notable size difference of the individual chambers between Brady's Recent and Trinidad's Early Cretaceous forms.

The *Reophax* form as here discussed and illustrated on Fig. 6.9–10 is a characteristic faunal component mainly in the Bon Accord Marl facies of the Cuche Formation. It has not been seen in other zones and therefore appears to be a good index, at least locally. From the above its eventual assignment to a new species seems justified.

REOPHAX SCORPIURUS Montfort
Figure **6.4–6**; 5

Type reference. *Reophax scorpiurus* Montfort 1808,
    Conchyliologie systématique et classification
    méthodique des coquilles, Paris, F. Schoell, 1,
    p. 331, textfig. on p. 330.
Trinidad reference. I: 15, 1/2–3.

The Trinidad specimens were placed by Bartenstein, Bettenstaedt & Bolli (1957) into *Reophax scorpiurus*, originally described from sands of the Adriatic Sea (Fig. 6.6). A comparison of the Trinidad forms (Fig. 6.4–5) with the original type however shows little in common. In fact the Trinidad specimens appear to be much closer to *Reophax subfusiformis* Earland another Recent form (Fig. 6.7). The assignment of the Early Cretaceous specimens to a Recent form remains questionable. According to Bartenstein, Bettenstaedt & Bolli (1957) the species occurs throughout the Cretaceous. In Trinidad it has so far been recorded from the *Lenticulina ouachensis ouachensis* Zone of the Cuche Formation.

REOPHAX SUBFUSIFORMIS Earland
Figure **6.7–8**; 5

Type reference. *Reophax subfusiformis* Earland
    1933, Discovery Reports, Cambridge, 7, p. 74,
    pl. 2, fig. 17.
Trinidad reference. I: 16, 2/34–35.

As pointed out under *Reophax scorpiurus*, the holotype of *Reophax subfusiformis* compares closely to the specimens from Trinidad described as *R. scorpiurus*. On the other hand the Trinidad specimens determined as *R. subfusiformis* (Fig. 6.8) differ from the Recent forms of Earland (Fig. 6.7) in the neck-like extended aperture being at a distinct angle against the longitudinal axis of the rectilinear uniserial test. These characteristic forms are in Trinidad restricted to the *Lenticulina ouachensis ouachensis* Zone of the Cuche Formation.

Family Lituolidae

AMMOBACULITES REOPHACOIDES
Bartenstein
Figure **6.17–19**; 5

Type reference. *Ammobaculites reophacoides* Bartenstein 1952, Senckenbergiana, 33/4–6, p. 307, 309, tf. 1a, b.
Trinidad references. IV: 546, 562–563, 1/10–13; V: 949, 1/31–32.

The illustrated Trinidad specimens (Fig. 6.17–18) compare in size approximately with the holotype (Fig. 6.19). They differ from it in the last chamber being less extended than is the case in the holotype. In fact the shape of the last chamber of these two Trinidad specimens is closer to that of the holotype of *Ammobaculites torosus* Loeblich & Tappan (Fig. 6.25) which is a larger sized species described from the Albian Walnut Formation of Texas that closely resembles *Ammobaculites reophacoides*. The specimens figured as *Ammobaculites reophacoides* from the *Hedbergella rohri* Zone (in Bartenstein & Bolli, 1986, pl. 1, figs. 31, 32) are strongly compressed in a tile-like overlapping of the chambers. They also do not show the initial spiral part as clearly as the specimens here illustrated on Fig. 6.17–18.

The holotype of *Ammobaculites reophacoides* has been described from the Middle-Late Barremian; the Trinidad specimens are from the Early Aptian *Leupoldina protuberans* and the Late Aptian-Early Albian *Hedbergella rohri* Zone. Bartenstein & Bolli (1977) quote *Ammobaculites reophacoides* as a common index form of the boreal and tethyan realms, ranging from Early Barremian to Middle Albian.

AMMOBACULITES TOROSUS Loeblich & Tappan
Figure **6.23–25**; 5

Type reference. *Ammobaculites torosus* Loeblich & Tappan 1949, J. Paleontol. 23, p. 251, pl. 46, fig. 7.
Trinidad reference. II: 139, 1/17–22, 44–49; III: 1/42–56.

The Trinidad specimens assigned to *Ammobaculites torosus* from the co-type locality of the Maridale Formation (Bartenstein & Bolli 1973) vary considerably in size, ranging up to 1.9 mm. In contrast to the holotype (Fig. 6.25) the last chamber in the Trinidad specimens is more extended longitudinally (Fig. 6.24), or more like the one of the holotype of *Ammobaculites reophacoides* (Fig. 6.19). Furthermore, the initial part of the test in the Trinidad specimens (Fig. 6.23) is more slender than that of the holotype (Fig. 6.25). Furthermore, the early planispiral arrangement of the chambers is only poorly recognizable in the Trinidad specimens. Bartenstein, Bettenstaedt & Bolli (1966) realized the differences and stated that these slender forms cannot with certainty be included in any known species.

The species is characteristic and frequent in the type locality material of the Upper Aptian Maridale Formation.

AMMOBACULITES TRINIDADENSIS
Bartenstein, Bettenstaedt & Bolli
Figure **6.15–16**; 5

Type reference. *Ammobaculites trinidadensis* Bartenstein, Bettenstaedt & Bolli 1957, p. 17, pl. 1, figs. 6–8 (7 Holotype).

The species so far known only from the *Lenticulina ouachensis ouachensis* Zone, Bon Accord Marl, of the Cuche Formation of Trinidad, is` characterized by its coarsely agglutinated, irregularly angular test with not more than two chambers forming the rectilinear portion. The initial planispiral part is usually difficult to recognize. Characteristic is the tube-shaped apertural part. There exists a considerable variability in test shape. The sutures between the coarsely grained chambers may be quite broad and distinctly incised, appearing as smooth bands.

TRIPLASIA EMSLANDENSIS ACUTA
Bartenstein & Brand
Figure **6.29–30**; 5

Type reference. *Triplasia emslandensis acuta*
    Bartenstein & Brand 1951, p. 274, pl. 3, fig. 68.
Trinidad reference. I: 2/26a, b, 38.

For annotations see *Triplasia emslandensis emslandensis* below.

TRIPLASIA EMSLANDENSIS
EMSLANDENSIS Bartenstein & Brand
Figure **6.31–33**; 5

Type reference. *Triplasia emslandensis emslandensis*
    Bartenstein & Brand 1951, p. 274, pl. 3,
    fig. 67.
Trinidad reference. I: 18, 1/21, 2/25.

Bartenstein & Brand (1951) described the two subspecies *Triplasia emslandensis emslandensis* and *T. emslandensis acuta*, of which the latter differs from the former in possessing more acute edges and a virtual absence of an initial spiral part. The two taxa are also distinguished in Trinidad.

Of the subspecies *emslandensis* were found only broken specimens, usually single chambers (Fig. 6.31) or a few chambers still attached to each other (Fig. 6.32). In comparing these Trinidad fragments with the holotype figure (Fig. 6.33) the height/width ratio of individual chambers is quite different, about 1: 1 against 1: 2 in the holotype. Further, there is almost no increase in chamber size in the Trinidad specimens (Fig. 6.32). This is another indication that one is probably dealing with two different species and that the Trinidad form does not belong to the *emslandensis* plexus at all. The illustrated Trinidad specimen of the subspecies *acuta* (Fig. 6.29) is on the other hand closer to the holotype figure of the subspecies (Fig. 6.30).

In 1971 Bartenstein, Bettenstaedt & Kovatchova placed the specimens from Trinidad illustrated as *Triplasia emslandensis emslandensis* and *T. emslandensis acuta* without further explanations in synonymy with *Triplasia georgsdorfensis* (Bartenstein & Brand), This is confirmed also in Trinidad (Bartenstein & Bolli 1986).

The *emslandensis* forms are known in Europe and North Africa (here described as *Frankeina djaffaensis* by Sigal 1952) from Late Valanginian to Late Hauterivian. The fairly frequent speci-

mens in the Middle to Late Barremian *Lenticulina ouachensis ouachensis* Zone of the Cuche Formation of Trinidad are thus a younger record. It is also for this reason that in particular the specimens illustrated from Trinidad as subspecies *emslandensis* may, as indicated above, belong to a different taxon.

Family Haplophragmiidae

HAPLOPHRAGMIUM cf. AEQUALE
(Roemer)
Figure **6.20–22**; 5

Type reference. *Spirulina aequalis* Roemer 1841,
    p. 98, pl. 15, fig. 27.
Trinidad reference. I: 18, 2/22–23.

Bartenstein, Bettenstaedt & Bolli (1957) named the Trinidad forms *Haplophragmium* cf. *aequale* and regarded them as closely related to the northwestern German species concerning test shape, arrangement of chambers and aperture. When comparing the Trinidad specimens on Fig. 6.20–21 with Roemer's type (Fig. 6.22), they appear quite different however. The Trinidad specimens are in comparison much more massive, compact and broader, with hardly any sutural incisions visible between the chambers. The initial planispiral part is difficult to see. Further, there is a tendency in the final chamber to become slightly smaller than the previous one. Roemer's specimen on the other hand appears more slender, and the sutures between chambers and the early planispiral part are distinct.

The characteristic Trinidad specimens are restricted to the *Lenticulina ouachensis ouachensis* Zone, Bon Accord Marl, Cuche Formation.

Family Cyclamminidae

CHOFFATELLA DECIPIENS Schlumberger
Figure **6.13–14**; 5

Type reference. *Choffatella decipiens* Schlumberger
    1905, Bull. Soc. Géol. France, Sér. 4, 4/6 (1904),
    p. 763, pl. 18, figs. 1–6.
Trinidad reference. IV: 549, 1/32.

In Trinidad the species occurs in the Tompire Formation, the *Lenticulina barri* Zone of the

Toco Formation, the *Lenticulina ouachensis ouachensis* and *Leupoldina protuberans* zones of the Cuche Formation, or from Early Barremian to Early Aptian.

Family Verneuilinidae

GAUDRYINA COMPACTA Grabert
Figure 7.5–8; 5

Type reference. *Gaudryina compacta* Grabert 1959, Senckenberg naturf. Ges., Abh., 498, p. 11, pl. 1, fig. 7.
Trinidad reference. V: 950, 1/40–41.

As the illustrations on Fig. 7.5–8 show, *Gaudryina compacta* is morphologically very close to *Gaudryina dividens*. Gradstein (1978) reported the two species as belonging to the same plexus. The Trinidad specimens of *Gaudryina compacta* differ from those of *Gaudryina dividens* in the shorter triserial initial stage and in the presence of a biserial stage.

Gaudryina dividens ranges in Trinidad from the *Leupoldina protuberans* to the *Hedbergella rohri* Zone, whereas the apparently phylogenetically younger *Gaudryina compacta* is restricted to the *Hedbergella rohri* Zone. *Gaudryina compacta* has been reported from the Early Albian (Alps, Sicily) and from the Late Aptian to Middle Albian in the subbottom of the western Atlantic.

GAUDRYINA DIVIDENS Grabert
Figure 7.1–4; 5

Type reference. *Gaudryina dividens* Grabert 1959, Senckenberg naturf. Ges., Abh., 498, p. 9, pl. 1, fig. 4.
Trinidad references. II: 141, 1/56–57; III: 390; IV: 562, 564; V: 951, 1/42–43.

The holotype (Fig. 7.4) is characterized by triserial and biserial growth stages. The Trinidad specimens (Fig. 7.1–2) as a rule are only represented by the triserial stage, similar to the paratype (Fig. 7.3).

For the range of the species see *Gaudryina compacta*.

GAUDRYINA REICHELI Bartenstein, Bettenstaedt & Bolli
Figure 7.17–22; 5

Type reference. *Gaudryina reicheli* Bartenstein, Bettenstaedt & Bolli 1966, p. 142–143, pl. 1, fig. 54.
Trinidad references. II: 142–143, 1/34–37, 50–55; III: 396; V: 951, 1/46–47.

The species described from the Maridale Formation is a very typical form, characterized by its slender test with 9 to 18 biserially arranged chambers, following the short triserial early stage, and occasionally with an uniserial end stage. Tests vary considerably in size, large ones often rotate up to 60 degrees in their longitudinal axis.

In Trinidad the species is frequent in the *Planomalina maridalensis* Zone. Fewer and less typical (wider) specimens occur also in the *Hedbergella rohri* Zone. Outside Trinidad the species was reported from the Late Aptian of Bulgaria.

GAUDRYINELLA HANNOVERIANA Bartenstein & Brand
Figure 7.13–14; 5

Type reference. *Gaudryinella hannoveriana* Bartenstein & Brand 1951, p. 276, pl. 4, fig. 329.
Trinidad reference. IV: 547, 562–563, 1/18–20.

The slender and flattened Trinidad specimen (Fig. 7.14) displays an indistinct triserial early stage, followed by a biserial arrangement of the chambers and a distinct uniserial end-stage consisting of one to two chambers. The aperture of the final chamber is positioned on a neck-like extension. The holotype (Fig. 7.13) is figured for comparison. It is distinctly more slender and smaller compared with the illustrated Trinidad specimen and does also not possess the neck-like apertural extension on the final chamber.

The species ranges world-wide from Late Valanginian to Early Aptian. In Trinidad it is present in the *Leupoldina protuberans* Zone.

GAUDRYINELLA SHERLOCKI
Bettenstaedt
Figure **7.23–25**; 5

Type reference. *Gaudryinella sherlocki* Bettenstaedt
1952, Senckenbergiana, 33, p. 268, pl. 1, fig. 1.
Trinidad references. II: 141, 1/38–40; III: 395, 2/22–
26; IV: 547, 562–563, 1/21–22; V: 951, 2/1–2.

Trinidad specimens (Fig. 7.23–24) appear broad
as a result of a flattening of the test. They are
dominated by the biserial stage, with the triserial
early one being strongly reduced. An uniserial
end-stage exists in occasional specimens. Fig.
7.25 shows the holotype for comparison.

The taxon is known to range world-wide from
Late Hauterivian to Early Albian. In Trinidad
the species occurs in the *Leupoldina protuberans*
and the *Hedbergella rohri* zones.

VERNEUILINOIDES NEOCOMIENSIS
Myatlyuk
Figure **7.15–16**; 5

Type reference. *Verneuilina neocomiensis* Myatlyuk
1939, Trans. Oil. Geol. Inst., Leningrad, Ser.
A, Fasc. 120, p. 50, pl. 1, fig. 12–13.
Trinidad reference. I: 19, 2/39–40.

The short early triserial stage is not well visible
in the holotype (Fig. 7.16). Trinidad specimens
(Fig. 7.15) are usually flattened.

The species has been reported from the
Berriasian of the Swiss Jura mountains and in
Northwest Germany from Middle Valendis to
Barremian. Its known paleogeographic distri-
bution extends from the Volga to northwestern
Germany, to Switzerland, and to Trinidad,
where it occurs in the *Lenticulina ouachensis
ouachensis* Zone of the Cuche Formation.

Family Tritaxiidae

TRITAXIA PLUMMERAE Cushman
Figure **7.9–10**; 5

Type reference. *Tritaxia plummerae* Cushman 1936,
Contrib. Cushman Lab. foramin. Res., Spec.
Publ. 6, p. 3, pl. 1, fig. 7.
Trinidad reference. V: 953, 2/16–18.

The species is characterized by the extended triserial stage, with only the last two or three cham-

bers becoming biserial to uniserial. Compared
with the holotype (Fig. 7.10) the illustrated
Trinidad specimen (Fig. 7.9) displays the long
triserial stage but not the characteristic biserial
and uniserial end stages.

The species is world-wide known from the
Middle and Late Albian. In Trinidad it is
recorded from the *Hedbergella rohri* Zone.

TRITAXIA PYRAMIDATA Reuss
Figure **7.11–12**; 5

Type reference. *Tritaxia pyramidata* Reuss 1863,
p. 32, pl. 1, fig. 9.
Trinidad references. I: 19, 2/37; IV: 546, 1/14; V:
977.

The Trinidad specimens (Fig. 7.11) compare
quite well with the type of Reuss (Fig. 7.12),
except that added chambers become progress-
ively wider.

The characteristic species is known to range
in Europe from the Hauterivian into the Late
Cretaceous. In Trinidad it is recorded from the
Lenticulina ouachensis ouachensis and *Leupoldina protuberans* zones of the Cuche Formation.

Family Eggerellidae

DOROTHIA FILIFORMIS (Berthelin)
Figure **7.38–39**; 5

Type reference. *Gaudryina filiformis* Berthelin 1880,
p. 25, pl. 1, fig. 8.
Trinidad references. II: 144, 1/43; III: 397, 2/32–35;
IV: 562, 564; V: 959, 1/37.

Compared with the holotype (Fig. 7.39) the
Trinidad specimens (Fig. 7.38) possess a less
extended biserial portion following the short triserial stage.

The distribution is world-wide from Early
Aptian to basal Albian, in Trinidad from the
Leupoldina protuberans to the *Hedbergella rohri*
Zone.

MARSSONELLA KUMMI Zedler
Figure **7.32–34**, 37; 5

Type reference. *Marssonella kummi* Zedler 1961,
Paläontol. Zeitschr., 35, p. 31, pl. 7, fig. 1.
Trinidad references. I: 20, 2/42–43 (as cf. *oxycona*);
IV: 548, 1/23–24.

The holotype (Fig. 7.32) has, following the short
triserial stage, a larger number of biserially
arranged chambers than is usually seen in the
Trinidad specimens (Fig. 7.33–34).

The species is restricted in Trinidad to the
Early Aptian *Leupoldina protuberans* Zone,
while in European sections it is more typical for
the Barremian where it disappears at its top. See
Bartenstein, Bettenstaedt & Kovatcheva (1971)
for further information.

MARSSONELLA OXYCONA (Reuss)
Figure **7.35–36**; 5

Type reference. *Gaudryina oxycona* Reuss 1860,
Sitzber. K. Akad. Wiss. Wien, math.-natwiss.
Cl., 40, p. 229, pl. 12, fig. 3.
Trinidad references. II: 144, 1/58–59 (as
praeoxycona); III: 396, 2/62; IV: 562, 564; V:
952, 2/6–7.

The general morphological aspect of the Trinidad forms (Fig. 7.36) appears close to or identical with the somewhat stylized illustration of this
taxon by Reuss (Fig. 7.35).

The species is known to occur world-wide in
the temperate and tethyan realms from Late
Aptian to Late Cretaceous. In Trinidad the
species is recorded from the *Leupoldina protuberans* to the *Hedbergella rohri* Zone.

MARSSONELLA PRAEOXYCONA
(Moullade)
Figure **7.26–29**; 5

Type reference. *Dorothia praeoxycona* Moullade
1966, Lyon, Univ. Fac. Sci. Lab. Géol. Doc.
no. 15, fasc. 1–2, p. 30–31, pl. 3, fig. 9.
Trinidad references. III: 397, 2/57–61; IV: 548, 562–
563, 1/25–26; V: 976.

Compared with the specimens from the *Leupoldina protuberans* Zone (Fig. 7.27–28) the holotype (Fig. 7.29) appears more slender, similar to
the Trinidad specimen figured from the Maridale
Formation (Fig. 7.26).

The species ranges in Europe from Barremian
to Middle (?Late) Aptian according to the
ammonite terminology.

MARSSONELLA SUBTROCHUS
Bartenstein
Figure **7.30–31**; 5

Type reference. *Marssonella subtrochus* Bartenstein
1962, Senckenbergiana Lethaea, 42, p. 137.

Holotype: Valvulina D 6, Hecht 1938. Abh.
Senckenberg Naturf. Ges., 434, pl. 6a, fig. 40.

Trinidad references. I: 20, 3/44–45 (as cf. *trochus*);
II: 144, 1/79–82; III: 397, 3/29–31; IV: 548, 1/
27–28; V: 952, 2/8–10.

The illustrated poorly preserved Trinidad specimen (Fig. 7.30) from the *Hedbergella rohri*
Zone, Maridale Formation, compares in its
characteristic wide conical test shape quite well
with the apparently somewhat stylized illustration of the holotype (Fig. 7.31).

Family Textulariidae

BIGENERINA CLAVELLATA Loeblich &
Tappan
Figure **6.34–36**; 5

Type reference. *Bigenerina clavellata* Loeblich &
Tappan 1946, J. Paleontol. 20., p. 245, pl. 35,
fig. 7.
Trinidad reference. III: 1/22–23; V: 1/35–36.

The generic assignment of the Trinidad specimens is somewhat problematic as the initial stage
required for the assignment to *Bigenerina* is
hardly detectable (Fig. 6.35–36) compared to
the holotype figure (Fig. 6.34). It is therefore
possible that the Trinidad forms would fit better
into the genus *Reophax*. The Trinidad specimens
are very close to *Reophax minuta* Tappan, a
species that occurs in the American and European Aptian-Albian. In Trinidad the species is
present in the two zones of the Maridale Formation, thus ranging from Late Aptian to basal
Albian. The species was originally described
from the Late Albian Washita Formation. In
Europe it has been reported from the Valanginian and Hauterivian.

TEXTULARIA BETTENSTAEDTI
Bartenstein & Oertli
Figure **6.26–28**; 5

Type reference. *Textularia bettenstaedti* Bartenstein
& Oertli 1977, Neues Jb. Geol. und Paläontol
Mh., 15–24, p. 16, fig, 3/1.
Trinidad reference. V: 953, 2/14–15.

As the illustrations show, the Trinidad specimens (Fig. 6.26–27) compare well in shape and
size with the northwestern German forms (Fig.
6.28, paratype). The species occurs in Trinidad
in the Late Aptian to Early Albian *Hedbergella
rohri* Zone of the Maridale Formation. Originally the species was described from the Early
Albian. World-wide it is known to range in the
boreal and tethyan realms from Late Aptian to
earliest Middle Albian. With its stratigraphic
restriction *Textularia bettenstaedti* can be
regarded as a good index form.

Family Involutinidae

TROCHOLINA INFRAGRANULATA
Noth
Figure **10. 29–30**; 5

Type reference. *Trocholina infragranulata* Noth
1951, Jb. geol. Bundesanstalt, spec. v. 3, p. 69,
pl. 1, fig. 32.
Trinidad reference. I: 44, 5/112, 6/141.

Compared with the side view of the holotype
(Fig.10.30*a*) the illustrated Trinidad specimen
(Fig. 10.29*a*) appears more low conical and not
nearly as hemispherical as is the case in the corresponding holotype illustration. However,
according to the statements given in Trinidad
I, these shapes fall within the variability of the
species.
    The species is regarded as an indicator for
shallow or reefal conditions. In Trinidad it is
frequent in the type locality of the Early Barremian *Lenticulina barri* Zone, Toco Formation.

Family Nodosariidae

DENTALINA BONACCORDENSIS
Bartenstein & Bolli
Figure **8.11–12**; 5

Type reference. *Dentalina bonaccordensis*
Bartenstein & Bolli 1986, p. 955, pl. 2, fig. 32.
Trinidad reference. I: 955, 2/32–34.

Reference is made to the original description by
Bartenstein & Bolli (1986) where the species is
said to have some similarities to *Nodosaria bifur-
cata* Tappan, without a bifurcation of the thin
ribs however.
    In Trinidad the species has only been recorded
from the Late Aptian-Early Albian *Hedbergella
rohri* Zone of the Maridale Formation.

NODOSARIA cf. CHAPMANI Tappan
Figure **8.3–4**; 5

Type reference. *Nodosaria chapmani* Tappan 1940,
J. Paleontol., 14, p. 103, pl. 16, fig. 9.
Trinidad reference. I: 36, 7/152.

The figured Trinidad specimen (Fig. 8.3) is only
a fragment consisting of four chambers which
are ornamented by six longitudinal ribs, with
only slightly incised sutures between the chambers. In contrast the test of the holotype (Fig.
8.4) appears more delicate and slender, with
longer chambers and only four longitudinal ribs.
Based on these differences which hardly fall
within the variability of a species the figured
Trinidad specimen is therefore better not
included in *Nodosaria chapmani*.

NODOSARIA cf. ZIPPEI Reuss
Figure **8.1–2**; 5

Type reference. *Nodosaria zippei* Reuss 1844,
Geognostische Skizzen aus Boehmen; II: Die
Kreidegebilde des westlichen Boehmens, ein
morphologischer Versuch. Prag, C. W. Medan,
2, p. 210, (fig. not given); see Reuss 1845, Die
Versteinerungen der böhmischen
Kreideformation, pt. 1, pl. 8, figs. 1–3.
Trinidad reference. I: 36, 6/125.

Bartenstein, Bettenstaedt & Bolli (1957) place
their illustrated specimen (Fig. 8.1) only with
reservation into *Nodosaria zippei*. The Trinidad
specimen consists of only three chambers and is

hardly comparable with the specimen originally figured by Reuss. The long and slender test of this specimen (Fig. 8.2) possesses in contrast 18 chambers and more strongly incised intercameral sutures.

The Trinidad specimen was found in the *Lenticulina barri* Zone, Toco Formation.

PSEUDONODOSARIA HUMILIS
(Roemer)
Figure **8.5–7**; 5

Type reference. *Nodosaria humilis* Roemer 1841, p. 95, pl. 15, fig. 6.
Trinidad references. I: 37, 7/153–155 (as *Pseudoglandulina*); II: 155, 3/246–247 (as *Rectoglandulina*); III: 406, 6/9–12 (as *Rectoglandulina*); V: 969, 5/36–37.

For discussions on the generic treatment of the species and its relationship to *Pseudonodosaria mutabilis* see the annotations on that species.

PSEUDONODOSARIA MUTABILIS
(Reuss)
Figure **8.8**; Figure **10.1–6**; 5

Type reference. *Glandulina mutabilis* Reuss 1863, p. 58, pl. 5, figs. 7–11.
Trinidad references. I: 37, 7/156 (as *Pseudoglandulina*); II: 154, 3/231–235 (as *Rectoglandulina*); III: 405 (as *Rectoglandulina*); V: 970, 6/1–3 (as *Pseudonodosaria*).

The described and illustrated specimens of the two species *mutabilis* and *humilis* assigned in Trinidad V to the genus *Pseudonodosaria* were in Trinidad I placed in *Pseudoglandulina* and in Trinidad II and III in *Rectoglandulina* with the previously applied genera placed in synonymy. It is obvious from this changing treatment on the generic level that specimens present in the various stratigraphic levels are either synonymous or closely related. Here we follow Bartenstein & Bolli (1986) who place the two species *mutabilis* and *humilis* in *Pseudonodosaria*.

The great variability in the specimens assigned to the two species *mutabilis* and *humilis* is already apparent in the description by Reuss of *mutabilis* where he stresses this fact concerning chamber shapes, growth rates of chambers and number of chambers. This is evident from his illustrations, three of which are reproduced here

(Fig. 10.3–5). He also places *Nodosaria humilis* Roemer (1841) in synonymy with his *mutabilis*. The specimen of Reuss here illustrated on Fig. 10.5 in fact compares closely with the form of Roemer.

In Trinidad specimens with poorly incised intercameral sutures resulting in more compact tests and assigned to *humilis* occur in the *Lenticulina barri* Zone, Toco Formation and in the *Lenticulina ouachensis ouachensis* Zone, Cuche Formation. Specimens placed in *Pseudonodosaria mutabilis* with partially more distinctly incised intercameral sutures and often irregular size of succeeding chambers have been recorded from the *Lenticulina ouachensis ouachensis* Zone of the Cuche Formation (as cf. *mutabilis*, Fig. 8.8). More typical *mutabilis* forms (Fig. 10.1, 2, 6) occur in the *Hedbergella rohri* Zone of the Maridale Formation. They differ, however, from those of Reuss (Fig. 10.3–5) in possessing much larger initial chambers.

LINGULINA PRAELONGA ten Dam
Figure **10.14–15**; 5

Type reference. *Lingulina praelonga* ten Dam 1946, J. Paleontol., 20, p. 576, pl. 88, fig. 12.
Trinidad reference. I: 38, 7/157–158.

Test shape and number of chambers (8) in the holotype compare closely with the larger of the illustrated Trinidad specimen (Fig. 10.15) Different are the higher arched intercameral sutures in the holotype compared with the Trinidad specimens.

Originally described from the Hauterivian of the Netherlands, the Trinidad specimens occur in the Early Barremian *Lenticulina barri* Zone, Toco Formation, and in the *Lenticulina ouachensis ouachensis* Zone of the Cuche Formation. This younger age may explain the different individual chamber shapes, high-arched in the Hauterivian forms, lower in the Barremian specimens of Trinidad.

FRONDICULARIA GAULTINA Reuss
Figure **8.13–15**; 5

Type reference. *Frondicularia gaultina* Reuss 1860,
Die Foraminiferen der westphaelischen
Kreideformation.-Sitzber. K. Akad. Wiss.
Wien, Math. Natwiss. Cl., 40, p. 194, pl. 5,
fig. 5.
Trinidad references. II: 156, 3/265, 270–272; III: 406,
6/1–2; V: 960, 3/42–42.

Compared with the original illustration of Reuss
(Fig.8.15), the Trinidad specimens (Fig. 8.13–
14) differ in that subsequent chambers increase
less rapidly in length, giving the specimen a less
rhomboidal shape. As considerable variability
exists in species of the genus *Frondicularia*, any
specific differentiation therefore depends on the
examination of a large number of specimens.
Forms here placed in *Frondicularia gaultina* consequently vary from subrhomboidal (Trinidad
specimens) to the more lanceolate ones of
Reuss.

The species has been reported world-wide
from Aptian to Cenomanian, in Trinidad from
the Early Aptian to Early Albian Maridale Formation, where they are rare.

Family Vaginulinidae

LENTICULINA CARIBICA Bartenstein &
Bolli
Figure **9.3–4**; 5

Type reference. *Lenticulina caribica* Bartenstein &
Bolli 1986, p. 962, pl. 5, fig. 10.
Trinidad reference. V: 962, 5/10–12.

This is a form regarded by its authors as transitional between the subgenera *Vaginulinopsis*
and *Marginulinopsis*. The species characterized
by a large number of longitudinal ribs (22–26)
shows no relationship to other species of the two
subgenera.

The occurrence in Trinidad is restricted to the
Late Aptian/earliest Albian *Hedbergella rohri*
Zone of the Maridale Formation.

LENTICULINA (LENTICULINA)
ANTILLICA Bartenstein & Bolli
Figure **9.5**; 5

Type reference. *Lenticulina (Lenticulina) antillica*
Bartenstein & Bolli 1986, p. 961, pl. 5, fig. 14.
Trinidad reference. V: 961, 5/13–14.

The species is characterized by numerous longitudinal ribs, robust and interrupted in the early
stage, more delicate and continuous in the last
chambers.

Like *Lenticulina caribica*, also with numerous
longitudinally arranged ribs, *Lenticulina (Lenticulina) antillica* was only observed in the Late
Aptian/earliest Albian *Hedbergella rohri* Zone
of the Maridale Formation.

LENTICULINA (LENTICULINA) BARRI
Bartenstein, Bettenstaedt & Bolli
Figure **9.1–2**; 5

Type reference. *Lenticulina (Lenticulina) barri*
Bartenstein, Bettenstaedt & Bolli 1957, p. 28,
pl. 5, fig. 97; pl. 6, fig. 118.
Trinidad reference. I: 28, 5/97; 6/118.

This is a characteristic, small, nearly circular
form with the last whorl consisting of 12–15
chambers. Typical for the taxon is the umbilical
area filled in by a distinct smooth calcite knob
which may continue to some extent as a rounded
rim between some of the chambers forming the
last two whorls.

The species is fairly frequent in the Early
Barremian of the Toco Formation.

LENTICULINA (LENTICULINA)
EICHENBERGI Bartenstein & Brand
Figure **9.22–23**; 5

Type reference. *Lenticulina (Lenticulina)
eichenbergi* Bartenstein & Brand 1951, p. 285,
pl. 5, fig. 118.
Trinidad references. I: 27, 3/51, 4/72–75; IV: 562,
564.

Compared with the holotype (Fig. 9.23) which
has an elongate equatorial outline, the illustrated Trinidad specimen (Fig. 9.22) is nearly
circular which is also the case for the specimens
figured in Trinidad I, pl. 4, figs. 72–75. Typical
for the species are the knobs along the intercameral sutures.

The species has been reported from Late Valanginian to Late Aptian. In Trinidad it occurs in the *Lenticulina ouachensis ouachensis* Zone of the Cuche Formation.

LENTICULINA (LENTICULINA) KUGLERI Bartenstein, Bettenstaedt & Bolli
Figure **9.6–7**; 5

Type reference. *Lenticulina (Lenticulina) kugleri* Bartenstein, Bettenstaedt & Bolli 1957, p. 27, pl. 5, fig. 96; pl. 6, fig. 116.
Trinidad references. I: 27, 5/95; 6/11. V: 551, 562–563.

Characteristics of this species are its small size, the numerous chambers forming the last whorl (11 in the holotype), and the deep umbilical area.

The species occurs in the Toco Formation to become more frequent in the *Lenticulina ouachensis ouachensis* Zone of the Cuche Formation and is scarce again and poorly preserved in the *Leupoldina protuberans* Zone.

LENTICULINA (LENTICULINA) MERIDIANA Bartenstein, Bettenstaedt & Kovatcheva
Figure **9.8–9**; 5

Type reference. *Lenticulina (Lenticulina) meridiana* Bartenstein, Bettenstaedt & Kovatcheva 1971, N. Jb. Geol. Paleontol. Abh., 139, p. 133, Abb. 1, fig. 15.
Trinidad references. III: 399, 3/40; IV: 562, 564; V: 964, 4/24.

The holotype (Fig. 9.8) and the paratypes figured in Bartenstein, Bettenstaedt & Kovatcheva (Abb. 1, figs. 16–21) are with the exception of their fig. 19 more circular in equatorial outline compared with the distinctly more elongate Trinidad specimens (Fig. 9.9) from the *Hedbergella rohri* Zone of the Maridale Formation and also the specimen illustrated in Trinidad III (pl. 3, fig. 40) from the *Planomalina maridalensis* Zone of the Maridale Formation. The species is characterized by circular or elongate knobs on the intercameral sutures and a narrow to wide umbilical area also surrounded by knobs. These knobs which are well developed in the original types are not or only poorly present in the Trinidad specimens.

LENTICULINA (LENTICULINA) NODOSA (Reuss)
Figure **9.10–14**; 5

Type reference. *Robulina nodosa* Reuss 1863, p. 78, pl. 9, fig. 6. Neotype: Bartenstein 1974, Eclog. geol. Helv., 67, p. 540, pl. 1, fig. 1.
Trinidad references. I: 24, 3/49; 4/66–67; IV: 550, 562–563, 1/34–36; V: 977.

Bartenstein (1974) and Aubert & Bartenstein (1976) discuss this world-wide important Early Cretaceous species in considerable detail. Bartenstein (1974) proposed a neotype (Fig. 9.14) which, compared with the original specimen of Reuss (Fig. 9.11) possesses distinctly more chambers in the last whorl, about 11 against 7 in the original specimen. The here illustrated Trinidad specimens, in particular the one on Fig. 9.10, are closer to the multichambered neotype.

LENTICULINA (LENTICULINA) cf. OUACHENSIS (Sigal)
Fig. **9.26**; 5

Trinidad reference. I: 27, 3/52.

Based on the general shape, forms regarded as cf. *ouachensis* show a similarity to the subspecies *ouachensis* (Fig. 6.15) except that the characteristic rim surrounding the umbilical area is absent. Specimens are generally small and it may be assumed therefore that they represent juvenile forms.

Specimens occur together with the subspecies *ouachensis* in the *Lenticulina ouachensis ouachensis* Zone of the Cuche Formation.

LENTICULINA (LENTICULINA) OUACHENSIS MULTICELLA Bartenstein, Bettenstaedt & Bolli
Figure **9.20–21**; 5

Type Reference. *Lenticulina (Lenticulina) ouachensis multicella* Bartenstein, Bettenstaedt & Bolli 1957, p. 26, pl. 3, fig. 47; pl. 4, fig. 68.
Trinidad reference. I: 26, 3/47; 4/68–69.

Compared with the subspecies *ouachensis, multicella* is more evolute and the last whorl is formed by distinctly more chambers, about 16 as against 8. The nearly circular equatorial outline and the characteristic umbilical features with its narrow circular rim are the same as in the subspecies *ouachensis* . Forms with a number of chambers intermediate between those typical for the two subspecies exist. Specimens with about 12 chambers in the last whorl occur together with typical representatives of the two subspecies.

So far the subspecies *multicella* has only been reported from Trinidad where it is restricted to the *Lenticulina ouachensis ouachensis* Zone of the Cuche Formation.

LENTICULINA (LENTICULINA) OUACHENSIS OUACHENSIS (Sigal)
Figure **9.15–17**; 5

Type reference. *Cristellaria ouachensis* Sigal 1952, 19me Congr. Géol. Int. Algers. Monogr. Rég. sér. 1, no. 26, p. 16, textfig. 10.
Trinidad references. I: 25, 3/50, 4/71, 76; IV: 550, 562–563, 1/37, 2/1–2; V: 977.

The taxon was originally poorly described and illustrated (Fig 9.17). The characteristic features of the subspecies are its close to circular outline in equatorial view, with about eight chambers forming the last whorl. Particularly typical is the umbilical area which is framed by a nearly circular rim. A peripheral keel may be present in better preserved specimens (Fig. 9.15).

World-wide the taxon has been reported from Middle Valanginian to Early Aptian. In Trinidad the subspecies occurs in the two zones of the Middle Barremian to Early Aptian Cuche Formation.

LENTICULINA (LENTICULINA) PRAEGAULTINA Bartenstein, Bettenstaedt & Bolli
Figure **9.18–19**; 5

Type reference. *Lenticulina (Lenticulina) praegaultina* Bartenstein, Bettenstaedt & Bolli 1957, p. 24, pl. 3, fig. 48; pl. 4, fig. 63.
Trinidad reference. I: 24, 3/48, 4/63–65.

The species, with about 10 chambers forming the last whorl, is characterized by its slightly elongate equatorial outline, a wide and delicate peripheral keel and the umbilical area filled in by a knob, often with a somewhat irregular surface.

The species is restricted in Trinidad to the *Lenticulina ouachensis ouachensis* Zone of the Cuche Formation.

LENTICULINA (LENTICULINA) SAXOCRETACEA Bartenstein
Figure **9.24–25**; 5

Type reference. *Cristellaria (Crist.) subalata* Reuss 1863, p. 76, pl. 8, fig. 10. *Lenticulina saxocretacea* Bartenstein 1954 is a new name for *Cristellaria subalata* (see Bartenstein, 1954, pp. 45–46).
Trinidad references. II: 146, 2/95–99; III: 399, 3/36–39; V: 965, 4/32–34.

As some specimens assigned to this species occasionally show weak pustules on the intercameral sutures, the taxon appears related to *Lenticulina (Lenticulina) meridiana* (Fig. 9.8–9).

The species occurs in Europe from Barremian to Albian. In Trinidad it is restricted to the two zones of the Late Aptian to earliest Albian Maridale Formation.

LENTICULINA (LENTICULINA) SUBGAULTINA Bartenstein
Figure **9.30–32**; 5

Type reference. *Lenticulina (Lenticulina) subgaultina* Bartenstein 1962, Senckenberg Lethaea, 43, p. 136, pl. 15, fig. 1.
Trinidad reference. II: 147, 2/128–129; III: 401, 4/30–31.

The smooth surface and the trend to an evolute end-stage are characteristic and give the test a distinct elongate shape.

The Trinidad specimens (Fig. 9.30–31) compare well with the holotype from northwestern Germany (Fig. 9.32), where the species occurs from Aptian to early Middle Albian. In Trinidad the species is known only from the Late Aptian *Planomalina maridalensis* Zone of the Maridale Formation.

LENTICULINA (LENTICULINA) VOCONTIANA Moullade
Fig. **9.27–29**; 5

Type reference. *Lenticulina vocontiana* Moullade 1966, Univ. Fac. Sci. Lab. Géol. Doc. Lyon, no. 15, fasc. 1–2, p. 56, pl. 5, fig. 12.
Trinidad references. III: 398, 3/32–35; V: 966, 5/7–9.

This characteristic species with its ribs following the coiling was originally described from the

Late Aptian of southern France (Fig. 9.27) and later reported from the Aptian of the Ukraine and the Crimea. In Trinidad (Fig. 9.28–29) it occurs in the Late Aptian to earliest Albian *Planomalina maridalensis* and *Hedbergella rohri* zones of the Maridale Formation.

LENTICULINA (MARGINULINOPSIS) ROBUSTA (Reuss)
Figure **8.32–35**; 5

Type reference. *Marginulina robusta* Reuss 1863, p. 63, pl. 6, fig. 5.
Trinidad reference. I: 32, 6/122 (as *Lenticulina (Marginulinopsis)* sp.); V: 965, 4/30–31.

The forms attributed to this species display considerable variability, in particular concerning number, thickness, and arrangement of the longitudinal ribs (Bartenstein & Brand, 1951). The original specimen of Reuss (Fig. 8.35) displays a broader test compared with the Trinidad specimens from the *Hedbergella rohri* Zone of the Maridale Formation (Fig. 8.33–34); on the other hand it is close to the specimen illustrated from the Toco Formation (Fig. 8.32).

The species is said to occur world-wide throughout the Cretaceous. Rare specimens are reported in Trinidad from the Early Barremian *Lenticulina barri* Zone of the Toco Formation and more frequent ones in the Late Aptian to Early Albian *Hedbergella rohri* Zone of the Maridale Formation.

LENTICULINA (MARGINULINOPSIS) SIGALI Bartenstein, Bettenstaedt & Bolli
Figure **8.30–31**; 5

Type reference. *Lenticulina (M.) sigali* Bartenstein, Bettenstaedt & Bolli 1957, p. 32, pl. 5, fig. 99; pl. 6, fig. 130.
Trinidad reference. I: 32, 5/99, 6/130–131.

Reference to Bartenstein, Bettenstaedt & Bolli (1957) is made for the original description of this characteristic species. Following its publication some authors regarded the species as synonymous with *Vaginulinopsis reticulosa* ten Dam, published from the Hauterivian of the Netherlands and/or with *Marginulinopsis djaffaensis* Sigal (1952) from the Hauterivian of Algeria.

Closer examination of the holotype figures and descriptions of the three taxa shows that compared with *sigali*, *reticulosa* is much smaller

(length 0.24 mm against 0.94 mm), has no distinct coiled early stage and possesses fewer longitudinal ribs (about 9 against 17). In comparison with *sigali* the holotype of *reticulosa* may represent a juvenile form. On the other hand the holotype illustrations of *sigali* and *djaffaensis* compare more closely.

Bartenstein, Bettenstaedt & Bolli (1957) also bring *Lenticulina (Marginulinopsis) robusta* into close relation to *Lenticulina (Marginulinopsis) sigali*.

World-wide forms identical or close to *sigali* have been reported mainly from the Hauterivian and Barremian. In Trinidad the species is restricted to the Middle-Late Barremian *Lenticulina ouachensis ouachensis* Zone of the Cuche Formation.

LENTICULINA (SARACENARIA) cf. BRONNI (Roemer)
Figure **8.42–43**: 5

Type reference. *Planularia bronni* Roemer 1841, p. 97, pl. 15, fig. 14.
Trinidad reference. I: 33, 3/61.

A comparison of the form originally illustrated by Roemer (Fig. 8.43) with the figured Trinidad specimen (Figure 8.42) shows considerable differences. The Trinidad specimen consists of fewer and broader chambers in the uncoiled stage which gives the test a less 'sickle'-like appearance than is typical in Roemer's specimen. Furthermore, the apertural face of the Trinidad specimen is distinctly broader in its lower part.

The species is frequent and characteristic in northwestern Germany where it occurs from Late Valendis to Aptian. In Trinidad it is present in the *Lenticulina barri* Zone, Early Barremian Toco Formation and in the *Lenticulina ouachensis ouachensis* Zone of the Middle to Late Barremian Cuche Formation.

LENTICULINA (SARACENARIA) FRANKEI ten Dam
Figure **8.36–38**; 5

Type reference. *Saracenaria frankei* ten Dam 1946, J. Paleontol., 20, p. 573, pl. 88, fig. 1.
Trinidad reference. I: 33, 3/60.; IV: 552, 2/9–10.

The Trinidad specimens figured from the two zones of the Cuche Formation (Fig. 8.36–37) are not as elongate and possess a broader and less high apertural face compared with the holotype (Fig. 8.38).

The species was originally described from the Hauterivian of Holland. European occurrences from various localities are reported from Hauterivian to Lower Aptian. Its presence in Trinidad from the Toco Formation to the Early Aptian *Leupoldina protuberans* Zone of the Cuche Formation support the limited range of the species in Europe.

LENTICULINA (SARACENARIA) SPINOSA (Eichenberg)
Figure **8.44–48**; 5

Type reference. *Saracenaria spinosa* Eichenberg 1935, Niedersaechs. Geol. Ver., Jahresber., 27, pl. 10, pl. 4, fig. 5.
Trinidad references. II: 151, 3/238–242, 256–259; III: 403, 5/11–15; IV: 553, 2/11–12; V: 997.

Based on the spines protruding on the umbilical side from the chamber edges, Bartenstein, Bettenstaedt & Bolli (1966) place the Trinidad specimens in this species which according to these authors displays considerable variability. It extends from elongate, quite slim forms as occur in the *Planomalina maridalensis* Zone of the Maridale Formation (Fig. 8.45–47) to short and broad forms as typified by the holotype itself (Fig. 8.48). The Trinidad specimen from the *Leupoldina protuberans* Zone (Fig. 8.44) is close to it.

Based on widespread occurrences in Europe, *Lenticulina (Saracenaria) spinosa* is regarded as a reliable index for the Aptian to lowermost Early Albian. In Trinidad it is restricted to the *Leupoldina protuberans* Zone of the Cuche Formation and the *Planomalina protuberans* Zone of the Maridale Formation.

LENTICULINA (ASTACOLUS) CREPIDULARIS (Roemer)
Figure **8.16–19**; 5

Type reference. *Planularia crepidularis* Roemer 1842, Neues Jb. Min. Geogn. Geol. Petrefacten-Kunde. Stuttgart, p. 273, pl. 78, fig. 4.
Trinidad references. I: 29, 3/55, 4/82–83; IV: 51, 562–563, 2/3–4; V: 976.

Lenticulina (Astacolus) crepidularis and *Lenticulina (Astacolus) tricarinella* are morphologically closely related and their stratigraphic ranges are also nearly identical. The former differs from the latter in the test being generally smaller and less robust; this applies in particular also to the more delicately developed three keels along the outer periphery. Stages of growth apparently also influence the above criteria. Forms intermediate between the two taxa may occur.

As far as the world-wide ranges of the two species are concerned, they are both reported predominantly from late Middle Jurassic (Dogger) to Early Aptian. In Trinidad *Lenticulina (Astacolus) crepidularis* occurs throughout the Early Barremian Toco- and Middle Barremian to Early Aptian Cuche formations, whereas *Lenticulina (Astacolus) tricarinella* was found only in the *Lenticulina ouachensis ouachensis* Zone, lower part of the Cuche Formation.

LENTICULINA (ASTACOLUS) cf. GRATA (Reuss)
Figure **8.23–24**; 5

Type reference. *Cristellaria (Cristellaria) grata* Reuss 1863, p. 70, pl. 7, fig. 14.
Trinidad reference. I: 30, 3/58, 4/87–88.

According to Bartenstein, Bettenstaedt & Bolli (1957) the specimens assigned by them to *Lenticulina (Astacolus)* cf. *grata* are very variable. Some, like the Trinidad specimen illustrated (Fig. 8. 23) possess a broad apertural face close to that typical for *Saracenaria*. This feature is less well developed in the specimen originally figured by Reuss (Fig. 8.24).

In Trinidad the species has only been reported from the *Lenticulina ouachensis ouachensis* Zone of the Cuche Formation.

LENTICULINA (ASTACOLUS)
MARIDALENSIS Bartenstein, Bettenstaedt &
Bolli
Figure **8.25–29**; 5

Type reference. *Lenticulina (A.) maridalensis*
Bartenstein, Bettenstaedt & Bolli 1966, p.148;
described as *Lenticulina (Astacolus)* sp. (n.
sp.), pl. 2, fig. 122.
Trinidad references. II: 148, 2/120–124, as
Lenticulina (Astacolus) sp. (n. sp.); III: 401, 4/
40; IV: 552, 2/6–8.

The species is characterized by deeply incised
sutures between the chambers which have the
tendency to become evolute in the final growth
stage of the specimen.

The characteristic species has so far only been
recorded from Trinidad where it occurs from
the upper part of the Cuche to the lower part
of the Maridale Formation and thus is typical
for the Aptian.

LENTICULINA (ASTACOLUS)
TRICARINELLA (Reuss)
Figure **8.20–22**; 5

Type reference. *Cristellaria (Cristellaria) tricarinella*
Reuss 1863, p. 68, pl. 7, fig. 9.
Trinidad references. I: 30, 3/56; IV: 562; V: 976.

For annotations see *Lenticulina (Astacolus) cre-*
pidularis (Roemer).

LENTICULINA (VAGINULINOPSIS)
EXCENTRICA (Cornuel)
Figure **8.39–40**; 5

Type reference. *Cristellaria excentrica* Cornuel 1848,
Soc. Géol. France. Mém., sér. 2, tome 3, pt.
1, no. 3, p. 254, pl. 3, figs. 11–13.
Trinidad references. I: 31, 3/59; 4/89–90 (described
as *Lenticulina (V.) prima* (d'Orbigny); II: 149,
2/136–138; III: 402, 4/32–36.

The specimen here illustrated from Trinidad
(Fig. 8.39) differs from Cornuel's (Fig. 8.40) in
the first place in consisting of considerably fewer
chambers. The illustrated Trinidad specimen
from the *Lenticulina ouachensis ouachensis*
Zone of the Cuche Formation possesses only 10
chambers in the last whorl and uncoiled part
compared to 18 in the primary type.

The other specimens identified in Trinidad as
Lenticulina (Vaginulinopsis) excentrica, in particular those from the co-type locality of the
Maridale Formation (Bartenstein & Bolli, 1973,
pl. 4, figs. 32–36) possess even fewer chambers,
are distinctly broader and with only the last
chamber showing a tendency for becoming uncoiled. The specimens illustrated in Bartenstein,
Bettenstaedt & Bolli (1966, pl. 2, figs. 136–138)
from the type locality of the Maridale Formation
are again closer to the one illustrated here (Fig.
8.39). Specimens showing strong variability have
been included as *Lenticulina (Vaginulinopsis)*
excentrica in the Trinidad publications by Bartenstein, Bettenstaedt & Bolli (1957, 1966) and
Bartenstein & Bolli (1973).

The specimens identified in Trinidad as *Len-*
ticulina (Vaginulinopsis) excentrica are from the
Middle to Late Barremian *Lenticulina ouachen-*
sis ouachensis Zone of the Cuche Formation and
the Late Aptian *Planomalina maridalensis* Zone
of the Maridale Formation.

LENTICULINA (VAGINULINOPSIS)
MATUTINA (d'Orbigny)
Figure **8.41**: 5

Type reference. *Cristellaria matutina* d'Orbigny
1850, Prodrome de paléontologie
stratigraphique universelle des animaux
mollusques et rayonés. Paris, V. Masson, 1,
p. 242, pl. 1, fig. 264.
Trinidad reference. I: 31, 6/120.

Compared with *Lenticulina (Vaginulinopsis)*
matutina originally illustrated by d'Orbigny from
the Liassic of Metz (France) the figured specimen from Trinidad differs in a less enrolled
initial stage and in the more incised sutures
dividing the chambers in the later, uncoiled
stage.

The form is restricted in Trinidad to the
Middle to Late Barremian *Lenticulina ouachensis*
ouachensis Zone of the Cuche formation.

CITHARINA ACUMINATA (Reuss)
Figure **8.9–10**; 5

Type reference. *Vaginulina acuminata* Reuss 1863,
p. 49, pl. 4, fig. 1.
Trinidad references. I: 39, 7/159; IV: 554, 562–563,
2/17; V: 976.

The original illustration of Reuss (Fig. 8.9) with
its fine longitudinal ribs compares well with the
figured Trinidad specimen (Fig. 8.10). A minor

difference between the two lies in the somewhat broader later chambers of the Trinidad specimen.

Citharina acuminata has been reported worldwide from Late Hauterivian to Early Albian. In Trinidad its scarce occurrence is restricted to the two zones of the Cuche Formation, Middle to Late Barremian to Early Aptian.

VAGINULINA GEINITZI Reuss
Figure **10.7–9**; 5

Type reference. *Vaginulina geinitzi* Reuss 1875, In: Geinitz, H. B. Das Elbtalgebirge in Sachsen. Palaeont. Beitr. Naturg. Cassel, Deutschland, 1875, 20, p. 91, pl. (II) 21, fig. 1.
Trinidad references. II: 156, 3/267–269; III: 406, 5/64–71.

The specimen described and illustrated by Reuss (Fig. 10.9) is a very small specimen from the Late Cretaceous. In comparison the illustrated Trinidad specimens (Fig. 10.7–8) show a less rapid broadening of the chambers as added, but others compare better with the specimen of Reuss.

In Trinidad the species is reported from the Late Aptian *Planomalina maridalensis* Zone, Maridale Formation. World-wide, specimens assigned to *Vaginulina geinitzi* are known from Albian to Turonian.

VAGINULINA KOCHI Roemer
Figure **10.10–11**; 5

Type reference. *Vaginulina kochi* Roemer 1841, p. 96, pl. 15, fig. 10.
Trinidad reference. I: 38, 5/105, 6/124.

The illustrated Trinidad specimen (Fig. 10.10) compares well with the robust form originally described from northern Germany (Fig. 10.11), whose chambers increase rapidly in size.

In Trinidad the species is present only in the Middle to Late Barremian *Lenticulina ouachensis ouachensis* Zone of the Cuche Formation. European occurrences are reported as rare in the Late Valanginian and Early Hauterivian, as frequent from Late Hauterivian to Barremian.

VAGINULINA PROCERA Albers
Figure **10.12–13**; 5

Type reference. *Vaginulina procera* Albers 1952, Geol. Staatsinst. Mitt. Hamburg, 1952, no. 21, p. 80, pl. 4, figs. 3, 9.
Trinidad references. I: 39, 5/102, 6/133; IV: 562, 564.

Reference is made to Albers (1952) who described and illustrated on numerous specimens the variability of the species. Compared with the holotype (Fig.10.13) the illustrated Trinidad specimen (Fig. 10.12) displays a more rapid broadening of the successive chambers, particularly in its later growth stage.

The species is regarded in Europe as an index for the Barremian. In Trinidad it is restricted to the Middle to Late Barremian *Lenticulina ouachensis ouachensis* Zone of the Cuche Formation.

Family Polymorphinidae

FALSOGUTTULINA VANDENBOLDI
(Bartenstein, Bettenstaedt & Bolli)
Figure **10.16**; 5

Type reference. *Guttulina vandenboldi* Bartenstein, Bettenstaedt & Bolli 1957, p. 40, pl. 7, fig. 163.
Trinidad references. I: 40, 7/163–164 (as *Guttulina*); II: 158, 3/309–314; III: 407; V: 972, 3/18–19.

The species was first described from the Middle to Late Barremian *Lenticulina ouachensis ouachensis* Zone of the Cuche Formation under the generic name *Guttulina*. No similar forms were described previously from the Early Cretaceous. Subsequently, the species was also reported and illustrated from the two zones of the Maridale Formation (Late Aptian to earliest Albian). Here, the specimens are however of a distinctly larger test size with a length of up to 0.5 mm as compared with 0.32 mm of the holotype.

Family Ceratobuliminidae

EPISTOMINA CARACOLLA (Roemer)
Figure **10.40–42**; 5

Type reference. *Gyroidina caracolla* Roemer 1841,
   p. 97, pl. 15, fig. 22.
Trinidad references. I: 46, 5/113–114, 6/142; IV: 557,
   2/34–36.

The Trinidad specimens compare well with the
forms known from Europe. They show the same
variability, particularly in the height of both the
spiral and umbilical sides. The Trinidad specimens are characterized by their more moderate
heights, both spirally and umbilically.

   The species is known to range in Europe from
Late Valanginian to about Early Barremian,
often in rich populations. In Trinidad the species
occurs in the Early Barremian *Lenticulina
ouachensis ouachensis* Zone of the Cuche Formation.

EPISTOMINA HECHTI Bartenstein,
Bettenstaedt & Bolli
Figure **10.43**; 5

Type reference. *Epistomina* D-7 Hecht 1938,
   Standardgliederung der Nordwest-deutschen
   Unterkreide nach Foraminiferen.
   Senckenberg. Naturf. Ges. Abh., 443, pl. 12b,
   fig. 13.
Trinidad reference. I: 46, 7/170.

The holotype of *Epistomina hechti* was selected
not from Trinidad but from Northwest Germany
where apparently identical forms were illustrated but not formally described by Hecht
(1938). Figured on his plate 12b are 29 specimens (figs. 1–29) as *Epistomina* D-7. Unfortunately all specimens are reproduced in very low
magnification and with only one view per specimen. As a result sufficient details of the individual specimens are not visible. The specimen
illustrated by Hecht as fig. 13 was re-illustrated
and selected by Bartenstein, Bettenstaedt &
Bolli (1957) as holotype for *Epistomina hechti*.
It is deposited at the Senckenberg Museum in
Frankfurt a. M.
   The species is characterized by the high intercameral sutures on the spiral side through which
the surfaces of the chambers, particularly those
of the older whorls, appear distinctly lowered.

In Northwest Germany the species is restricted to the highest Early Barremian to middle
Middle Barremian and thus makes an excellent
index for this short stratigraphic interval. *Epistomina hechti* occurs in Trinidad in the Middle to
Late Barremian *Lenticulina ouachensis ouachensis* Zone of the Cuche Formation.

EPISTOMINA ORNATA (Roemer)
Figure **10.44–45**; 5

Type reference. *Planulina ornata* Roemer 1841,
   p. 98, pl. 15, fig. 25.
Trinidad references. I: 46, 5/110, 115, 6/143; IV: 557,
   2/37.

The illustrations of Roemer's original specimen
are at such low magnification that no details are
recognizable, nor is a sufficient description
given.
   Epistomina ornata seems to be closely related
to *Epistomina caracolla* from which it differs in
the stronger development of the intercameral
sutures, a feature which might be environmentally controlled.
   World-wide the two species also have nearly
the same stratigraphic range. *Epistomina ornata*
ranges from Late Valanginian to ?Early Aptian.
In Trinidad it occurs in the Middle to Late
Barremian *Lenticulina ouachensis ouachensis*
Zone and in the Early Aptian *Leupoldina protuberans* Zone, both Cuche Formation.

Family Bolivinidae

BOLIVINA TEXTILARIOIDES Reuss
Figure **10.23–24**; 5

Type reference. *Bolivina textilarioides* Reuss 1863,
   p. 81, pl. 10, fig. 1.
Trinidad reference. I: 42, 5/108, 6/140.

The species occurs in Trinidad together with
Trocholina infragranulata and some megafossils
considered to have lived in a shallow water
environment (reefal area in a wider sense). On
the other hand the presence of, for instance,
several *Lenticulina* species at the type locality of
the Toco Formation rather point to a deeper
water environment. A heterogeneous association of shallow and deeper water taxa at the

type locality of the *Lenticulina barri* Zone can therefore not be excluded.

World-wide *Bolivina textilarioides* is considered to range from Late Jurassic to Early Barremian. The rare specimens found in Trinidad (Fig. 10.23) are restricted to the Early Barremian *Lenticulina barri* Zone. The holotype (Fig. 10.24) is illustrated for comparison.

Family Siphogenerinoididae

ORTHOKARSTENIA SHASTAENSIS
Dailey
Figure **10.19–22**; 5

Type reference. *Orthocarstenia shastaensis* Dailey 1970, Contrib. Cushman Found. foramin. Res., 21, p. 107, pl. 12, fig. 8.
Trinidad reference. IV: 554, 2/18–19; V: 969, 5/35.

As noted in Trinidad IV, p. 554–555, and also shown here, the Trinidad specimens (Fig. 10.21–22) are slimmer as compared to the holotype (Fig. 10.20) but close in this respect to the paratype (Fig. 10.19). Further, the chambers in both the holotype and the paratype are not as high as in the Trinidad specimens.

So far *Orthokarstenia shastaensis* is known only from the western hemisphere (California, Grand Banks, Blake Plateau, Trinidad) from where it has been reported from uppermost Barremian to Cenomanian. The species occurs in Trinidad in the Early Aptian *Leupoldina protuberans* Zone of the Cuche Formation and very scarce in the Late Aptian to earliest Albian *Hedbergella rohri* Zone of the Maridale Formation.

Family Pleurostomellidae

'NODOSARELLA' ROHRI Bartenstein,
Bettenstaedt & Bolli
Figure **10.17–18**; 5

Type reference. *Nodosarella rohri* Bartenstein, Bettenstaedt & Bolli 1957, p. 43, pl. 6, fig. 126.
Trinidad reference. I: 43, 6/126–127.

The authors placed the entirely biserial species into the genus *Nodosarella* which is defined as uniserial throughout. We could not find a genus

in the family Pleurostomellidae nor in others like Fursenkoinidae that would accommodate this characteristic and very slender biserial form. It is therefore likely to belong to a new genus.

So far the species has only been reported from the Middle to Late Barremian *Lenticulina ouachensis ouachensis* Zone of the Cuche Formation.

Family Bagginidae

VALVULINERIA LOETTERLEI (Tappan)
Figure **10.25–28**; 5

Type reference. *Gyroidina loetterlei* Tappan 1940, J. Paleontol., 14, p. 120, pl. 19, fig. 10.
Trinidad references. II: 161, 4/354–355; III: 410, 6/66–77; IV: 556, 2/30–33; V: 975, 6/23–26.

There exists a considerable variability amongst specimens described world-wide as *Valvulineria loetterlei*; it concerns general test shape and number of chambers forming the last whorl. There also exist a number of synonyms (Trinidad III, p. 410).

The Trinidad specimens identified as *Valvulineria loetterlei* from the Early Aptian *Leupoldina protuberans* Zone of the Cuche Formation have five well-inflated chambers in the last whorl, with distinctly incised intercameral sutures (Fig.10.26–28). In contrast the holotype of *Valvulineria loetterlei* (Fig.10.25), originally placed in *Gyroidina*, shows eight chambers in the last whorl and with intercameral sutures hardly incised, giving the test a circular outline in equatorial view. While in the illustrated spiral view of a Trinidad specimen (Fig. 10.26) the chambers of the earlier whorl are visible, they appear in the holotype (Fig.10.25) to be covered by a knob.

Family Globorotalitidae

CONOROTALITES BARTENSTEINI APTIENSIS (Bettenstaedt)
Figure **11.3–6**; 5

Type reference. *Conorotalites bartensteini aptiensis* Bettenstaedt 1952, p. 278, 282, pl. 3, fig. 32.
Trinidad references. II: 162–163, 4/357–359; III: 411, 6/62–65; IV: 558, 562, 3/6–10; V: 974, 6/11–13.

Reference is made to *Conorotalites bartensteini intercedens* for a discussion of the two subspecies *intercedens* and *aptiensis*.

CONOROTALITES BARTENSTEINI INTERCEDENS (Bettenstaedt)
Figure **11.1–2**; 5

Type reference. *Globorotalites bartensteini intercedens* Bettenstaedt 1952, p. 280, pl. 3, fig. 30.
Trinidad references. I: 48, 8/171 *a–c*, 172–202; IV: 562, 564.

Bettenstaedt established a phylogenetic sequence *Conorotalites bartensteini bartensteini – C. bartensteini intercedens – C. bartensteini aptiensis*. Of these the subspecies *intercedens* and *aptiensis* occur also in Trinidad: the former in the Middle to Late Barremian *Lenticulina ouachensis ouachensis* and the Early Aptian *Leupoldina protuberans* zones of the Cuche Formation; the latter – the youngest of the three subspecies – from the *Leupoldina protuberans* Zone to the Late Aptian-earliest Albian *Hedbergella rohri* Zone of the Maridale Formation and rarely up to the Early Cenomanian.

Illustrated on Fig. 11 are the holotypes of the two subspecies *intercedens* and *aptiensis* (Fig. 11.2, 2*a*, 2*b* and 6, 6*a*, 6*b*). In side view they show the different and characteristic outlines: on the umbilical side rounded in *intercedens* (Fig. 11.2*a*), conical in *aptiensis* (Fig. 11.6*a*). This reflects the main features characterizing the phylogenetic sequence, i.e. a gradual change from a strongly inflated, almost hemispherical, to a pointed conical stage. Corresponding Trinidad specimen are shown on Fig.11.1, 1*a–b* for the subspecies *intercedens*, on Fig. 11.3–5 for *aptiensis*.

For a detailed description of the morphologi-

cal characters and differentiation of the three subspecies *bartensteini*, *intercedens* and *aptiensis* reference is made to the original descriptions by Bettenstaedt (1952), and also to Bartenstein, Bettenstaedt & Bolli (1957).

The subspecies of *Conorotalites bartensteini* have been recorded from many parts of Europe, always in the same stratigraphic sequence. They can thus be regarded as reliable index forms.

Family Gavelinellidae

GAVELINELLA BARREMIANA Bettenstaedt
Figure **10.35–39**; 5

Type reference. *Gavelinella barremiana* Bettenstaedt 1952, p. 275, pl. 2, fig. 27 (holotype), pl. 2, figs. 26, 28, 29 (paratypes).
Trinidad references. I: 47, 7/168–169; IV: 558, 562, 564, 2/38, 3/1–3.

Gavelinella barremiana is considered the ancestor of *G. intermedia* from which it differs in the more strongly backwards drawn intercameral sutures, resulting in narrower chambers (see also Trinidad I, p. 47).

Bartenstein, Bettenstaedt & Bolli (1957), illustrate two *Gavelinella barremiana* specimens from Trinidad. One, here reproduced on Fig. 10.36, consists of 7½ chambers in the last whorl. In equatorial view the test is distinctly elongate and the periphery shows slightly lobate last chambers. This specimen compares well with the holotype and paratype illustrated in Bettenstaedt (1952, his pl. 2, figs 27a-c and 26a-c respectively). The paratype is here reproduced on Fig. 10.37, together with the Trinidad specimen (Fig. 10.36). The reason for here reproducing Bettenstaedt's paratype and not his holotype is that the holotype illustrations, in particular the spiral view, are very poor.

The other Trinidad specimen (Fig. 10.38) consists of nine chambers in the last whorl. In equatorial view it is more circular and the last chambers do not form a lobate periphery as is the case with the other illustrated Trinidad specimens. The specimen compares well with that illustrated by Bettenstaedt (1952, pl. 2, figs. 29a-b), reproduced here on Fig. 10.39.

From the above discussed differences between the two groups of specimens one could consider

Fig. 5. Distribution of selected benthic foraminifera in the Early Cretaceous of Trinidad (illustrated on Figs. 6–11).

					AGE	Distribution of selected benthic foraminifera in the Early Cretaceous of Trinidad, W.I.
Early Barremian	Middle/Late Barremian	Early Aptian	Late Aptian	Earliest Albian/Late Aptian		
Toco	Cuche		Maridale		FM	
Lenticulina barri	Lenticulina ouachensis ouachensis	Leupoldina protuberans	Planomalina maridalensis	Hedbergella rohri	ZONE	

Species list (alphabetical, with figure number):

Ammobaculites reophacoides	52
Ammobaculites torosus	62
Ammobaculites trinidadensis	15
Bigenerina clavellata	65
Bolivina textilarioides	1
Choffatella decipiens	8
Citharina acuminata	38
Conorotalites bartensteini aptiensis	53
Conorotalites bartensteini intercedens	39
Dentalina bonaccordensis	71
Dorothia filiformis	54
Epistomina caracolla	9
Epistomina hechti	16
Epistomina ornata	40
Falsoguttulina vandenboldi	46
Frondicularia gaultina	66
Gaudryina compacta	72
Gaudryina dividens	55
Gaudryina reicheli	67
Gaudryinella hannoveriana	48
Gaudryinella sherlocki	56
Gavelinella barremiana	41
Gavelinella intermedia	57
Glomospirella gaultina	68
Haplophragmium cf. aequale	17
Lenticulina (A) cf. grata	19
Lenticulina (A) crepidularis	10
Lenticulina (A) maridalensis	49
Lenticulina (A) tricarinella	25
Lenticulina (L) antillica	73
Lenticulina (L) barri	2
Lenticulina (L) cf. ouachensis	22
Lenticulina (L) eichenbergi	18
Lenticulina (L) kugleri	42
Lenticulina (L) meridiana	58
Lenticulina (L) nodosa	12
Lenticulina (L) ouachensis multicella	21
Lenticulina (L) ouachensis ouachensis	43
Lenticulina (L) praegaultina	23
Lenticulina (L) saxocretacea	69
Lenticulina (L) subgaultina	63
Lenticulina (L) vocontiana	70
Lenticulina (M) robusta	14
Lenticulina (M) sigali	24
Lenticulina (S) cf. bronni	5
Lenticulina (S) frankei	11
Lenticulina (S) spinosa	50
Lenticulina (V) excentrica	45
Lenticulina (V) matutina	20
Lenticulina caribica	74
Lenticulina sp.	26
Lingulina praelonga	6
Marssonella kummi	13
Marssonella oxycona	59
Marssonella praeoxycona	51
Marssonella subtrochus	47
"Nodosarella" rohri	27
Nodosaria cf. chapmani	28
Nodosaria cf. zippei	3
Orthocarstina shastaensis	60
Pseudonodosaria cf. mutabilis	29
Pseudonodosaria humilis	7
Pseudonodosaria mutabilis	75
Reophax guttifer	30
Reophax scorpiurus	31
Reophax subfusiformis	32
Textularia bettenstaedti	76
Triplasia emslandensis acuta	33
Triplasia emslandensis emslandensis	34
Tritaxia plummerae	77
Tritaxia pyramidata	44
Trocholina infragranulata	4
Vaginulina geinitzi	64
Vaginulina kochi	35
Vaginulina procera	36
Valvulinera loetterlei	61
Verneuilinoides neocomiensis	37

Numbered species list (order of appearance in chart):

1	Bolivina textilarioides
2	Lenticulina (L) barri
3	Nodosaria cf. zippei
4	Trocholina infragranulata
5	Lenticulina (S) cf. bronni
6	Lingulina praelonga
7	Pseudonodosaria humilis
8	Choffatella decipiens
9	Epistomina caracolla
10	Lenticulina (A) crepidularis
11	Lenticulina (S) frankei
12	Lenticulina (L) nodosa
13	Marssonella kummi
14	Lenticulina (M) robusta
15	Ammobaculites trinidadensis
16	Epistomina hechti
17	Haplophragmium cf. aequale
18	Lenticulina (L) eichenbergi
19	Lenticulina (A) cf. grata
20	Lenticulina (V) matutina
21	Lenticulina (L) ouachensis multicella
22	Lenticulina (L) cf. ouachensis
23	Lenticulina (L) praegaultina
24	Lenticulina (M) sigali
25	Lenticulina (A) tricarinella
26	Lenticulina sp.
27	"Nodosarella" rohri
28	Nodosaria cf. chapmani
29	Pseudonodosaria cf. mutabilis
30	Reophax guttifer
31	Reophax scorpiurus
32	Reophax subfusiformis
33	Triplasia emslandensis acuta
34	Triplasia emslandensis emslandensis
35	Vaginulina kochi
36	Vaginulina procera
37	Verneuilinoides neocomiensis
38	Citharina acuminata
39	Conorotalites bartensteini intercedens
40	Epistomina ornata
41	Gavelinella barremiana
42	Lenticulina (L) kugleri
43	Lenticulina (L) ouachensis ouachensis
44	Tritaxia pyramidata
45	Lenticulina (V) excentrica
46	Falsoguttulina vandenboldi
47	Marssonella subtrochus
48	Gaudryinella hannoveriana
49	Lenticulina (A) maridalensis
50	Lenticulina (S) spinosa
51	Marssonella praeoxycona
52	Ammobaculites reophacoides
53	Conorotalites bartensteini aptiensis
54	Dorothia filiformis
55	Gaudryina dividens
56	Gaudryinella sherlocki
57	Gavelinella intermedia
58	Lenticulina (L) meridiana
59	Marssonella oxycona
60	Orthocarstina shastaensis
61	Valvulinera loetterlei
62	Ammobaculites torosus
63	Lenticulina (L) subgaultina
64	Vaginulina geinitzi
65	Bigenerina clavellata
66	Frondicularia gaultina
67	Gaudryina reicheli
68	Glomospirella gaultina
69	Lenticulina (L) saxocretacea
70	Lenticulina (L) vocontiana
71	Dentalina bonaccordensis
72	Gaudryina compacta
73	Lenticulina (L) antillica
74	Lenticulina caribica
75	Pseudonodosaria mutabilis
76	Textularia bettenstaedti
77	Tritaxia plummerae

them as two distinct species. The reason why Bettenstaedt (1952), and Bartenstein, Bettenstaedt & Bolli (1957), place these forms into one species is their great variability with apparently morphologically intermediate forms.

World-wide, *Gavelinella barremiana* is recorded from Middle Barremian to Early Aptian. Its distribution in Trinidad is Middle to Late Barremian *Lenticulina ouachensis ouachensis* Zone to the Early Aptian *Leupoldina protuberans* Zone, both Cuche Formation, from where on the taxon gets replaced by the younger *Gavelinella intermedia* (Fig. 10.31–34).

GAVELINELLA INTERMEDIA (Berthelin)
Figure **10.31–34**; 5

Type reference. *Anomalina intermedia* Berthelin 1880, p. 67, pl. 4, fig. 14.
Trinidad references. II: 161, 4/340–353; III: 410, 6/ 48–49; IV: 558, 562, 564, 3/4–5; V: 975, 6/ 14–22.

Regarding the world-wide distribution and the position of *Gavelinella intermedia* within an evolutionary sequence of species reference is made to Trinidad V, p. 975. Where the general test shape and number of chambers forming the last whorl are concerned, the illustrated Trinidad specimens placed into the species *intermedia* (Fig. 10.31–33) compare reasonably well with the original figures of Berthelin (Fig. 10.34).

World-wide the range of the species is given as Early Aptian to Cenomanian. In Trinidad it ranges from the Early Aptian *Leupoldina protuberans* Zone, Cuche Formation, to the Late Aptian to earliest Albian *Hedbergella rohri* Zone, Maridale Formation.

Fig. 6.

1–3 *Glomospirella gaultina* (Berthelin)
1, 2 from the *Hedbergella rohri* Zone, Maridale Fm. (Td.V:1.7–8), × 40. **3** Holotype, × 90.

4–6 *Reophax scorpiurus* Montfort
4, 5 from the *Lenticulina ouachensis ouachensis* Zone, Cuche Fm. (Td. I: 1.2–3), × 40. **6** Holotype, "une demi-ligne de longeur".

7–8 *Reophax subfusiformis* Earland
7 Primary type, × 19. **8** from the *Lenticulina ouachensis ouachensis* Zone, Cuche Fm. (Td. I: 2.35), × 30.

9–12 *Reophax guttifer* Brady
9, 10 from the *Lenticulina ouachensis ouachensis* Zone, Cuche Fm. (Td. I: 1:13, 15), × 40. **11, 12** Primary types, × 50.

13–14 *Choffatella decipiens* Schlumberger
13 from the *Leupoldina protuberans* Zone, Cuche Fm. (Td. IV: 1.32), × 55. **14** from the Tomon Fm. between Rio Frio and Uribante bridge, Tachira, Venezuela. From Maync (1949), pl. 11, fig. 13, × 30.

15–16 *Ammobaculites trinidadensis* Bartenstein, Bettenstaedt & Bolli
15, 16 from the *Lenticulina ouachensis ouachensis* Zone, Cuche Fm. (Td. I: 1.7, 8). **15** Holotype, **16** Paratype, both × 40.

17–19 *Ammobaculites reophacoides* Bartenstein
17, 18 from the *Leupoldina protuberans* Zone, Cuche Fm. (Td. IV: 1.12, 13), × 55. **19** Holotype, × 80.

20–22 *Haplophragmium* cf. *aequale* (Roemer)
20, 21 from the *Lenticulina ouachensis ouachensis* Zone, Cuche Fm. (Td. I: 2.22, 23), × 30. **22** Holotype, magnification not given.

23–25 *Ammobaculites torosus* Loeblich & Tappan
23, 24 from the *Planomalina maridalensis* Zone, Maridale Fm. (Td. II: 1.19; III: 1.55). **23** × 20, **24** × 25. **25** Holotype, × 50.

26–28 *Textularia bettenstaedti* Bartenstein & Oertli
26, 27 from the *Hedbergella rohri* Zone, Maridale Fm. (Td. V: 2.14, 15), × 40. **28** Paratype, × 40.

29–30 *Triplasia emslandensis acuta* Bartenstein & Brand
29 from the *Lenticulina ouachensis ouachensis* Zone, Cuche Fm. (Td. I: 2.38), × 30. **30** Holotype, × 25.

31–33 *Triplasia emslandensis emslandensis* Bartenstein & Brand
31, 32 from the *Lenticulina ouachensis ouachensis* Zone, Cuche Fm. (Td. I: 1.21a, b, 2.25a). **31** × 40, **32** × 30. **33** Holotype, × 25.

34–36 *Bigenerina clavellata* Loeblich & Tappan
34 Holotype, × 46. **35, 36** from the *Hedbergella rohri* Zone, Maridale Fm. (Td. V: 1.35, 36), × 40.

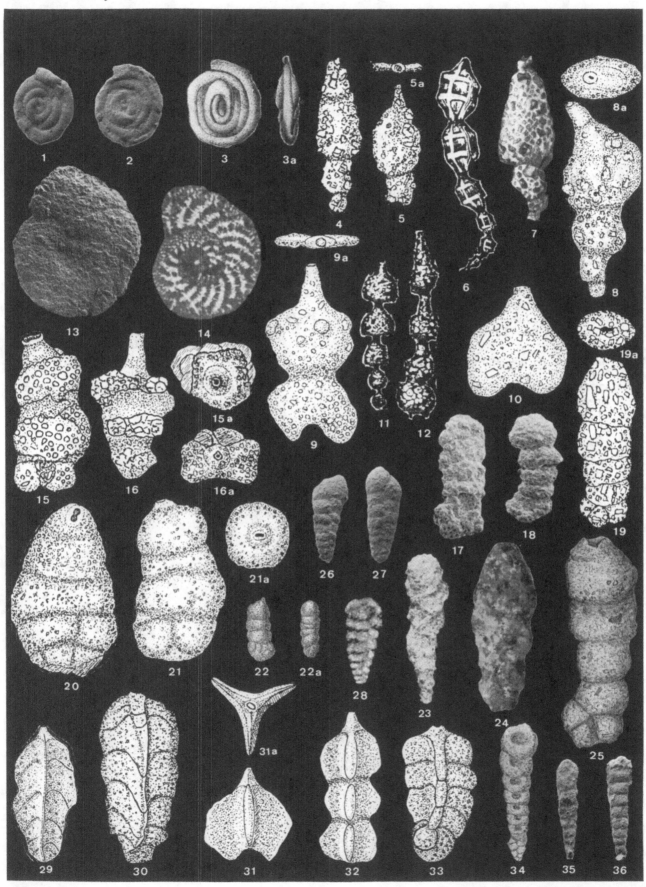

Fig. 7.

1–4 *Gaudryina dividens* Grabert
   1, 2 from the *Hedbergella rohri* Zone, Maridale Fm.
   (Td. V: 1/42, 43), × 40. **3** Paratype, **4** Holotype,
   × 50.

5–8 *Gaudryina compacta* Grabert
   5, 6 from the *Hedbergella rohri* Zone, Maridale Fm.
   (Td. V: 1740, 41), × 40. **7** Paratype, **8** Holotype,
   × 50.

9–10 *Tritaxia plummerae* Cushman
   9 from the *Hedbergella rohri* Zone, Maridale Fm.
   (Td. V: 1/40,41), × 40. **10** Holotype, × 28.

11–12 *Tritaxia pyramidata* Reuss
   11 *Lenticulina ouachensis ouachensis* Zone, Maridale Fm. (Td. I: 2/37a), × 30. **12** Holotype, × 18.

13–14 *Gaudryinella hannoveriana* Bartenstein & Brand
   13 Holotype, × 40. **14** from the *Leupoldina protuberans* Zone Cuche Fm. (Td. IV: 1/20), × 55.

15–16 *Verneuilinoides neocomiensis* (Myatlyuk)
   15 from the *Lenticulina ouachensis ouachensis*
   Zone, Cuche Fm. (Td. I: 2/40), × 40. **16** Holotype,
   × 50.

17–22 *Gaudryina reicheli* Bartenstein, Bettenstaedt &
   Bolli
   17, 18 Holotype, **19** from the *Planomalina maridalensis* Zone, Maridale Fm. (Td. II: 1/53–55), ×
   20. **20–22** from Bartenstein 1985, pl. 1, figs. 19–
   21, × 22.

23–25 *Gaudryinella sherlocki* Bettenstaedt
   23, 24 from the *Hedbergella rohri* Zone, Maridale
   Fm. (Td. V: 2/1–2), × 40. **25** Holotype, × 40.

26–29 *Marssonella praeoxycona* Moullade
   26 from the *Planomalina maridalensis* Zone, Maridale Fm. (Td. III: 2/61), × 25. **27–28** from the
   Leupoldina protuberans Zone, Cuche Fm. (Td. IV:
   1/25–26), × 55. **29** Holotype, × 40.

30–31 *Marssonella subtrochus* Bartenstein
   30 from the *Hedbergella rohri* Zone, Maridale Fm.
   (Td. V: 2/10), × 40. **31** from the *Lenticulina
   ouachensis ouachensis* Zone, Cuche Fm. (Td. I: 3/
   45a, b), × 40.

32–34, *Marssonella kummi* Zedler
   37 **32** Holotype, × 90. **33, 34** from the *Leupoldina
   protuberans* Zone, Cuche Fm. (Td. IV: 1/23, 24),
   × 55. **37** from the *Lenticulina ouachensis ouachensis*
   Zone, Cuche Fm. (Td. I:2/42), × 40.

35–36 *Marssonella oxycona* (Reuss)
   35 Holotype, magnification not given. **36** from the
   Hedbergella rohri Zone, Maridale Fm. (Td. V: 2/
   7), × 40.

38–39 *Dorothia filiformis* (Berthelin)
   38 from the *Hedbergella rohri* Zone, Maridale Fm.
   (Td. V: 1/37), × 40. **39** Holotype, × 60.

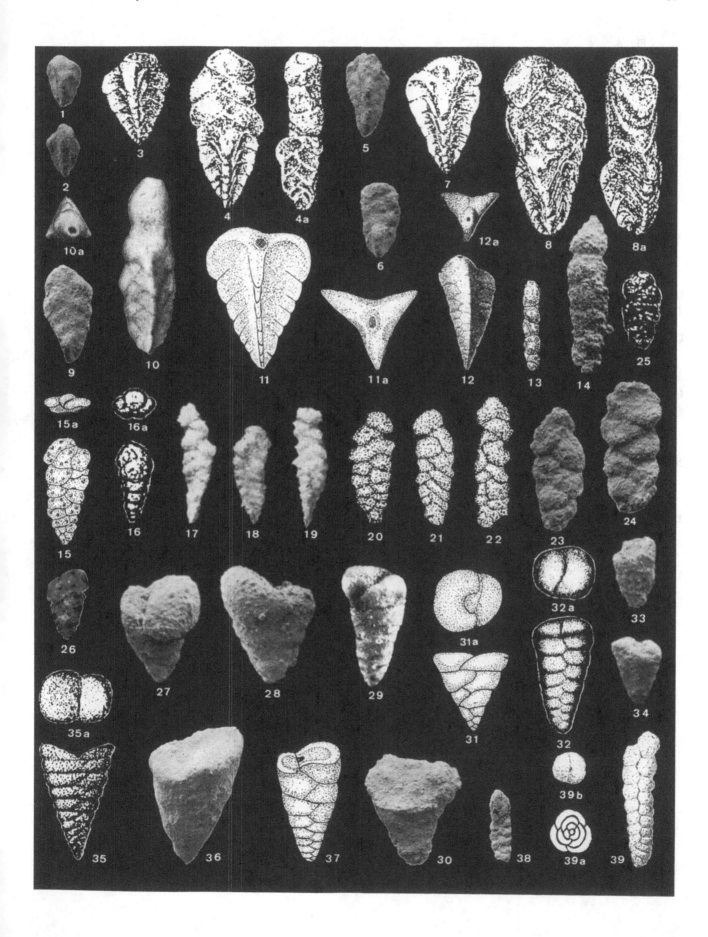

Fig. 8.

1–2 *Nodosaria* cf. *zippei* Roemer
1 from the *Lenticulina barri* Zone, Toco Fm. (Td. I: 6/125), × 50. **2** Holotype.

3–4 *Nodosaria* cf. *chapmani* Tappan
3 from the *Lenticulina ouachensis ouachensis* Zone, Cuche Fm. (Td. I: 7/152), × 40. **4** Holotype, × 44.

5–7 *Pseudonodosaria humilis* (Roemer)
from the *Lenticulina ouachensis ouachensis* Zone, Cuche Fm. (Td. I: 7/153–155), × 40.

8 *Pseudonodosaria* cf. *mutabilis* (Reuss)
from the *Lenticulina ouachensis ouachensis* Zone, Cuche Fm. (Td. I: 7/156), × 40.

9–10 *Citharina acuminata* (Reuss)
9 Holotype, length 1.9 mm. **10** from the *Lenticulina ouachensis ouachensis* Zone, Cuche Fm. (Td. I: 7/159), × 40.

11–12 *Dentalina bonaccordensis* Bartenstein & Bolli
11 Holotype, **12** Paratype, both from the *Hedbergella rohri* Zone, Maridale Fm. (Td. V: 2/32, 34), × 40.

13–15 *Frondicularia gaultina* Reuss
13, **14** from the *Planomalina maridalensis* Zone, Maridale Fm. (Td. II: 3/265, 270), × 20. **15** Holotype, × 62.

16–19 *Lenticulina (Astacolus) crepidularis* (Roemer)
16, **19** from the *Lenticulina ouachensis ouachensis* Zone, Cuche Fm. (Td. I: 3/55, 4/82), × 27. **17**, **18** from the *Leupoldina protuberans* Zone, Cuche Fm. (Td. IV: 2/3, 4), × 50.

20–22 *Lenticulina (Astacolus) tricarinella* (Reuss)
20, **22** from the *Lenticulina ouachensis ouachensis* Zone, Cuche Fm. (Td. I: 4/85, 3/56), × 27. **21** Holotype, length 0.77 mm.

23–24 *Lenticulina (Astacolus)* cf. *grata* (Reuss)
23 from the *Lenticulina ouachensis ouachensis* Zone, Cuche Fm. (Td. I: 3/58), × 27. **24**: Holotype, × 79.

25–29 *Lenticulina (Astacolus) maridalensis* Bartenstein & Bolli
25, **27** from the *Leupoldina protuberans* Zone, Cuche Fm. (Td. IV: 2/8, 7). **26** and **28** are drawings of **25** and **27**, × 50. **29** Holotype (Td. II: 2/122), × 20.

30–31 *Lenticulina (Marginulinopsis) sigali* Bartenstein, Bettenstaedt & Bolli
30 Paratype, (Td. I: 6/131), × 40. **31** Holotype (Td. 5/99, 6/130), × 35. Both from the *Lenticulina ouachensis ouachensis* Zone, Cuche Fm.

32–35 *Lenticulina (Marginulinopsis) robusta* (Reuss)
32 from the *Lenticulina ouachensis ouachensis* Zone, Cuche Fm. (Td. I: 6/122), × 40. **33**, **34** from the *Hedbergella rohri* Zone, Maridale Fm. (Td. V: 4/30), × 60. **35** Holotype, × 40.

36–38 *Lenticulina (Saracenaria) frankei* ten Dam
36 from the *Lenticulina ouachensis ouachensis* Zone, Cuche Fm. (Td. I: 3/60), × 27. **37** from the *Leupoldina protuberans* Zone, Cuche Fm. (Td. IV: 2/9), × 55. **38** Holotype, × 84.

39–40 *Lenticulina (Vaginulinopsis) excentrica* (Cornuel)
39 from the *Lenticulina ouachensis ouachensis* Zone, Cuche Fm. (Td. I: 3/59), × 27. **40** Holotype, × 28.

41 *Lenticulina (Vaginulinopsis) matutina* (d'Orbigny)
from the *Hedbergella rohri* Zone, Maridale Fm. (Td. I: 6/120), × 35.

42–43 *Lenticulina (Saracenaria)* cf. *bronni* (Roemer)
42 from the *Lenticulina ouachensis ouachensis* Zone, Cuche Fm. (Td. I: 3/61), × 27. **43** Holotype, no magnification given.

44–48 *Lenticulina (Saracenaria) spinosa* (Eichenberg)
44 from the *Leupoldina protuberans* Zone, Cuche Fm. (Td. IV: 2/11), × 55. **45–47** from the *Planomalina maridalensis* Zone, Maridale Fm. (Td. II: 3/238, 242, 257), × 20. **48** Holotype, × 53.

Fig. 9.

1–2 *Lenticulina (Lenticulina) barri* Bartenstein, Bettenstaedt & Bolli
from the *Lenticulina barri* Zone, Toco Fm. **1** photograph of holotype, × 35, (Td. I: 5/97). **2** drawing of holotype, × 40 (Td. I: 6/118).

3–4 *Lenticulina caribica* Bartenstein & Bolli
3, 4 from the *Hedbergella rohri* Zone, Maridale Fm. **3** Holotype (Td. V: 5/10). **4** Paratype (Td. V: 5/11) Both × 40.

5 *Lenticulina (Lenticulina) antillica* Bartenstein & Bolli
from the *Hedbergella rohri* Zone, Maridale Fm., Holotype. (Td. V: 5/14), × 40.

6–7 *Lenticulina (Lenticulina) kugleri* Bartenstein, Bettenstaedt & Bolli
from the *Lenticulina ouachensis ouachensis* Zone, Cuche Fm. **6** photograph of holotype (Td. I: 5/95), × 35. **7** drawing of holotype (Td. I: 6/116), × 40.

8–9 *Lenticulina (Lenticulina) meridiana* Bartenstein, Bettenstaedt & Kovatcheva
8 Holotype, × 35. **9** from the *Hedbergella rohri* Zone, Maridale Fm. (Td. V: 4/24), × 40.

10–14 *Lenticulina (Lenticulina) nodosa* (Reuss)
10, 13 from the *Lenticulina ouachensis ouachensis* Zone, Cuche Fm. (Td. I: 3/49, 4/66), × 27. **11** Holotype, × 50. **12** from the *Leupoldina protuberans* Zone, Cuche Fm. (Td. IV: 1/136), × 55. **14** Neotype, × 50.

15–17 *Lenticulina (Lenticulina) ouachensis ouachensis* (Sigal)
15 from the *Lenticulina ouachensis ouachensis* Zone, Cuche Fm. (Td. I: 3/50), × 27. **16** from the *Leupold-*

ina protuberans Zone, Cuche Fm. (Td. IV: 1/37), × 55. **17** Holotype, × 36.

18–19 *Lenticulina (Lenticulina) praegaultina* Bartenstein, Bettenstaedt & Bolli
from the *Lenticulina ouachensis ouachensis* Zone, Cuche Fm. **18** drawing of holotype (Td. I: 3/48). **19** photograph of holotype (Td. I: 4/63), × 27.

20–21 *Lenticulina (Lenticulina) ouachensis multicella* Bartenstein, Bettenstaedt & Bolli
from the *Lenticulina ouachensis ouachensis* Zone, Cuche Fm. **20** Paratype (Td. I: 4/69). **21** Holotype (Td. 3/47). Both × 27.

22–23 *Lenticulina (Lenticulina) eichenbergi* Bartenstein & Brand
22 from the *Lenticulina ouachensis ouachensis* Zone, Cuche Fm. (Td. I: 3/51), × 27. **23** Holotype, × 35.

24–25 *Lenticulina (Lenticulina) saxocretacea* Bartenstein (new name for *Cristellaria subalata* Reuss, 1863, pl. 8, fig.10, pl 9. fig.1).
24 Holotype, × 32. **25** from the *Hedbergella rohri* Zone, Maridale Fm. (Td. V: 4/34), × 40.

26 *Lenticulina (Lenticulina)* cf. *ouachensis* (Sigal)
from the *Lenticulina ouachensis ouachensis* Zone, Cuche Fm. (Td. I: 3/52), × 27.

27–29 *Lenticulina (Lenticulina) vocontiana* Moullade
27 Holotype, × 50. **28, 29** from the *Hedbergella rohri* Zone, Maridale Fm. (Td. V: 5/8, 9), × 40.

30–32 *Lenticulina (Lenticulina) subgaultina* Bartenstein
30, 31 from the *Planomalina maridalensis* Zone, Maridale Fm. (Td. II: 2/128, 129), × 20. **32** Holotype, × 25.

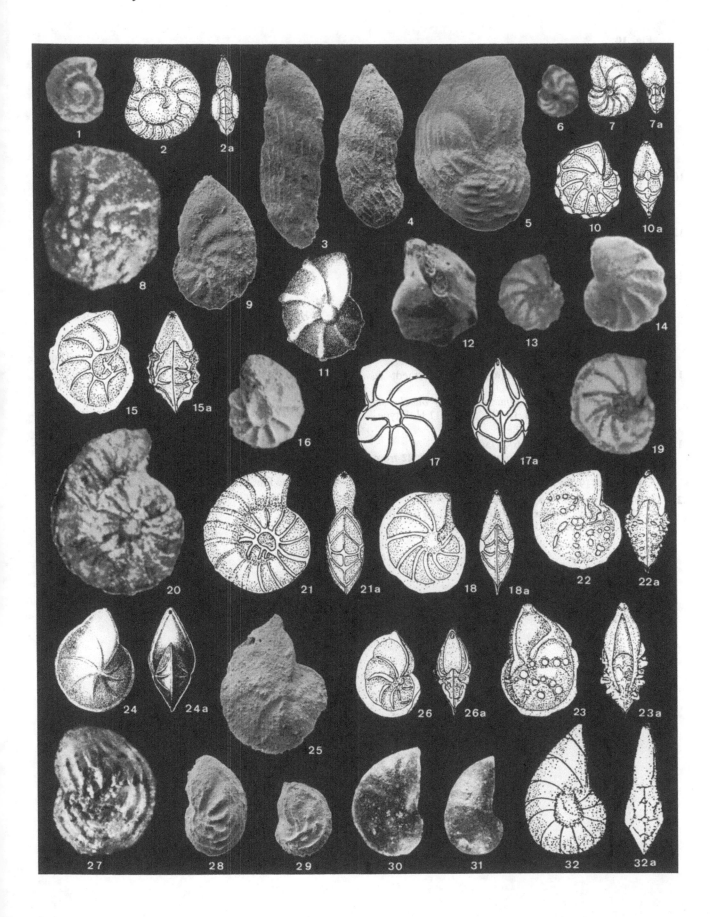

Fig. 10.

1–6 *Pseudonodosaria mutabilis* (Reuss)
1, 2, 6 from the *Hedbergella rohri* Zone, Maridale Fm. (Td. V: 5/36, 6/1, 2), × 40. **3–5** primary types of Reuss 1863 (his pl. 5, figs. 7–9).

7–9 *Vaginulina geinitzi* Reuss
7, 8 from the *Planomalina maridalensis* Zone, Maridale Fm. (Td. III: 5/64, 65), × 25. **9** Holotype, no magnification given.

10–11 *Vaginulina kochi* Roemer
10 from the *Lenticulina ouachensis ouachensis* Zone, Cuche Fm. (Td. I: 5/105), × 35. **11** Holotype, no magnification given.

12–13 *Vaginulina procera* Albers
12 from the *Lenticulina ouachensis ouachensis* Zone, Cuche Fm. (Td. I: 5/102), × 20. **13** Holotype, × 26.

14–15 *Lingulina praelonga* ten Dam
14, 15 from the *Lenticulina ouachensis ouachensis* Zone, Cuche Fm. (Td. I: 7/157, 158), × 40.

16 *Falsoguttulina vandenboldi* (Bartenstein, Bettenstaedt & Bolli)
from the *Lenticulina ouachensis ouachensis* Zone, Cuche Fm. (Td. I: 7/163), Holotype, × 40.

17–18 '*Nodosarella*' *rohri* Bartenstein, Bettenstaedt & Bolli
both from the *Lenticulina ouachensis ouachensis* Zone, Cuche Fm. (Td. I: 6/126, 127). **17** Holotype, **18** Paratype. Both × 40.

19–22 *Orthokarstenia shastaensis* Dailey
19 Paratype, × 35, **20** Holotype, × 40. **21, 22** from the *Leupoldina protuberans* Zone, Cuche Fm. (Td. IV: 2/18, 19), × 50.

23–24 *Bolivina textilarioides* Reuss
23 from the *Lenticulina barri* Zone, Toco Fm. (Td. I: 5/108), × 35. **24** Holotype, × 50.

25–28 *Valvulineria loetterlei* (Tappan)
25 Holotype, × 82. **26–28** from the *Leupoldina protuberans* Zone, Cuche Fm. (Td. IV: 2/30–32), × 100.

29–30 *Trocholina infragranulata* Noth
29 from the *Lenticulina barri* Zone, Toco Fm. (Td. I: 5/112a, b; 6/141), × 35. **30** Holotype, × 40.

31–34 *Gavelinella intermedia* (Berthelin)
31–33 from the *Hedbergella rohri* Zone, Maridale Fm. (Td. V: 6/17, 19, 14b), × 40. **34** Holotype, × 50.

35–39 *Gavelinella barremiana* Bettenstaedt
35 from the *Leupoldina protuberans* Zone, Cuche Fm. (Td. IV: 3/1), × 80. **36** from the *Lenticulina ouachensis ouachensis* Zone, Cuche Fm. (Td. I: 7/168), × 50. **37** Paratype, × 40 (pl. 2, figs. 26a, b in Bettenstaedt 1952). **38** from the *Lenticulina ouachensis ouachensis* Zone, Cuche Fm. (Td. I: 7/169), × 50. **39** Paratype, × 40 (pl. 2, figs. 29a, b in Bettenstaedt, 1952).

40–42 *Epistomina caracolla* (Roemer)
40, 41 from the *Leupoldina protuberans* Zone, Cuche Fm. (Td. IV: 2/35, 34), × 75. **42** from the *Lenticulina ouachensis ouachensis* Zone, Cuche Fm. (Td. I: 6/142), × 40.

43 *Epistomina hechti* Bartenstein, Bettenstaedt & Bolli
from the *Lenticulina ouachensis ouachensis* Zone, Cuche Fm. (Td. I: 170a–c), × 50.

44–45 *Epistomina ornata* (Roemer)
44 from the *Lenticulina ouachensis ouachensis* Zone, Cuche Fm. (Td. I: 6/243), × 40. **45** from the *Leupoldina protuberans* Zone, Cuche Fm. (Td. IV: 2/37), × 25.

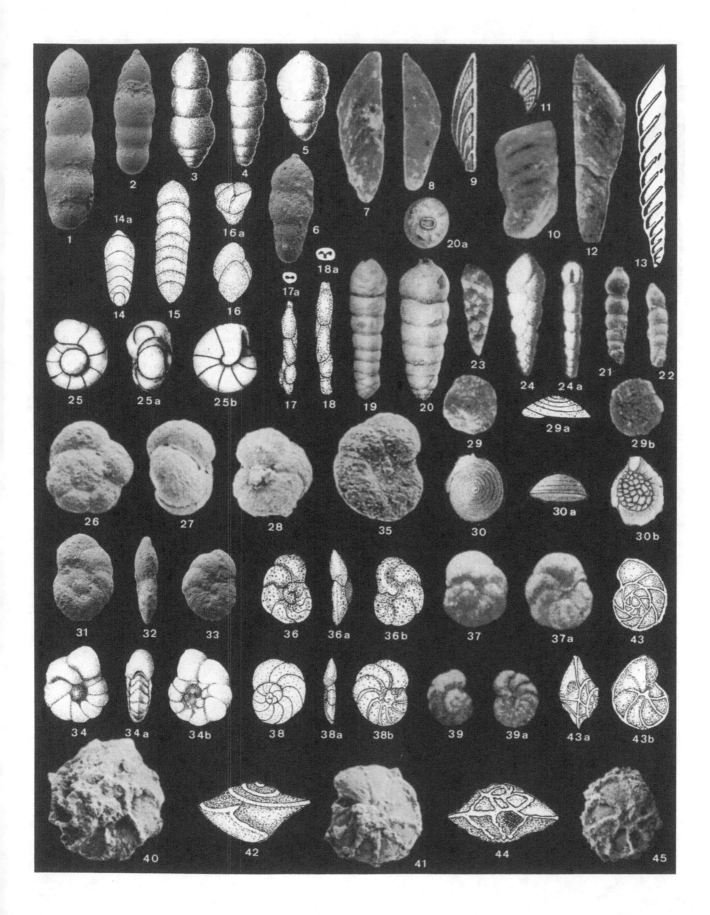

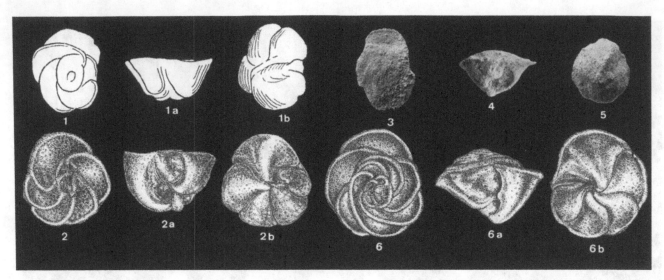

Fig. 11.

1–2 *Conorotalites bartensteini intercedens* (Bettenstaedt)
1, **1a**, **1b** from the *Lenticulina ouachensis ouachensis*
Zone, Cuche Fm. (Td. I: 8/171a–c), × 55. **2**, **2a**,
2b Holotype, × 70.

3–6 *Conorotalites bartensteini aptiensis* (Bettenstaedt)
3–5 from the *Hedbergella rohri* Zone, Maridale Fm.
(Td. V: 6/11–13), × 40. **6**, **6a**, **6b** Holotype, × 70.

Late Albian to Early Eocene

Geologic-stratigraphic background

This section deals with the Middle Cretaceous (Late Albian) to Early Eocene benthic foraminifera of Central and South Trinidad. This time interval includes, from bottom to top, the Gautier, Naparima Hill, Guayaguayare and Lizard Springs formations, as defined in the Stratigraphic Lexicon of Trinidad (Kugler 1956). Their position within the stratigraphic framework of the southern Caribbean area is shown on the regional correlation charts (Fig. 2 and Fig. 3). This complex of four formations consists of a marine sequence of calcareous to argillaceous rocks which has a compound thickness of approximately 1500 metres; its actual thickness, however, may vary considerably due to local or regional hiatuses and/or tectonic disturbances. It lies above a Lower Cretaceous group of formations, of which the youngest is the Maridale Formation (Late Aptian-Early Albian) and is, in turn, overlain by the Navet Formation (mostly of Middle to Late Eocene age). The underlying and overlying formations form the subject of the preceding and the following chapter, respectively, of this volume.

The rocks which nowadays represent the Gautier, Naparima Hill, Guayaguayare and Lizard Springs formations have for a long time been known from relatively small and isolated surface outcrops such as those described by Wall & Sawkins (1860), Waring (1926), Liddle (1928; 2nd edition 1946), Jarvis (1929) and Skelton (1929). These authors have reported various megafossils, usually indicative of a Cretaceous age. Gradually, attention was also focussed on microfossils, usually foraminifera, particularly since the pioneer papers on the Lizard Springs area by Cushman & Jarvis (1928, 1932). Microfaunas from other parts of Trinidad became increasingly well known in the following years (Kugler 1936, Vaughan & Cole 1941, Renz 1942, Cushman & Renz 1946 and 1947a, Kugler 1950). These microfossils, like the associated mega-invertebrate fauna, were generally regarded as supporting a Cretaceous age, although the possibility was left open that outcrops in the Lizard Springs area could represent the lowermost Tertiary.

The major breakthrough in the understanding of the biostratigraphy of these formations is mainly due to two reasons: (a) the drilling by various oil companies of deep exploratory wells allowing a more or less continuous sampling of the Cretaceous-Early Tertiary rock sequence, and (b) the establishment and application of a biostratigraphic zonal scheme based on planktic foraminifera (Bolli 1951, 1957a, 1957b).

The first decisive arguments for an Early Tertiary age of the Lizard Springs Formation, relying on the planktic foraminifera, were presented by Grimsdale (1947) and Bolli (1950); they met the initial opposition of adherents to the traditional stratigraphy like J. A. Cushman and H. H. Renz, but final convincing support was given in particular by Bolli (1952). The succession of foraminiferal biostratigraphic zones, as founded in its essence in Bolli (1957a, b), later underwent minor revisions by Bolli (1966) and Premoli Silva & Bolli (1973). A comprehensive discussion

AGE		PLANKTIC FORAMINIFERAL ZONES			FORMATION
		BOLLI et al. (ed. 1985)		SAUNDERS & BOLLI (1985)	
EOCENE	EARLY	Morozovella aragonensis	P8	Globorotalia aragonensis	UPPER LIZARD SPRINGS
		Morozovella formosa formosa	P7	Globorotalia formosa formosa	
		Morozovella subbotinae	P6	Globorotalia subbotinae	
		Morozovella edgari		Globorotalia edgari	
PALEOCENE	LATE	Morozovella velascoensis	P5	Globorotalia velascoensis	LOWER LIZARD SPRINGS
		Planorotalites pseudomenardii	P4	Globorotalia pseudomenardii	
	MIDDLE	Planorotalites pusilla pusilla	P3	Globorotalia pusilla pusilla	
		Morozovella angulata		Globorotalia angulata	
		Morozovella uncinata	P2	Globorotalia uncinata	
	EARLY	Morozovella trinidadensis	c	Globorotalia trinidadensis	
		Morozovella pseudobulloides	P1 b	Globorotalia pseudobulloides	
		Globigerina eugubina	a	Globigerina eugubina — Rz	
CRETACEOUS	MAASTRICHTIAN	Abathomphalus mayaroensis		Abathomphalus mayaroensis	GUAYAGUAYARE
		Gansserina gansseri		Globotruncana gansseri	
		Globotruncana aegyptiaca		Globotruncana lapparenti tricarinata	
		Globotruncanella havanensis			
	CAMPANIAN	Globotruncana calcarata			not recorded
		Globotruncana ventricosa		Globotruncana stuarti	NAPARIMA HILL
		Globotruncanita elevata			
	SANTONIAN	Dicarinella asymetrica		Globotruncana fornicata	
	CONIACIAN	Dicarinella concavata		Globotruncana concavata	
		Dicarinella primitiva		Globotruncana renzi	
	TURONIAN	Marginotruncana sigali		Globotruncana inornata	
		Helvetoglobotruncana helvetica			
		Whiteinella archaeocretacea			not recorded
	CENOMANIAN	Rotalipora cushmani			
		Rotalipora reicheli			
		Rotalipora brotzeni		Rotalipora appenninica appenninica	GAUTIER
	ALBIAN	Rotalipora appenninica		Favusella washitensis	
		Rotalipora ticinensis		Rotalipora ticinensis ticinensis	

Fig. 12. Late Albian to Early Eocene biostratigraphy: ages, planktic foraminiferal zones and formations. The planktic foraminiferal zones to the left (Bolli *et al.* 1985) represent an international standard; the subdivision to the right (Saunders & Bolli 1985) is the local zonation for Trinidad used in this volume.

of various planktic microfossil groups and their bio-stratigraphy was presented by Bolli *et al.* (1985). The biostratigraphic scheme used throughout the present volume relies on an updated version of Bolli's original zonation which was proposed by Saunders & Bolli (1985). This zonation serves as a local biostratigraphic background for calibrating the stratigraphic ranges of the benthic foraminifera described in the following pages with the planktic foraminiferal sequence observed in Trinidad. Fig. 12 shows the zonation of Saunders & Bolli (1985), its revised ages and its corre-lation with the more generally applicable bio-stratigraphic subdivisions of the Cretaceous and Tertiary proposed in Bolli *et al.* (1985).

The standardisation of the biostratigraphy has also led to a better understanding of the lithostratigraphic units (formations), as they are presently used (see Fig. 12). As presently defined, all the formation boun-daries coincide with biostratigraphic zonal boun-daries. The lithology and the fossil content of the three Cretaceous formations dealt with in the present chapter, the Gautier, Naparima Hill and Guayaguay-are formations, are discussed in Kugler & Bolli (1967). They were all deposited in an open marine environment. The dark grey calcareous shales of the Gautier Formation (first named by Bolli 1951; Late Albian to Early Cenomanian in age), are separated from the underlying Maridale Formation (Late Aptian to Early Albian; see Bartenstein & Bolli 1986) by a hiatus. Between the Gautier Formation and the next higher formation, the Naparima Hill Formation (Kugler 1950; Turonian to Campanian), there is, at least on the Trinidad mainland, a major stratigraphic gap which corresponds to a considerable portion of the Cenomanian as well as the lower part of the Turonian. At the top of the Naparima Hill Formation, the Late Campanian (*Globotruncana calcarata* Zone) has so far not been recognized, and there is also a distinct lithologic change from a hard calcareous shale to a more argillaceous facies of the Guayaguayare Formation (Bolli 1952; Maastrichtian). In the lower part of this formation, a thin but distinct facies fauna with *Siphogenerinoides* ('Marabella facies') is found locally.

In the basal Tertiary, possibly also at the top of the Cretaceous, a non-calcareous facies (Chaudiere Formation, 'Upper Tarouba', 'Arenaceous Lizard Springs', *Rzehakina epigona* Zonule) is frequently encountered. This non-calcareous interval or, where it is not present, the youngest Cretaceous (normally the Guayaguayare Formation) is followed by the calcareous-argillaceous rocks of the Lizard Springs Formation. This formation is a typical representative

of the 'Velasco type' deep-water facies as defined in Berggren & Aubert (1975). The fauna of the Lizard Springs Formation (Paleocene to Early Eocene) was first described by Cushman & Jarvis (1928). The sep-aration of this formation into a lower and upper part, proposed by Cushman & Renz (1946), is justified bio-stratigraphically more than lithologically. Near the base of the Tertiary, the normal Lizard Springs facies is locally replaced by a shallower water 'Midway type' (Berggren & Aubert 1975) intercalation here termed the 'Rochard facies'. A slightly younger (Late Paleo-cene) fauna with a distinct neritic 'Midway' aspect is that of the Soldado Formation, described by Vaughan & Cole (1941), Cushman & Renz (1942) and Kugler & Caudri (1975) from the small island of Soldado Rock. The boundary of the Lizard Springs Formation against the overlying Navet Formation (latest Early Eocene to Late Eocene) is characterised by a change into a lighter grey colour of the sediment and a distinct increase of its calcium carbonate content (Bolli 1957*b*, p. 64).

Previous studies in Trinidad

The classical taxonomic papers on the smaller benthic foraminifera of the stratigraphic interval covered in this chapter are those by Cushman & Jarvis (1928, 1932) and Cushman & Renz (1946, 1947*a*). The first three give a record of the fauna in the type area of the Lizard Springs Formation (Southeast Trinidad), formerly included in the Late Cretaceous but now-adays dated as Paleocene to Early Eocene. The separ-ation of the Lizard Springs Formation into a lower and an upper zone, attempted by Cushman & Renz (1946), was based on the presence of *Rzehakina epi-gona lata* in the former and its absence in the latter. However, as the samples of the upper zone may con-tain *Rzehakina epigona minima*, a subspecies which normally coexists with *R. epigona lata*, the original biostratigraphic argument for such a definition of the upper zone is not valid (see also Beckmann 1960, p. 59). The distinction of the lower and upper Lizard Springs Formation is maintained at the level originally suggested by Cushman & Renz (Bolli 1957*b*), but is now defined at the top of the *Globorotalia velascoensis* Zone. This level is marked by a distinct change of both the planktic and benthic foraminiferal faunas. The last paper mentioned above (Cushman & Renz 1947*a*) deals with five isolated localities in Central Trinidad (see comments in Bolli 1951, p. 193 and Beckmann 1960, p. 59). Two of these localities (Chaudiere shale and Upper Tarouba member) show a foraminiferal fauna consisting entirely of agglutinated species whose exact age is usually difficult to determine. The remaining three localities (Pointe-à-Pierre Railway Cut marl, Lantern marl and lower member of the Tarouba Formation) contain faunas with both calcareous and arenaceous species which, with the possible exception of the Lower Tarouba member, are not homogeneous. We must therefore conclude that the faunas described by Cushman & Jarvis and Cushman & Renz, while being a useful source of taxonomic information, are of limited biostratigraphic value. The only other significant publication is that by Kaminski *et al.* (1988). It is based on about 60 samples from the Lizard Springs Formation and two samples of the Upper Cretaceous Guayaguayare Formation. Its taxonomic part deals exclusively with the agglutinated species. It also includes discussions of modern aspects of the paleoecology and sedimentology.

It can be seen that the published record on the Middle Cretaceous to Early Eocene foraminifera of Trinidad is most uneven, and for the greater part of the Cretaceous portion is patchy and poorly documented. The benthic foraminiferal faunas of the Gautier Formation (Late Albian to possibly Early Cenomanian) and of the greater part of the Naparima Hill Formation (Turonian to Santonian) have so far never been described, although some genera and species are listed in the paleoecological paper by Koutsoukos & Merrick (1985). From the younger part of the Cretaceous section, the uppermost part of the Naparima Hill Formation and the Guayaguayare Formation (Campanian to Maastrichtian), a few species have been figured by Cushman & Renz (1947*a*) and by Kaminski *et al.* (1988). For the Paleocene-Early Eocene, the record is considerably better, particularly for the agglutinated species, but the calcareous species, especially their stratigraphic distribution, is still incompletely known. The faunas recorded here from the Lizard Springs Formation are derived for the greater part from samples which were not available to the earlier authors; for this reason they include a considerable number of species not previously described. On the other hand, some samples used earlier were not available during the present investigations (parts of the Lizard Springs type section; some surface and well samples used by Kaminski *et al.* 1988), and the presence of some species recorded earlier could therefore not be confirmed.

Source of the Trinidad material

The material available for study consists of both surface samples and well samples (mostly cores). The original samples and documentation are stored in the Geological Laboratory of the Trinidad and Tobago Oil Company, Pointe-à-Pierre, Trinidad, West Indies. Duplicates of some of the samples or of representative faunas are deposited in the Natural History Museum, Basel, Switzerland. The following list shows the localities of the surface outcrops and well sites (see also the locality map, Fig. 13).

A. Surface

Bn 61: Pickering Quarry. NE side of Naparima Hill, San Fernando.

Bo 521: North entrance of the town of San Fernando; outcrop on eastern bank of Pointe-à-Pierre Road, behind a house some 60 feet (18 m) from the north end of the bridge across the Vista Bella River (same locality as for Rz 413).

G 3644: Outcrop on right bank of Gautier River (right side branch of Cunapo River), about 1100 feet (335 m) southwest above junction of waterfall branch, 1¼ miles (2 km) southeast of Mamon, Guaico-Tamana Road, Chert Hill area, eastern Central Range (same general area as locality JS 1019/KR 8385). See Bolli (1957*a*), p. 59.

Hk 1831: Left bank tributary of Cascas River, 180 feet (55 m) from its junction with the Cascas River, in a ravine which enters the Cascas River 3400 feet (1035m) from its confluence with the Moriquite River, about 1½ miles (2.4 km) west of the point where the Moriquite River crosses the Moruga Road (between the 14 (22.5 km) and 14¼ mile (23 km) posts), South Trinidad. Type locality of the *Globorotalia subbotinae* Zone; see Bolli (1957*b*), p. 64.

Hk 1832: Same locality as for HK 1831.

JS 1019: In the Gautier river, a right side branch of the Cunapo River, at junction of waterfall branch, 1¼ miles (2 km) southeast of Mamon, Guaico-Tamana Road, north of Chert Hill, eastern Central Range. Same locality as for KR 8385.

K 9415: On left side of disused railway track, south of Pointe-à-Pierre Railway Station, about 730 feet (222 m) from the level crossing of Station

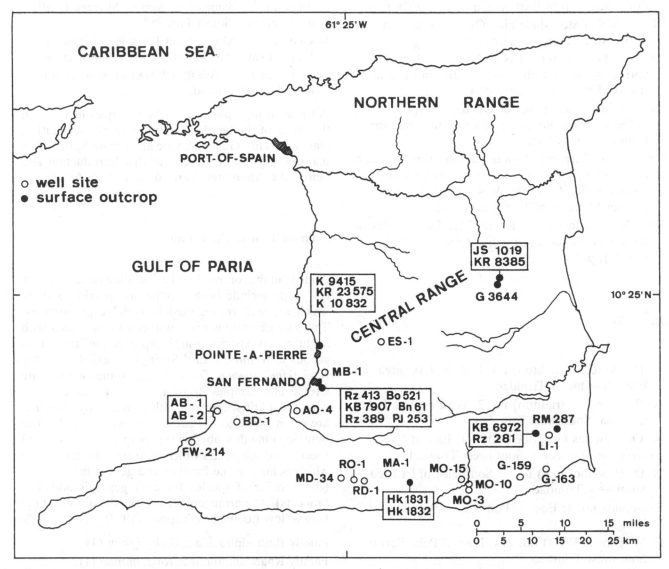

Fig. 13. Map of Trinidad showing the Late Albian to Early Eocene sample localities. Abbreviated well locations: AB, Antilles Brighton; AO, Antilles Oropouche; BD, Boodoo-singh; ES, Esmeralda; FW, Fortin West; G, Guayaguayare; LI, Lizard; MA, Marac; MB, Marabella; MD, Morne Diablo; MO, Moruga; RD, Rock Dome; RO, Rochard.

Road, Pointe-à-Pierre (very close to Locality KR23575). Co-type locality of the *Globorotalia velascoensis* Zone; see Bolli (1957*b*), p. 64.

K 10832: Northeast bank of Tank Farm at the Old Club Site, Pointe-à-Pierre. Type locality of the *Globorotalia pseudomenardii* Zone; see Bolli (1957*b*), p. 64.

KB 6972: Ravine in eastern tributary of Ampelu River, about 150 feet (45 m) from its junction with the Ampelu River; Lizard Springs area, about 1¼ miles (2 km) southeast of the road junction of the Rio Claro-Guayaguayare Road (8¾ mile (14 km) post) and the old Trinidad Central

Oilfields Road leading to the abandoned Lizard Springs Oilfield; Southeast Trinidad. See Cushman & Renz (1946) for description and map.

KB 7907: Outcrop of 'Lower Tarouba' marly-silty clay, 18th Street, San Fernando.

KR 8385: Same locality as for JS 1019. See Bolli (1957*a*), p. 52.

KR 23575: West side of railway track, south of the Pointe-à-Pierre Railway Station, about 560 feet (170 m) from the level crossing of Station Road, Pointe-à-Pierre. Designated as type locality of the *Globorotalia uncinata* Zone (Bolli 1957*b*, p. 64) but now recognized as representing the *G. angulata*

Zone; see map in Bartenstein *et al.* (1957), p. 8.

PJ 253: Usine Ste. Madeleine Quarry, southeast side of Naparima Hill, San Fernando.

RM 287: Head waters of Navette River, Lizard Springs area. Test pit in shale inlier in Miocene (Lower Cruse Formation) clay.

Rz 281: Same locality as for KB 6972. Type locality of the *Globorotalia formosa formosa* Zone; see Bolli (1957*b*, p. 64).

Rz 389: San Fernando Power Station, north of Carib Road, eastern cliff opposite high level water reservoir. See Cushman & Renz (1947*a*), p. 32 (Lower Member of Tarouba Formation).

Rz 413: Same locality as for Bo 521. Type locality of the *Globorotalia aragonensis* Zone; see Bolli (1957*b*), p. 64.

B. Wells

A. B. (Antilles Brighton) no. 1: Pitch Lake area, La Brea, Southwest Trinidad.

A. B. (Antilles Brighton) no. 2: West of Pitch Lake, La Brea, Southwest Trinidad.

A. O. (Antilles Oropouche) no. 4: East of Oropouche Lagoon, Southwest Trinidad.

B. D. (Boodoosingh) no. 1: Southeast of La Brea, Southwest Trinidad.

Esmeralda no. 1: East of Poonah, eastern Central Range.

F. W. (Fortin West) no. 214: East of Point Fortin, Southwest Trinidad.

Guayaguayare no. 159: West of Guayaguayare Field, Southeast Trinidad.

Guayaguayare no. 163: Guayaguayare Field, Southeast Trinidad.

Lizard no. 1 (and no. 1x redrilled): East of Lizard Springs, Southeast Trinidad.

Marabella no. 1: Marabella, north of San Fernando, West-Central Trinidad.

Marac no. 1: Along Moreau Road, south of junction with Penal Rock Road, South Trinidad.

Morne Diablo no. 34: Along Penal Rock Road near crossing with Cooramata River, South Trinidad.

Moruga no. 3: South of Moruga Bouffe, South Trinidad.

Moruga no. 10: Same area, north of Moruga no. 3 well location, South Trinidad.

Moruga no. 15: Same area, west of Moruga Bouffe mud volcano, South Trinidad.

Rochard no. 1: Along Penal Rock Road, east of Morne Diablo no. 34 well location, South Trinidad.

Rock Dome no. 1: West of Karamat mud volcano area, South Trinidad.

A list in stratigraphic order of all samples included in the study of the benthic foraminifera of the Gautier, Naparima Hill, Guayaguayare and Lizard Springs formations is given at the end of this Introduction and before the Annotated Taxonomic List (p. 63).

Composition of the fauna

Almost all the foraminiferal associations examined for this study include both benthic and planktic species which are well represented in variable proportions. The only exceptions are a non-calcareous fauna with agglutinated species only, representing the arenaceous lowermost Lizard Springs (referred to as the *Rzehakina epigona* Zonule), and some of the Late Cretaceous samples from the San Fernando area where planktic specimens, although usually reported, are often rare and not present in every sample. The benthic fauna described in this chapter consists of 553 species and subspecies which represent 189 genera and which belong to the families and genera listed below (the number of species for each genus is added in brackets). The arrangement and the definitions of the taxa follow Loeblich & Tappan (1987).

Family Bathysiphonidae: *Bathysiphon* (4)

Family Rhabdamminidae: *Rhizammina* (1), *Dendrophrya* (1)

Family Saccamminidae: *Lagenammina* (1), *Saccammina* (3), *Thurammina?* (1)

Family Hippocrepinidae: *Hyperammina* (3)

Family Ammodiscidae: *Ammodiscus* (5), *Ammolagena* (1), *Glomospira* (3), *Glomospirella* (2), *Repmanina* (2)

Family Rzehakinidae: *Rzehakina* (2), *Silicosigmoilina* (2)

Family Aschemocellidae: *Aschemocella* (1)

Family Hormosinidae: *Hormosinella* (2), *Reophax* (5), *Subreophax* (3), *Hormosina* (1)

Family Haplophragmoididae: *Asanospira* (1), *Cribrostomoides* (1), *Evolutinella* (2), *Haplophragmoides* (11), *Reticulophragmoides* (1)

Family Lituotubidae: *Lituotuba* (1),

Paratrochamminoides (1), *Trochamminoides* (3)

Family Lituolidae: *Ammobaculites* (3), *Ammomarginulina* (1), *Sculptobaculites* (1), *Triplasia* (2)

Family Ammosphaeroidinidae: *Ammosphaeroidina* (1), *Recurvoides* (3), *Thalmannammina* (1), *Bulbobaculites* (2)

Family Cyclamminidae: *Popovia* (1), *Quasicyclammina* (1), *Reticulophragmium* (1)

Family Spiroplectamminidae: *Bolivinopsis* (2), *Quasispiroplectammina* (5), *Spiroplectammina?* (1), *Spiroplectinella* (2), *Vulvulina* (1)

Family Textulariopsidae: *Textulariopsis* (2)

Family Trochamminidae: *Ammoglobigerina* (4), *Trochammina* (4)

Family Conotrochamminidae: *Conotrochammina* (2)

Family Prolixoplectidae: *Karrerulina* (4), *Plectina* (1), *Verneuilinella* (1)

Family Verneuilinidae: *Gaudryinopsis* (3), *Verneuilinoides* (3), *Belorussiella* (1), *Gaudryina* (12), *Gaudryinella* (1), *Pseudogaudryinella* (1)

Family Tritaxiidae: *Tritaxia* (1)

Family Ataxophragmiidae: *Arenobulimina* (2)

Family Globotextulariidae: *Remesella* (1)

Family Eggerellidae: *Bannerella* (1), *Dorothia* (6), *Marssonella* (6), *Eggerina* (1), *Karreriella* (1), *Martinottiella?* (1)

Family Textulariidae: *Textularia* (1)

Family Pseudogaudryinidae: *Clavulinoides* (4), *Pseudoclavulina* (6)

Family Patellinidae: *Patellina* (1)

Family Hauerinidae: *Quinqueloculina* (4)

Family Nodosariidae: *Chrysalogonium* (5), *Dentalina* (4), *Laevidentalina* (13), *Nodosaria* (4), *Pseudonodosaria* (3), *Pyramidulina* (8), *Gonatosphaera* (1), *Lingulina* (1), *Frondicularia* (7), *Tristix* (1)

Family Vaginulinidae: *Cristellariopsis* (2), *Dimorphina?* (1), *Lenticulina* (22), *Marginulinopsis* (6), *Percultazonaria* (1), *Pravoslavlevia* (1), *Saracenaria* (4), *Neoflabellina* (11), *Palmula* (2), *Astacolus* (10), *Hemirobulina* (9), *Vaginulinopsis* (2), *Citharina* (2), *Planularia* (2), *Psilocitharella* (1), *Vaginulina* (2)

Family Lagenidae: *Lagena* (8), *Reussoolina* (2)

Family Polymorphinidae: *Globulina* (7), *Guttulina* (1), *Palaeopolymorphina* (1), *Polymorphina* (1), *Polymorphinella?* (1), *Pseudopolymorphina* (1),

Pyrulina (2), *Pyrulinoides* (2), *Ramulina* (6)

Family Ellipsolagenidae: *Buchnerina* (3), *Exsculptina* (1), *Favulina* (1), *Lagnea* (1), *Oolina* (1), *Fissurina* (2), *Palliolatella* (1)

Family Glandulinidae: *Glandulina* (4)

Family Heterohelicidae: *Bifarina* (1)

Family Bolivinidae: *Brizalina* (1), *Gabonita* (2), *Loxostomoides* (2), *Tappanina* (2)

Family Bolivinoididae: *Bolivinoides* (7)

Family Loxostomatidae: *Aragonia* (3), *Zeauvigerina* (1)

Family Cassidulinidae: *Globocassidulina* (1)

Family Eouvigerinidae: *Eouvigerina* (1)

Family Turrilinidae: *Neobulimina* (2), *Praebulimina* (9), *Pseudouvigerina* (2), *Pyramidina* (4), *Sitella* (1), *Turrilina* (2)

Family Stainforthiidae: *Hopkinsina* (1), *Stainforthia* (1)

Family Siphogenerinoididae: *Siphogenerinoides* (2), *Orthokarstenia* (3)

Family Buliminidae: *Bulimina* (9), *Globobulimina?* (1), *Praeglobobulimina* (1)

Family Buliminellidae: *Buliminella* (1)

Family Uvigerinidae: *Euuvigerina* (1)

Family Fursenkoinidae: *Coryphostoma* (5)

Family Pleurostomellidae: *Ellipsodimorphina* (3), *Ellipsoglandulina* (2), *Ellipsoidella* (4), *Ellipsoidina* (2), *Ellipsopolymorphina* (2), *Nodosarella* (5), *Pleurostomella* (8), *Bandyella* (1), *Czarkowyella* (1)

Family Stilostomellidae: *Orthomorphina* (1), *Siphonodosaria* (4), *Stilostomella* (3)

Family Conorbinidae: *Conorbina* (1)

Family Bagginidae: *Rotamorphina* (1), *Valvulineria* (2)

Family Eponidae: *'Eponides'* (4)

Family Mississippinidae: *Stomatorbina?* (1)

Family Parrelloididae: *Cibicidoides* (8)

Family Cibicididae: *Cibicidina* (2), *Falsoplanulina* (2)

Family Epistomariidae: *Nuttallides* (2), *Nuttallinella* (4)

Family Nonionidae: *Spirotecta?* (1), *Nonion* (1), *Nonionella* (3), *Pullenia* (6)

Family Chilostomellidae: *Allomorphina* (1), *Chilostomella* (1), *Globimorphina* (1), *Pallaimorphina* (1), *Quadrimorphinella* (1)

Family Quadrimorphinidae: *Quadrimorphina* (9)

Family Alabaminidae: *Alabamina* (3)

Family Globorotaliidae: *Conorotalites* (1), *Globorotalites* (3)

Family Osangulariidae: *Charltonina* (1), *Osangularia* (7)

Family Heterolepidae: *Anomalinoides* (8), *Heterolepa* (1)

Family Gavelinellidae: *Gyroidinoides* (11), *Stensioeina* (1), *Angulogavelinella* (1), *Gavelinella* (16), *Gyroidina* (6), *Linaresia* (1)

Family Rotaliidae: *Rotalia*? (1)

The 553 species making up this list are distributed among the suborders of Loeblich & Tappan (1987) as follows:

Textulariina	162 species
Spirillinina	1 species
Miliolina	4 species
Lagenina	171 species
Globigerinina	1 species
Rotaliina	214 species

101 of the 214 species representing the Rotaliina are here attributed to the families Bolivinidae to Stilostomellidae; they form a distinct group which is practically identical with the 'Buliminidea' of Glaessner (1945); the remaining 113 species would represent the 'Rotaliidea' of this author. This separation has been considered useful for the construction of the frequency table of the major groups of foraminifera (Fig. 14).

The overall proportions of the families and of the suborders are representative for the greater part of the stratigraphic section from Late Albian to Early Eocene. The most noticeable divergences are a relatively lower total diversity of the Textulariina (agglutinants) in the Naparima Hill Formation and their slightly higher diversity in the Lower Lizard Springs Formation.

The two locally restricted facies faunas briefly mentioned above (Marabella facies, Rochard facies), however, contain some species which are not, or only exceptionally, found elsewhere. The first one occurs in Marabella well no. 1, core at 3512–3540 feet (probably Early Maastrichtian, *Globotruncana lapparenti tricarinata* Zone) and includes the following restricted or nearly restricted species:

Arenobulimina puschi (Reuss)

Bannerella biformis (Finlay)

Clavulinoides disjunctus (Cushman)

Frondicularia clarki Bagg

Neoflabellina cf. *permutata* Koch

Neoflabellina sp. A

Loxostomoides applinae (Plummer)

Praebulimina cf. *spinata* (Cushman & Campbell)

Pyramidina cushmani (Brotzen)

Hopkinsina dorreeni (Weiss)

Stainforthia cf. *navarroana* (Cushman)

Siphogenerinoides bramlettei Cushman

Orthokarstenia cretacea (Cushman)

Orthokarstenia cf. *parva* (Cushman)

Coryphostoma cf. *gemma* (Cushman)

'Eponides' cf. *huaynai* Frizzell

Stomatorbina(?) sp. A

Gavelinella spissicostata (Cushman)

The other facies fauna is found in Rochard well no. 1, cores at 8556–8565 and 8565–8571 feet (Early Paleocene, probably *Globorotalia pseudobulloides* Zone) and is characterized by the following species:

Textularia midwayana Lalicker

Lenticulina midwayensis carinata (Plummer)

Palmula toulmini ten Dam & Sigal

Planularia advena Cushman & Jarvis

Vaginulina plummerae (Cushman)

Loxostomoides applinae (Plummer)

Hopkinsina dorreeni (Weiss)

Stainforthia cf. *navarroana* (Cushman)

Cibicidina newmanae (Plummer)

Cibicidina cf. *simplex* (Brotzen)

Allomorphina cretacea (Reuss)

Gavelinella midwayensis trochoidea (Plummer)

Gavelinella rochardensis Beckmann

Other age-equivalent but more distinct faunas representing a distinctly different facies in the Paleocene are included in other formations (Chaudiere Formation, Soldado Formation) and are not further discussed in this work.

The preservation of the individual specimens is mostly good, with the exception of a few samples in the lower three zones of the Naparima Hill Formation, where pyrite impregnations or other kinds of mineralization are sometimes observed. Solution effects on the specimens are only rarely seen, but compressed or deformed specimens, particularly among those with an agglutinated wall, are not uncommon in some samples.

There are 27 species or subspecies described in the present chapter which are also recorded in the underlying Maridale or older formations. For the overlying Navet Formation, there is no up-to-date and reasonably complete compilation of the benthic foraminifera, and no reliable counts or estimates are therefore possible. In all cases, any available information concerning an extended stratigraphic range for a particular taxon is included in the Annotated Taxonomic List.

Comparisons with other areas

When the Trinidad faunas discussed here were first published by Cushman & Jarvis (1928), they were immediately found to be closely comparable to those of the Velasco shale of Mexico (Cushman 1926, White 1928, 1929). The term 'Velasco type' fauna was suggested by Berggren & Aubert (1975) for such deepwater benthic faunas, as a counterpart to the 'Midway type' faunas, which are typically represented in the early Tertiary of the Gulf Coast area of the United States. The 'Velasco type' faunas were not extensively discussed by Berggren & Aubert at that time but have since been well exemplified by Tjalsma & Lohmann (1983) and Morkhoven et al. (1986). The occurrence of faunas of Velasco aspect in other places in North America and Europe was already noticed by Cushman (1926) and commented upon in more detail by Hay (1959). Subsequent investigations (particularly Koutsoukos & Merrick 1985, Kaminski et al. 1988) as well as the present study have confirmed that a Velasco faunal type is predominantly represented in Trinidad, not only in the latest Cretaceous and Paleocene but also, with some modifications, in the earlier Cretaceous of the Gautier and Naparima Hill formations.

Examples of similar faunas have been found on all continents and of course, more recently, at the various sites drilled since 1969 within the Deep Sea Drilling Project and Ocean Drilling Program. The following discussion of some comparable faunas outside Trinidad does not imply that they were all deposited in exactly the same environment as in Trinidad. Some of the authors cited below do not give a paleobathymetrical interpretation at all. Others are rather indefinite or, where they give an interpretation, it is usually suggestive of a bathyal or, more rarely, abyssal depth (see also the paragraph on Paleoecology below). In addition to the Mexican faunas referred to above, examples on the American continent are found in the papers by Dailey (1973; California, Aitken member, Albian), Graham & Church (1963; California, Campanian), Sliter (1968; California, La Jolla

fauna, Campanian-Maastrichtian), McGugan (1964; western Canada, Campanian-Maastrichtian), Natland *et al.* (1974; Magellanes Basin, Maastrichtian to Paleocene), Sellier de Civrieux (1952, Socuy member, western Venezuela, Campanian?-Maastrichtian) and Graham & Classen (1955; California, Late Paleocene). In eastern Venezuela, the Santa Anita Group (Cushman 1947, Galea-Alvarez 1985; Maastrichtian to Early Eocene) contains a mixture of deeper (Velasco type) and shallower (Midway type) faunas.

In Europe, the following authors may be cited: Gandolfi (1942; Albian-Cenomanian, southern Switzerland), Noth (1951; Austria, Albian to Senonian), Hanzlikova (1972; Moravia, Turonian to Maastrichtian), Gawor-Biedowa (1980; Poland, Turonian-Coniacian), Cati (1964; northern Italy, Campanian), Beckmann *et al.* (1982; northern Italy, Campanian to Paleocene), Glaessner (1937; Caucasus, Campanian to Paleocene), Schreiber (1980; Austria, Maastrichtian), Hillebrandt (1962; Bavaria, Paleocene-Early Eocene). Some faunas from the Late Cretaceous and Early Paleogene of Russia (Subbotina 1964, Myatlyuk 1970) are rich in very similar agglutinated species but the calcareous foraminifera are in part somewhat different. Faunas from other areas include those described by Klasz *et al.* (1987b; Senegal, Danian), Yasuda 1986 (Japan, Campanian to Paleocene) and Belford (1960; Australia, Santonian to Campanian).

An increasing number of similar faunas has been recovered in recent years by the Deep Sea Drilling Project (DSDP) and Ocean Drilling Program (ODP). Similar faunas in the Atlantic Ocean are included in the descriptions by Berggren (1972; Leg 12, North Atlantic, Early Eocene), Sliter (1977; Leg 39, Southwest Atlantic, Albian to Maastrichtian), Boersma (1977, Leg 39, Southwest Atlantic, Early Eocene), Proto Decima & Bolli (1978; Leg 40, Southeast Atlantic, Paleogene), Schnitker (1979; Bay of Biscay, Paleocene to Middle Eocene), Sliter (1980; Leg 50, Moroccan Basin, redeposited faunas, Albian-Cenomanian), Basov & Krasheninnikov (1983; Leg 71, Southwest Atlantic, Albian), Dailey (1983; Leg 72, Rio Grande Rise, Paleocene), Clark & Wright (1984; Leg 73, South Atlantic, Paleocene), Hemleben & Troester (1984; Leg 78, north of Barbados, Campanian-Maastrichtian), Saint-Marc (1987; Leg 93, northwestern Atlantic, Paleocene, mixed Velasco-Midway type faunas), Hulsbos (1987; Leg 93, northwestern Atlantic, Eocene), Miller & Katz (1987; Leg 95, New Jersey Transect, Early to Middle Eocene). A few faunas, which are characterized by a relative scarcity of agglutinants and nodosariids but are otherwise similar, are found in the Late Cretaceous of the Angola Basin and Walvis Ridge area of the southeastern Atlantic (Todd 1970; Beckmann 1978, Leg 40; McNulty 1984, Leg 75).

In the Pacific, Indian and Antarctic Oceans, comparable faunas are described by Douglas (1973; Leg 17, northeastern Pacific, Campanian and Paleogene), Vincent *et al.* (1974; Leg 24, western Indian Ocean, Paleocene to Early Eocene), Scheibnerova (1974; Leg 27, off West Australia, Aptian-Albian), Sliter (1985; Leg 89, East Mariana Basin, Coniacian to Maastrichtian) and Thomas (1990; Leg 113, Weddell Sea, Maastrichtian to Early Eocene).

For the two facies faunas separately mentioned above (Marabella and Rochard facies), a few similar examples can also be cited from the literature. The Marabella facies (Late Cretaceous) has a close equivalent in the Colon shale (Late Campanian?–Maastrichtian) of northern Colombia and western Venezuela (Cushman & Hedberg 1941, Fuenmayor 1989, Martinez 1989). Other somewhat similar faunas, also with *Siphogenerinoides/Orthokarstenia*, are described by Frizzell (1943; Peru), Colom (1947; former Spanish Sahara), Cushman (1949; Arkadelphia marl, Arkansas, USA), Petri (1962; Sergipe, Brazil) and Martin (1964; California). The Rochard facies (Early Paleocene) is comparable to some Midway-type faunas like those described by Plummer (1926; Texas), Jennings (1936; New Jersey, USA), Brotzen (1948; Sweden) and Martin (1964; California). It differs from the Soldado fauna of Trinidad (Vaughan & Cole 1941, Cushman & Renz 1942, Kugler & Caudri 1975) in lacking larger foraminifera and other more distinctly shallow shelf indicators).

A general analysis and comparison of the faunas mentioned in this paragraph is beyond the scope of the present article, but some specific aspects of the composition of the faunas, the biostratigraphic distribution and the paleoecology of the taxonomic groups are discussed in the following two sections.

Biostratigraphy

The biostratigraphic framework for the subdivision and dating the Late Albian to Early Eocene sequence of Trinidad is summarized on Fig. 12. On this chart, the plankton zonation of Trinidad, compiled essentially by Bolli (1957*a,b*) and emended in Saunders & Bolli (1985), is calibrated with an international zonation (Bolli *et al.* 1985) based on the paleomagnetic time scale and on biostratigraphic subdivisions by

other microfossil groups. An important aspect of the Trinidad subdivision is that it is normally not possible to rely on continuously exposed stratigraphic sections but rather on isolated surface outcrops or spot cores in wells. A refined biostratigraphic resolution within each zone is therefore in general not attempted. The total stratigraphic range of each species in Trinidad is indicated for each species in the Annotated Taxonomic List, and the ranges of selected species are shown on three range charts (Figs. 15–17).

Figure 14 is a summary of the first and last occurrences of the major groups and of the total diversity of benthic foraminifera for each zone. It gives an indication of the changes in faunal composition and total diversity in time for each zone. For a better understanding of details of the categories listed, the reader is referred to the legend. The main conclusions which can be drawn from this table are:

1. There is a rather high percentage of first appearances in the three lowermost zones (Late Albian to Early Cenomanian; Gautier Formation). The major hiatus between the *Rotalipora appenninica appenninica* Zone (Early Cenomanian) and the *Globotruncana inornata* Zone (Middle Turonian) is presumably the cause of the high number of last occurrences in the former zone.

2. The overall diversity increases gradually in the interval represented by the *Globotruncana inornata* to *G. stuarti* zones (Naparima Hill Formation), but the proportion of agglutinated species remains relatively low. The Early Campanian turnover observed by Kuhnt & Moullade (1991) for the agglutinants in the North Atlantic Ocean is not indicated for Trinidad but a similar change seems to have taken place near the base of the Maastrichtian.

3. The relatively high diversity in the three uppermost Cretaceous zones (Guayaguayare Formation), up to the Cretaceous/Tertiary boundary, is shared by all major groups.

4. The rate of extinctions in the highest Cretaceous zone (*Abathomphalus mayaroensis* Zone) is moderate in the Textulariina and Lagenina (12.5 and 15%) but is distinctly higher in the Rotaliina ('Buliminidea' with 37.5% and 'Rotaliidea' with 30.5%). A similar phenomenon is noted by Widmark & Malmgren (1992) at DSDP Sites 525 and 527 in the South Atlantic Ocean. The overall figure for the extinction rate of all groups is 20%.

5. Through most of the Paleocene, the first and last occurrences and the total diversity remain fairly constant. An increase in the Early Paleocene may have taken place but is not well documented.

ZONE	AGGLUTINANTS			NODOSARIINA			BULIMINIDEA			ROTALIIDEA			others			ALL TAXA		
	first	total	last	first	total	last	first	total	last	first	total	last	first	total	last	first	total	last
G. aragonensis	1	?35	?1	0	?14	?0	0	?20	?0	0	?20	?0	0	?1	?0	1	?90	?1
G. f. formosa	4	55	20	4	42	28	3	31	11	2	27	9	1	0	0	14	155	68
G. subbotinae	8	65	14	3	47	9	5	33	5	6	29	4	0	0	1	22	174	33
G. edgari	5	58	1	0	48	4	4	28	0	4	24	1	1	0	0	14	158	6
G. velascoensis	1	71	18	1	69	21	1	28	4	1	31	11				4	199	54
G. pseudomenardii	7	88	18	11	79	11	7	32	5	1	32	2				26	231	36
G. p. pusilla	2	86	5	2	77	9	2	28	3	8	36	5				14	227	22
G. angulata	0	84	0	5	78	3	2	28	2	1	29	1				8	219	6
G. uncinata	5	84	0	0	74	1	1	26	0	1	28	0				7	212	1
G. trinidadensis	14	81	2	5	76	2	7	25	0	6	31	3				32	213	7
G. pseudobulloides	1	68	1	4	75	4	0	21	2	6	31	7				11	195	14
R. epigona	11	67	0													11	67	0
A. mayaroensis	16	64	8	17	85	13	8	32	12	7	36	11				48	217	44
G. gansseri	7	51	3	0	68	1	3	27	3	9	31	2				19	177	9
G. l. tricarinata	27	51	7	15	77	9	17	37	13	8	28	6				67	193	35
G. stuarti	11	27	3	21	68	6	14	22	2	8	29	9				54	146	20
G. fornicata	8	24	8	10	57	10	7	14	6	14	27	5				39	122	29
G. concavata	2	16	0	12	53	5	4	10	3	1	15	3	0	1	1	19	95	12
G. renzi	1	16	2	2	53	2	0	11	5	2	15	1	0	1	0	5	96	10
G. inornata	2	18	3	12	46	5	6	13	2	9	15	2	0	1	0	29	93	12
R. a. appenninica	12	29	13	6	43	8	7	9	2	9	14	8	1	1	0	35	96	31
F. washitensis	9	17	0	14	41	3	1	3	1	3	7	2	2	3	2	29	71	8
G. t. ticinensis	4	4	0	7	28	2	2	2	0	4	5	1	0	1	0	17	40	3

Fig. 14. Frequency distribution of benthic foraminiferal taxa in the Late Albian to Early Eocene of Trinidad. The horizontal rows correspond to the biostratigraphic zones listed on Fig. 12. The columns represent the major faunal groups as explained in the text. The three figures in each rectangle show, from left to right, the number of first appearances, the total number of taxa and the number of last occurrences. The figure in the middle (total diversity) includes all taxa whose stratigraphic range falls within a particular zone, not only the taxa actually observed in each particular case. The question marks in the top row (*Globorotalia aragonensis* Zone) reflect both the relatively poor representation of this zone (one locality only) and the incomplete faunal record in the overlying Navet Formation.

6. There is an increase in the extinction rates to 16% and 27%, respectively, in the upper two zones of the Paleocene, which is shared by the major groups except the 'Buliminidea'. In the uppermost zone of the Paleocene (*Globorotalia velascoensis* Zone), these extinctions are accompanied by a drastic decrease of first appearances.

7. A gradual recovery seems to take place in the Early Eocene, but it is obscured in Trinidad by the uneven number of samples and by the insufficient faunal documentation at the transition to the overlying Navet Formation.

In principle, the data on Fig. 14 confirm the conclusions drawn earlier (Beckmann 1960) on the faunal changes at the top of Cretaceous and of the Paleocene but, due mainly to subsequent taxonomic refinements, some of the actual figures are not exactly the same. The different proportions, particularly at the Cretaceous/Tertiary boundary, of the changes in the Textulariina and Lagenina, on the one hand, and the Rotaliina ('Buliminidea' and Rotaliidea') on the other hand have now also become more apparent than they were before. A review of the recent literature on the degree of change in the deep-water benthic foraminifera at the tops of the Cretaceous and of the Paleocene indicates no major disagreement between the various authors as long as there is a reasonable continuity in deposition and environmental conditions (Dailey 1983, Tjalsma & Lohmann 1983, Bignot 1984, Thomas 1990). However, if this is not so, the recorded figures may show a wide variation (see Beckmann *et al.* 1982, p. 109).

Several authors have attempted to establish assemblage zones based on the stratigraphic distribution of benthic species. These zonations usually depend on a combination of biostratigraphic, paleoecologic, paleoclimatic or sedimentological factors and may or may not be easily comparable with each other. Zones created in a relatively stable deep-water environment (Kaiho 1988, Berggren & Miller 1989, Thomas 1990) probably have the best chance of being widely applicable and, in fact, are those which are most easily compatible with the Trinidad faunal succession. Other zonations such as Caldwell *et al.* (1978), McNeil & Caldwell (1981), Bailey *et al.* (1983) and Jenkins & Murray (1989) are less suitable for direct comparisons. The zonal sequence for the Late Cretaceous and Paleocene of western Venezuela suggested by Fuenmayor (1989) is of some interest as it concerns an area relatively close to Trinidad; it is therefore not surprising that his zonation has quite a few elements reflecting the faunal compositions in Trinidad, not only those of the regular Lizard Springs-Guayaguayare type but also of the 'Marabella' and 'Rochard' facies faunas.

Palaeoecology

Generalized statements about the presumed depth of deposition and depositional environment are already included in early papers on the Cretaceous-Early Paleogene interval of Trinidad stratigraphy and are found sporadically in later general papers on general geology and stratigraphy of the island. In particular, the early conclusion by Cushman & Jarvis (1928) that the Lizard Springs Formation is a deep-water deposit has never been seriously challenged. Koutsoukos & Merrick (1985) have surveyed the entire Cretaceous section of Trinidad, mostly relying on the subsurface information provided by various oil companies. Unfortunately, their well-presented micropaleontological data are not backed up by a documentation of the source of the examined material; this leaves some open questions concerning sampling localities and levels, taxonomy, biostratigraphic subdivisions and age determinations. The second important contribution is by Kaminski *et al.* (1988), who deal mostly with the latest Cretaceous to Early Tertiary agglutinated foraminifera and present a paleoecological interpretation supported by a factor analysis of some Early Paleocene samples.

A detailed statistical analysis of the Trinidad faunas is beyond the scope of the present chapter. However, an attempt is made here to compare the general composition of the faunas in Trinidad with the results of selected recent studies of closely similar associations, both at Deep Sea Drilling Sites and from surface outcrops. The background of these studies is formed by the lists of genera and species presented by Sliter (1972), Sliter & Baker (1972) and Olsson (1977). The Trinidad faunas, at least up to the Paleocene, are mostly comparable with the general lists compiled by these authors for a bathyal depth, and in particular with that of the La Jolla fauna of Sliter (1968) cited as a typical example. In the following discussion, an evaluation of the palaeoecology of the sequence of the formations in Trinidad in stratigraphic order is attempted.

Conclusions on the faunal type of the Gautier Formation (Late Albian to Early Cenomanian) are in fairly good agreement with each other. Sliter (1977) finds in the Albian of Site 356 (Southwest Atlantic) a similar fauna indicative of a middle bathyal environment (500–1500 m). Haig (1979) defines a Middle Cretaceous '*Marssonella* association', in which his zones 3 and 4 (upper to middle slope) have the most similar faunal composition. Displaced faunas of the Gautier type in the Moroccan Basin (Sliter 1980) are placed in a middle to lower bathyal environment (500–2500 m). Moullade (1984) defines in the mid-Cretaceous eight faunal types, of which types 3 (epibathyal, North Atlantic), 5 (mesobathyal, North Atlantic) and 10 (mesobathyal, South Atlantic), with *Praebulimina nannina* and *Valvulineria parva*, show a fairly good agreement with the Trinidad faunas. The only possible irregularity is the presence of *Patellina subcretacea* in faunal type 8 (outer shelf, South Atlantic). In Koutsoukos and Merrick (1985), a 'Vraconian' fauna in Trinidad, which appears to correspond to that of the Gautier Formation, is placed at a depth between deep neritic and middle/lower bathyal (about 1500 m) in relatively warm water; their most similar faunal list is that indicative of upper bathyal. The average of the records just cited indicates an upper to middle bathyal environment for the Gautier type faunas.

Faunas comparable to those of the Naparima Hill Formation (Turonian to Campanian) are also recorded at some Deep Sea Drilling sites. Sliter (1977; Southwest Atlantic) mentions them in the Santonian of Site 356 (lower bathyal, 1500–2500 m), in the Santonian of Site 357 (middle to lower bathyal, 1000–1500 m) and in the Campanian of the same Site (lower bathyal, with some dissolution). In his standard succession described on p. 672, the Trinidad faunas would best fit the middle bathyal assemblage. Dailey (1983) places a very similar fauna from the Santonian of the Rio Grande Rise in the upper to middle bathyal depth zone (200–1500 m). In Basov and Krasheninnikov (1983; Southwest Atlantic), the recorded Coniacian-Santonian faunas have a greater proportion of

Fig. 15. Stratigraphic distribution of selected taxa (part 1): species and subspecies from Trinidad that are restricted to the Cretaceous.

			CRETACEOUS						PALEOCENE			EOCENE			CRETACEOUS (ALBIAN) TO EARLY EOCENE BENTHIC FORAMINIFERA FROM TRINIDAD, W. I.

Left-hand alphabetical index:

Aragonia materna kugleri	41
Bandyella greatvalleyensis	42
Charltonina australis	28
Citharina angustissima	57
Conorotalites aptiensis	8
Conorotalites subconicus	49
Dorothia hyperconica	3
Ellipsoidella binaria	58
Evolutinella flagleri	31
Evolutinella renzi	9
Gabonita kugleri	50
Gaudryina frankei	54
Gaudryina laevigata	32
Gaudryinopsis bosquensis	10
Gaudryinopsis canadensis	29
Gavelinella cenomanica	11
Gavelinella dakotensis	35
Gavelinella drycreekensis	21
Gavelinella eriksdalensis	39
Gavelinella schloenbachi	43
Gyroidina bandyi	5
Gyroidina mauretanica	6
Gyroidina praeglobosa	45
Gyroidinoides grahami	60
Gyroidinoides quadratus	47
Gyroidinoides subglobosus	22
Haplophragmoides formosus	12
Lenticulina exarata	13
Lenticulina lepida morugaensis	36
Lenticulina lobata	18
Lenticulina secans	33
Lingulina taylorana	48
Marginulinopsis curvisepta	51
Marginulinopsis lituola	44
Marginulinopsis praetschoppi	52
Neobulimina irregularis	30
Neobulimina primitiva	7
Neoflabellina projecta	55
Nodosarella wintereri	19
Osangularia californica	23
Osangularia insigna secunda	24
Patellina subcretacea	25
Planularia complanata	20
Praebulimina nannina	26
Praebulimina reussi	53
Praebulimina wyomingensis	1
Pseudoclavulina californica	27
Pyramidina rudita	59
Pyramidina szajnochae	37
Quadrimorphina allomorphinoides	46
Quadrimorphina camerata camerata	56
Quadrimorphina camerata umbilicata	34
Saracenaria cretacea	16
Tappanina eouvigeriniformis	38
Textulariopsis losangica	4
Tristix excavata	14
Tritaxia tricarinata	15
Trochammina globolaevigata	17
Turrilina carseyae	40
Valvulineria parva	2

AGE columns: Albian, Cenomanian, Turonian, Coniacian, Santonian, Campanian, Maastrichtian, Early, Middle, Late (PALEOCENE), Early (EOCENE)

FORMATION: GAUTIER, NAPARIMA HILL, GUAYA-GUAYARE, Lower LIZARD SPRINGS, Upper LIZARD SPRINGS

ZONE (Saunders & Bolli 1985)

Stratigraphic distribution of selected species (part 1)

Species columns (left to right): Rotalipora ticinensis ticinensis, Favusella washitensis, Rotalipora app. appenninica, Globotruncana inornata, Globotruncana renzi, Globotruncana concavata, Globotruncana fornicata, Globotruncana stuarti, Globotruncana lapp. tricarinata, Globotruncana gansseri, Abathomphalus mayaroensis, Rzehakina epigona, Globorotalia pseudobulloides, Globorotalia trinidadensis, Globorotalia uncinata, Globorotalia angulata, Globorotalia pusilla pusilla, Globorotalia pseudomenardii, Globorotalia velascoensis, Globorotalia edgari, Globorotalia subbotinae, Globorotalia formosa formosa, Globorotalia aragonensis

Right-hand numbered list:

1	Praebulimina wyomingensis
2	Valvulineria parva
3	Dorothia hyperconica
4	Textulariopsis losangica
5	Gyroidina bandyi
6	Gyroidina mauretanica
7	Neobulimina primitiva
8	Conorotalites aptiensis
9	Evolutinella renzi
10	Gaudryinopsis bosquensis
11	Gavelinella cenomanica
12	Haplophragmoides formosus
13	Lenticulina exarata
14	Tristix excavata
15	Tritaxia tricarinata
16	Saracenaria cretacea
17	Trochammina globolaevigata
18	Lenticulina lobata
19	Nodosarella wintereri
20	Planularia complanata
21	Gavelinella drycreekensis
22	Gyroidinoides subglobosus
23	Osangularia californica
24	Osangularia insigna secunda
25	Patellina subcretacea
26	Praebulimina nannina
27	Pseudoclavulina californica
28	Charltonina australis
29	Gaudryinopsis canadensis
30	Neobulimina irregularis
31	Evolutinella flagleri
32	Gaudryina laevigata
33	Lenticulina secans
34	Quadrimorphina camerata umbilicata
35	Gavelinella dakotensis
36	Lenticulina lepida morugaensis
37	Pyramidina rudita
38	Tappanina eouvigeriniformis
39	Gavelinella eriksdalensis
40	Turrilina carseyae
41	Aragonia materna kugleri
42	Bandyella greatvalleyensis
43	Gavelinella schloenbachi
44	Marginulinopsis lituola
45	Gyroidina praeglobosa
46	Quadrimorphina allomorphinoides
47	Gyroidinoides quadratus
48	Lingulina taylorana
49	Conorotalites subconicus
50	Gabonita kugleri
51	Marginulinopsis curvisepta
52	Marginulinopsis praetschoppi
53	Praebulimina reussi
54	Gaudryina frankei
55	Neoflabellina projecta
56	Quadrimorphina camerata camerata
57	Citharina angustissima
58	Ellipsoidella binaria
59	Pyramidina szajnochae
60	Gyroidinoides grahami

Fig. 16. Stratigraphic distribution of selected taxa (part 2): species and subspecies from Trinidad that occur mostly in the latest Cretaceous and Paleocene.

CRETACEOUS (ALBIAN) TO EARLY EOCENE BENTHIC FORAMINIFERA FROM TRINIDAD, W. I.

Stratigraphic distribution of selected species (part 2)

	CRETACEOUS						PALEOCENE			EOCENE	A G E	FORMA-TION	ZONE (Saunders & Bolli 1985)
	Albian	Cenomanian	Turonian	Coniacian	Santonian	Campanian	Maas-trichtian	Early	Middle	Late	Early			

Formations: GAUTIER, NAPARIMA HILL, GUAYA-GUAYARE, Lower LIZARD SPRINGS, Upper LIZARD SPRINGS

Species columns (left to right):
Rotalipora ticinensis ticinensis; Favusella washitensis; Rotalipora app. appenninica (not exposed); Globotruncana inornata; Globotruncana renzi; Globotruncana concavata; Globotruncana formicata; Globotruncana stuarti; Globotruncana lapp. tricarinata (not exposed); Globotruncana gansseri; Abathomphalus mayaroensis; Rzehakina epigona; Globorotalia pseudobulloides; Globorotalia trinidadensis; Globorotalia uncinata; Globorotalia angulata; Globorotalia pusilla pusilla; Globorotalia pseudomenardii; Globorotalia velascoensis; Globorotalia edgari; Globorotalia subbotinae; Globorotalia formosa formosa; Globorotalia aragonensis

Angulogavelinella avnimelechi	119
Anomalinoides rubiginosus	88
Anomalinoides welleri	63
Aragonia velascoensis	70
Bolivinoides draco draco	117
Bolivinoides draco miliaris	103
Bolivinoides draco verrucosus	120
Bolivinoides granulatus emersus	79
Bulbobaculites jarvisi	111
Bulimina midwayensis	89
Bulimina taylorensis	105
Citharina barcoensis	73
Clavulinoides asper	112
Coryphostoma incrassatum	80
Czarkowiella torta	82
Dorothia bulbosa	110
Dorothia subretusa	92
Frondicularia jarvisi	101
Frondicularia mucronata	65
Gaudryina cretacea	109
Gaudryina ingens	74
Gavelinella beccariiformis	93
Gavelinella dayi	113
Gavelinella spissicostata	104
Globimorphina trochoides	102
Globobulimina ? suteri	94
Guttulina trigonula	99
Gyroidina bollii	95
Gyroidina megastoma	66
Gyroidinoides globosus	96
Gyroidinoides nitidus	61
Gyroidinoides subangulatus	100
Hormosinella ovulum	69
Lagnea perfectocostata	106
Lenticulina velascoensis	67
Marginulinopsis subrecta	90
Marssonella oxycona trinitatensis	97
Neoflabellina numismalis	71
Neoflabellina praereticulata	76
Neoflabellina thalmanni	77
Nodosaria aspera parvavermiculata	85
Nuttallides truempyi truempyi	116
Nuttallinella coronula	75
Nuttallinella florealis	68
Osangularia texana	62
Praebulimina kickap. kickapooensis	83
Praebulimina kickapooensis pingua	72
Praebulimina petroleana	78
Praebulimina ? quadrata	86
Pseudogaudryinella compacta	118
Pseudopolymorphina rudis	98
Pyramidulina multicostata	87
Remesella varians	114
Rzehakina epigona	64
Rzehakina minima	91
Silicosigmoilina akkeshiensis	81
Sitella colonensis	107
Spiroplectinella dentata	84
Trochammina ruthvenmurrayi	115
Turrilina cushmani	108

61	Gyroidinoides nitidus
62	Osangularia texana
63	Anomalinoides welleri
64	Rzehakina epigona
65	Frondicularia mucronata
66	Gyroidina megastoma
67	Lenticulina velascoensis
68	Nuttallinella florealis
69	Hormosinella ovulum
70	Aragonia velascoensis
71	Neoflabellina numismalis
72	Praebulimina kickapooensis pingua
73	Citharina barcoensis
74	Gaudryina ingens
75	Nuttallinella coronula
76	Neoflabellina praereticulata
77	Neoflabellina thalmanni
78	Praebulimina petroleana
79	Bolivinoides granulatus emersus
80	Coryphostoma incrassatum
81	Silicosigmoilina akkeshiensis
82	Czarkowiella torta
83	Praebulimina kickap. kickapooensis
84	Spiroplectinella dentata
85	Nodosaria aspera parvavermiculata
86	Praebulimina ? quadrata
87	Pyramidulina multicostata
88	Anomalinoides rubiginosus
89	Bulimina midwayensis
90	Marginulinopsis subrecta
91	Rzehakina minima
92	Dorothia subretusa
93	Gavelinella beccariiformis
94	Globobulimina ? suteri
95	Gyroidina bollii
96	Gyroidinoides globosus
97	Marssonella oxycona trinitatensis
98	Pseudopolymorphina rudis
99	Guttulina trigonula
100	Gyroidinoides subangulatus
101	Frondicularia jarvisi
102	Globimorphina trochoides
103	Bolivinoides draco miliaris
104	Gavelinella spissicostata
105	Bulimina taylorensis
106	Lagnea perfectocostata
107	Sitella colonensis
108	Turrilina cushmani
109	Gaudryina cretacea
110	Dorothia bulbosa
111	Bulbobaculites jarvisi
112	Clavulinoides asper
113	Gavelinella dayi
114	Remesella varians
115	Trochammina ruthvenmurrayi
116	Nuttallides truempyi truempyi
117	Bolivinoides draco draco
118	Pseudogaudryinella compacta
119	Angulogavelinella avnimelechi
120	Bolivinoides draco verrucosus

Fig. 17. Stratigraphic distribution of selected taxa (part 3): species and subspecies from Trinidad that are found mainly in the latest Cretaceous and Early Tertiary.

Chart column groups (top): CRETACEOUS — Albian, Cenomanian, Turonian, Coniacian, Santonian, Campanian, Maastrichtian; PALEOCENE — Early, Middle, Late; EOCENE — Early | AGE | Stratigraphic distribution of selected species (part 3) | CRETACEOUS (ALBIAN) TO EARLY EOCENE BENTHIC FORAMINIFERA FROM TRINIDAD, W.I.

Formations: GAUTIER | NAPARIMA HILL | GUAYA-GUAYARE | Lower LIZARD SPRINGS | Upper LIZARD SPRINGS | FORMA-TION | ZONE (Saunders & Bolli 1985)

Species columns (left to right): Rotalipora ticinensis ticinensis · Favusella washitensis · Rotalipora app. appenninica (not exposed) · Globotruncana inornata · Globotruncana renzi · Globotruncana concavata · Globotruncana fornicata · Globotruncana stuarti · Globotruncana lapp. tricarinata (not exposed) · Globotruncana gansseri · Abathomphalus mayaroensis · Rzehakina epigona · Globorotalia pseudobulloides · Globorotalia trinidadensis · Globorotalia uncinata · Globorotalia angulata · Globorotalia pusilla pusilla · Globorotalia pseudomenardii · Globorotalia velascoensis · Globorotalia edgari · Globorotalia subbotinae · Globorotalia formosa formosa · Globorotalia aragonensis

Species (alphabetical)	No.
Ammodiscus latus	174
Anomalinoides dorri aragonensis	162
Aragonia aragonensis	171
Asanospira walteri	155
Bolivinoides delicatulus curtus	143
Bolivinoides delicatulus delicatulus	124
Bulbobaculites lueckei	133
Bulimina excavata	144
Bulimina semicostata	177
Bulimina stokesi	170
Bulimina trinitatensis	150
Bulimina truncanella	176
Bulimina tuxpamensis	168
Buliminella grata	178
Cibicidoides proprius	151
Cibicidoides succedens	132
Clavulinoides trilaterus	160
Conorotalites michelinianus	128
Coryphostoma crassum	129
Coryphostoma midwayense	146
Cribrostomoides trinitatensis	134
Dorothia beloides	152
Dorothia cylindracea	169
Eggerina subovata	121
Ellipsoidella kugleri	138
Eouvigerina subsculptura	130
Falsoplanulina acuta	142
Falsoplanulina waltonensis	164
Gaudryina pyramidata	135
Gaudryina rutteni	180
Glomospira serpens	156
Haplophragmoides porrectus	139
Haplophragmoides trinitatensis	140
Karrerulina conversa	127
Lenticulina insulsa	153
Marssonella indentata	145
Neoflabellina delicatissima	131
Neoflabellina jarvisi	158
Neoflabellina paleocenica	163
Neoflabellina semireticulata	157
Nuttallides truempyi crassaformis	154
Osangularia expansa	161
Osangularia mexicana	179
Osangularia velascoensis	147
Popovia beckmanni	141
Praeglobobulimina ovata	172
Pseudoclavulina amorpha	165
Pseudoclavulina trinitatensis	173
Pseudouvigerina naheolensis	166
Pseudouvigerina plummerae	122
Quasispiroplectammina jarvisi	148
Repmanina charoides favilla	125
Rotamorphina cushmani	126
Stensioeina excolata	123
Subreophax velascoensis	137
Tappanina selmensis	159
Triplasia fundibularis	149
Triplasia inaequalis	175
Vaginulina anitana	136
Vaginulinopsis hemicylindrica	167

No.	Species (numbered)
121	Eggerina subovata
122	Pseudouvigerina plummerae
123	Stensioeina excolata
124	Bolivinoides delicatulus delicatulus
125	Repmanina charoides favilla
126	Rotamorphina cushmani
127	Karrerulina conversa
128	Conorotalites michelinianus
129	Coryphostoma crassum
130	Eouvigerina subsculptura
131	Neoflabellina delicatissima
132	Cibicidoides succedens
133	Bulbobaculites lueckei
134	Cribrostomoides trinitatensis
135	Gaudryina pyramidata
136	Vaginulina anitana
137	Subreophax velascoensis
138	Ellipsoidella kugleri
139	Haplophragmoides porrectus
140	Haplophragmoides trinitatensis
141	Popovia beckmanni
142	Falsoplanulina acuta
143	Bolivinoides decoratus curtus
144	Bulimina excavata
145	Marssonella indentata
146	Coryphostoma midwayensis
147	Osangularia velascoensis
148	Quasispiroplectammina jarvisi
149	Triplasia fundibularis
150	Bulimina trinitatensis
151	Cibicidoides proprius
152	Dorothia beloides
153	Lenticulina insulsa
154	Nuttallides truempyi crassaformis
155	Asanospira walteri
156	Glomospira serpens
157	Neoflabellina semireticulata
158	Neoflabellina jarvisi
159	Tappanina selmensis
160	Clavulinoides trilaterus
161	Osangularia expansa
162	Anomalinoides dorri aragonensis
163	Neoflabellina paleocenica
164	Falsoplanulina waltonensis
165	Pseudoclavulina amorpha
166	Pseudouvigerina naheolensis
167	Vaginulinopsis hemicylindrica
168	Bulimina tuxpamensis
169	Dorothia cylindracea
170	Bulimina stokesi
171	Aragonia aragonensis
172	Praeglobobulimina ovata
173	Pseudoclavulina trinitatensis
174	Ammodiscus latus
175	Triplasia inaequalis
176	Bulimina truncanella
177	Bulimina semicostata
178	Buliminella grata
179	Osangularia mexicana
180	Gaudryina rutteni

agglutinated species and are believed to have been deposited at rather greater depth than coeval faunas in Trinidad, probably near the calcite compensation depth (CCD). In the Early Campanian, the assemblages are more similar, although with some dissolution, and are estimated to be deposited at 1500–2500 m (lower bathyal). Similar Coniacian-Santonian faunas from the eastern Marianas Basin (Sliter 1985) are regarded as middle bathyal. Turonian samples from Trinidad (Koutsoukos & Merrick 1985) are believed to have lived in deep neritic, rather cool water, at a depth of about 150 m. The Coniacian-Santonian samples (presumably representing the *Globotruncana renzi* and *G. concavata* zones) would indicate somewhat warmer water in an upper bathyal environment (200–350 m). Considering the faunal lists given by these authors, their surprisingly shallow depth interpretations for the Turonian to Santonian are difficult to explain. Combining all data, the Trinidad faunas of the Naparima Hill Formation would fall into a rather wide depth range from deep neritic to lower bathyal. It was mentioned above that these faunas contain a relatively lower percentage (16–27 %) of agglutinated species than those of the other formations (Fig. 14). According to Moullade (1984), these percentages would best fit into an upper to middle bathyal depth.

Records of faunas resembling those of the Guayaguayare Formation (Maastrichtian) are relatively more numerous. Sliter (1977) postulates a calcite compensation depth (CCD) at 3000 to 4000 m in the Late Cretaceous. His Maastrichtian faunas in the Southwest Atlantic Sites 356 and 357, which closely resemble the Trinidad faunas, indicate a lower bathyal depth (1500–2500 m). A *Praebulimina* fauna (Site 356, core 34) believed to indicate the bottom of the oxygen minimum zone (1000–1500 m) might indicate a similar depth for the Marabella facies fauna locally present in the Early Maastrichtian in Trinidad. Dailey (1983) would prefer a depth of 500–2000 m (middle to shallow lower bathyal) for a Guayaguayare type fauna from the Rio Grande Rise, and Basov & Krasheninnikov (1983) would locate corresponding faunas at Site 511 in the Southwest Atlantic at the lower bathyal. A depth estimate by Nyong & Olsson (1983) would be in the order of 500–2000 m (middle to lower bathyal). The Late Campanian-Maastrichtian faunas of Hemleben & Troester (1984; western Atlantic east of the Antillean Arc) lack certain important genera like *Neoflabellina*, *Bolivinoides* and *Stensoeina* and are almost exclusively benthic. They indicate a depth of deposition greater than that of the Trinidad equivalents, estimated by the authors at 3500–4000 m, close

to the CCD, before they were presumably displaced into abyssal depths. Sliter (1985; East Mariana Basin) records displaced Maastrichtian faunas of this type which indicate an original depth of 1500–3500 m (lower bathyal). From surface outcrops in eastern Venezuela, Galea-Alvarez (1985) suggests a depth of more than 900 m but above the CCD for a calcareous fauna of the San Antonio Formation; the associated arenaceous faunas are not depth-specific. According to Koutsoukos & Merrick (1985), Late Campanian to Middle Maastrichtian faunas in Trinidad are believed to indicate warm waters at depths of 350–3000 m, mostly well oxygenated but also with intermittent phases of restricted circulation; the most similar one of their faunal lists is attributed to the middle to lower bathyal zone. Kaminski *et al.* (1988) assume a bathyal environment for the Guayaguayare Formation in Trinidad where the lower part of the Formation is probably somewhat shallower than the upper part. In conclusion, the depositional depth of the Guayaguayare Formation is mostly regarded as lower to middle bathyal, possibly locally slightly shallower in the lower part of the formation.

For faunas comparable to those of the Lower Lizard Springs Formation (Paleocene), depth estimates by Vincent *et al.* (1974; Indian Ocean) suggest an equivalent fauna in their Assemblage 3 (with *Nuttallides truempyi*, *Gavelinella beccariiformis*, etc.; lower bathyal, depth 600–2500 m). The underlying Assemblage 2 (Midway type; lower neritic to upper bathyal, depth 50–600 m) includes the *Anomalina midwayensis* group and *Loxostomoides applinae* which are characteristic components of the Rochard facies fauna (Early Paleocene) in Trinidad. A distinctly Lower Lizard Springs type fauna is recorded by Dailey (1983; Rio Grande Rise) and interpreted as lower bathyal (1500–2000 m). Saint-Marc (1986) has reached the same conclusions for very similar Paleocene faunas in the area of the Sierra Leone Rise (lower bathyal, about 2000 m). The Paleocene fauna described by the same author (Saint-Marc 1987) from the Northwest Atlantic Site 605 is regarded as being slightly shallower (middle to lower bathyal, 600–1800 m), because of the more noticeable presence of Midway-type species, some of which (*Ceratobulimina perplexa*, *Cibicidoides alleni*, *Gavelinella danica*) have not been found in the Lizard Springs Formation. Among the Paleocene surface outcrops, the Danian of Senegal is indeed very similar in its faunal contents to the Lizard Springs and is regarded by Klasz *et al.* (1987a) as a middle bathyal deposit (about 2000 m water depth). In eastern Venezuela, the Vidoño Formation (latest Maastrichtian to Early Eocene) is believed by Galea-Alvarez (1985) to be deposited in a lower slope environment, in spite of the presence of some (redeposited?) 'Midway type' elements; the depth would be 2000 m or more but above the CCD which is estimated at 4000–4500 m. In Trinidad, Koutsoukos & Merrick (1985), without any supporting evidence, propose an abyssal environment close to the CCD for the Lizard Springs Formation. More recently, Kaminski *et al.* (1988), have contributed a paleoecological study which is mostly concerned with the agglutinated foraminifera of the lowermost part of the Lizard Springs Formation (equivalent to the *Globorotalia pseudobulloides* and *G. trinidadensis* zones) from a well section (Guayaguayare well no. 287). In a Q-mode factor analysis of this interval, they produce a rather complex combination of autochthonous and redeposited faunal elements apparently deposited in various environments from upper bathyal to probably abyssal. This includes presumably the deepest part of the Lizard Springs Formation, which otherwise reflects deposition at upper bathyal to lower bathyal depths. A combination of the data just reviewed for the Velasco-type Lower Lizard Springs (Paleocene) faunas of Trinidad would suggest a fairly deep (lower bathyal) average environment with a water depth of very approximately 2000 m, with possible extensions into the shallower bathyal and the abyssal ranges. In this connection, it should be noted that a relatively shallow palaeodepth of about 900 m produced by backtracking has been recorded in Tjalsma & Lohmann (1983) for three surface localities of the Lower Lizard Springs Formation (Bolli 1957*b*); unfortunately, the exact source of this depth information is not indicated.

In spite of the faunal changes at the Paleocene/Early Eocene boundary, leading to the disappearance of the *Gavelinella beccariformis* assemblage of Tjalsma & Lohmann (1983), there is still enough continuity, also in Trinidad, to trace the major faunal groups upwards into the Upper Lizard Springs Formation (Early Eocene). The fundamentals of the *Pleurostomella-Nuttallides* fauna of Bandy (1970), classified by this author as abyssal, are certainly valid. In the Indian Ocean DSDP sites of Vincent *et al.* (1974), their Assemblage 4 is mostly a continuation of the Paleocene Assemblage 3, but with the replacement of *Gavelinella beccariiformis* by *Anomalinoides dorri aragonensis*. The indicated depth range of 600 to 2500 m (lower bathyal) has still to be justified for the equivalent Trinidad faunas, but a detailed comparison becomes more difficult as Assemblage 4 contains an increasing number of species which are unknown in Trinidad. In the California Coast Ranges (Berggren & Aubert in Brabb, (ed.) 1983), an essentially similar fauna of the middle part of the Lodo Formation (Zones P5 to P8) is regarded as middle to upper bathyal with an upper depth limit of 500–600 m. In the Early Eocene sections off New Jersey described by Miller & Katz (1987) a paleodepth of 1000–2000 m (*Lenticulina-Bulimina-Osangularia* assemblage) is estimated. Kaminski *et al.* (1988) observe no major palaeoecologic change at the Lower/Upper Lizard Springs boundary but see indications of shallowing in the *Globorotalia aragonensis* Zone. On the whole, the examined data indicate a middle to lower bathyal palaeoenvironment most probably between about 500 and 2000 m, for the Upper Lizard Springs Formation.

Summarizing the discussion above, the average depth for the time interval between the Late Albian and Early Eocene might therefore be within a range of upper to middle bathyal for the Gautier and Naparima Hill formations and of middle to lower bathyal for the Guayaguayare and Lizard Springs formations, with probable sporadic extensions upwards or downwards. The method used here does admittedly involve some rather crude generalizations and does take little account of redeposition, backtracking and complex statistical tools; it should therefore be accepted with caution.

Sample list

This list shows the stratigraphic arrangement and the ages of all samples included in this study of benthic foraminifera from the Gautier, Naparima Hill, Guayaguayare and Lizard Springs formations.

Upper Lizard Springs Formation

Surface samples:
Bo 521, Rz 413 (*Globorotalia aragonensis* Zone).
KB 6972, Rz 281 (*Globorotalia formosa formosa* Zone).
Hk 1831 (*Globorotalia subbotinae* Zone).
Hk 1832 (*Globorotalia edgari* Zone).

Well samples:
Boodoosingh no. 1, core at 6892–6909 feet; Guayaguayare no. 159, core at 3707–3713 feet; Oropouche no. 4, core at 7829–7849 feet (all *Globorotalia subbotinae* Zone).

Lower Lizard Springs Formation

Surface samples:
K4915 (*Globorotalia velascoensis* Zone).
K 10832, RM 287 (*Globorotalia pseudomenardii* Zone).
KR 23575 (*Globorotalia angulata* Zone).

Well samples:
Guayaguayare no. 159, cores at 3813–3825, 3827–3839 feet (*Globorotalia velascoensis* Zone).
Guayaguayare no. 159, core at 3986–4000 feet; Moruga no. 15, core at 3796–3816 feet (both *Globorotalia pseudomenardii* Zone).

Guayaguayare no. 159, cores at 4524–4536, 4778–4790 feet (*Globorotalia pusilla pusilla* Zone).
Moruga no. 15, core at 4296–4310 feet (*Globorotalia uncinata* Zone).

Moruga no.3, cores at 10 258–10 264, 11 428–11 448 feet; Moruga no. 15, core at 4401–4421 feet (all *Globorotalia trinidadensis* Zone).
Rochard no. 1, cores at 8556–8565, 8565–8571 feet (both probably *Globorotalia pseudobulloides* Zone, Rochard facies).
Moruga no. 15, 4617–4637 feet (*Rzehakina epigona* Zonule).

Guayaguayare Formation

Surface samples:
G 3643 (*Abathomphalus mayaroensis* Zone or possibly *Globotruncana gansseri* Zone).
Bn 61 (probably *Abathomphalus mayaroensis* Zone or *Globotruncana gansseri* Zone).

Well samples:
Guayaguayare no. 163, core at 5588–5598 feet (*Abathomphalus mayaroensis* Zone).
F. W. no. 214, core at 10 290–10 310 feet (*Abathomphalus mayaroensis* Zone or *Globotruncana gansseri* Zone).
Guayaguayare no. 163, screen samples at 5672–5752 feet (*Globotruncana gansseri* Zone).

A. B. no. 1, core at 4400–4005 feet (*Globotruncana gansseri* Zone or *Globotruncana lapparenti tricarinata* Zone).
Guayaguayare no. 163, core at 5882–5902 feet; Moruga no. 10, core sample at 7520 feet; Moruga no. 15, core at 4831–4849 feet; Rock Dome no. 1,

core at 6087–6141 feet (all *Globotruncana lapparenti tricarinata* Zone).
Marabella no. 1, cores at 3489–3512, 3512–3540 feet (probably *Globotruncana lapparenti tricarinata* Zone, Marabella facies).

Naparima Hill Formation

Surface samples:
KB 7907, PJ 253, Rz 389 (*Globotruncana stuarti* Zone, partly possibly *Globotruncana lapparenti tricarinata* Zone).

Well samples:
Guayaguayare no. 163, cores at 6046–6052, 6052–6079, 6248–6260 feet; Lizard no. 1x, cores at 6333–6353, 6373–6393 feet; Marabella no. 1, cores at 3600–3629, 4210–4232 feet; Moruga no. 10, core sample at 7987 feet; Moruga no. 15, core at 5195–5210 feet; Rock Dome no. 1, core at 6142–6206 feet (all *Globotruncana stuarti* Zone).
A. B. no. 2, cores at 6460–6480, 6725–6740 feet (both *Globotruncana stuarti* Zone or *G. fornicata* Zone).

Guayaguayare no. 163, cores at 7358–7393, 7641–7656 feet; Lizard no. 1x, core at 7089–7099 feet; Marabella no. 1, cores at 4465–4476, 4711–4731 feet; Marac no.1, core at 7947–7967 feet; Morne Diablo no. 34, cores at 12 578–12 602, 12 602–12 622, 12 622–12 641, 12 641–12 660, 12 660–12 680, 12 680–12 700, 12 700–12 720, 12 720–12 750 feet; Moruga no. 15, cores at 5329–5336, 5410–5415, 5427–5452, 5647–5671 feet (all *Globotruncana fornicata* Zone).
Moruga no. 15, core at 6121–6140 feet (*Globotruncana fornicata* Zone or *G. concavata* Zone).

Esmeralda no. 1, core at 10 678–10 694 feet; Marabella no. 1, cores at 5018–5038, 5316–5329 feet; Marac no. 1, cores at 8076–8129, 8129– 8180, 8180–8237, 8260–8305, 8305–8332, 8332–8363, 8657–8698, 8965–9017 feet; Moruga no. 15, core at 6330–6355 feet (all *Globotruncana concavata* Zone).

Marac no. 1, cores at 9017–9075, 9347–9403 feet; Morne Diablo no. 34, cores at 12 911–12 913, 13 205–13 206, 13 260–13 300, 13 300–13 347 feet; Moruga no. 15, cores at 6519–6544, 6802–6827 feet (all *Globotruncana renzi* Zone).

Lizard no. 1x, cores at 7186–7198, 7198–7200, 7378–7390, 7390– 7402 feet; Morne Diablo no. 34, cores at 13 387–13 430, 13 729–13 774, 13 774–13 821, 13 842–13 889 feet; Moruga no. 15, core at 6980– 7005 feet (all *Globotruncana inornata* Zone).

Gautier Formation

Surface samples:
JS 1019, KR 8385 (both *Rotalipora appenninica appenninica* Zone).

Well samples:
Lizard no. 1x, cores at 7612–7624, 7827–7830, 7837–7848, 7848–7875 feet; Marac no. 1, core at 9578–9621 feet; Morne Diablo no. 34, cores at 13 889–13 936, 13 936–13 965, 13 965–13 984, 13 984–14 018 feet (all *Rotalipora appenninica appenninica* Zone).

Marac no. 1, core at 9853–9891 feet (*Favusella washitensis* Zone).

Marac no. 1, screen samples at 11 959–11 989, 12 035–12 040 feet (*Rotalipora ticinensis ticinensis* Zone).

Annotated taxonomic list

The following list includes 553 species and subspecies of Late Albian to Early Eocene age, which are illustrated on Figs. 18 to 47. The arrangement of this fauna into families and genera follows the classification by Loeblich & Tappan (1987). It includes 13 new taxa which were described and named separately (Beckmann 1991). The synonymy lists for each taxon include the original references as well as references to verified occurrences in Trinidad. The designation of the ages and zones corresponds to that of Fig. 12.

Family Bathysiphonidae

BATHYSIPHON cf. DISCRETUS (Brady)
Figure **18**.1–2

Rhabdammina discreta Brady 1881, Quart. J. Micr. Sci., 21, p. 489. – Brady 1884, Report Sci. Results HMS Challenger, Zoology, 9, p. 258, pl. 22, figs. 11–13.

Rhabdammina discreta H. B. Brady, Cushman & Jarvis 1932, p. 4, pl. 1, figs. 1, 2. – Cushman & Renz 1946, p. 12, pl. 1, fig. 1.
Rhabdammina discreta H. B. Brady, var., Cushman & Renz 1946, p. 12, pl. 1, figs. 2, 3.
Rhabdammina cf. *discreta* H. B. Brady, Cushman & Renz 1948, p. 4, pl. 1, fig. 3.
Rhabdammina ex gr. *discreta* Brady, Kaminski et al. 1988, p. 183, pl. 1, figs. 8, 9.

Similar fragments have been named *Rhabdammina cylindrica* Glaessner and *Rh. lineariformis* Myatlyuk. They are tubular, not compressed, fine sandy/sugary in appearance, occasionally with slight thickenings or compressions. Diameter of fragments 0.17–0.4 mm.

Stratigraphic range: Maastrichtian (*Globotruncana gansseri* Zone) to Early Eocene (*Globorotalia aragonensis* Zone). Fairly common. Also found in the Eocene of the overlying Navet Formation.

BATHYSIPHON EOCENICUS Cushman & Hanna
Figure **18**.3

Bathysiphon eocenica Cushman & Hanna 1927, Proc. California Acad. Sci., ser. 4, 16, p. 210, pl. 13, figs. 2, 3.

The wall has often a chalky appearance and contains fine sand grains. Diameter of tubes 0.25–0.8 mm.

Stratigraphic range: Early Paleocene (*Rzehakina epigona* Zonule) to Early Eocene (*Globorotalia aragonensis* Zone). Common. Also found in the Eocene of the overlying Navet Formation.

BATHYSIPHON ROBUSTUS (Grzybowski)
Figure **18**.4

Dendrophrya robusta Grzybowski 1898, p. 273, pl. 10, fig. 7.
? *Dendrophrya latissima* Grzybowski, Kaminski et al. 1988, p. 182, pl. 1, fig. 6.

Highly variable in size, sometimes relatively large. Wall finely granular. Diameter 0.2–1.9 mm.

Stratigraphic range: Albian (*Rotalipora ticinensis ticinensis* Zone?, *Favusella washitensis* Zone) to Maastrichtian (*Globotruncana gansseri* Zone). Common.

BATHYSIPHON sp. A
Figure **18.5–6**

?Bathysiphon?dubia (White), Cushman &
    Renz 1946, p. 12, pl. 1, figs. 4, 5.
Bathysiphon sp., Kaminski *et al.* 1988, p. 182,
    pl. 1, figs. 2, 3.

The tubes show nodes or slightly constricted
zones which may indicate areas of relative weak-
ness. The wall is amorphous and smooth, often
glossy. Diameter 0.12–0.4 mm.

Stratigraphic range: Late Maastrichtian
(*Abathomphalus mayaroensis* Zone) to Early
Eocene (*Globorotalia aragonensis* Zone).
Fairly common.

Family Rhabdamminidae

RHIZAMMINA (?) INDIVISA Brady
Figure **18.7**

Rhizammina indivisa Brady 1884, Reports Sci.
    Results *HMS Challenger*, Zoology, 9, p. 277,
    pl. 28, figs. 5–7.
Rhizammina indivisa Brady, Kaminski *et al.*
    1988, p. 183, pl. 1, figs. 10–13.

Diameter of tubes 0.2–0.55 mm.

Stratigraphic range: Maastrichtian (*Globotruncana
gansseri* Zone and *Abathomphalus mayaroensis*
Zone)?; Early Paleocene (*Globorotalia
trinidadensis* Zone) to Early Eocene
(*Globorotalia aragonensis* Zone). Fairly
common.

DENDROPHRYA (?) EXCELSA
Grzybowski
Figure **18.8**

Dendrophrya excelsa Grzybowski 1898, p. 16,
    pl. 10, figs. 2–4.
Dendrophrya ex gr. *excelsa* Grzybowski,
    Kaminski *et al.* 1988, p. 182, pl. 1, figs. 4,
    5.

Diameter of tubes 0.08–0.55 mm.

Stratigraphic range: Late Maastrichtian
(*Abathomphalus mayaroensis* Zone) to Early
Eocene (*Globorotalia aragonensis* Zone).
Rather scarce.

Family Saccamminidae

LAGENAMMINA GRZYBOWSKII
(Schubert)
Figure **18.9–10**

Reophax difflugiiformis Brady, Grzybowski
    1898, p. 255, pl. 10, figs. 11–12.
Reophax grzybowskii Schubert 1901, Beitr.
    Palaeontol. Geol. Oesterreich-Ungarn, 14,
    p. 20, pl. 1, fig. 13.
Lagenammina grzybowskii (Schubert),
    Kaminski *et al.* 1988, p. 182, pl. 2, fig. 7.

The shape varies from elongated flask-shaped to
almost spherical. Usually compressed. Length
0.41–1.08, width 0.21–0.66 mm.

Stratigraphic range: Late Maastrichtian
(*Abathomphalus mayaroensis* Zone) to Late
Paleocene (*Globorotalia pseudomenardii*
Zone). Rather scarce.

SACCAMMINA COMPLANATA (Franke)
Figure **18.11**

Pelosina complanata Franke 1912, Jahrb. K.
    preuss. geol. Landesanstalt, 32, p. 107, pl. 3,
    fig. 1.
Saccammina complanata (Franke), Kaminski *et
    al.* 1988, p. 183, pl. 2, fig. 8.

Compressed-spherical, with a small aperture.
Length 0.3–1.1, width 0.25–1.0 mm.

Stratigraphic range: Campanian (*Globotruncana
stuarti* Zone) to Early Eocene (*Globorotalia
aragonensis* Zone). Rather scarce.

SACCAMMINA GLOBOSA Crespin
Figure **18.12–13**

Saccammina globosa Crespin 1963, Bull.
    Bureau Min. Res. Geol. Geoph., Canberra,
    66, p. 21, pl. 1, figs. 13–17.
Pelosina complanata Franke, Cushman &
    Jarvis 1932, p. 5, pl. 1, fig. 5 (not figs. 4, 6).
    – Cushman & Renz 1946, p. 13, pl. 1, fig. 8.

A few specimens with a relatively wide neck
resemble *Saccammina alexanderi* (Loeblich &
Tappan; *Proteonina a.*), but generally the shape
is more spherical, and the wall material is finer.
The test is usually less distinctly compressed than

in other representatives of the Saccamminidae in Trinidad. Length 0.51–0.92, width 0.36–0.76 mm.

Stratigraphic range: Early Cenomanian (*Rotalipora appenninica appenninica* Zone) to Turonian (*Globotruncana inornata* Zone). Rather scarce.

SACCAMMINA PLACENTA (Grzybowski)
Figure **18.14–15**

Reophax placenta Grzybowski 1898, p. 276, pl. 10, figs. 9, 10.
Pelosina complanata Franke, Cushman & Jarvis 1932, p. 5, pl. 1, figs. 4, 6 (not fig. 5).
Saccammina rhumbleri (Franke), Cushman & Renz 1946, p. 13, pl. 1, figs. 6, 7.
Saccammina placenta (Grzybowski), Kaminski *et al.* 1988, p. 183, pl. 2, fig. 9.

Typical specimens are moderately compressed and have a small aperture, but there are also numerous tests without a visible aperture; occasional specimens are almost spherical. Diameter 0.28–0.95 mm.

Stratigraphic range: Late Maastrichtian (*Abathomphalus mayaroensis* Zone) to Early Eocene (*Globorotalia subbotinae* Zone). Common. Also found in the Eocene of the overlying Navet Formation.

THURAMMINA (?) spp.
Figure **18.16–17**

Spherical, more or less compressed specimens with a finely granular, non-calcareous wall showing a distinct reticulation. There is no visible aperture. Similar forms are described by Subbotina (1964). Although they superficially resemble radiolarian skeletons, their wall material has the same aspect as that of other Saccamminidae in our samples. The more finely reticulated specimens (Fig.18.17) are very similar to *Thurammina magnalveolata* Bulatova in Subbotina (1964). Diameter 0.45–0.9 mm.

Stratigraphic range: Paleocene (*Rzehakina epigona* Zonule to *Globorotalia pseudomenardii* Zone). Very rare.

Family Hippocrepinidae

HYPERAMMINA ELONGATA Brady
Figure **18.18–19**

Hyperammina elongata Brady 1878, Ann. Mag. Nat. Hist., ser. 5, 1, p. 433, pl. 20, fig. 2.
Hyperammina elongata H. B. Brady, Cushman & Renz 1928, p. 86, pl. 12, fig. 1. – Cushman & Jarvis 1932, p. 6, pl. 1, figs. 7, 8. – Cushman & Renz 1946, p. 13, pl. 1, figs. 9, 10. – Cushman & Renz 1948, p. 5, pl. 1, figs. 10, 11. – Kaminski *et al.* 1988, p. 184, pl. 1, figs. 14, 15.

A few large specimens (proloculus 0.85–1.05, tube 0.7–0.8 mm in diameter) occur in the Late Cretaceous Guayaguayare Formation; externally they resemble *H. primitiva* Myatlyuk. In the Tertiary, the corresponding dimensions are 0.27–0.55 and 0.15–0.4 mm.

Stratigraphic range: Early Maastrichtian (*Globotruncana lapparenti tricarinata* Zone) to Early Eocene (*Globorotalia formosa formosa* Zone). Rather scarce. Also found in the Eocene of the overlying Navet Formation.

HYPERAMMINA ERUGATA Sliter
Figure **18.20**

Hyperammina erugata Sliter 1968, p. 41, pl. 1, fig. 6.

The specimens are much compressed and smaller than those of *H. elongata*. The wall is finely agglutinated, sometimes amorphous and slightly glossy. Diameter of the proloculus 0.12–0.28 mm, of the tube 0.1–0.22 mm.

Stratigraphic range: Maastrichtian (*Globotruncana gansseri* Zone) to Early Eocene (*Globorotalia aragonensis* Zone). Fairly common.

HYPERAMMINA cf. SUBNODOSIFORMIS Grzybowski
Figure **18.21–22**

Hyperammina subnodosiformis Grzybowski 1898, p. 274, pl. 10, fig. 5.
Hyperammina (?) sp., Cushman & Jarvis 1932, p. 6, pl. 1, fig. 9.

Hyperammina ex gr. *subnodosiformis*
    Grzybowski, Kaminski *et al.* 1988, p. 184,
    pl. 1, figs. 16, 17.
? *Hyperammina arenifera* Bermudez 1963,
    p. 36, pl. 1, figs. 6, 7.

The morphology and taxonomy is discussed in
Kaminski *et al.* (1988).
    Length 0.55–2.05, diameter 0.18–0.5 mm.

Stratigraphic range: Early Paleocene (*Globorotalia
trinidadensis* Zone) to Early Eocene
(*Globorotalia formosa formosa* Zone). Scarce.
Also found in the Eocene of the overlying
Navet Formation.

Family Ammodiscidae

AMMODISCUS GLABRATUS Cushman &
Jarvis
Figure **18.23–24**

Ammodiscus glabratus Cushman & Jarvis 1928,
    p. 86, pl. 12, fig. 6.
Ammodiscus glabratus Cushman & Jarvis,
    Cushman & Jarvis 1932, p. 8, pl. 2, fig. 1. –
    Cushman & Renz 1946, p. 14, pl. 1, fig. 26.
    – Cushman & Renz 1948, p. 6, pl. 1, fig. 16.
    – Kaminski *et al.* 1988, p. 184, pl. 3, fig. 8.
Ammodiscus peruvianus Berry, Kaminski *et al.*
    1988, p. 185, pl. 3, figs. 11, 12.
? *Ammodiscus cretaceus* (Reuss), Kaminski *et
    al.* 1988, p. 184, pl. 3, fig. 7.

This appears to be a homogenous group which,
from the *Globorotalia velascoensis* Zone
upwards, contains an increasing number of
specimens with a slight tendency towards
developing a wider spiral in the last two whorls.
Compressed specimens with an elliptical outline
('*A. peruvianus*' Berry) occur throughout the
range of the species. The possibility of a synonymy with *Operculina cretacea* Reuss and *Cor-
nuspira polygyra* Reuss needs further
consideration and is here left open. Diameter
0.45–1.1 mm.

Stratigraphic range: Early Maastrichtian
(*Globotruncana lapparenti tricarinata* Zone)
to Early Eocene (*Globorotalia aragonensis*
Zone). Common. Also found in the Eocene of
the overlying Navet Formation.

AMMODISCUS LATUS Grzybowski
Figure 18.25

Ammodiscus latus Grzybowski 1898, p. 282,
    pl. 10, figs. 27, 28.
Lituotuba eocenica Cushman & Renz 1948,
    p. 7, pl. 1, figs. 20, 21.
Lituotuba navetensis Cushman & Renz, new
    name, 1950, Contrib. Cushman Found.
    foramin. Res., 1, p. 45.

Diameter 0.45–0.8 mm, thickness of spiral tube
about 0.12 mm.

Stratigraphic range: Early Eocene (*Globorotalia
subbotinae* Zone to *G. formosa formosa* Zone).
Rather scarce. Also found in the Eocene of the
overlying Navet Formation.

AMMODISCUS MACILENTUS (Myatlyuk)
Figure **18.26–27**

Grzybowskiella macilenta Myatlyuk 1970,
    p. 72, pl. 13, figs. 2, 3.

More robust than *A. planus*, with some transitional specimens to *A. glabratus*. Diameter
0.24–0.5, thickness 0.04–0.07 mm.

Stratigraphic range: Late Paleocene (*Globorotalia
pseudomenardii* Zone) to Early Eocene
(*Globorotalia aragonensis* Zone). Rather
scarce. Also found in the Eocene of the
overlying Navet Formation.

AMMODISCUS PENNYI Cushman & Jarvis
Figure **18.28**, 36

Ammodiscus pennyi Cushman & Jarvis 1928,
    p. 87, pl. 12, figs. 4, 5.
Ammodiscus pennyi Cushman & Jarvis,
    Cushman & Jarvis 1932, p. 9, pl. 2, figs. 2,
    3. – Cushman & Renz 1946, p. 14, pl. 1,
    fig. 27. – Kaminski *et al.* 1988, p. 184, pl. 3,
    figs. 9, 10.
? *Ammodiscus cretaceus* (Reuss) var. *rugosa*
    Schijfsma 1946, Meded. Geol. Stichting,
    Haarlem, ser. C, sec. 5, 7, p. 28, pl. 6, fig. 2.

Two intergrading morphological types can be
recognized. The larger specimens (here named
type A, see Fig. 18.28) seem to correspond to
the (lost) holotype from the Lower Lizard
Springs Formation; a slightly smaller and rather

more regular form (type B, see Fig. 18.36) is
confined to the Lower Lizard Springs Formation. Diameter of type A 1.0–3.0 mm (thickness
of spiral chamber 0.15–0.24 mm). Diameter of
type B 0.5–1.25 mm (spiral chamber 0.1–
0.15 mm).

Stratigraphic range: Type A: Early Maastrichtian
(*Globotruncana lapparenti tricarinata* Zone)
to Late Paleocene (*Globorotalia
pseudomenardii* Zone). Type B: Early
Paleocene (*Rzehakina epigona* Zonule) to Late
Paleocene (*Globorotalia velascoensis* Zone).
Fairy common.

AMMODISCUS PLANUS Loeblich
Figure **18.29**

Ammodiscus planus Loeblich 1946, J.
Paleontol., 20, p. 133, pl. 22, fig. 2.
Ammodiscus planus Loeblich, Kaminski *et al.*
1988, p. 185, pl. 3, fig. 13.
? *Ammodiscus asanoi* Yoshida 1963, J.
Hokkaido Gakugei Univ., Sapporo (Japan),
13, p. 219, pl. 1, fig. 12.

Thin, often transparent. Diameter 0.12–
0.45 mm, thickness 0.02–0.05 mm.

Stratigraphic range: Early Maastrichtian
(*Globotruncana lapparenti tricarinata* Zone)
to Early Eocene (*Globorotalia aragonensis*
Zone). Fairly common.

AMMOLAGENA CLAVATA (Jones &
Parker)
Figure **18.32**

Trochammina irregularis (d'Orbigny) var.
clavata Jones & Parker 1860, Quart. J. Geol.
Soc. London, 16, p. 304.
Webbina clavata (Jones & Parker), Brady
1884, Report Sci. Results HMS Challenger,
Zoology, 9, p. 349, pl. 41, figs. 12–15.
Ammolagena clavata (Jones & Parker),
Cushman & Jarvis 1928, p. 90, pl. 12, fig. 14.
– Cushman & Jarvis 1932, p. 11, pl. 2, fig. 12.
– Cushman & Renz 1946, p. 15, pl. 1, fig. 38.
– Cushman & Renz 1948, p. 7, pl. 2, fig. 2.
– Kaminski *et al.* 1988, p. 185, pl. 3, fig. 24.

Length 0.45–1.6 or more, diameter of chamber
0.2–0.75 mm.

Stratigraphic range: Early Maastrichtian
(*Globotruncana lapparenti tricarinata* Zone)
to Early Eocene (*Globorotalia subbotinae*
Zone). Rather scarce. Also found in the
Eocene of the overlying Navet Formation.

GLOMOSPIRA GORDIALIS
DIFFUNDENS Cushman & Renz
Figure **18.30**

Glomospira gordialis (Jones & Parker),
Cushman & Jarvis 1928, p. 87, pl. 12, figs.
7, 8. – Cushman & Jarvis 1932, p. 9, pl. 2,
figs. 6, 7.
Glomospira gordialis (Jones & Parker) var.
diffundens Cushman & Renz 1946, p. 15, pl. 1,
fig. 30. – Cushman & Renz 1948, p. 7, pl. 1,
fig. 17.
Glomospira diffundens (Cushman & Renz),
Kaminski *et al.* 1988, p. 185, pl. 3, figs. 18,
19.

Diameter 0.45–1.05 mm.

Stratigraphic range: Late Maastrichtian
(*Abathomphalus mayaroensis* Zone) to Late
Paleocene (*Globorotalia pseudomenardii*
Zone). Rather scarce. Also found in the Eocene
of the overlying Navet Formation.

GLOMOSPIRA IRREGULARIS
(Grzybowski)
Figure **18.33**

Ammodiscus irregularis Grzybowski 1898,
p. 285, pl. 11, figs. 2, 3.
Glomospira irregularis (Grzybowski),
Kaminski *et al.* 1988, p. 185, pl. 3, figs. 20,
21.
? *Glomospira glomerata* (Grzybowski),
Kaminski *et al.* 1988, p. 185, pl. 3, fig. 16.

Length 0.45–1.25 mm, diameter of tube 0.06–
0.25 mm.

Stratigraphic range: Maastrichtian (*Globotruncana
gansseri* Zone)?, Early Paleocene (*Globorotalia
trinidadensis* Zone) to Early Eocene
(*Globorotalia subbotinae* Zone). Rather scarce.

GLOMOSPIRA SERPENS (Grzybowski)
Figure **18.31**

Ammodiscus serpens Grzybowski 1898, p. 285,
pl. 10, fig. 31.
Glomospira sp. A, Beckmann 1960, p. 62,
fig. 1.
Glomospira serpens (Grzybowski), Kaminski *et
al.* 1988, p. 185, pl. 3, figs. 22, 23.

Length 0.45–1.4, width 0.2–0.55 mm.

Stratigraphic range: Middle Paleocene (*Globorotalia
uncinata* Zone) to Early Eocene (*Globorotalia
aragonensis* Zone). Fairly common. Also found
in the Eocene of the overlying Navet
Formation.

GLOMOSPIRELLA GAULTINA
CONFUSA (Zaspelova)
Figure **19.1–2**

Glomospira gaultina (Berthelin) var. *confusa*
Zaspelova 1948, Microfauna oil fields
USSR, VNIGRI, Leningrad, Sbornik 1,
p. 196, pl. 1, fig. 3.

The smaller Trinidad specimens (Fig. 19.1), with
irregular initial coiling, seem to correspond to
the type, but others have a better developed
planispiral stage (Fig. 19.2). Diameter 0.35–
1.2 mm.

Stratigraphic range: Late Maastrichtian
(*Abathomphalus mayaroensis* Zone; here most
typical) to Late Paleocene (*Globorotalia
pseudomenardii* Zone). Rather scarce.

GLOMOSPIRELLA GAULTINA
GAULTINA (Berthelin), emend. Bartenstein
1954
Figure **18.34–35**

Ammodiscus gaultinus Berthelin 1880, p. 19,
pl. 1, fig. 3.
Glomospira gaultina Berthelin 1880,
Bartenstein 1954, Senckenbergiana Lethea,
Frankfurt, 35, p. 38.
Ammodiscus gaultinus Berthelin 1880,
Bartenstein & Bolli 1973, p. 394, pl. 2, figs.
1–13.
Glomospirella gaultina (Berthelin 1880),
Bartenstein & Bolli 1986, p. 947, pl. 1, figs.
7, 8.

The wall is white and glossy; more rarely it has
a fine sandy appearance (Fig. 18.35). Specimens
with a reduced planispiral stage resemble *Glomospira gordialis* (Jones & Parker). Diameter
0.12–0.85 mm.

Stratigraphic range: Late Maastrichtian
(*Abathomphalus mayaroensis* Zone) to Early
Eocene (*Globorotalia aragonensis* Zone).
Rather scarce. Also recorded in Trinidad from
the Late Aptian to Early Albian Maridale
Formation.

REPMANINA CHAROIDES CORONA
(Cushman & Jarvis)
Figure **19.3**

Glomospira charoides (Jones & Parker) var.
corona Cushman & Jarvis 1928, p. 89,
pl. 12, figs. 9–11.
Glomospira charoides (Jones & Parker) var.
corona Cushman & Jarvis, Cushman & Jarvis
1932, p. 10, pl. 2, figs. 8–10.
Glomospira charoides (Jones & Parker),
Kaminski *et al.* 1988, p. 185, pl. 3, figs. 14,
15.

Diameter 0.15–0.44 mm.

Stratigraphic range: Early Paleocene (*Rzehakina
epigona* Zonule) to Early Eocene (*Globorotalia
aragonensis* Zone). Common. Also found in
the Eocene of the overlying Navet Formation.

REPMANINA CHAROIDES FAVILLA
(Emiliani)
Figure **19.4–6**

Glomospira favilla Emiliani 1954, Palaeontogr.
Italica, 48, p. 133, pl. 23, fig. 15.
Glomospira charoides (Jones & Parker) var.
corona Cushman & Renz 1946, p. 15, pl. 1,
fig. 31.

This subspecies differs from *corona* in the larger
size and the hemispheric to hood-shaped outline
in side view. Diameter 0.33–0.75 mm.

Stratigraphic range: Maastrichtian (*Globotruncana
gansseri* Zone) to Late Paleocene (*Globorotalia
pseudomenardii* Zone). Fairly common.

Family Rzehakinidae

RZEHAKINA EPIGONA (Rzehak)
Figure 19.7–8

Silicina epigona Rzehak 1895, Ann. Haturhist.
Hofmuseum, Vienna, 10, p. 214, pl. 6,
fig. 1.
Rzehakina epigona (Rzehak) var. *lata* Cushman
& Jarvis 1928, p. 93, pl. 13, fig. 11.
Rzehakina epigona (Rzehak) var. *lata* Cushman
& Jarvis, Cushman & Jarvis 1932, p. 20,
pl. 6, fig. 1. – Cushman & Renz 1946, p. 23,
pl. 3, fig. 6.
Rzehakina epigona (Rzehak), Kaminski *et al.*
1988, p. 186, pl. 7, figs. 6, 7.

A considerable range of variation (see Hiltermann 1974) is also observed in Trinidad. Length
0.3–1.2, width 0.16–1.0 mm.

Stratigraphic range: Santonian (*Globotruncana
fornicata* Zone) to Late Paleocene (*Globorotalia
velascoensis* Zone). Common.

RZEHAKINA MINIMA Cushman & Renz
Figure 19.9

Rzehakina epigona (Rzehak) var. *minima*
Cushman & Renz 1946, p. 24, pl. 3, fig. 5.
Rzehakina minima (Cushman & Renz),
Kaminski *et al.* 1988, p. 186, pl. 7, figs. 8, 9.

Length 0.25–0.75, width 0.15–0.5 mm.

Stratigraphic range: Campanian (*Globotruncana
stuarti* Zone) to Late Paleocene (*Globorotalia
pseudomenardii* Zone). Common.

SILICOSIGMOILINA AKKESHIENSIS
Yoshida
Figure 19.10–11

Silicosigmoilina (Bramletteia) akkeshiensis
Yoshida 1963, J. Hokkaido Gakugei Univ.,
13, p. 64, pl. 5, figs. 1, 2.

S. californica Cushman & Church seems to have
a more acute periphery but is otherwise very
similar. Length 0.4–0.75, width 0.35–0.5, thickness 0.22–0.3 mm.

Stratigraphic range: Campanian (*Globotruncana
stuarti* Zone) to Maastrichtian (*Globotruncana
gansseri* Zone). Scarce.

SILICOSIGMOILINA ANGUSTA Turenko
Figure 19.12–14

Silicosigmoilina angusta Turenko 1983,
Paleontol. Zhurnal, Moscow, 1983/3, p. 13,
pl. 15, figs. 5, 6 (Paleontol. J., translation,
17/3, p. 14, pl. 1, figs. 5, 6).

The aperture may or may not have a tooth-like
constriction. Turenko's type description mentions a rounded aperture, but his illustration
seems to show a small tooth. The size, length/
width ratio, symmetry in side view and peripheral angle show a considerable variability, and
relatively large and flat specimens resemble *S.
futabaensis* Asano. Length 0.33–0.55, width
0.11–0.3, thickness 0.07–0.13 mm.

Stratigraphic range: Early Maastrichtian
(*Globotruncana lapparenti tricarinata* Zone)
to Early Paleocene (*Globorotalia trinidadensis*
Zone). Rare slightly broader specimens (up to
0.42 mm) occur in the earliest Eocene
(*Globorotalia edgari* Zone to *G. subbotinae*
Zone). Rather scarce.

Family Aschemocellidae

ASCHEMOCELLA MONILIFORMIS
(Neagu)
Figure 19.15–16

Aschemonella moniliformis Neagu 1964,
Rocznik Polsk. Towarz. Geol., Krakow, 34,
p. 586, pl. 27, figs. 4–7, textfigs. 4/7–8, 5/1–
7.
? *Aschemonella* ex gr. *grandis* (Grzybowski),
Kaminski *et al.* 1988, p. 186, pl. 2, figs. 11–
13.

Compressed, thin-walled fragments, subcylindrical to almost spherical, consisting of single
chambers, which are sometimes slightly smaller
than the types from the Late Cretaceous of
Romania. Extreme variants may have the shape
of *A. carpathica* Neagu or 'Reophax' *grandis*
Grzybowski. Length of fragments 1.25–4.2,
width (compressed) 0.45–1.5 mm.

Stratigraphic range: Early Maastrichtian
(*Globotruncana lapparenti tricarinata* Zone)
to Early Eocene (*Globorotalia formosa
formosa* Zone). Fairly common.

Family Hormosinidae

HORMOSINELLA OVULUM (Grzybowski)
Figure **19.17–19**

Reophax ovulum Grzybowski 1896, p. 276,
  pl. 8, figs. 8, 9.
Reophax ovuloides Grzybowski 1901, p. 268,
  pl. 7, fig. 3.
Hormosina ovulum ovulum (Grzybowski),
  Kaminski *et al*. 1988, p. 186, pl. 2, fig. 10.
Hormosina ovuloides (Grzybowski), Kaminski
  et al. 1988, p. 186, pl. 2, figs. 3, 4.

Ovoid/fusiform to almost globular chambers,
which appear to be fragments of a multilocular
form, with an extended, often neck-like aperture. Two-chambered specimens or specimens
with only one aperture occur very rarely.
Smaller subspherical specimens (Fig. 19.19) are
predominant in the Early Eocene samples. The
synonymy is discussed in Kaminski *et al*. (1988).
Length of chambers 0.25–1.23, diameter 0.24–
0.7 mm.

Stratigraphic range: Santonian (*Globotruncana
fornicata* Zone; rare), Early Maastrichtian
(*Globotruncana lapparenti tricarinata* Zone) to
Early Eocene (*Globorotalia subbotinae*
Zone). Fairly common.

HORMOSINELLA sp. A
Figure **19.20**

Compressed, more or less spherical single chambers with two distinct apertural tubes of variable
length. The wall is fairly coarsely arenaceous.
Length 0.9–1.15, width 0.65–1.1 mm.

Stratigraphic range: Paleocene (*Rzehakina epigona*
Zonule to *Globorotalia pseudomenardii*
Zone). Rare.

REOPHAX ELONGATUS Grzybowski
Figure **19.21**

Reophax elongata Grzybowski 1898, p. 279,
  pl. 10, figs. 19, 20.

Always compressed, with more or less fusiform
chambers. Length 0.75–2.00 (incomplete),
width 0.3–0.55 mm.

Stratigraphic range: Early Eocene (*Globorotalia
subbotinae* Zone to *G. formosa formosa* Zone).
Rare.

REOPHAX cf. FOLKESTONENSIS (Khan)
Figure **19.22**

Hormosina globulifera Brady, Chapman 1892,
  J. Royal Micr. Soc. London, p. 326, pl. 6,
  fig. 10.
Hormosina folkestonensis Khan 1950, J. Royal
  Micr. Soc. London, ser. 3, 70, p. 268.

The regular series of globular chambers is
characteristic, but the Trinidad specimens are
about twice the size of the types from the English
Albian. Length 0.51–1.0, width 0.18–0.34 mm.

Stratigraphic range: Early Paleocene (*Globorotalia
trinidadensis* Zone) to Early Eocene
(*Globorotalia formosa formosa* Zone). Rather
rare.

REOPHAX GLOBOSUS Sliter
Figure **19.23–24**

Reophax globosus Sliter 1968, p. 43, pl. 1,
  fig. 12.
Reophax globosus Sliter, Kaminski *et al*. 1988,
  p. 187, pl. 3, fig. 4.

Fairly typical specimens (Fig. 19.24) occur in a
probable outer shelf facies which is found in
Marabella well no. 1 (probably *Globotruncana
lapparenti tricarinata* Zone). A smaller type
(Fig. 19.23) is stratigraphically more widespread
and is found sporadically in a deeper water facies. Length 0.85–1.64, width 0.5–0.81 mm for
the large type, 0.58–1.05 and 0.28–0.55 mm
respectively for the smaller type.

Stratigraphic range: Early Cenomanian (*Rotalipora
appenninica appenninica* Zone) to Early
Paleocene (*Globorotalia trinidadensis* Zone).
Rare.

REOPHAX PROLATUS Sliter
Figure **19.25**

Reophax prolatus Sliter 1968, p. 43, pl. 1,
  fig. 11.

Length 0.56–0.98, width 0.2–0.41 mm.

Stratigraphic range: Late Maastrichtian
(*Abathomphalus mayaroensis* Zone), Early
Eocene (*Globorotalia formosa formosa* Zone).
Very rare.

REOPHAX cf. TROYERI Tappan
Figure **19.26**

Reophax troyeri Tappan 1960, Bull. Am.
 Assoc. Petroleum Geol., 44, p. 291, pl. 1, figs.
 10–12.
Reophax sp. 2, Kaminski *et al.* 1988, p. 187,
 pl. 3, figs. 2, 3.

As for other species of *Reophax*, the Trinidad
form is larger than the type, which is described
from the Albian of Alaska. Length 0.58–1.03,
width 0.21–0.36 mm.

Stratigraphic range: Santonian (*Globotruncana
fornicata* Zone) to Late Paleocene (*Globorotalia
velascoensis* Zone). Rather scarce.

SUBREOPHAX PSEUDOSCALARIUS
(Samuel)
Figure **19.27**

Reophax pseudoscalaria Samuel 1977, Zapadne
 Karpaty, Bratislava, ser. Paleontol., 2–3,
 p. 36, pl. 3, fig. 4.
Subreophax pseudoscalaria (Samuel),
 Kaminski *et al.* 1988, p. 187, pl. 3, figs. 5,
 6.

Length up to 2.0, diameter 0.3–0.75 mm.

Stratigraphic range: Maastrichtian (*Globotruncana
lapparenti tricarinata* Zone?, *Globotruncana
gansseri* Zone) to Early Eocene (*Globorotalia
aragonensis* Zone). Rather rare.

SUBREOPHAX SCALARIUS (Grzybowski)
Figure **19.28–29**

Reophax guttifera Brady var. *scalaria*
 Grzybowski 1896, p. 277, pl. 8, fig. 26.
Reophax sp. (?), Cushman & Jarvis 1928, p. 86,
 pl. 12, fig. 2. – Cushman & Jarvis 1932, p. 7,
 pl. 1, fig. 13. – Cushman & Renz 1946,
 p. 14, pl. 1, figs. 14, 25.
Subreophax scalaria (Grzybowski), Kaminski *et
al.* 1988, p. 187, pl. 2, figs. 16, 17.

Subglobular chambers in irregular linear

arrangement. Length up to 1.7 mm (incomplete), width 0.16–0.5 mm.

Stratigraphic range: Early Maastrichtian
(*Globotruncana lapparenti tricarinata* Zone)
to Early Eocene (*Globorotalia aragonensis*
Zone). Fairly common. Also found in the
Eocene of the overlying Navet Formation.

SUBREOPHAX VELASCOENSIS
(Cushman)
Figure **19.38**

Nodosinella velascoensis Cushman 1926, p. 583,
 pl. 20, fig. 9.
Nodellum velascoensis Cushman, Cushman &
 Jarvis 1932, p. 8, pl. 1, figs. 15–17.
Nodellum velascoense Cushman, Cushman &
 Renz 1946, p. 14, pl. 1, figs. 20–24. –
 Kaminski *et al.* 1988, p. 187, pl. 1, figs. 21,
 22.

Length up to 2.0 mm (incomplete), diameter
0.22–0.5 mm.

Stratigraphic range: Late Maastrichtian
(*Abathomphalus mayaroensis* Zone) to Early
Eocene (*Globorotalia formosa formosa* Zone).
Fairly common. Also found in the Eocene of the
overlying Navet Formation.

HORMOSINA TRINITATENSIS Cushman
& Renz
Figure **19.30–31, 36–37**

Hormosina globulifera H. B. Brady var.
 trinitatensis Cushman & Renz 1946, p. 14,
 pl. 1, figs. 15–19.
Hormosina globulifera H. B. Brady, Cushman
 & Jarvis 1928, p. 86, pl. 12, fig. 3. –
 Cushman & Jarvis 1932, p. 7, pl. 1, fig. 14.
Hormosina globulifera H. B. Brady var.
 trinitatensis Cushman & Renz, Cushman &
 Renz 1948, p. 6, pl. 1, figs. 8, 9.
Hormosina trinitatensis Cushman & Renz,
 Kaminski *et al.* 1988, p. 187, pl. 3, fig. 1.
Reophax duplex (Grzybowski), Kaminski *et al.*
 1988, p. 187, pl. 2, fig. 15.

Apart from the globular chambers, the test
shows a great variability in shape (subcylindrical
to tapering), number of chambers (1 to 4, rarely
up to 7), apertural features (depressed, with or
without a slight neck). A variant with a tubular

aperture occurs in some samples of the Lower Lizard Springs Formation (Paleocene). Length 0.5–4.0, width 0.24–1.0 mm.

Stratigraphic range: Santonian (*Globotruncana fornicata* Zone) to Early Eocene (*Globorotalia aragonensis* Zone). Abundant. Also found in the Eocene of the overlying Navet Formation.

Family Haplophragmoididae

ASANOSPIRA WALTERI (Grzybowski)
Figure **19.33–34, 40**

Trochammina walteri Grzybowski 1898, p. 290, pl. 11, fig. 31.
Haplophragmoides sp. (?), Cushman & Jarvis 1928, p. 91, pl. 12, fig. 16.
Haplophragmoides excavata Cushman & Waters, Cushman & Jarvis 1932, p. 12, pl. 3, fig. 1.
Haplophragmoides walteri (Grzybowski), Kaminski *et al.* 1988, p. 190, pl. 5, figs. 14, 15.

Diameter 0.3–0.7, thickness 0.16–0.19 mm.

Stratigraphic range: Middle Paleocene (*Globorotalia uncinata* Zone) to Early Eocene (*Globorotalia aragonensis* Zone). Fairly common. Also found in the Eocene of the overlying Navet Formation.

CRIBROSTOMOIDES TRINITATENSIS
Cushman & Jarvis
Figure **19.35, 41**

Cribrostomoides trinitatensis Cushman & Jarvis 1928, p. 91, pl. 12, fig. 12.
Cribrostomoides trinitatensis Cushman & Jarvis, Cushman & Jarvis 1932, p. 12, pl. 3, fig. 3. – Cushman & Renz 1946, p. 19, pl. 2, figs. 4, 5. – Kaminski *et al.* 1988, p. 188, pl. 6, figs. 1, 2.

Diameter 0.35–0.8 mm.

Stratigraphic range: Late Maastrichtian (*Abathomphalus mayaroensis* Zone) to Early Eocene (*Globorotalia aragonensis* Zone). Rather scarce.

EVOLUTINELLA FLAGLERI (Cushman & Hedberg)
Figure **19.42–43**

Haplophragmoides flagleri Cushman & Hedberg 1941, Contrib. Cushman Lab. foramin. Res., 17, p. 82, pl. 21, fig. 2.

Diameter 0.36–0.75, thickness 0.25–0.4 mm.

Stratigraphic range: Early Cenomanian (*Rotalipora appenninica appenninica* Zone) to Santonian (*Globotruncana fornicata* Zone). Fairly common.

EVOLUTINELLA RENZI Beckmann
Figure **20.1–3**

Evolutinella renzi Beckmann 1991, p. 882, pl. 1, figs. 1–3.

Planispiral, evolute, with slightly deepened umbilicus. 8–12 chambers in the final whorl which increase very gradually in size, varying in shape from not inflated and almost unrecognizable to subglobular. Sutures radial, flush with the chamber wall surface to moderately depressed. Wall fairly coarsely arenaceous. Aperture semicircular, at the base of the apertural face, but sometimes hardly visible. Diameter 0.55–1.05, thickness 0.16–0.3 mm.

In the older samples (*Favusella washitensis* Zone), the specimens usually have a depressed umbilical area and not more than 10 chambers in the final whorl (Fig. 20.1); the final chambers are often distinctly inflated but sometimes secondarily collapsed. The stratigraphically younger specimens (Fig. 20.3) have a less depressed umbilicus and up to 12 chambers which are often less distinctly inflated.

The superficially similar species *Haplophragmoides fraseri* Wickenden and *H. gilberti* Eicher are considerably smaller. *H. yukonensis* Chamney has more (14) chambers. All three species show more involute coiling than the present species.

Stratigraphic range: Late Albian to Early Cenomanian (*Favusella washitensis* Zone to *Rotalipora appenninica appenninica* Zone). Rare.

HAPLOPHRAGMOIDES CAUCASICUS
Shutskaya
Figure **20.4–5**

Haplophragmoides caucasicus Shutskaya 1956,
   Trudy Akad. Nauk SSSR, Inst. geol. Nauk,
   Geol. ser., vyp. 164, 71, p. 85, pl. 1, figs. 5,
   6.
? *Haplophragmoides apenninica* Montanaro
   Gallitelli 1958, Accad. Sci. Lett. Arti
   Modena, Atti Mem., ser. 5, 16, p. 132, pl. 1,
   fig. 8.

Diameter 0.3–0.6, thickness 0.16–0.38 mm.

Stratigraphic range: Early Maastrichtian
(*Globotruncana lapparenti tricarinata* Zone)
to Early Eocene (*Globorotalia aragonensis*
Zone). Fairly common. Also found in the
Eocene of the overlying Navet Formation.

HAPLOPHRAGMOIDES CONCAVUS
(Chapman)
Figure **20.6, 12**

Trochammina concava Chapman 1892, J.
   Royal Micr. Soc. London, 1892, p. 327, pl. 6,
   fig. 14.
Haplophragmoides concavus (Chapman 1892),
   Bartenstein & Bolli 1973, p. 393, pl. 3, figs.
   2–26. – Bartenstein & Bolli 1986, p. 952,
   pl. 1, figs. 48–53.

Always irregularly compressed with collapsed
chambers (5–6 ½ in the last whorl). Diameter
0.36–0.66 mm.

Stratigraphic range: Late Albian (*Favusella washitensis* Zone) to Santonian (*Globotruncana fornicata* Zone). Fairly common. Also recorded in
Trinidad from the Late Aptian to Early Albian
Maridale Formation.

HAPLOPHRAGMOIDES FORMOSUS
Takayanagi
Figure **20.7**

Haplophragmoides formosus Takayanagi 1960,
   p. 71, pl. 1, fig. 22.
Haplophragmoides stelcki Hanzlikova 1966,
   Foram. Lhoty Schichten, Cas. Morav. Univ.,
   Vedy prirodni, Brno, 51, p. 115, pl. 3, fig. 5.

Diameter 0.45–0.95, thickness 0.21–0.25 mm.

Stratigraphic range: Late Albian to Early
Cenomanian (*Favusella washitensis* Zone to
Rotalipora appenninica appenninica Zone).
Scarce.

HAPLOPHRAGMOIDES HORRIDUS
(Grzybowski)
Figure **20.8–9**

Haplophragmium horridum Grzybowski 1901,
   p. 54, pl. 7, fig. 12.
Haplophragmoides horridus (Grzybowski),
   Kaminski *et al.* 1988, p. 189, pl. 5, fig. 11.

The two figures show the range of variation
within this species. Diameter 0.8–1.2, thickness
0.55–0.65 mm.

Stratigraphic range: Campanian (*Globotruncana
stuarti* Zone) to Early Maastrichtian
(*Globotruncana lapparenti tricarinata* Zone).
Very rare.

HAPLOPHRAGMOIDES KIRKI Wickenden
Figure **20.10**

Haplophragmoides kirki Wickenden 1932,
   Trans. Proc. Royal Soc. Canada, ser. 3, 26/4,
   p. 85, pl. 1, fig. 1.

A rather poorly defined group of more of less
compressed specimens with 4½–6 moderately
inflated chambers. They may become somewhat
bigger than the type from Canada. Diameter
0.25–0.65, thickness 0.22–0.34 mm.

Stratigraphic range: Early Cenomanian (*Rotalipora
appenninica appenninica* Zone) to Campanian
(*Globotruncana stuarti* Zone); Early Paleocene
(*Globorotalia trinidadensis* Zone) to Early
Eocene (*Globorotalia aragonensis* Zone).
Rather rare. Also found in the Eocene of the
overlying Navet Formation.

HAPLOPHRAGMOIDES MJATLIUKAE
Maslakova
Figure **20.11**

Haplophragmoides mjatliukae Maslakova 1955,
   Materialy po biostrat. zapadnykh oblastei
   USSR, Moscow, p. 48, pl. 3, figs. 7, 8.

Always compressed. 4–5½ chambers which are
hardly lobate. Diameter 0.15–0.4 mm.

Stratigraphic range: Middle to Late Paleocene (*Globorotalia pusilla pusilla* Zone to *G. pseudomenardii* Zone); possibly also Early Eocene (*Globorotalia subbotinae* Zone). Rare.

HAPLOPHRAGMOIDES NEOCHAPMANI Beckmann
Figure **20.13–14**

Haplophragmium latidorsatum Chapman 1892, Foram. Gault Folkestone, pt. 2, J. Royal Micr. Soc, p. 323, pl. 5, fig. 12 (non *Nonionina latidorsata* Bornemann).
Haplophragmoides chapmani Morozova 1948, Byull. Obshch, Ispyt. Prirody, n. ser., 53, Otdel Geol. 23, p. 33, pl. 1, figs. 2, 3 (primary homonym of *H. chapmani* Crespin, 1944).
Haplophragmoides neochapmani Beckmann 1991, new name, p. 823, pl. 1, fig. 12.

4½–6 chambers in the final whorl. Periphery not lobate. Wall fine-grained, often glossy. Diameter 0.46–0.68 mm.

Stratigraphic range: Late Albian (*Favusella washitensis* Zone) to Santonian (*Globotruncana fornicata* Zone). Fairly common.

HAPLOPHRAGMOIDES PORRECTUS Maslakova
Figure **20.15–16**

Haplophragmoides porrectus Maslakova 1955, Materialy po biostrat. zapadnych oblastei URRS, Moscow, p. 47, pl. 3, figs. 5, 6.
Haplophragmoides porrectus Maslakova, Kaminski *et al.* 1988, p. 189, pl. 5, figs. 7, 8.

Usually compressed, with 5½–6 chambers. Diameter 0.35–0.55 mm.

Stratigraphic range: Early to Middle Paleocene (*Rzehakina epigona* Zonule to *Globorotalia pusilla pusilla* Zone). Very rare.

HAPLOPHRAGMOIDES RETROSEPTUS (Grzybowski)
Figure **20.17**

Cyclammina retrosepta Grzybowski 1896, p. 284, pl. 9, figs. 7, 8.
Haplophragmoides retroseptus (Grzybowski), Kaminski *et al.* 1988, p. 189, pl. 5, figs. 9, 10.

Often deformed, with 5–7 chambers. Similar specimens are sometimes assigned to *H. eggeri* Cushman, probably a closely related, if not synonymous species. Diameter 0.48–1.3, thickness 0.35–0.77 mm.

Stratigraphic range: Santonian (*Globotruncana fornicata* Zone) to Late Paleocene (*Globorotalia pseudomenardii* Zone). Common.

HAPLOPHRAGMOIDES cf. SUBORBICULARIS (Grzybowski)
Figure **20.18**

Cyclammina suborbicularis Grzybowski 1896, p. 24, pl. 9, figs. 5, 6.
Haplophragmoides ex gr. *suborbicularis* (Grzybowski), Kaminski *et al.* 1988, p. 189, pl. 5, figs. 12, 13.

Similar to *H. retroseptus* but more regular and with a smoother wall. The number of chambers (6–7½) is less than in Grzybowski's types. Diameter 0.5–0.98, thickness 0.36–0.56 mm.

Stratigraphic range: Early to Middle Paleocene (*Rzehakina epigona* Zonule to *Globorotalia pusilla pusilla* Zone). Rather rare.

HAPLOPHRAGMOIDES TRINITATENSIS Cushman & Renz
Figure **20.19–20**

Haplophragmoides flagleri Cushman & Hedberg var. *trinitatensis* Cushman & Renz 1946, p. 18, pl. 2, figs. 2, 3.
Budashevaella trinitatensis Cushman & Hedberg, Kaminski *et al.* 1988, p. 188, pl. 5, fig. 2; pl. 10, figs. 2, 3.

Planispiral, sometimes deformed. Diameter 0.45–0.95, thickness 0.22–0.46 mm.

Stratigraphic range: Early Paleocene (*Rzehakina epigona* Zonule) to Late Paleocene (*Globorotalia pseudomenardii* Zone). Rather scarce.

RETICULOPHRAGMOIDES JARVISI
(Thalmann)
Figure **19.32, 39**

Nonion cretaceum Cushman & Jarvis 1932,
    p. 41, pl. 12, fig. 12.
Nonion jarvisi Thalmann, n. name, 1932,
    Eclog. geol. Helv., 25, p. 313.
Nonion jarvisi Thalmann, Cushman & Renz
    1946, p. 35, pl. 5, fig. 31.
Haplophragmoides (?) *jarvisi* (Thalmann),
    Kaminski *et al.* 1988, p. 190, pl. 7, figs. 1,
    2; pl. 10, figs. 8, 9.
Reticulophragmoides jarvisi (Thalmann)
    emend., Gradstein & Kaminski 1989, p. 81,
    pl. 7, figs. 1–8; textfig. 4.

The morphology and taxonomy of this strange
form is briefly discussed in Kaminski *et al.*
(1988); Gradstein & Kaminski (1989) reviewed
the occurrences of this species and made it the
type species of the new genus *Reticulophragmoides*. Diameter 0.4–0.7, thickness 0.18–
0.23 mm.

Stratigraphic range: Late Paleocene (*Globorotalia
pseudomenardii* Zone). Scarce.

Family Lituotubidae

LITUOTUBA LITUIFORMIS (Brady)
Figure **20.21–22**

Trochammina lituiformis Brady 1879, Quart. J.
    Microsc. Soc., 19, p. 59, pl. 5, fig. 16.
Lituotuba lituiformis (H. B. Brady), Cushman
    & Jarvis 1928, p. 90, pl. 12, fig. 15. –
    Cushman & Jarvis 1932, p. 10, pl. 2, fig. 11.
    – Cushman & Renz 1946, p. 15, pl. 1, figs.
    32, 33. – Kaminski *et al.*, 1988, p. 190, pl. 4,
    figs. 14, 15.
Lituotuba cf. *lituiformis* (H. B. Brady),
    Cushman & Renz 1948, p. 7, pl. 1, fig. 22.
Ammovertella retrorsa Cushman & Stainforth,
    Cushman & Renz 1948, p. 7, pl. 2, fig. 1.

The coiling of the early part of the test varies
from irregular to almost planispiral. Length 0.5–
2.1 mm, diameter of the coiled early portion
0.3–1.05 mm.

Stratigraphic range: Early Maastrichtian
(*Globotruncana lapparenti tricarinata* Zone)
to Early Eocene (*Globorotalia aragonensis*
Zone). Fairly common. Also found in the
Eocene of the overlying Navet Formation.

PARATROCHAMMINOIDES
IRREGULARIS (White)
Figure **20.23–24**

Trochamminoides irregularis White 1928,
    p. 307, pl. 42, fig. 1.
Trochamminoides irregularis White, Kaminski
    et al. 1988, p. 191, pl. 4, fig. 18.

Diameter 0.5–1.15 mm.

Stratigraphic range: Late Maastrichtian
(*Abathomphalus mayaroensis* Zone) to Early
Eocene (*Globorotalia subbotinae* Zone). Rare.

TROCHAMMINOIDES DUBIUS
(Grzybowski)
Figure **20.27–28**

Ammodiscus dubius Grzybowski 1901, p. 274,
    pl. 8, figs. 12, 14.
Trochamminoides dubius (Grzybowski),
    Kaminski *et al.* 1988, p. 191, pl. 4, figs. 16,
    17.

Often compressed. The coiling may be very
irregularly to almost regularly planispiral, and
the chamber separation varies from very distinct
to almost non-existent. *T. dubius* and *T. velascoensis* Cushman represent two end members of
this series of morphotypes. It is possible that
some of the specimens included here are initial
stages of *Lituotuba lituiformis*. Diameter 0.5–
1.2 mm.

Stratigraphic range: Early Maastrichtian
(*Globotruncana lapparenti tricarinata* Zone)
to Early Eocene (*Globorotalia aragonensis*
Zone). Common. Also found in the Eocene of
the overlying Navet Formation.

TROCHAMMINOIDES PROTEUS (Karrer)
Figure **20.25**

Trochammina proteus Karrer 1866, Sitzungsb.
    K. Akad. Wiss., Vienna, math.-phys. Kl.,
    52 (1865), Abth. 1/9, p. 494, pl. 1, fig. 8.
Trochamminoides proteus (Karrer), Kaminski
    et al. 1988, p. 192, pl. 4, fig. 20.
Trochamminoides sp., Cushman & Renz 1948,
    p. 10, pl. 2, fig. 4.

Diameter 0.6–1.25 (1.75?), thickness 0.15–0.4 mm.

Stratigraphic range: Early Maastrichtian (*Globotruncana lapparenti tricarinata* Zone) to Late Paleocene (*Globorotalia velascoensis* Zone). Rather rare. Also found in the Eocene of the overlying Navet Formation.

TROCHAMMINOIDES SUBCORONATUS (Grzybowski)
Figure **20.26**

Trochammina subcoronata Grzybowski 1896, p. 283, pl. 9, fig. 3.
Haplophragmoides coronata (H. B. Brady), Cushman & Jarvis 1928, p. 90, pl. 12, fig. 17. – Cushman & Jarvis 1932, p. 11, pl. 2, figs. 13–15. – Cushman & Renz 1946, p. 18, pl.1, figs. 36, 37.
Trochamminoides cf. *irregularis* White, Cushman & Renz 1948, p. 10, pl. 2, fig. 3.
Trochamminoides subcoronatus (Grzybowski), Kaminski *et al.* 1988, p. 192, pl. 4, fig. 19.

Diameter 0.48–1.6, thickness 0.25–0.9 mm.

Stratigraphic range: Early Maastrichtian (*Globotruncana lapparenti tricarinata* Zone) to Early Eocene (*Globorotalia aragonensis* Zone). Common. Also found in the Eocene of the overlying Navet Formation.

Family Lituolidae

AMMOBACULITES FRAGMENTARIUS
Cushman
Figure **20.31–33**

Ammobaculites fragmentarius Cushman 1927, Trans Royal Soc. Canada, 3rd ser., 21, p. 130, pl. 1, fig. 8.
Ammobaculites coprolithiformis (Schwager), Cushman & Jarvis 1932, p. 13, pl. 3, figs. 4, 5. – ? Cushman & Renz 1946, p. 19, pl. 2, fig. 6.
Ammobaculites sp. 1, Kaminski *et al.* 1988, p. 188, pl. 4, figs. 1, 2.

The illustrated specimens give an impression of the considerable variability of this species in Trinidad. The shape resembles that of the paratype in the Cushman Collection rather than that

of the flaring holotype. Small specimens with non-inflated chambers resemble *A. alexanderi* Cushman. Length 0.5–1.6, width 0.25–0.55, thickness 0.1–0.4 mm.

Stratigraphic range: Santonian (*Globotruncana fornicata* Zone) to Middle Paleocene (*Globorotalia pusilla pusilla* Zone). Rather scarce.

AMMOBACULITES LACERTAE
Beckmann
Figure **21.1–4**

Ammobaculites sp. 2, Kaminski *et al.* 1988, p. 188, pl. 4, fig. 3.
Ammobaculites lacertae Beckmann 1991, p. 823, pl. 1, figs. 4, 5, 10, 11.

Test elongated, rounded to oval in cross section but occasionally compressed. Initial coil with 5–6 subglobular chambers in the last whorl, mostly with an open umbilicus. Uniserial part consisting of up to 5 chambers which are broader than high, usually all about equal in shape and slightly to distinctly inflated; more rarely they show a very gradual increase in size. Wall fairly coarsely arenaceous. Aperture terminal, often on a short neck. Length 0.4–1.55, width 0.30–0.55, thickness 0.2–0.53 mm; diameter of spiral 0.3–0.65 mm.

Some transitional forms to *A. fragmentarius* Cushman are present and may indicate a possible relationship or ancestry in this direction. *Ammobaculites chiranus* Cushman & Stone and *A. subagglutinans* Bandy resemble some extreme variants of the present species; the former has a greater number of chambers whereas the latter is distinctly cylindrical with indistinct sutures.

Stratigraphic range: Early Eocene (*Globorotalia subbotinae* Zone to *G. formosa formosa* Zone). Rather rare.

AMMOBACULITES SUBCRETACEUS
Cushman & Alexander
Figure **20.34**

Ammobaculites subcretaceus Cushman & Alexander 1930, Contrib. Cushman Lab. foramin. Res., 6, p. 6, pl. 2, figs. 9, 10.
Ammobaculites subcretaceus Cushman & Alexander 1930, Bartenstein & Bolli 1973,

p. 394, pl. 2, figs. 63, 64. – Bartenstein & Bolli 1986, p. 950, pl. 1, figs. 33, 34.

The Trinidad specimens have only one or two uniserial chambers but otherwise correspond to the types. Length 0.3–0.8, thickness 0.08–0.12 mm; width of spiral 0.2–0.5 mm.

Stratigraphic range: Albian to Early Cenomanian (*Rotalipora ticinensis ticinensis* Zone to *R. appenninica appenninica* Zone). Very rare. Also recorded in Trinidad from the Late Aptian to Early Albian Maridale Formation

AMMOMARGINULINA cf. EXPANSA (Plummer)
Figure **21.5**

Ammobaculites expansus Plummer 1933, Texas Univ. Bull., 3201, p. 65, pl. 5, figs. 4–6.

Compressed, consisting of a spiral with 5–8 (9?) chambers, followed by up to three uniserial chambers. Chambers sometimes separated by depressed sutures, sometimes hardly recognizable. Wall distinctly arenaceous. Aperture not clearly visible, possibly terminal. Length 0.45–1.1, width 0.4–0.65, thickness 0.14–0.32 mm.

The specimens are very close to *A. expansa* but are somewhat bigger and have a more lobate periphery.

Stratigraphic range: Early Eocene (*Globorotalia edgari* Zone). Scarce.

SCULPTOBACULITES BARRI Beckmann
Figure **21.6–8, 15**

Ammobaculites sp. 3, Kaminski *et al.* 1988, p. 188, pl. 4, figs. 5, 6.
Sculptobaculites barri Beckmann 1991, p. 823, pl. 1, figs. 6–9.

Early part of the test discoid, planispiral, evolute, with 7–11 chambers in the last whorl; periphery rounded. This may be followed by up to three uniserial chambers. Chambers about as broad as high, moderately inflated, in the early spiral increasing very gradually in size but later remaining constant. Sutures radial, straight or very slightly curved, depressed but sometimes obscured by abundant or coarse wall material. Wall variable from coarse to fine grained, resulting accordingly in a very rough to relatively smooth chamber surface. Aperture terminal,

apparently a simple rounded opening. Length 0.65–1.35, diameter of spiral part 0.5–1.1, thickness 0.13–0.36 mm.

Specimens from the Upper Lizard Springs Formation (Early Eocene, Fig. 21.8) often have somewhat coarser wall material than those from the Paleocene. A variant with a rather thick fine grained wall (Fig. 21.15) is found in the Late Paleocene *Globorotalia pseudomenardii* Zone. Planispiral specimens may resemble small specimens of *Trochamminoides proteus*, but normally the wall material is distinctly coarser.

Stratigraphic range: Early Paleocene (*Rzehakina epigona* Zonule) to Early Eocene (*Globorotalia formosa formosa* Zone). Rather scarce.

TRIPLASIA FUNDIBULARIS (Harris & Jobe)
Figure **21.9–10**

Frankeina fundibularis Harris & Jobe 1951, Transcript Press, Norman (Oklahoma), p. 7, pl. 1, fig. 8.
Frankeina sp. A, Beckmann 1960, p. 62, fig. 3.

The uniserial chambers (up to 7) of our specimens are rather less flaring than in the holotype, but there is little doubt that we are dealing with the same species (see also Loeblich & Tappan 1952). Length 0.8–2.4, width 0.55–1.0 mm.

Stratigraphic range: Paleocene (*Globorotalia trinidadensis* Zone to *G. velascoensis* Zone). Rare.

TRIPLASIA INAEQUALIS Hagn
Figure **21.11**

Triplasia inaequalis Hagn 1956, Palaeontographica, Stuttgart, A 107, p. 112.
Frankeina sp. A, Israelsky 1951, p. 12, pl. 3, figs. 5–8.
Triplasia sp., Loeblich & Tappan 1952, Smithsonian Misc. Coll., 117, no. 15, p. 42, pl. 7, fig. 12.

Length 0.6–1.7, width 0.4–0.75 mm.

Stratigraphic range: Early Eocene (*Globorotalia subbotinae* Zone to *G. formosa formosa* Zone). Rare. Also found in the Eocene of the overlying Navet Formation.

Family Ammosphaeroidinidae

AMMOSPHAEROIDINA PSEUDOPAUCILOCULATA (Myatlyuk)
Figure 21.12–13

Cystamminella pseudopauciloculata Myatlyuk 1966, Voprosy Mikropaleont., Moscow, 10, p. 264, pl. 1, figs. 5–7; pl. 2, fig. 6; pl. 3, fig. 3.
Ammosphaeroidina pseudopauciloculata (Mjatliuk), Kaminski *et al.* 1988, p. 193, pl. 8, figs. 3–5.

Always compressed. Diameter 0.2–1.03 mm.

Stratigraphic range: Maastrichtian (*Globotruncana gansseri* Zone) to Early Eocene (*Globorotalia aragonensis* Zone). Fairly common. Also found in the Eocene of the overlying Navet Formation.

RECURVOIDES DEFLEXIFORMIS (Noth)
Figure 21.14

Trochammina deflexiformis Noth 1912, Beitr. Palaeont. Geol. Oesterr.-Ungarn Orient., Vienna, 25, p. 14, pl. 1, fig. 10.
Recurvoides deflexiformis (Noth), Kaminski *et al.* 1988, p. 190, pl. 6, fig. 3; pl. 10, fig. 15.

6–8 chambers in the final whorl. The sutures are hardly, if at all, depressed. The arrangement of the chambers is shown in the drawings by Kaminski *et al.* (1988). Diameter 0.35–0.9 mm.

Stratigraphic range: Early Maastrichtian (*Globotruncana lapparenti tricarinata* Zone) to Late Paleocene (*Globorotalia pseudomenardii* Zone). Fairly common.

RECURVOIDES cf. LOCZYI (Majzon)
Figure 21.16

Haplophragmoides loczyi Majzon 1943, Magyar K. Földt. Int., Evk., Budapest, 37, p. 156, pl. 2, fig. 13.

4 to 5 chambers in the last whorl. The Trinidad specimens appear to be more distinctly streptospiral than the type (see also Grün *et al.* 1964). A very similar form has been described by Sandulescu (1973) under the name *Thalmannammina*

subturbinata (Grzybowski). Diameter 0.27–0.5 mm.

Stratigraphic range: Early Cenomanian (*Rotalipora appenninica appenninica* Zone) to Turonian (*Globotruncana inornata* Zone). Rare.

RECURVOIDES cf. WALTERI (Grzybowski)
Figure 21.17–18

Haplophragmium walteri Grzybowski 1898, p. 280, pl. 10, fig. 24.
Recurvoides sp. 2, Kaminski *et al.* 1988, p. 191, pl. 6, figs. 10, 11.

4½ to 7 chambers in the last whorl. The early chambers are hardly visible, and it is uncertain whether the aperture is basal or slightly areal. Diameter 0.3–1.2 mm.

Stratigraphic range: Maastrichtian (*Globotruncana gansseri* Zone)?, Early Paleocene (*Rzehakina epigona* Zonule) to Early Eocene (*Globorotalia aragonensis* Zone). Fairly common. Also found in the Eocene of the overlying Navet Formation.

THALMANNAMMINA SUBTURBINATA (Grzybowski)
Figure 21.19–20

Haplophragmium subturbinatum Grzybowski 1898, p. 280, pl. 10, fig. 23.
? *Recurvoides* cf. *subturbinatus* (Grzybowski), Kaminski *et al.* 1988, p. 191, pl. 6, figs. 8, 9.

4½–6 chambers in the last whorl. Sutures flush with the surface to moderately depressed. Wall finely arenaceous, often smooth. The group also includes specimens which resemble *Recurvoides* sp. 1 of Kaminski *et al.* (1988). Diameter 0.35–0.75 mm.

Stratigraphic range: Early Paleocene (*Rzehakina epigona* Zonule) to Early Eocene (*Globorotalia aragonensis* Zone). Fairly common. Also found in the Eocene of the overlying Navet Formation.

BULBOBACULITES JARVISI (Cushman & Renz)
Figure **21.25**

Ammobaculites jarvisi Cushman & Renz 1946, p. 19, pl. 2, figs. 8, 9.
Ammobaculites jarvisi Cushman & Renz, Kaminski *et al.* 1988, p. 188, pl. 4, fig. 4.

Length 1.2–2.7, thickness 0.7–1.05 mm.

Stratigraphic range: Maastrichtian (*Globotruncana lapparenti tricarinata* Zone to *G. gansseri* Zone); very rare. Middle to Late Paleocene (*Globorotalia uncinata* Zone to *G. velascoensis* Zone); rather rare.

BULBOBACULITES LUECKEI (Cushman & Hedberg)
Figure **21.26–27**

Ammobaculites lueckei Cushman & Hedberg 1941, p. 83, pl. 21, fig. 4.
Ammobaculites cf. *lueckei* Cushman & Hedberg, Cushman & Renz 1946, p. 19, pl. 2, fig. 7.

Length 0.4–1.1, thickness 0.12–0.35 mm.

Stratigraphic range: Late Maastrichtian (*Abathomphalus mayaroensis* Zone) to Late Paleocene (*Globorotalia velascoensis* Zone). Fairly common.

Family Cyclamminidae

POPOVIA BECKMANNI (Kaminski & Geroch)
Figure **20.29–30**

Phenacophragma beckmanni Kaminski & Geroch 1987, Micropaleontology, 33, p. 98, pl. 1, figs. 1–7; textfigs. 1–4.
Ammomarginulina sp. A, Beckmann 1960, p. 62, fig. 2.
Phenacophragma beckmanni Kaminski & Geroch, Kaminski *et al.* 1988, p. 190, pl. 4, figs. 8, 9; pl. 10, figs. 6, 7.

Charnock & Jones (1990) have shown for the related species '*Phenacophragma' elegans* Kaminski & Geroch that the position of the hemisepta and the aperture place this group in the family Cyclamminidae. The illustrations of *P.*

beckmanni (see synonymy) indicate that this is also the case for *P. beckmanni*. Length 0.3–1.2, thickness 0.12–0.22 mm; diameter of spiral 0.3–0.8 mm.

Stratigraphic range: Early Paleocene (*Rzehakina epigona* Zonule) to Late Paleocene (*Globorotalia pseudomenardii* Zone). Rather scarce.

QUASICYCLAMMINA? sp. A
Figure **21.21–22**

Planispiral/nautiloid, sometimes slightly asymmetrical; periphery rounded. 7–11 chambers in the last whorl. Sutures may show short forward projections in the direction of coiling. Wall finely arenaceous, smooth. Aperture not well visible, apparently at the base of the apertural face. Diameter 0.35–0.6, thickness 0.15–0.21 mm.

Stratigraphic range: Early Eocene (*Globorotalia formosa formosa* Zone to *G. aragonensis* Zone). Rare. Also found in the Eocene of the overlying Navet Formation.

RETICULOPHRAGMIUM cf. GARCILASSOI (Frizzell)
Figure **21.23–24, 30–31**

Cyclammina garcilassoi Frizzell 1943, J. Paleontol., 17, p. 338, pl. 55, fig. 11.
Cyclammina cf. *garcilassoi* Frizzell, Cushman & Renz 1946, p. 19, pl. 2, fig. 11. – Beckmann 1960, p. 62, fig. 4.
Cyclammina garcilassoi Frizzell, Cushman & Renz 1948, p. 11, pl. 2, fig. 11.
Reticulophragmium cf. *garcilassoi* (Frizzell), Kaminski *et al.* 1988, p. 192, pl. 7, figs. 3–5; pl. 10, fig. 5.
? *Cyclammina? intermedia* Myatlyuk 1970, p. 89, pl. 21, fig. 6; pl. 28, fig. 1.

The problems with this group are summarized in Kaminski *et al.* (1988). The Paleocene morphotype (Fig. 21.23–24) appears to be connected with *Reticulophragmoides jarvisi* (see above) through some intermediate specimens; it is gradually replaced by the Early Eocene morphotype (Fig. 21.30–31). Both types show 8 to 12 chambers in the last whorl. Diameter 0.3–0.85, thickness 0.16–0.28 mm in the Paleocene Lower Lizard Springs Formation; the corresponding dimensions in the Upper Lizard Springs Formation are 0.46–0.95 and 0.19–0.32 mm.

Stratigraphic range: Paleocene (*Globorotalia pseudomenardii* Zone); Early Eocene (*Globorotalia subbotinae* Zone to *G. aragonensis* Zone). Rather scarce. Also found in the Eocene of the overlying Navet Formation.

Family Spiroplectamminidae

BOLIVINOPSIS SPECTABILIS (Grzybowski)
Figure 21.32–33

Spiroplecta spectabilis Grzybowski 1898, p. 293, pl. 12, fig. 12.
Spiroplectoides clotho (Grzybowski), Cushman & Jarvis 1928, p. 101, pl. 14, figs. 13, 14. – Cushman & Jarvis 1932, p. 43, pl. 13, figs. 5, 6.
Spiroplectammina grzybowskii Frizzell, Cushman & Renz 1946, p. 20, pl. 5, figs. 34–38.
Bolivinopsis directa Cushman & Siegfus, Cushman & Renz 1948, p. 23, pl. 5, fig. 6.
Spiroplectammina spectabilis (Grzybowski), Kaminski *et al.* 1988, p. 193, pl. 7, figs. 16–18.

Length 0.25–0.6, width 0.16–0.4, thickness 0.06–0.22 mm.

Stratigraphic range: Early Maastrichtian (*Globotruncana lapparenti tricarinata* Zone) to Early Eocene (*Globorotalia aragonensis* Zone). Common. Also found in the Eocene of the overlying Navet Formation.

BOLIVINOPSIS TRINITATENSIS (Cushman & Renz)
Figure 21.34–35

Spiroplectammina trinitatensis Cushman & Renz 1948, p. 11, pl. 2, figs. 13, 14.

This species is broader and has a more distinctly agglutinated wall than *B. spectabilis*. Length 0.45–1.45, width 0.3–0.5, thickness 0.12–0.21 mm.

Stratigraphic range: Late Paleocene (*Globorotalia pseudomenardii* Zone). Fairly common.

QUASISPIROPLECTAMMINA CHICOANA (Lalicker)
Figure 21.28–29

Spiroplectammina chicoana Lalicker 1935, p. 7, pl. 1, figs. 8, 9.

In the youngest occurrence (Campanian), the specimens become rather more massive and broader (Fig. 21.29). Length 0.25–0.8, width 0.2–0.4, thickness 0.08–0.18 mm.

Stratigraphic range: Coniacian/Early Santonian (*Globotruncana concavata* Zone) to Campanian (*Globotruncana stuarti* Zone). Rather scarce.

QUASISPIROPLECTAMMINA cf. EMBAENSIS (Myatlyuk)
Figure 21.36–38

Spiroplectammina embaensis Myatlyuk, in Vasilenko 1961, Trudy VNIGRI, Leningrad, 171, p. 14, pl. 1, figs. 5–7.

The Trinidad specimens are rather smaller than the types. They often show a ragged periphery as on Fig. 21.36, but in others the periphery may be almost straight. Length 0.25–0.65, width 0.18–0.33, thickness 0.07–0.24 mm.

Stratigraphic range: Late Turonian/Coniacian (*Globotruncana renzi* Zone). Rather rare.

QUASISPIROPLECTAMMINA cf. GANDOLFII (Carbonnier)
Figure 21.39

Spiroplectammina gandolfii Carbonnier 1952, Bull. Soc. géol. France, ser. 6, 2, p. 112, pl. 5, fig. 2.

Apart from being somewhat larger and having more distinctly parallel sides, the specimens from Trinidad are comparable to the type. Length 0.3–0.6, width 0.2–0.4, thickness 0.06–0.09 mm.

Stratigraphic range: Early Cenomanian (*Rotalipora appenninica appenninica* Zone). Rare.

QUASISPIROPLECTAMMINA JARVISI
(Cushman)
Figure 22.1–2

Spiroplectammina jarvisi Cushman 1939,
   Contrib. Cushman Lab. foramin. Res., 15,
   p. 90, pl. 16, fig. 1.
Spiroplectammina anceps (Reuss) var.,
   Cushman & Jarvis 1932, p. 14, pl. 3, fig. 8.
Spiroplectammina jarvisi Cushman, Cushman
   & Renz 1946, p. 20, pl. 2, fig. 12.

The holotype is relatively broader than the average Trinidad specimens. *S. israelskyi* Hillebrandt (1962) may be a junior synonym. Length 0.35–0.6, width 0.2–0.5, thickness 0.12–0.24 mm.

Stratigraphic range: Paleocene (*Globorotalia trinidadensis* Zone to *G. velascoensis* Zone). Rather scarce.

QUASISPIROPLECTAMMINA NUDA
(Lalicker)
Figure 22.3

Spiroplectammina nuda Lalicker 1935, p. 4,
   pl. 1, fig. 6, 7.

Length 0.27–0.42, width 0.15–0.29, thickness 0.08–0.14 mm.

Stratigraphic range: Late Albian (*Favusella washitensis* Zone) to Santonian (*Globotruncana fornicata* Zone). Rare.

SPIROPLECTAMMINA (?)
NAVARROANA (Cushman)
Figure 22.10–12

Spiroplectammina navarroana Cushman 1932,
   Contrib. Cushman Lab. foramin. Res., 8,
   p. 96, pl. 11, fig. 14.
? *Textularia concinna* Reuss, Cushman & Jarvis
   1928, p. 91, pl. 13, fig. 1. – Cushman & Jarvis
   1932, p. 15, pl. 4, fig. 1 (not fig. 2).
Gaudryina foeda (Reuss), Cushman & Renz
   1946, p. 21, pl. 2, fig. 18. – Cushman & Renz
   1948, p. 12, pl. 2, fig. 17.
Spiroplectammina navarroana Cushman,
   Kaminski *et al.* 1988, p. 193, pl. 7, figs. 13–15.

While the presence of a small initial spiral is indicated in some specimens, in others, especially those with a pointed base and a very small proloculus, the chamber arrangement is difficult to see and may be somewhat irregular. The assignment of this species to *Spiroplectammina* is explained in Kaminski *et al.* (1988) and Gradstein & Kaminski (1989). Length 0.48–1.25, width 0.21–0.4, thickness 0.2–0.32 mm.

Stratigraphic range: Late Maastrichtian (*Abathomphalus mayaroensis* Zone) to Early Eocene (*Globorotalia aragonensis* Zone). Fairly common. Also found in the Eocene of the overlying Navet Formation.

SPIROPLECTINELLA DENTATA (Alth)
Figure 22.4–7

Textularia dentata Alth 1850, Haidinger's
   Naturw. Abhandl., Vienna, 3, p. 262,
   pl. 13, fig. 13.
? *Spiroplectammina dentata* (Alth), Cushman
   & Jarvis 1932, p. 14, pl. 3, fig. 7.
Spiroplectammina dentata (Alth), Cushman &
   Renz 1947a, p. 37, pl. 11, fig. 3.
Spiroplectammina sp. aff. *dentata* (Alth),
   Kaminski *et al.* 1988, p. 192, pl. 7, figs. 10, 11.
Spiroplectammina denticuligera Myatlyuk 1970,
   p. 99, pl. 5, fig. 8; pl. 10, figs. 16, 17; pl. 32,
   fig. 9.

One gets the impression that two intergrading morphotypes are predominant in the Trinidad faunas, one with relatively convex sides (Fig. 22.4–5) and one which is slightly more rhombic in cross section and has rather more distinct spines (Fig. 22.6). Length 0.35–1.15, width 0.3–0.7, thickness 0.15–0.33 mm.
   A slightly larger variety (length up to 1.3 mm; see Fig. 22.7) occurs in the Late Cretaceous shallower water facies encountered in Marabella well no. 1.

Stratigraphic range: Campanian (*Globotruncana stuarti* Zone) to Late Maastrichtian (*Abathomphalus mayaroensis* Zone). Fairly common.

SPIROPLECTINELLA EXCOLATA
(Cushman)
Figure **22.8, 17**

Textularia excolata Cushman 1926, p. 585,
   pl. 15, fig. 9.
Spiroplectammina excolata (Cushman),
   Cushman & Jarvis 1932, p. 14, pl. 3, figs. 9,
   10. – Cushman & Renz 1946, p. 20, pl. 2,
   fig. 13. – Kaminski *et al.* 1988, p. 192, pl. 7,
   fig. 12.

Length 0.39–0.95, width 0.25–0.85, thickness
0.24–0.5 mm.

Stratigraphic range: Paleocene (*Globorotalia
trinidadensis* Zone to *G. velascoensis* Zone).
Rather scarce.

VULVULINA COLEI Cushman
Figure **22.13**

Vulvulina colei Cushman 1932, Contrib.
   Cushman Lab. foramin. Res., 8, p. 84, pl. 10,
   figs. 21, 22.
Vulvulina cf. *jarvisi* Cushman, Cushman &
   Renz 1946, p. 20, pl. 2. fig. 15.

Length 0.55–0.9, width 0.35–0.5, thickness
0.18–0.23 mm.

Stratigraphic range: Early Eocene (*Globorotalia
subbotinae* Zone to *G. formosa formosa* Zone).
Rare. Also found in the Eocene of the
overlying Navet Formation.

Family Textulariopsidae

TEXTULARIOPSIS cf.
HIKAGEZAWENSIS (Takayanagi)
Figure **22.14–16**

Textularia hikagezawensis Takayanagi 1960,
   p. 78, pl. 2, figs. 19–21.

This species shows a considerable variability.
The Trinidad form is practically identical with
the types from Japan, except that its sutures are
slightly raised in about one quarter of the specimens. Length 0.25–0.6, width 0.3–0.6, thickness
0.18–0.33 mm.

Stratigraphic range: Late Albian to Early
   Cenomanian (*Favusella washitensis* Zone to
   Rotalipora appenninica appenninica Zone).
   Rather scarce.

TEXTULARIOPSIS LOSANGICA
(Loeblich & Tappan)
Figure **22.9, 18**

Textularia washitensis Carsey, Tappan 1943, J.
   Paleontol., 17, p. 486, pl. 78, figs. 5–9.
Textularia losangica Loeblich & Tappan 1951,
   in: Lozo, F. E. & Perkins, B. F. 1951,
   Southern Methodist Univ. Press, Dallas,
   p. 82, pl. 21, figs. 4, 5.

The holotype is more elongated, but other specimens figured by Loeblich & Tappan (1951, see
synonymy; 1982) appear to be identical with the
Trinidad form. Length 0.26–0.7, width 0.25–
0.45, thickness 0.24–0.33 mm.

Stratigraphic range: Albian to Early Cenomanian
   (*Rotalipora ticinensis ticinensis* Zone to *R.
   appenninica appenninica* Zone). Scarce.

Family Trochamminidae

AMMOGLOBIGERINA BOEHMI (Franke)
Figure **22.19**

Trochammina böhmi Franke 1928, p. 174,
   pl. 15, fig. 24.

Compressed tests with 3½ to 5 subglobular chambers in the final whorl. Wall roughly finished.
Diameter 0.35–1.1 mm.
Stratigraphic range. Early Cenomanian (*Rotalipora
   appenninica appenninica* Zone) to Early
   Maastrichtian (*Globotruncana lapparenti
   tricarinata* Zone). Rare.

AMMOGLOBIGERINA ALTIFORMIS
(Cushman & Renz)
Figure **22.20–22**

Trochammina globigeriniformis (Parker &
   Jones) var. *altiformis* Cushman & Renz
   1946, p. 24, pl. 3, figs. 7–11.
Trochammina globigeriniformis (Parker &
   Jones), Cushman & Jarvis 1928, p. 95, pl. 13,
   fig. 12. – Cushman & Jarvis 1932, p. 21, pl. 6,
   figs. 2–5.
Trochammina altiformis Cushman & Renz,
   Kaminski *et al.* 1988, p. 193, pl. 8, figs. 1, 2.

Highly variable, globular to secondarily compressed; usually with 3½ to 4 chambers in the
final whorl. Wall material fine to fairly coarse.
Diameter 0.35–1.55 mm.

Stratigraphic range: Early Maastrichtian (*Globotruncana lapparenti tricarinata* Zone) to Late Paleocene (*Globorotalia velascoensis* Zone). Common.

AMMOGLOBIGERINA cf. ALTIFORMIS (Cushman & Renz)
Figure 22.23

Specimens which are always compressed and are slightly smaller than the typical form occur throughout the Lizard Springs Formation. Diameter 0.3–0.82 mm.

Stratigraphic range: Early Paleocene (*Globorotalia trinidadensis* Zone) to Early Eocene (*Globorotalia aragonensis* Zone). Fairly common.

AMMOGLOBIGERINA ALTIFORMIS
Cushman & Renz), forma A
Figure 22.24–25

This form has a coarse wall surface and rather truncated chambers, which may give it a subspherical to subquadrangular appearance. Although this form is quite characteristic at least in parts of the Early Eocene, it is difficult to keep it clearly separated within the *altiformis* group. Diameter 0.4–0.9 mm.

Stratigraphic range: Early Eocene (*Globorotalia subbotinae* Zone). Rather scarce.

TROCHAMMINA (?) cf. DEFORMIS
Grzybowski
Figure 22.26–27

Trochammina deformis Grzybowski 1898, p. 288, pl. 11, figs. 20–22.

Subspherical but always compressed in various directions, with 5–6 chambers in the final whorl. The coiling is probably trochospiral. Diameter 0.22–0.7 mm.

Stratigraphic range: Turonian (*Globotruncana inornata* Zone) to Santonian (*Globotruncana fornicata* Zone). Rather scarce.

TROCHAMMINA GLOBOLAEVIGATA
Beckmann
Figure 22.30–34

Trochammina globolaevigata Beckmann 1991, p. 824, pl. 1, figs. 13–19.

Test subspherical, with tendencies towards becoming either somewhat flattened or slightly conical. Chambers moderately inflated, arranged trochospirally, mainly triserial but sometimes with four chambers in the initial coil and occasionally becoming irregularly biserial in the final stage. Sutures depressed, those of the larger specimens sometimes slightly sinuous. Wall smooth to finely hispid, finely agglutinated, slightly calcareous. Aperture basal, varying from a broad low slit to a low arch, rarely with a thin raised rim. Maximum diameter 0.3–0.85 mm
The oldest specimens (*Favusella washitensis* Zone) are the largest; they decrease gradually in size in the younger zones.
Praecystammina globigerinaeformis Krasheninnikov is superficially very similar but has an areal aperture.

Stratigraphic range: Late Albian (*Favusella washitensis* Zone) to Late Turonian/Coniacian (*Globotruncana renzi* Zone). Rather scarce.

TROCHAMMINA RUTHVENMURRAYI
Cushman & Renz
Figure 22.28–29

Trochammina ruthven-murrayi Cushman & Renz 1946, p. 24, pl. 3, fig. 13.
Trochammina ruthven-murrayi Cushman & Renz, Kaminski *et al.* 1988, p. 193, pl. 8, fig. 6.

Diameter 0.4–0.65, height 0.2–0.35 mm.

Stratigraphic range: Early Maastrichtian (*Globotruncana lapparenti tricarinata* Zone); Paleocene (*Rzehakina epigona* Zonule to *Globorotalia velascoensis* Zone). Fairly common.

TROCHAMMINA WICKENDENI Loeblich
Figure 22.35

Trochammina wickendeni Loeblich, 1946, J. Paleontol., 20, p. 138, pl. 22, fig. 17.

Always compressed. Diameter 0.2–0.5 mm.

Stratigraphic range: Albian (*Rotalipora ticinensis ticinensis* Zone) to Santonian (*Globotruncana fornicata* Zone). Fairly common.

Family Conotrochamminidae

CONOTROCHAMMINA WHANGAIA
Finlay
Figure **22.36–37**

Conotrochammina whangaia Finlay 1940, p. 448, pl. 62, figs. 1, 2.
Trochammina sp. A, Beckmann 1960, p. 62, fig. 7.
Conotrochammina whangaia Finlay, Kaminski *et al.* 1988, p. 193, pl. 7, figs. 19, 20.

The shape varies from distinctly to moderately conical (Beckmann 1960; Kaminski *et al.* 1988) to rather low conical (Fig. 22.37). Diameter 0.5–0.85, height 0.4–0.95 mm.

Stratigraphic range: Late Paleocene (*Globorotalia pseudomenardii* Zone). Very rare.

CONOTROCHAMMINA (?) cf. DEPRESSA Finlay
Figure **22.38–39**

Conotrochammina depressa Finlay 1947, New Zealand J. Sci. Technol., sec. B, 28, p. 260, pl. 1, figs. 1–4.

The specimens in Trinidad resemble the types from New Zealand quite closely, although they may be slightly more loosely coiled. Diameter 0.45–0.85, thickness 0.27–0.44 mm.

Stratigraphic range: Early Eocene (*Globorotalia subbotinae* Zone). Rather scarce.

Family Prolixoplectidae

KARRERULINA BORTONICA Finlay
Figure **23.1–2**

Karrerulina bortonica Finlay 1940, p. 451, pl. 62, figs. 18–20.

The initial chambers are indistinct but probably multiserial, except for some relatively thin specimens which have a triserial beginning. A short

apertural neck is usually present in the final uniserial chamber. Length 0.55–1.15, thickness 0.15–0.32 mm.

Stratigraphic range: Early Eocene (*Globorotalia edgari* Zone to *G. formosa formosa* Zone). Rare.

KARRERULINA CONIFORMIS
(Grzybowski)
Figure **23.3–4**

Gaudryina coniformis Grzybowski 1898, p. 295, pl. 12, fig. 7.
Karreriella coniformis (Grzybowski), Kaminski *et al.* 1988, p. 195, pl. 9, figs. 15, 16.

Length 0.4–1.42, thickness 0.22–0.38 mm.

Stratigraphic range: Early Paleocene (*Globorotalia trinidadensis* Zone) to Early Eocene (*Globorotalia aragonensis* Zone). Fairly common. Also found in the Eocene of the overlying Navet Formation.

KARRERULINA CONVERSA (Grzybowski)
Figure **23.5–8**

Gaudryina conversa Grzybowski 1901, p. 224, pl. 8, figs. 15, 16.
Gaudrina bentonensis (Carman), Cushman & Renz 1946, p. 21, pl. 2, fig. 19.
Karreriella conversa (Grzybowski), Kaminski *et al.* 1988, p. 196, pl. 9, figs. 17, 18.

Compared to the otherwise similar *K. smokyensis* (see below), the multiserial-triserial stage of the test is longer and the biserial stage is relatively short. Details of the chamber arrangement vary considerably, and the terminal final chamber is often absent. As shown by Grzybowski, the shape and thickness of the test are also variable. Relatively thick specimens with a coarse wall and indistinct chambers, like those referred to *K. horrida* Myatlyuk by Kaminski *et al.* 1988, seem to be intermediate forms between *K. bentonensis* and *K. coniformis*. *Dorothia hokkaidoana* Takayanagi may be practically identical but apparently lacks the pseudouniserial final chamber. Length 0.28–1.05, thickness 0.09–0.33 mm.

Stratigraphic range: Maastrichtian (*Globotruncana gansseri* Zone?, *Abathomphalus mayaroensis* Zone) to Early Eocene (*Globorotalia aragonensis* Zone). Abundant (except in the Early Eocene, where it is scarcer and less typical).

KARRERULINA SMOKYENSIS (Wall)
Figure **23.9–10**

Dorothia smokyensis Wall 1960, Bull. Alberta Research Council, 6, p. 23, pl. 4, figs. 22–28.

The test includes a relatively long biserial portion. A pseudouniserial final chamber with subterminal aperture, but without a distinct neck, is occasionally present. Length 0.4–1.05, thickness 0.14.-0.3 mm.

Stratigraphic range: Coniacian-Santonian (*Globotruncana concavata* Zone) to Maastrichtian (*Globotruncana lapparenti tricarinata* Zone, possibly also *G. gansseri* Zone). Common.

PLECTINA KUGLERI Beckmann
Figure **23.11–12**

Plectina kugleri Beckmann 1991, p. 825, pl. 1, figs. 20, 21.

Test subcylindrical to slightly conical; thick oval or even subquadrangular in cross section. Early part rounded, multiserial but usually without distinct chamber sutures, then changing fairly abruptly to a biserial chamber arrangement. Chambers in the biserial part of the test somewhat inflated with slightly depressed and often slightly oblique sutures, rather broader than high except for the last two or three which may become either more globular or, in the last stage, obliquely truncated. Wall rough, including rather coarse sand grains. Apertural face obliquely sloping; aperture variable, occasionally semicircular and situated at the basal sutures, but more often oval to circular with a tendency to migrate into an areal position.

Comparable species like *P. ruthenica* (Reuss; *Gaudryina r.*) or *P. watersi* Cushman both have higher chambers and a more pointed and subtriangular early portion.

Length 0.5–1.25, width 0.36–0.8, thickness 0.38–0.74 mm.

Stratigraphic range: Campanian (*Globotruncana stuarti* Zone) to Maastrichtian (*Abathomphalus mayaroensis* Zone). Rare.

VERNEUILINELLA sp. A
Figure **23.13**

Small, quadriserial, with an indistinct but probably low slitlike aperture. Length 0.3–0.42, thickness 0.2–0.3 mm.

Stratigraphic range: Early Eocene (*Globorotalia formosa formosa* Zone). Very rare.

Family Verneuilinidae

GAUDRYINOPSIS BOSQUENSIS Loeblich & Tappan
Figure **23.14–15**

Gaudryina bosquensis Loeblich & Tappan 1946, J. Paleontol., 20, p. 245, pl. 35, figs. 12–13; textfig. 3.
Gaudryina propria Botwinnik, 1974, Paleontol. Sbornik, Lvov, 11, p. 10, pl. 1, figs. 3, 4.

The test varies considerably in length; on the average, the Trinidad specimens seem to be slightly wider than the types and to have a slightly larger early triserial stage. Length 0.6–1.35, width 0.2–0.45, thickness 0.18–0.32 mm.

Stratigraphic range: Late Albian to Early Cenomanian (*Favusella washitensis* Zone to *Rotalipora appenninica appenninica* Zone). Rare.

GAUDRYINOPSIS CANADENSIS (Cushman)
Figure **23.16**

Bigenerina angulata Cushman 1927, Trans. Royal Soc. Canada, 3d ser., 21, p. 131, pl. 1, fig. 10 (non *Gaudryina triangularis* var. *angulata* Cushman).
Gaudryina canadensis new name, Cushman 1943, Contrib. Cushman Lab. foramin. Res., 19, p. 27, pl. 6, figs. 7, 8.

The test is often deformed, and the shape and size as well as the coarseness of the wall material vary considerably. *G. cobbani* (Fox; *Gaudryina c.*) appears to be closely related or may even be synonymous. Length 0.21–1.18, thickness 0.09–0.33 mm.

Stratigraphic range: Early Cenomanian (*Rotalipora appenninica appenninica* Zone) to Turonian (*Globotruncana inornata* Zone). Rare.

GAUDRYINOPSIS GRADATA (Berthelin)
Figure **23.17**

Gaudryina gradata Berthelin 1880, p. 24, pl. 1, fig. 6.

Although Berthelin described this species as having a triserial early stage, most subsequent authors have referred it to *Dorothia*. As the Trinidad form is tri-biserial, this usage is not followed here (see also Haig 1980). Length 0.6–0.75, thickness 0.4 mm.

Stratigraphic range: Early Cenomanian (*Rotalipora appenninica appenninica* Zone). Very rare.

VERNEUILINOIDES POLYSTROPHA (Reuss)
Figure **23.18–19**

Bulimina polystropha Reuss 1846, p. 109, pl. 24, fig. 53.
Verneuilina polystropha (Reuss), Cushman & Jarvis 1932, p. 15, pl. 4, fig. 3. – Cushman & Renz 1946, p. 20, pl. 2, fig. 16. – Cushman & Renz 1948, p. 12, pl. 2, fig. 18.
Verneuilinoides polystrophus (Reuss), Kaminski *et al.* 1988, p. 194, pl. 8, fig. 8.

The concept of this species is that of Cushman & Jarvis (1932) and Hagn (1953), not necessarily that of Reuss. Length 0.45–1.1, thickness 0.3–0.5 mm.

Stratigraphic range: Early Maastrichtian (*Globotruncana lapparenti tricarinata* Zone) to Paleocene (*Globorotalia pusilla pusilla* Zone). Rather scarce.

VERNEUILINOIDES TAILLEURI Tappan
Figure **23.20–21**

Verneuilinoides tailleuri Tappan 1957, p. 208, pl. 66, figs. 19–22.

Length 0.35–0.75, thickness 0.18–0.3 mm.

Stratigraphic range: Santonian (*Globotruncana fornicata* Zone); possibly, less typical, in the Maastrichtian (*Abathomphalus mayaroensis* Zone). Rare.

VERNEUILINOIDES TRIFORMIS (Sliter)
Figure **23.22–23**

Trochammina triformis Sliter 1968, p. 47, pl. 3, fig. 2.

Beside typical specimens, this species in Trinidad also includes some rather more elongated tests (Fig. 23.23). Length 0.6–1.25, thickness 0.5–0.85 mm.

Stratigraphic range: Paleocene (*Globorotalia trinidadensis* Zone to *G. velascoensis* Zone). Rather scarce.

BELORUSSIELLA cf. OSSIPOVAE (Bykova)
Figure **23.24–25**

Gaudryina ossipovae Bykova 1953, Trudy VNIGRI, Leningrad, 73, p. 239, pl. 3, figs. 9–12; pl. 4, figs. 1, 2.

This species from the Early Eocene of Central Asia is quite variable in its size, in the shape of the aperture and in its wall material. The Trinidad form is usually smaller, very fine-walled and always more or less compressed; it resembles the smallest and most fine-walled specimens of Bykova. The aperture is vertically elongated and often slightly comma-shaped. Length 0.18–0.39, width 0.06–0.16 mm.

Stratigraphic range: Paleocene (*Globorotalia trinidadensis* Zone to *G. pusilla pusilla Zone*), possibly also Early Eocene (*Globorotalia subbotinae* Zone). Rare.

GAUDRYINA CRETACEA (Karrer)
Figure **23.26**

Verneuilina cretacea Karrer 1870, Jahrb. geol. Reichsanst., Vienna, p. 164, pl. 1, fig. 1.
Gaudryina ex gr. *cretacea* Karrer, Kaminski *et al.* 1988, p. 194, pl. 9, fig. 3.

Similar specimens are often referred to *G. laevigata*. However, the form recorded here is bigger than *G. laevigata* and has a flatter, more truncated apertural face. Its edges, particularly in the triserial part, are rather less serrate than mentioned by Karrer. A specimen which is practically identical is figured by Cushman (1937*a*, pl. 6, fig. 5); the same applies to a 'paratype' of *Verneuilina cretacea* Karrer in the Cushman collection. Length 0.35–1.38, width 0.35–1.0 mm.

Stratigraphic range: Early Maastrichtian
  (*Globotruncana lapparenti tricarinana* Zone)
  to Early Paleocene (*Globorotalia trinidadensis*
  Zone). Rather rare.

GAUDRYINA EXPANSA Israelsky
Figure **23.27–28**

Gaudryina (Gaudryina) expansa Israelsky
  1951, p. 16, pl. 5, figs. 21–24.

Apart from the rapidly expanding typical form,
there are some specimens with more inflated
chambers which may be transitional to *G.
phenocrysta* Israelsky. Length 0.45–1.1, greatest
width 0.4–0.85 mm.

Stratigraphic range: Early Eocene (*Globorotalia
formosa formosa* Zone). Rare.

GAUDRYINA FRANKEI Brotzen
Figure **23.29–30**

Gaudryina frankei Brotzen 1936, p. 33, pl. 1,
  fig. 7; textfig. 5.

There is some variability in shape between subangular and subrounded in the early triserial
part. The aperture may be low semicircular to
almost circular but is always connected with the
basal suture. Length 0.43–1.45, width 0.25–
0.6 mm.

Stratigraphic range: Santonian (*Globotruncana
fornicata* Zone) to Campanian (*Globotruncana
stuarti* Zone). Rather rare.

GAUDRYINA sp. aff. G. INFLATA
Israelsky
Figure **23.31–32**

Gaudryina (Gaudryina) inflata Israelsky 1951,
  p. 16, pl. 6, figs. 1–12.

The test is distinctly flaring with a pointed and
relatively short early triserial part. The resemblance to *G. inflata* is very general, and no direct
relationship is implied. *G. inflata* is rather
stouter and has more distinctly inflated chambers, probably also a finer wall. A possible
successor in Trinidad is *G.* sp. aff. *G. nitida* (see
below). Length 0.65–1.1, width 0.45–0.8, thickness 0.3–0.45 mm.

Stratigraphic range: Late Paleocene (*Globorotalia
velascoensis* Zone) to Early Eocene
(*Globorotalia subbotinae* Zone). Rare.

GAUDRYINA INGENS Voloshina
Figure **23.33–34**

Gaudryina ingens Voloshina 1972, Trudy
  UkrNIGRI, Lvov, 27, p. 55, pl. 1, fig. 1.
Dorothia cf. *glabrella* Cushman, Beckmann
  1960, p. 62, fig. 6.

This species appears to be the youngest member
of a group which also includes *G. healyi* Finlay
and *G. pulvina* Belford. The test is distinctly
broader than in these two species. Length 0.6–
1.55, width 0.55–1.0, thickness 0.5–0.8 mm.

Stratigraphic range: Campanian (*Globotruncana
stuarti* Zone) to Early Maastrichtian
(*Globotruncana lapparenti tricarinata* Zone).
Rather rare.

GAUDRYINA LAEVIGATA Franke
Figure **23.35–37**

Gaudryina laevigata Franke 1914, Zeitschr.
  deutsch. geol. Ges., 66, p. 431, pl. 27, figs. 1,
  2.

The specimens from Trinidad are slightly smaller
than the types but agree in the other characters.
The biserial part has a tendency to become
slightly more angular and broader (see Fig.
23.37) in the higher stratigraphic occurrences of
the species; this change is accompanied by a
moderate size increase of the whole test. Small
specimens can hardly be distinguished from *G.
praepyramidata* Hercogova (1987). Length 0.2–
0.85, width 0.18–0.58 mm.

Stratigraphic range: Early Cenomanian (*Rotalipora
appenninica appenninica* Zone) to Santonian
(*Globotruncana fornicata* Zone). Rather rare.

GAUDRYINA cf. MODICA Bermudez
Figure **23.38–40**

Gaudryina modica Bermudez 1949, p. 76, pl. 3,
  figs. 53–54.

The initial part of the test is usually more
rounded than in the types, and the biserial chambers are less rapidly expanding and less inflated.
Length 0.5–1.1, width 0.45–0.75, thickness
0.38–0.67 mm.

Stratigraphic range: Middle Paleocene (*Globorotalia uncinata* Zone) to Early Eocene (*Globorotalia subbotinae* Zone). Rare. Also found in the Eocene of the overlying Navet Formation.

GAUDRYINA sp. aff. G. NITIDA Haque
Figure **24.1–3**

Gaudryina (Pseudogaudryina) pyramidata
   Cushman var. *nitida* Haque 1956, Mem. geol.
   Survey Pakistan, Palaeontol. Pakistanica, 1,
   p. 41, pl. 9, fig. 2.

This form gradually replaces the group here referred to *G.* sp. aff. *G. inflata* (see above). Intermediate specimens, with a more triangular shape and with less inflated chambers (Fig. 24.1), occur in the *Globorotalia subbotinae* Zone. In the overlying *Globorotalia formosa formosa* Zone, the same tendency continues (Fig. 24.2–3), without reaching the distinctly triangular and triserial test shape of *G. nitida*. Length 0.45–0.8, maximum width 0.3–0.5 mm.

Stratigraphic range: Early Eocene (*Globorotalia subbotinae* Zone to *G. formosa formosa* Zone). Rare.

GAUDRYINA PYRAMIDATA Cushman
Figure **24.4–6**

Gaudryina laevigata var. *pyramidata* Cushman
   1926, p. 587, pl. 16, fig. 8.
Gaudryina laevigata Franke var. *pyramidata*
   Cushman, Cushman & Jarvis 1928, p. 92,
   pl. 13, fig. 6. – Cushman & Jarvis 1932, p. 18,
   pl. 5, fig. 3.
Gaudryina (Pseudogaudryina) pyramidata
   Cushman, Cushman & Renz 1946, p. 21, pl. 2,
   fig. 21.
Gaudryina pyramidata Cushman, Kaminski *et al.* 1988, p. 194, pl. 8, fig. 7.
Gaudryina pyramidata var. *tumeyensis*
   Israelsky 1951, p. 18, pl. 7, figs. 8–12.

There is a remarkable difference in size among the specimens in Trinidad. The larger ones (Figs. 24.4–5) have relatively few biserial chambers; in the smaller specimens (Fig. 24.6), the biserial part is often better developed. The same two morphologies seem to be represented by the holotype and the paratype of *G. pyramidata tumeyensis*, a probable synonym of *G. pyra-*

midata. Length 0.35–1.25, greatest width 0.22–1.1 mm.

Stratigraphic range: Late Maastrichtian (*Abathomphalus mayaroensis* Zone) to Late Paleocene (*Globorotalia velascoensis* Zone). Common.

GAUDRYINA cf. RUGOSA d'Orbigny
Figure **24.7–8**

Gaudryina rugosa d'Orbigny 1840, p. 44, pl. 4,
   figs. 20–21.

The Trinidad specimens have the typical rugose wall surface (sometimes showing embedded minute tests of calcareous foraminiferal shells). They differ from *G. rugosa* mainly in their non-inflated chambers with nearly invisible sutures. Length 0.52–1.55, width 0.4–0.75 mm.

Stratigraphic range: Paleocene (*Globorotalia trinidadensis* Zone to *G. pusilla pusilla* Zone). Rare.

GAUDRYINA RUTTENI Cushman & Bermudez
Figure **24.9–10**

Gaudryina rutteni Cushman & Bermudez 1936,
   Contrib. Cushman Lab. foramin. Res., 12,
   p. 56, pl. 10, figs. 15–16.
Gaudryina (Pseudogaudryina) pyramidata
   Cushman, Cushman & Renz 1948, p. 13, pl. 3,
   fig. 4.

Apart from fairly typical specimens (Fig. 24.9), there are some which have a relatively larger biserial portion. This second type may be represented by a hypotype of *G. pyramidata* in the Cushman Collection from the Navet Formation of Trinidad (see synonymy). Length 0.45–1.15, width 0.4–0.77, thickness 0.33–0.6 mm.

Stratigraphic range: Late Eocene (*Globorotalia aragonensis* Zone). Rare. Also found in the Eocene of the overlying Navet Formation.

GAUDRYINA sp. A
Figure **24.11–13**

This species superficially resembles *Dorothia hyperconica* but the test is distinctly narrower. Specimens with a smaller proloculus (microspheric?) have a small triserial early stage. Others, with a larger proloculus, are entirely biserial; these can then hardly be distinguished from those of some other species like *Textularia rioensis* Carsey or *Spiroplectammina (?) linki* Petri, whose initial stages may require a re-examination. Length 0.3–0.95, width 0.15–0.35, thickness 0.13–0.28 mm.

Stratigraphic range: Albian to Early Cenomanian (*Rotalipora ticinensis ticinensis* Zone?, *Favusella washitensis* Zone to *R. appenninica appenninica* Zone). Rare.

GAUDRYINELLA PUSILLA
Magniez-Jannin
Figure **24.14**

Gaudryinella pusilla Magniez-Jannin 1975, Cahiers Paléontol., Paris, p. 68, pl. 5, figs. 15–22.

Apart from being slightly bigger and thinner-walled (and therefore compressed), the Trinidad specimens are in good agreement with the types. *G. mollis* Cushman is also similar but has less inflated chambers. Length 0.28–0.6, width 0.12–0.24 mm.

Stratigraphic range: Turonian (*Globotruncana inornata* Zone) to Santonian (*Globotruncana fornicata* Zone). Rare.

PSEUDOGAUDRYINELLA COMPACTA
ten Dam & Sigal
Figure **24.15–16**

Pseudogaudryinella compacta ten Dam & Sigal 1950, p. 33, pl. 2, fig. 12.

Beside the typical form (Fig. 24.15), there is a variant in the Late Cretaceous shallower water facies of Marabella well no. 1 (Fig. 24.16); this has a more sandy wall and a more elongated and pointed triserial portion. Length 0.5–1.45, width 0.4–0.65 mm.

Stratigraphic range: Maastrichtian (*Globotruncana lapparenti tricarinata* Zone? (Marabella facies), *Globotruncana gansseri* Zone to *Abathomphalus mayaroensis* Zone). Rare.

Family Tritaxiidae

TRITAXIA TRICARINATA (Reuss)
Figure **24.17**

Textularia tricarinata Reuss 1844, Geognost. Skizzen Böhmen, Prague, 2, p. 215. – Reuss 1845, p. 39, pl. 8, fig. 60.

Apart from typical specimens with a terminal final chamber and aperture, there are others with a vertical slitlike aperture which is attached to the basal suture. Length 0.35–0.95, maximum width 0.3–0.6 mm.

Stratigraphic range: Late Albian to Early Cenomanian (*Favusella washitensis* Zone to *Rotalipora appenninica appenninica* Zone). Rare.

Family Ataxophragmiidae

ARENOBULIMINA PUSCHI (Reuss)
Figure **24.18**

Bulimina puschi Reuss 1851, p. 57, pl. 3, fig. 6.

The specimens in Trinidad usually have 3 chambers in the last whorl. They are shorter than the holotype but resemble the topotype figured by Cushman (1937*b*, pl. 4, fig. 23) very much. Length 0.27–0.58, thickness 0.18–0.36 mm.

Stratigraphic range: Early Maastrichtian (*Globotruncana lapparenti tricarinata* Zone?, Marabella facies). Rather rare.

ARENOBULIMINA TRUNCATA (Reuss)
Figure **24.19–20**

Bulimina truncata Reuss 1845, p. 37, pl. 8, fig. 73.
Arenobulimina truncata (Reuss), Kaminski *et al.* 1988, p. 194, pl. 8, fig. 10.

Normally about 4 chambers to a whorl. Sutures hardly if at all depressed. Aperture virguline. Some short, fat specimens resemble *A. malkinae* Jennings but have a more rounded base. Length 0.24–0.42, thickness 0.18–0.32 mm.

Stratigraphic range: Early Paleocene (*Globorotalia trinidadensis* Zone) to Early Eocene (*Globorotalia subbotinae* Zone). Rare.

Family Globotextulariidae

REMESELLA VARIANS (Glaessner)
Figure **24.22–24**

Textulariella? *varians* Glaessner 1937,
  Problems Palaeontol., Moscow, 2–3, p. 366,
  pl. 2, fig. 15.
Gaudryina retusa Cushman, Cushman & Jarvis
  1928, p. 92, pl. 13, figs. 3, 4. – Cushman &
  Jarvis 1932, p. 17, pl. 4, figs. 7–10.
Textulariella trinitatensis Cushman & Renz
  1946, p. 23. pl. 3, figs. 1–3.
Textulariella trinitatensis Cushman & Renz var.
  subcylindrica Cushman & Renz 1946, p. 23,
  pl. 3, fig. 4.
Matanzia varians (Glaessner), Kaminski *et al.*
  1988, p. 196, pl. 9, fig. 14; pl. 10, fig. 14.

Microspheric form: Length 0.44–1.8, thickness
0.4–1.1 mm. Megalospheric form: Length 0.25–
1.25, thickness 0.32–0.7 mm.

Stratigraphic range: Early Maastrichtian
  (*Globotruncana lapparenti tricarinata* Zone)
  to Late Paleocene (*Globorotalia velascoensis*
  Zone). Fairly common.

Family Eggerellidae

BANNERELLA BIFORMIS (Finlay)
Figure **24.25–27**

Dorothia biformis Finlay 1939, Trans. Proc.
  Royal Soc. New Zealand, 69/3, p. 313,
  pl. 25, figs. 26–28.

The two phenotypes, fusiform and subcylindrical, described by Finlay and interpreted as
micro- and megalospheric, can also be distinguished in Trinidad. Their common characteristics are the short initial multispiral stage, the
relatively high, hardly inflated biserial chambers
and the fairly coarse wall material and sandy test
surface. They occur together in the relatively
shallower water facies developed in the Late
Cretaceous of Marabella well no. 1. Length 0.5–
1.45, thickness 0.35–0.95 mm for thick fusiform
specimens, 0.55–1.0 and 0.3–0.45 mm respectively for subcylindrical specimens.

Stratigraphic range: Probably Early Maastrichtian
  (*Globotruncana lapparenti tricarinata* Zone,
  Marabella facies). Rather scarce.

DOROTHIA BELOIDES Hillebrandt
Figure **24.21, 31**

Dorothia beloides Hillebrandt 1962, p. 39, pl. 2,
  figs. 8–14; pl. 15, figs. 12–13; textfig. 3.
Dorothia beloides Hillebrandt, Kaminski *et al.*
  1988, p. 195, pl. 9, figs. 4, 5.

Length 0.4–1.25, thickness 0.35–0.55 mm.

Stratigraphic range: Paleocene (*Globorotalia
  uncinata* Zone to *G. velascoensis* Zone).
  Rather scarce. Very rare and less typical also
  in the Early Eocene (*Globorotalia subbotinae*
  Zone and *G. formosa formosa* Zone).

DOROTHIA BULBOSA Israelsky
Figure **24.28**

Dorothia bulbosa Israelsky 1951, p. 23, pl. 9,
  figs. 33–35; pl. 11, fig. 17.
Dorothia retusa (part), Kaminski *et al.* 1988,
  p. 195, pl. 9, fig. 6 (not fig. 11).

Among the thick fusiform Dorothias which are
commonly combined under names like *D.
bulletta*, *D. retusa*, *D. pupa* and others, Israelsky
(1951) has recognized three morphotypes of
Dorothia, which he named *D. alticamerata* (see
D. pupa), *D. bulbosa* and *D. subretusa*. These
three types are morphologically similar and have
similar stratigraphic ranges but nevertheless can
easily be distinguished in Trinidad. Length
0.48–1.65, thickness 0.45–1.0 mm.

Stratigraphic range: Early Maastrichtian
  (*Globotruncana lapparenti tricarinata* Zone)
  to Late Paleocene (*Globorotalia
  pseudomenardii* Zone). Fairly common.

DOROTHIA CYLINDRACEA Bermudez
Figure **24.32–34**

Dorothia cylindracea Bermudez 1963, p. 26,
  pl. 2, figs. 10, 11.

The species resembles *D. beloides* but is easily
distinguished by its consistently smaller size and,
at least in Trinidad, by its slightly flattened apertural face. Length 0.33–0.87, thickness 0.24–
0.42 mm.

Stratigraphic range: Late Paleocene (*Globorotalia
  pseudomenardii* Zone) to Early Eocene
  (*Globorotalia aragonensis* Zone). Fairly
  common. Also found in the Eocene of the
  overlying Navet Formation.

DOROTHIA HYPERCONICA Risch
Figure 24.35–36

Dorothia hyperconica Risch 1971, Palaeontographica, Stuttgart, p. 37, pl. 1, figs. 18, 19.

The Trinidad specimens show the typical inflated final two chambers. Length 0.35–0.8, width 0.22–0.65, thickness 0.18–0.54 mm.

Stratigraphic range: Albian to Early Cenomanian (*Rotalipora ticinensis ticinensis* Zone to *R. appenninica appenninica* Zone). Rare.

DOROTHIA PUPA (Reuss)
Figure 24.29

Textilaria pupa Reuss 1860, p. 232, pl. 13, fig. 4 (not fig. 5).
Dorothia alticamerata Israelsky 1951, p. 24, pl. 10, figs. 3–6; pl. 11, figs. 1–3.

This species is distinguished from *D. bulbosa* by its relatively short multi/triserial stage and the higher and more inflated biserial chambers. *Dorothia retusa* Cushman is possibly a synonym. Length 0.55–1.5, thickness 0.45–0.8 mm.

Stratigraphic range: Early Maastrichtian (*Globotruncana lapparenti tricarinata* Zone) to Late Paleocene (*Globorotalia velascoensis* Zone). Fairly common.

DOROTHIA SUBRETUSA Israelsky
Figure 24.30

Dorothia subretusa Israelsky 1951, p. 23, pl. 9, figs. 36–38; pl. 11, figs. 4, 5.
Dorothia retusa (part), Kaminski *et al.* 1988, p. 195, pl. 9, fig. 11 (not fig. 6).

Length 0.35–1.6, thickness 0.45–0.95 mm.

Stratigraphic range: Campanian (*Globotruncana stuarti* Zone) to Late Paleocene (*Globorotalia velascoensis* Zone). Fairly common.

MARSSONELLA INDENTATA (Cushman & Jarvis)
Figure 24.37

Gaudryina indentata Cushman & Jarvis 1928, p. 92, pl. 13, fig. 7.
Gaudryina indentata Cushman & Jarvis, Cushman & Jarvis 1932, p. 17, pl. 4, fig. 11.

Marssonella indentata (Cushman & Jarvis), Cushman & Renz 1946, p. 23, pl. 2, fig. 28.
Dorothia indentata (Cushman & Jarvis), Kaminski *et al.* 1988, p. 195, pl. 9, figs. 7, 8.

Length 0.4–1.55, thickness 0.3–1.05 mm.

Stratigraphic range: Paleocene (*Globorotalia trinidadensis* Zone to *G. pseudomenardii* Zone). Fairly common.

MARSSONELLA cf. NAMMALENSIS
Haque
Figure 24.38

Marssonella oxycona (Reuss) var. *nammalensis* Haque 1956, p. 50, pl. 3, fig. 4.

This may be a flat variant of *Dorothia cylindracea*, not related to *Marssonella oxycona*. It has the size and general shape of *M. nammalensis*, but the terminal face is rather flatter or even slightly concave. *Dorothia* ('*Textularia*') *nacataensis* White is relatively thicker. Length 0.4–0.85, thickness 0.34–0.45 mm.

Stratigraphic range: Paleocene (mostly *Globorotalia pseudomenardii* Zone); probably also in the *Globorotalia trinidadensis* Zone and in the Early Eocene (*Globorotalia subbotinae* Zone and *G. formosa formosa* Zone). Rare.

MARSSONELLA OXYCONA
FLORIDANA Applin & Jordan
Figure 25.1–2

Marssonella oxycona floridana Applin & Jordan 1945, J. Paleontol. 19, p. 135.
Gaudryina trochoides Marsson, White 1928, J. Paleontol., 2, p. 314, pl. 42, fig. 11.
Marssonella sp. A, Beckmann 1960, p. 62, fig. 5.
Dorothia cf. *trochoides* Marsson, Kaminski *et al.* 1988, p. 195, pl. 9, fig. 10.

This form was not figured by Applin & Jordan, but reference was made to a small type from Mexico (*Gaudryina trochoides* in White, 1928, size 0.35 × 0.3 mm) and a larger type from Trinidad (Cushman & Jarvis 1928, pl. 5, fig. 1; size 1.1 × 0.7 mm.). Applin and Jordan gave no reason for uniting two such dissimilar morphologies; possibly they implied that they were megalospheric and microspheric generations. The smaller type is deposited in the Cushman

Collection, Washington D.C., and appears to be conspecific with the smaller, more or less barrel-shaped form described here (Fig. 25.1) and also mentioned by other authors (see synonymy). In Trinidad, it usually has a fairly large proloculus followed by a very short and often indistinct multiserial stage, so that triserial coiling is almost immediately reached. In the Trinidad samples, this small form is often accompanied by a larger one (Fig. 25.2). This one is rather difficult to distinguish from *M. oxycona trinitatensis*, although it is sometimes relatively slender and the last chambers are often slightly inflated; it has a small proloculus and an initial coil of 4 to 5 subglobular chambers and is here tentatively assigned to *M. o. floridana*. Small form: Length 0.2–0.67, thickness 0.24–0.38 mm; large form: length 0.4–1.1, thickness 0.3–0.65 mm.

Stratigraphic range: Paleocene (*Globorotalia trinidadensis* Zone to *G. velascoensis* Zone). Rather scarce. Very rare and less typical also in the Campanian-Maastrichtian (*Globotruncana stuarti* Zone to *Abathomphalus mayaroensis* Zone). Rather scarce.

MARSSONELLA OXYCONA OXYCONA (Reuss)
Figure 25.5–6

Gaudryina oxycona Reuss 1860, p. 229, pl. 12, fig. 3.
Marssonella oxycona (Reuss 1860), Bartenstein & Bolli 1973, p. 396, pl. 2, fig. 62. – Bartenstein & Bolli 1986, p. 952, pl. 2, figs. 6, 7.

The apical end varies in shape from almost pointed to subrounded. Subacute specimens predominate in the Gautier Formation (*Rotalipora ticinensis ticinensis* Zone to *R. appenninica appenninica* Zone), whereas the specimens in the overlying Naparima Hill Formation have mostly a more distinctly subrounded base. Length 0.24–0.65, thickness 0.2–0.57 mm.

Stratigraphic range: Albian (*Rotalipora ticinensis ticinensis* Zone) to Campanian (*Globotruncana stuarti* Zone). Common. Also recorded in Trinidad from the Late Aptian to Early Albian Maridale Formation.

MARSSONELLA OXYCONA TRINITATENSIS Cushman & Renz
Figure 25.3–4

Marssonella oxycona (Reuss) var. *trinitatensis* Cushman & Renz 1946, p. 22, pl. 2, fig. 29.
Gaudryina oxycona Reuss, Cushman & Jarvis 1932, p. 18, pl. 5, figs. 1, 2.
Dorothia oxycona (Reuss), Kaminski *et al.* 1988, p. 195, pl. 9, fig. 9.

The highly arched aperture, also mentioned by Cushman & Renz as a diagnostic character of this subspecies, is present in the greater part of the specimens in Trinidad. Length 0.33–1.25, thickness 0.35–0.88 mm.

Stratigraphic range: Campanian (*Globotruncana stuarti* Zone) to Late Paleocene (*Globorotalia velascoensis* Zone). Fairly common.

MARSSONELLA TROCHOIDES (Marsson)
Figure 25.7

Gaudryina crassa Marsson var. *trochoides* Marsson 1878, p. 158, pl. 3, fig. 27.

Length 0.35–0.9 (?1.2), thickness 0.33–0.62 (?0.9) mm.

Stratigraphic range: Campanian (*Globotruncana stuarti* Zone) to Maastrichtian (*Globotruncana gansseri* Zone). Very rare.

EGGERINA SUBOVATA Beckmann
Figure 25.9–10

Eggerina subovata Beckmann 1991, p. 825, pl. 1, fig. 22–24.
Bulimina sp. A Beckmann 1960, p. 62, fig. 12.

Test ovoid to fusiform with the initial part rounded to subpointed. Chambers hardly to distinctly inflated, somewhat embracing, arranged in a triserial manner. Sutures steeply inclined, sometimes curved, often depressed but occasionally almost flush with the surface and then hardly visible. Wall smooth, fine grained, slightly calcareous, not glossy. Aperture an arched slit at the basal suture of the final chamber. Length 0.35–0.57, thickness 0.2–0.33 mm.

Eggerina cylindrica Toulmin 1941 from the Early Eocene differs in being somewhat larger; its shape is more cylindrical with a more distinctly pointed base, and the chambers are rather more inflated.

Stratigraphic range: Maastrichtian (*Globotruncana gansseri* Zone to *Abathomphalus mayaroensis* Zone). Rare.

KARRERIELLA LODOENSIS Israelsky
Figure **25.11–12**

Karreriella? *lodoensis* Israelsky 1951, p. 24, pl. 10, figs. 10–14; pl. 11, figs. 6–9.

The initial stage of the test is multiserial, rarely possibly triserial. Occasionally, the final one or two chambers of the biserial part are more loosely attached; this trend seems to continue and to lead to the development of a more distinctly uniserial final stage (see *Martinottiella*? sp. A). Length 0.34–0.66, thickness 0.16–0.28 mm.

Stratigraphic range: Early Eocene (*Globorotalia edgari* Zone to *G. formosa formosa* Zone). Rare.

MARTINOTTIELLA? sp. A
Figure **25.13**

These specimens combine a multiserial-biserial stage, like that of *Karreriella lodoensis*, with a final stage of up to 4 pseudo-uniserial to practically uniserial chambers. Length 0.6–0.69, thickness 0.18–0.23 mm.

Stratigraphic range: Early Eocene (*Globorotalia formosa formosa* Zone). Very rare.

Family Textulariidae

TEXTULARIA MIDWAYANA Lalicker
Figure **25.14**

Textularia midwayana Lalicker 1935, p. 49, pl. 6, figs. 7–9.

Length 0.4–1.0 or more (incomplete), width 0.22–0.3, thickness 0.15–0.22 mm.

Stratigraphic range: Early Paleocene (probably *Globorotalia pseudobulloides* Zone, Rochard facies). Rare.

Family Pseudogaudryinidae

CLAVULINOIDES ASPER (Cushman)
Figure **25.8, 15–16**

Clavulina trilatera Cushman var. *aspera* Cushman 1926, p. 589, pl. 17, fig. 3.
Clavulina trilatera Cushman var. *aspera* Cushman, Cushman & Jarvis 1928, p. 93, pl. 13, fig. 5.
Clavulina aspera (Cushman), Cushman & Jarvis 1932, p. 19, pl. 5, fig. 4.
Clavulina aspera Cushman var. *whitei* Cushman & Jarvis 1932, p. 19, pl. 5, figs. 6–8.
Clavulinoides aspera (Cushman), Cushman & Renz 1946, p. 22, pl. 2, fig. 25. – Kaminski *et al.* 1988, p. 194, pl. 8, figs. 11, 12.
Clavulinoides aspera (Cushman) var. *whitei* (Cushman & Jarvis), Cushman & Renz 1946, p. 22, pl. 2, fig. 26.
Clavulinoides rugulosa ten Dam & Sigal 1950, p. 32, pl. 2, figs. 8–10.
? *Clavulinoides globulifera* (ten Dam & Sigal), Kaminski *et al.* 1988, p. 194, pl. 8, figs. 14, 15.

The cylindrical form (*aspera*-type) and the larger, broader form (interpreted by Cushman & Jarvis as microspheric form of the *whitei*-type) are here regarded as one species (see also Kaminski *et al.*, 1988). Length up to 2.6 mm; thickness of megalospheric? form 0.4–0.7, of microspheric? form 0.6–0.9 mm.

Stratigraphic range: Early Maastrichtian (*Globotruncana lapparenti tricarinata* Zone) to Late Paleocene (*Globorotalia velascoensis* Zone). Fairly common. Specimens whose uniserial chambers have a rounded cross section (Fig. 25.15; see '*Clavulinoides globulifera*' in Kaminski *et al.* 1988) occur mostly in the Late Paleocene *Globorotalia pseudomenardii* and *G. velascoensis* zones.

CLAVULINOIDES DISJUNCTUS
(Cushman)
Figure **25.17–18**

Clavulina plummerae Sandidge, Cushman 1932, J. Paleontol., 6, p. 333, pl. 50, fig. 1.
Clavulina disjuncta Cushman 1933, new name, Contrib. Cushman Lab. foramin. Res., 9, p. 21.

The test shows some variability, similar to that of the probably closely related *C. midwayensis* Cushman although not quite as extreme. Collapsed final chambers, which produce a horizontal ridge like that of the holotype, are also observed. The aperture is rounded to oval, with or without a raised rim.
Length 0.55–1.8, thickness 0.33–0.5 mm.

Stratigraphic range: Probably Early Maastrichtian (*Globotruncana lapparenti tricarinata* Zone, Marabella facies). Rare.

CLAVULINOIDES PLUMMERAE
(Sandidge)
Figure **25.19–20**

Clavulina plummerae Sandidge 1932, p. 270, pl. 41, fig. 17–18.

The specimens in Trinidad show the typical three vertical ridges and moderately inflated chambers. Length 0.55–1.35, thickness 0.28–0.48 mm.

Stratigraphic range: Late Maastrichtian (*Abathomphalus mayaroensis* Zone). Very rare.

CLAVULINOIDES TRILATERUS
(Cushman)
Figure **25.21**

Clavulina trilatera Cushman 1926, p. 588, pl. 17, fig. 2.
Clavulina trilatera Cushman, Cushman & Jarvis 1928, p. 93, pl. 13, fig. 8. – Cushman & Jarvis 1932, p. 18, pl. 5, fig. 5.
Clavulinoides trilatera (Cushman), Cushman & Renz 1946, p. 22, pl. 2, fig. 24. – Kaminski *et al.* 1988, p. 195, pl. 9, fig. 2.

Length 0.6–1.5, thickness 0.25–0.45 mm.

Stratigraphic range: Paleocene (*Globorotalia pusilla pusilla* Zone to *G. pseudomenardii* Zone). Rare.

PSEUDOCLAVULINA AMORPHA
(Cushman)
Figure **25.22**

Clavulina amorpha Cushman 1926, p. 589, pl. 17, fig. 5.
Clavulina amorpha Cushman, Cushman & Jarvis 1928, p. 93, pl. 13, fig. 9.

Pseudoclavulina amorpha (Cushman), Cushman & Renz, 1946, p. 21, pl. 2, fig. 23.
Clavulinoides amorpha (Cushman), Kaminski *et al.* 1988, p. 194, pl. 8, fig. 13.

Length 0.65–1.5, thickness 0.4–0.58 mm.

Stratigraphic range: Late Paleocene (*Globorotalia pseudomenardii* Zone). Rare.

PSEUDOCLAVULINA ANGLICA
Cushman
Figure **25.23–24**

Pseudoclavulina anglica Cushman 1936, Cushman Lab. foramin. Res., Special Publ. 6, p. 18, pl. 3, fig. 5.

A few specimens with an almost cylindrical shape resemble *P. amorpha incrustata* Cushman, but usually the tests show a subtriangular early stage and weakly to moderately incised sutures in the uniserial portion. Length 0.65–1.6, thickness 0.35–0.75 mm.

Stratigraphic range: Early Eocene (*Globorotalia formosa formosa* Zone). Rather rare.

PSEUDOCLAVULINA CALIFORNICA
Cushman & Todd
Figure **25.25–26**

Pseudoclavulina californica Cushman & Todd 1948, Contrib. Cushman Lab. foramin. Res., 24, p. 92, pl. 16, fig. 6.
Clavulina gabonica Le Calvez, deKlasz & Brun 1971, Rev. Espan. Micropaleontol., 3, p. 308, pl. 1, figs. 7, 9.

The shape of the triserial early stage varies from rounded triangular (*californica*-type) to fairly sharply triangular (*gabonica*-type). Length 0.85 mm or more (broken), thickness of the uniserial part 0.18–0.32 mm.

Stratigraphic range: Early Cenomanian (*Rotalipora appenninica appenninica* Zone). Rare.

PSEUDOCLAVULINA CARINATA
(Neagu)
Figure **25.27**

Clavulinoides gaultinus carinatus Neagu 1962, Rocznik Polsk. Towarz. Geol., Krakow, 32, p. 420, pl. 40, figs. 14–20.

This form may be more closely related to *P. californica/gabonica* than to *P. gaultina* as suggested by Neagu. Length 0.62–1.05, thickness (uniserial part) 0.3–0.35 mm.

Stratigraphic range: Early Cenomanian (*Rotalipora appenninica appenninica* Zone). Very rare.

PSEUDOCLAVULINA CHITINOSA
(Cushman & Jarvis)
Figure **25.28**

Clavulina chitinosa Cushman & Jarvis 1932,
  p. 20, pl. 5, figs. 9–11.
Pseudoclavulina chitinosa (Cushman & Jarvis),
  Cushman & Renz 1946, p. 21, pl. 2, fig. 22.

Length 0.77–0.9, thickness 0.18–0.25 mm.

Stratigraphic range: Late Maastrichtian (*Abathomphalus mayaroensis* Zone). Very rare.

PSEUDOCLAVULINA TRINITATENSIS
Cushman & Renz
Figure **25.29**

Pseudoclavulina trinitatensis Cushman & Renz
  1948, p. 13, pl. 3, fig. 5.

The chambers are often slightly collapsed; this results in a horizontal rim around the chamber. *P. maqfiensis* LeRoy may be a synonym. Length 0.36–1.5, thickness 0.12–0.3 mm.

Stratigraphic range: Early Eocene (*Globorotalia edgari* Zone to *G. aragonensis* Zone). Rather scarce. Also found in the Eocene of the overlying Navet Formation.

Family Patellinidae

PATELLINA SUBCRETACEA Cushman & Alexander
Figure **25.30–32**

Patellina subcretacea Cushman & Alexander
  1930, Contrib. Cushman Lab. foramin.
  Res., 6, p. 10, pl. 3, fig. 1.
Patellina subcretacea Cushman & Alexander
  1930, Bartenstein & Bolli 1986, p. 976.

Two to three chambers in the last whorl. The ventral side, figured by Tappan (1943) and Hanzlikova (1972), is rather different from that

of the type species of *Patellina*, *P. corrugata*; and may possibly suggest a closer relationship to the rather poorly defined genus *Palaeopatellina* (placed in the family Placentulinidae by Loeblich & Tappan, 1987). Diameter 0.2–0.36, height 0.07–0.19 mm.

Stratigraphic range: Early Cenomanian (*Rotalipora appenninica appenninica* Zone). Fairly common. Also recorded in Trinidad from the Barremian Toco Formation, the Early Aptian of the Cuche Formation and the Late Aptian to Early Albian Maridale Formation.

Family Hauerinidae

QUINQUELOCULINA ANTIQUA
(Franke)
Figure **25.33**

Miliolina antiqua Franke 1928, p. 126, pl. 11,
  figs. 25, 26.

The angles of the test are often somewhat abraded. The type has a slightly projecting aperture but seems to be otherwise the same. *Quinqueloculina moremani* Cushman may be a junior synonym; its holotype in the Cushman collection is larger, but a paratype is practically identical. Length 0.3–0.35, width 0.24–0.26, thickness 0.2 mm.

Stratigraphic range: Late Albian (*Favusella washitensis* Zone). Very rare.

QUINQUELOCULINA EOCENICA
Cushman
Figure **25.34–35**

Quinqueloculina eocenica Cushman 1939,
  Contrib. Cushman Lab. foramin. Res., 15,
  p. 51, pl. 9, fig. 12.

Most of the Trinidad specimens are partially abraded but they seem to fit the original description. The wall is smooth to slightly roughened. Length 0.35–0.52, width 0.18–0.32, thickness 0.16–0.22 mm.

Stratigraphic range: Early Eocene (*Globorotalia formosa formosa* Zone to *G. aragonensis* Zone). Rare.

QUINQUELOCULINA SANDIEGOENSIS Sliter
Figure 25.36–37

Quinqueloculina sandiegoensis Sliter 1968, p. 52, pl. 4, fig. 7.

A few specimens in the Trinidad material have a short apertural neck, but otherwise they appear to be identical with the holotype in particular (Cushman collection). The aperture may show a trace of a tooth or flap. Length 0.25–0.55, width 0.16–0.35, thickness 0.11–0.26 mm.

Stratigraphic range: Early Cenomanian (*Rotalipora appenninica appenninica* Zone) to Coniacian/ Santonian (*Globotruncana concavata* Zone). Rare.

QUINQUELOCULINA sp. A
Figure 25.38–39

Triangular in cross section, with rounded edges. Aperture round, without a distinct tooth. *Quinqueloculina avelinoi* Petri is similar in shape but is smaller and has a relatively larger aperture with a distinct tooth. Length 0.28–0.5, width 0.24–0.4, thickness 0.15–0.27 mm.

Stratigraphic range: Late Albian (*Favusella washitensis* Zone). Very rare.

Family Nodosariidae

CHRYSALOGONIUM CRETACEUM
Cushman & Church
Figure 26.1–2

Chrysalogonium cretaceum Cushman & Church 1929, p. 513, pl. 39, figs. 23, 24.
? *Dentalina filiformis* Reuss (?), Cushman & Jarvis 1932, p. 29, pl. 9, figs. 6, 7.
Chrysalogonium cretaceum Cushman & Church, Cushman & Renz 1946, p. 30, pl. 5, fig. 4.

Length up to 2.7 mm or more (fragments), thickness 0.15–0.33 mm.

Stratigraphic range: Santonian (*Globotruncana fornicata* Zone) to Early Eocene (*Globorotalia subbotinae* Zone). Fairly common.

CHRYSALOGONIUM ELONGATUM
Cushman & Jarvis
Figure 26.3–4

Chrysalogonium elongatum Cushman & Jarvis 1934, p. 73, pl. 10, figs. 10, 11.

Fragments of small slender specimens are difficult to distinguish from *Ch. cretaceum*. Length up to 3.0 mm or more (broken), thickness 0.2–0.6 mm.

Stratigraphic range: Campanian (*Globotruncana stuarti* Zone) to Early Eocene (*Globorotalia formosa formosa* Zone). Fairly common. Also found in the Eocene of the overlying Navet Formation.

CHRYSALOGONIUM EOCENICUM
Cushman & Todd
Figure 26.5

Chrysalogonium eocenicum Cushman & Todd 1946, Contrib. Cushman Lab. foramin. Res., p. 53, pl. 9, figs. 3–5.

The larger specimens in Trinidad are transitional to the Middle Eocene species *Ch. laeve* Cushman & Bermudez. Length 0.85–1.6, thickness 0.16–0.22 mm.

Stratigraphic range: Late Paleocene (*Globorotalia pseudomenardii* Zone) to Early Eocene (*Globorotalia aragonensis* Zone). Rare. Also found in the Eocene of the overlying Navet Formation.

CHRYSALOGONIUM LANCEOLUM
Cushman & Jarvis
Figure 26.6–7

Chrysalogonium lanceolum Cushman & Jarvis 1934, p. 75, pl. 10, fig. 16.
? *Dentalina lorneiana* d'Orbigny, Cushman & Jarvis 1932, p. 31, pl. 10, fig. 2.
Chrysalogonium lanceolum Cushman & Jarvis, Cushman & Renz 1948, p. 20, pl. 4, figs. 6, 7?, 8.

Length up to 1.8 mm or more (fragments), thickness 0.16–0.33 mm.

Stratigraphic range: Late Maastrichtian (*Abathomphalus mayaroensis* Zone) to Early

Eocene (*Globorotalia aragonensis* Zone). Rather rare. Also found in the Eocene of the overlying Navet Formation.

CHRYSALOGONIUM VELASCOENSE (Cushman)
Figure **26.8–9**

Nodosaria fontannesi (Berthelin) var. *velascoensis* Cushman 1926, p. 594, pl. 18, fig. 12.
Nodosaria velascoensis Cushman, Cushman & Jarvis 1928, p. 97, pl. 13, figs. 15, 16. – Cushman & Jarvis 1932, p. 35, pl. 11, figs. 1–4. – Cushman & Renz 1946, p. 29, pl. 4, fig. 34.
? *Chrysalogonium tenuicostatum* Cushman & Bermudez, Cushman & Renz 1948, p. 20, pl. 4, fig. 9.

The variability of this species in Trinidad is shown by Cushman & Jarvis (1932). The holotype, together with specimens of the same age sometimes assigned to *Ch. tenuicostatum* Cushman & Bermudez in the literature, should be assigned to the genus *Chrysalogonium*. Length up to 3.3 mm or more (broken), thickness 0.2–0.5 mm.

Stratigraphic range: Late Maastrichtian (*Abathomphalus mayaroensis* Zone) to Early Eocene (*Globorotalia aragonensis* Zone). Fairly common. Also found in the Eocene of the overlying Navet Formation.

DENTALINA CONFLUENS (Reuss)
Figure **26.10**

Nodosaria (Dentalina) confluens Reuss 1862, Sitzungsber. K. Akad. Wiss., Vienna, math.-naturwiss. Kl., 44 (1861), p. 335, pl. 7, fig. 5.
Dentalina confluens Reuss, Cushman & Jarvis 1932, p. 30, pl. 9, figs. 10–12. – Cushman & Renz 1946, p. 28, pl. 4, figs. 14–16.

The species is rather variable in Trinidad (see Cushman & Jarvis, 1932). In some specimens the early chambers are somewhat inflated, and the basal spine may be missing. Length 0.8–2.15, thickness 0.17–0.45 mm.

Stratigraphic range: Late Paleocene (*Globorotalia velascoensis* Zone). Rare.

DENTALINA FRONTIERENSIS Petersen
Figure **26.11**

Dentalina frontierensis Petersen 1953, Bull. Utah Geol. Min. Survey, 47, p. 37, pl. 1, figs. 25, 26.
? *Nodosaria* cf. *marcki* Reuss, Cushman & Jarvis 1928, p. 97, pl. 14, fig. 4. – Cushman & Jarvis 1932, p. 34, pl. 10, fig. 12.

The available specimens are somewhat smaller than the types but appear so be otherwise identical. Length 0.7–0.82, thickness 0.13–0.2 mm.

Stratigraphic range: Early Cenomanian (*Rotalipora appenninica appenninica* Zone). Very rare.

DENTALINA RANCOCASENSIS Olsson
Figure **26.21**

Dentalina rancocasensis Olsson 1960, p. 15, pl. 3, fig. 7.

Length 0.6–1.25 mm or more (broken), thickness 0.15–0.32 mm.

Stratigraphic range: Early Paleocene (*Globorotalia trinidadensis* Zone) to Early Eocene (*Globorotalia formosa formosa* Zone). Rare. Also found in the Eocene of the overlying Navet Formation.

LAEVIDENTALINA CATENULA (Reuss)
Figure **26.12–13**

Dentalina catenula Reuss 1860, p. 185, pl. 3, fig. 6.
Dentalina catenula Reuss 1860, Bartenstein & Bolli 1986, p. 956, pl. 2, fig. 35.
Dentalina catenula Reuss, Cushman & Renz 1947a, p. 41, pl. 11, fig. 8.

The test consists of 2–5 chambers and usually carries a basal spine. *L. oligostegia* (Reuss; *Nodosaria (Dentalina) o.*) differs only in the lesser number of chambers (2–3). Length 0.43–1.45, thickness 0.14–0.45 mm.

Stratigraphic range: Early Cenomanian (*Rotalipora appenninica appenninica* Zone) to Late Paleocene (*Globorotalia velascoensis* Zone). Fairly common. Also recorded in Trinidad from the Late Aptian to Early Albian Maridale Formation.

LAEVIDENTALINA COMMUNIS
(d'Orbigny)
Figure **26.29**

Nodosaria (Dentalina) communis d'Orbigny
1826, p. 254.
Dentalina communis d'Orbigny, d'Orbigny
1840, p. 13, pl. 1, fig. 4.
Dentalina communis Orbigny 1826,
Bartenstein, Bettenstaedt & Bolli 1966,
p. 153, pl. 3, figs. 195–199. – Bartenstein &
Bolli 1973, p. 404, pl. 5, figs. 45–56. –
Bartenstein & Bolli 1986, p. 956, pl. 2, figs.
36, 37.
Dentalina legumen (Reuss), Cushman & Jarvis
1932, p. 30, pl. 9, fig. 9. – Cushman & Renz
1946, p. 27, pl. 4, fig. 10.
? *Dentalina* cf. *mucronata* Neugeboren,
Cushman & Renz 1948, p. 19, pl. 4, fig. 3.

Length 0.4–1.3, thickness 0.13–0.3 mm.

Stratigraphic range: Early Cenomanian (*Rotalipora
appenninica appenninica* Zone) to Early Eocene
(*Globorotalia aragonensis* Zone). Fairly
common. In Trinidad also recorded from the
Barremian Cuche Formation and the Late
Aptian to Early Albian Maridale Formation.
Found also in the Eocene Navet Formation.

LAEVIDENTALINA CYLINDROIDES
(Reuss)
Figure **26.15**

Dentalina cylindroides Reuss 1860, p. 185,
pl. 1, fig. 8.
Dentalina cylindroides Reuss 1860,
Bartenstein, Bettenstaedt & Bolli 1966,
p. 153, pl. 3, figs. 200–202, 218–219. –
Bartenstein & Bolli 1973, p. 405, pl. 5, figs.
25–35. – Bartenstein & Bolli 1986, p. 957,
pl. 2, figs. 38, 39.
Dentalina cylindroides Reuss, Cushman &
Renz 1946, p. 28, pl. 4, figs. 17, 18.

Specimens with or without an initial spine occur
throughout the stratigraphic range of this
species; the latter could be included in *Dentalina
ghorabi* Said & Kenawy. Length 0.4–2.3, thickness 0.16–0.6 mm.

Stratigraphic range: Late Turonian/Coniacian
(*Globotruncana renzi* Zone) to Late
Paleocene (*Globorotalia velascoensis* Zone).
Fairly common. Also recorded in Trinidad from

the Late Aptian to Early Albian Maridale
Formation.

LAEVIDENTALINA DISTINCTA (Reuss)
Figure **26.26**

Dentalina distincta Reuss 1860, p. 184, pl. 2,
fig. 5.
Dentalina distincta Reuss 1860, Bartenstein,
Bettenstaedt & Bolli 1966, p. 153, pl. 3, figs.
203, 204, 209, 217. – Bartenstein & Bolli
1973, p. 405, pl. 5, figs. 36–42.

Length 0.54–0.81, thickness 0.15–0.2 mm.

Stratigraphic range: Late Albian to ?Early
Cenomanian (*Favusella washitensis* Zone;
Rotalipora appenninica appenninica Zone?).
Very rare. Also recorded in Trinidad from the
Late Aptian to Early Albian Maridale
Formation.

LAEVIDENTALINA GRACILIS
(d'Orbigny)
Figure **26.14**

Nodosaria (Dentalina) gracilis d'Orbigny 1840,
p. 14, pl. 1, fig. 5.
Dentalina gracilis d'Orbigny 1839, Bartenstein,
Bettenstaedt & Bolli 1966, p. 153, pl. 3, figs.
187–194, 208. – Bartenstein & Bolli 1986,
p. 958, pl. 3, figs. 4–6.

Length 0.55–1.7, thickness 0.12–0.2 mm.

Stratigraphic range: Campanian (*Globotruncana
stuarti* Zone) to Early Eocene (*Globorotalia
formosa formosa* Zone). Rather rare. Also
recorded in Trinidad from the Late Aptian to
Early Albian Maridale Formation. Found also
in the Eocene Navet Formation.

LAEVIDENTALINA HAEGGI (Brotzen)
Figure **26.16**

Marginulina häggi Brotzen 1936, p. 69, pl. 4,
fig. 17; textfig. 23.
? *Dentalina catenula* Reuss, Cushman & Jarvis
1932, p. 29, pl. 9, fig. 8.

The initial chamber may or may not have a
spine. Length 0.63–0.82, thickness 0.22–
0.24 mm.

Stratigraphic range: Late Paleocene (*Globorotalia*

pseudomenardii Zone to *G. velascoensis* Zone). Very rare.

LAEVIDENTALINA HAMMENSIS
(Franke)
Figure **26.17**

Marginulina hammensis Franke 1928, p. 77, pl. 7, figs. 9, 10.

The size increase of the chambers, their arrangement (curved or straight) and the wall surface (hispid or almost smooth) vary between the extremes represented by the species *hammensis* and *aequalis (Marginulina aequalis* Franke). Length 0.32–0.93, thickness 0.16–0.35 mm.

Stratigraphic range: Turonian (*Globotruncana inornata* Zone) to Paleocene (*Globorotalia pusilla pusilla* Zone). Very rare.

LAEVIDENTALINA HAVANENSIS
(Cushman & Bermudez)
Figure **26.18**

Dentalina havanensis Cushman & Bermudez 1937, p. 11, pl. 1, figs. 39, 40.

Length 0.42–0.9, thickness 0.13–0.2 mm.

Stratigraphic range: Late Maastrichtian (*Abathomphalus mayaroensis* Zone) to Early Eocene (*Globorotalia formosa formosa* Zone). Rare.

LAEVIDENTALINA LEGUMEN (Reuss)
Figure **26.19–20**

Nodosaria (Dentalina) legumen Reuss 1845, p. 28, pl. 13, figs. 23, 24.

Some of the Tertiary specimens sometimes have slightly more oblique sutures than those from the Cretaceous, approaching *L. inornata* (d'Orbigny; *Dentalina i.*). Length 0.6–1.7, thickness 0.12–0.3 mm.

Stratigraphic range: Late Maastrichtian (*Abathomphalus mayaroensis* Zone) to Early Eocene (*Globorotalia formosa formosa* Zone). Common. In Trinidad also found in the Eocene of the overlying Navet Formation.

LAEVIDENTALINA LUMA (Belford)
Figure **26.22–23**

Dentalina luma Belford 1960, p. 34, pl. 10, figs. 6–11.

Length 0.92–1.65, thickness 0.15–0.22 mm.

Stratigraphic range: Early Maastrichtian (*Globotruncana lapparenti tricarinata* Zone) to Late Paleocene (*Globorotalia velascoensis* Zone). Rare.

LAEVIDENTALINA MEGALOPOLITANA (Reuss)
Figure **26.24–25**

Dentalina megalopolitana Reuss 1855, Zeitschr. deut. geol. Ges., 7, p. 267, pl. 8, fig. 10.
Dentalina megalopolitana Reuss, Cushman & Jarvis 1932, p. 29, pl. 9, fig. 5. – Cushman & Renz 1946, p. 28, pl. 4, fig. 11.

The initial spine, characteristic of the type specimens of this species, is very weak or absent in the present material. The early portion of the test often resembles that of *L. basiplanata* (Cushman; *Dentalina b.*), but this species has distinctly inflated chambers. The specimens of the Upper Lizard Springs Formation have a tendency to become slightly thinner and a little more compressed (Fig. 26.25). Length 0.7–1.8, thickness 0.12–0.35 mm.

Stratigraphic range: Early Maastrichtian (*Globotruncana lapparenti tricarinata* Zone) to Early Eocene (*Globorotalia formosa formosa* Zone). Fairly common. In Trinidad also found in the Eocene of the overlying Navet Formation.

LAEVIDENTALINA PARADOXA
(Hussey)
Figure **26.27**

Dentalina paradoxa Hussey 1949, J. Paleontol., 23, p. 126, pl. 26, fig. 23.

Relatively thin and elongated specimens resemble *L. pseudonana* (ten Dam; *Dentalina p.*). Length 0.45–0.95, thickness 0.22–0.32 mm.

Stratigraphic range: Coniacian-Santonian (*Globotruncana concavata* Zone) to Late Paleocene (*Globorotalia velascoensis* Zone). Rare.

LAEVIDENTALINA REUSSI (Neugeboren)
Figure **26.28**

Dentalina reussi Neugeboren 1856, Denkschr. K. Akad. Wiss., Vienna, math.-naturw. Kl., 12/2, p. 85, pl. 3, figs. 6, 7, 17.

Length 0.65–1.4, thickness 0.12–0.25 mm.

Stratigraphic range: Santonian (*Globotruncana fornicata* Zone) to Late Maastrichtian (*Abathomphalus mayaroensis* Zone). Rare.

LAEVIDENTALINA SPINESCENS (Reuss)
Figure **38.19–20**

Dentalina spinescens Reuss 1851, Zeitschr. Deutsche Geol. Ges., 3, p. 62, pl. 3, fig. 10.

Length up to 1.45, thickness 0.12–0.23 mm.

Stratigraphic range: Late Paleocene (*Globorotalia pseudomenardii* Zone) to Early Eocene (*Globorotalia formosa formosa* Zone). Rather rare.

NODOSARIA ASPERA ASPERA Reuss
Figure **26.30–31**

Nodosaria (Nodosaria) aspera Reuss 1845, p. 26, pl. 13, figs. 14, 15.
Nodosaria aspera Reuss, Cushman & Jarvis 1932, p. 35, pl. 11, fig. 5. – Cushman & Renz 1946, p. 29, pl. 4, fig. 30.

The shape of the test and of the chambers is variable in Trinidad, and the distinction from *N. spinosa* Olsson is therefore difficult. Slightly piriform chambers are found from the Late Paleocene (*Globorotalia velascoensis* Zone) upwards. Length 0.32–1.35, thickness 0.15–0.4 mm.

Stratigraphic range: Coniacian-Santonian (*Globotruncana renzi* Zone) to Early Eocene (*Globorotalia formosa formosa* Zone). Rare. In Trinidad also found in the Eocene of the overlying Navet Formation.

NODOSARIA ASPERA PARVAVERMICULATA Beckmann
Figure **26.32**

Nodosaria aspera parvavermiculata Beckmann 1991, p. 826, pl. 2, figs. 1, 2.

Test short, thick, subcylindrical, consisting of 3–4 chambers; greatest thickness either at the proloculus or at the final chamber. Base rounded. Chambers spherical but in the middle of the test rather short. Sutures moderately depressed. Wall surface covered with a mixture of pustules and short irregular costae. Aperture rounded, on a short neck.

This subspecies differs from *N. aspera* in its relatively short and thick test and its characteristic wall surface. Length 0.88–1.65, thickness 0.38–0.85 mm.

Stratigraphic range: Campanian (*Globotruncana stuarti* Zone) to Middle Paleocene (*Globorotalia angulata* Zone). Rare.

NODOSARIA LIMBATA d'Orbigny
Figure **26.33–34**

Nodosaria limbata d'Orbigny 1840, p. 12, pl. 1, fig. 1.
Nodosaria concinna Reuss, Cushman & Jarvis 1928, p. 97, pl. 14, figs. 5, 11. – Cushman & Jarvis 1932, pl. 31, pl. 10, fig. 4.
Nodosaria limbata d'Orbigny, Cushman & Jarvis 1932, p. 32, pl. 10, fig. 5. – Cushman & Renz 1946, p. 29, pl. 4, figs. 24, 31.
Nodosaria limbata d'Orbigny var. *tumidata* Cushman & Jarvis 1932, p. 32, pl. 10, fig. 6. – Cushman & Renz 1946, p. 29, pl. 4, fig. 32.

The largest specimens are found in the Campanian-Early Maastrichtian; they also show the typical limbate sutures mentioned by d'Orbigny. Higher up, the sutures often get narrower until they form an acute angle. Occasional deformed specimens resembling the variety *tumidata* of Cushman & Jarvis (1932) are found in the Late Paleocene (*Globorotalia pseudomenardii* and *G. velascoensis* zones). Length 0.7–3.0 mm or more (broken), thickness 0.35–0.9 mm.

Stratigraphic range: Coniacian-Santonian (*Globotruncana concavata* Zone) to Late Paleocene (*Globorotalia velascoensis* Zone). Fairly common.

NODOSARIA (?) LONGISCATA d'Orbigny
Figure **26.35–36**

Nodosaria longiscata d'Orbigny 1846, Foramin. foss. bassin tert. Vienne, (Paris), p. 32, pl. 1, figs. 10–12.

The aperture is not of the radiate *Nodosaria* type. In the Trinidad specimens it is very rarely seen but appears to be a simple opening with a slightly thickened rim; it may be slightly asymmetric, althouth not distinctly crescentic as in the morphologically similar *Ellipsonodosaria exilis* Cushman. Nevertheless, a relationship to the *Stilostomellidae* cannot be excluded. Length up to 2.2 mm (broken), thickness 0.1–0.3 mm, diameter of initial chamber 0.12–0.35 mm.

Stratigraphic range: Late Maastrichtian (*Abathomphalus mayaroensis* Zone) to Early Eocene (*Globorotalia aragonensis* Zone). Fairly common. In Trinidad also found in the Eocene of the overlying Navet Formation.

PSEUDONODOSARIA CLEARWATERENSIS Mellon & Wall
Figure 26.37

Pseudonodosaria clearwaterensis Mellon & Wall 1956, Report Alberta Res. Council, 72, p. 23, pl. 2, figs. 15–17.

This species appears to be closely related to *P. mutabilis* (Reuss; *Glandulina m.*), from which it differs only in the lesser inflation of the chambers. Length 0.45–0.58, thickness 0.15–0.23 mm.

Stratigraphic range: Late Albian (*Favusella washitensis* Zone). Very rare.

PSEUDONODOSARIA (?) CYLINDRICA
(Neugeboren)
Figure 26.38

Glandulina cylindrica Neugeboren 1850, Verhandl. Mitt. Siebenbürg. Ver. Naturw., Hermannstadt, 1, p. 53, pl. 1, fig. 10.

The test consists of 3 to 6 uniserial chambers. Examination of the pointed base in transparent light suggests that it may possibly include a minute multiserial initial stage. Length 0.32–0.65, thickness 0.15–0.21 mm.

Stratigraphic range: Turonian (*Globotruncana inornata* Zone) to Middle Paleocene (*Globorotalia pusilla pusilla* Zone). Very rare.

PSEUDONODOSARIA HUMILIS
(Roemer)
Figure 26.39

Nodosaria humilis Roemer 1841, Verstein. d. norddeut. Kreidegebirges, Hannover, p. 95, pl. 15, fig. 6.
Rectoglandulina humilis (Roemer 1841), Bartenstein, Bettenstaedt & Bolli 1966, p. 155, pl. 3, figs. 246–247.
Pseudonodosaria humilis (Roemer 1841), Bartenstein & Bolli 1986, p. 969, pl. 5, figs. 36, 37.

Length 0.24–0.55, thickness 0.2–0.32 mm.

Stratigraphic range: Late Albian (*Favusella washitensis* Zone) to Santonian (*Globotruncana fornicata* Zone). Rare. Also recorded in Trinidad from the Barremian Cuche und Toco formations and from the Late Aptian to Early Albian Maridale Formation.

PYRAMIDULINA LATEJUGATA
(Guembel)
Figure 27.3–4

Nodosaria latejugata Guembel 1868, Abhandl. K. Bayer. Akad. Wiss., math.-physik. Kl., 10/2, p. 619, pl. 1, fig. 32.
Nodosaria affinis Reuss, Cushman & Renz 1942, p. 6, pl. 1, figs. 8, 9, ?10.
? *Nodosaria affinis* Reuss, Cushman & Jarvis 1932, p. 34, pl. 10, fig. 13.
Nodosaria aff. *affinis* Reuss, Cushman & Renz 1946, p. 30, pl. 5, fig. 1.

Length up to 1.6 mm or more (fragments), thickness 0.36–0.75 mm.

Stratigraphic range: Early Paleocene (*Globorotalia trinidadensis* Zone) to Early Eocene (*Globorotalia subbotinae* Zone). Rather rare.

PYRAMIDULINA MULTICOSTATA
(d'Orbigny)
Figure 27.5–6

Nodosaria (Dentalina) multicostata d'Orbigny 1840, p. 15, pl. 1, figs. 14, 15.
Nodosaria sp. A, Beckmann 1960, p. 62, fig. 9.

For this species, the concept of Cushman (1940, p. 80, *Dentalina multicostata*) is followed. Late Cretaceous specimens: Length 1.1–1.9, thickness 0.53–0.85 mm. Paleocene specimens: Length up to 0.7, thickness 0.25–0.42 mm.

Stratigraphic range: Campanian (*Globotruncana stuarti* Zone) to Maastrichtian (*Abathomphalus mayaroensis* Zone); Middle Paleocene (*Globorotalia pusilla pusilla* Zone). Rare

PYRAMIDULINA OBSCURA (Reuss)
Figure **27.1–2**

Nodosaria (Nodosaria) obscura Reuss 1845, p. 26, pl. 13, figs. 7–9.
Nodosaria obscura Reuss 1863, Bartenstein & Bolli 1986, p. 968, pl. 5, figs. 25, 26.
Nodosaria orthopleura Reuss, Cushman & Jarvis 1932, p. 33, pl. 10, fig. 10. – Cushman & Renz 1946, p. 30, pl. 4, fig. 35.
Nodosaria cf. *gracilitatis* Cushman, Cushman & Renz 1947*a*, p. 42, pl. 11, figs. 10, 11.

This group includes specimens with 7–9 (exceptionally 10) costae and whose chambers are little if at all inflated. *Nodosaria bighornensis* Young is hardly distinguishable and may be a junior synonym. Length up to 3.1 mm or more (broken), thickness 0.25–0.56 mm.

Stratigraphic range: Campanian (*Globotruncana stuarti* Zone) to Early Eocene (*Globorotalia formosa formosa* Zone). Rather rare. Also recorded in Trinidad from the Barremian Cuche Formation and from the Late Aptian to Early Albian Maridale Formation.

PYRAMIDULINA OBSOLESCENS (Reuss)
Figure **27.7**

Nodosaria (Dentalina) obsolescens Reuss 1875, Palaeontographica, Cassel, 20/2 (1872–1875), p. 83, pl. 20, fig. 14.

The tests have 8–11 vertical costae. Length up to 2.6 mm or more (broken), thickness 0.22–0.54 mm.

Stratigraphic range: Early Maastrichtian (*Globotruncana lapparenti tricarinata* Zone) to Late Paleocene (*Globorotalia velascoensis* Zone). Rare.

PYRAMIDULINA ORTHOPLEURA
(Reuss)
Figure **27.8–9**

Nodosaria orthopleura Reuss 1863, p. 89, pl. 12, fig. 5.
Nodosaria orthopleura Reuss 1863, Bartenstein, Bettenstaedt & Bolli 1966, p. 152, pl. 3, fig. 230. – Bartenstein & Bolli 1986, p. 968, pl. 5, figs. 27–29.

This species includes specimens which have straight sides and 5–6 costae. Length up to 1.8 mm or more (broken), thickness 0.15–0.4 mm.

Stratigraphic range: Coniacian-Santonian (*Globotruncana concavata* Zone) to Late Maastrichtian (*Abathomphalus mayaroensis* Zone). Rare. Also recorded in Trinidad from the Late Aptian to Early Albian Maridale Formation.

PYRAMIDULINA PAUPERCULA (Reuss)
Figure **27.10–11**

Nodosaria (Nodosaria) paupercula Reuss 1845, p. 26, pl. 12, fig. 12.
Nodosaria paupercula Reuss 1845, Bartenstein & Bolli 1986, p. 968, pl. 5, figs. 30–31.
Nodosaria paupercula Reuss, Cushman & Jarvis 1932, p. 33, pl. 10, figs. 14, 15. – Cushman & Renz 1946, p. 30, pl. 5, figs. 2, 3.

The present concept of the species is that of Cushman (1946*b*). It includes specimens with 14–25 costae and relatively broad sutures. Length up to 1.9 mm or more (broken), thickness 0.42–0.9 mm.

Stratigraphic range: Campanian (*Globotruncana stuarti* Zone) to Late Paleocene (*Globorotalia pseudomenardii* Zone). Fairly common. Also recorded in Trinidad from the Barremian Toco Formation and from the Late Aptian to Early Albian Maridale Formation.

PYRAMIDULINA PRISMATICA (Reuss)
Figure 27.12–13

Nodosaria prismatica Reuss 1860, p. 180, pl. 2, fig. 2.

Length up to 0.8 mm or more (broken), thickness 0.25–0.33 mm.

Stratigraphic range: Santonian (*Globotruncana fornicata* Zone). Very rare.

PYRAMIDULINA TETRAGONA (Reuss)
Figure 27.14

Nodosaria tetragona Reuss 1860, p. 181, pl. 2, fig. 1

Length up to 0.9 mm or more (broken), thickness 0.15–0.26 mm.

Stratigraphic range: Coniacian to Santonian (*Globotruncana concavata* Zone to *G. fornicata* Zone). Very rare.

GONATOSPHAERA cf. PRINCIPIENSIS
Cushman & Bermudez
Figure 27.15

Gonatospharea principiensis Cushman & Bermudez 1936, Contrib. Cushman Lab. foramin. Res., 12, p. 59, pl. 11, figs. 7–9.

The Trinidad form differs from the types only in the costae, which have the tendency to break up and alternate. Length 0.65–1.0, thickness 0.42–0.6 mm.

Stratigraphic range: Early Eocene (*Globorotalia formosa formosa* Zone). Very rare.

LINGULINA TAYLORANA Cushman
Figure 27.16–17, 24

Lingulina taylorana Cushman 1938, Contrib. Cushman Lab. foramin. Res., 14, p. 43, pl. 7, fig. 9.

This species shows a considerable variability in the shape of the test and of the chambers; a keel may or may not be present. It includes specimens which closely resemble *L. californiensis* Trujillo and *L. lucillea* Trujillo but these variants cannot be clearly separated morphologically or stratigraphically. Length 0.25–0.72, width 0.13–0.22, thickness 0.07–0.14 mm.

Stratigraphic range: Turonian (*Globotruncana inornata* Zone) to Late Maastrichtian (*Abathomphalus mayaroensis* Zone). Rare.

FRONDICULARIA CLARKI Bagg
Figure 27.18–19

Frondicularia clarki Bagg 1895, Circ. John Hopkins Univ., 15, no. 121, p. 11. – Bagg 1898, Bull. U.S. geol. Survey, 88, p. 48, pl. 3, fig. 4.

The proloculus varies in size and in a few cases may be bigger than described by Bagg or by Plummer (1931) (see Fig. 27.18); it carries 1 to 3 short vertical ribs (the greater part of the specimens have 3). Length 0.5 to about 4.0 mm (usually damaged), width 0.4–1.2, thickness 0.07–0.1 (proloculus 0.12–0.17) mm.

Stratigraphic range: Santonian (*Globotruncana fornicata* Zone); Early Maastrichtian (*Globotruncana lapparenti tricarinata* Zone?, Marabella facies). Rare.

FRONDICULARIA JARVISI Cushman
Figure 27.20–21

Frondicularia jarvisi Cushman 1939, Contrib. Cushman Lab. foramin. Res., 15, p. 91, pl. 16, fig. 6.
Frondicularia elongata White (?), Cushman & Jarvis 1928, p. 98, pl. 14, fig. 1. – Cushman & Jarvis 1932, p. 39, pl. 12, fig. 3.
Frondicularia jarvisi Cushman, Cushman & Renz 1946, p. 31, pl. 5, fig. 16.

Length 0.7–1.6, breadth 0.3–0.44, maximum thickness 0.18–0.3 mm.

Stratigraphic range: Campanian (*Globotruncana stuarti* Zone)?, Maastrichtian (*Globotruncana gansseri* Zone) to Late Paleocene (*Globorotalia velascoensis* Zone). Rather rare.

FRONDICULARIA MUCRONATA Reuss
Figure 27.22–23

Frondicularia mucronata Reuss 1845, p. 31, pl. 13, figs. 43–44.
Frondicularia mucronata Reuss, Cushman & Renz 1946, p. 34, pl. 5, fig. 17.

Length 0.8–5.0, breadth 0.38–2.5, thickness 0.11–0.4 mm.

Stratigraphic range: Santonian (*Globotruncana fornicata* Zone) to Late Paleocene (*Globorotalia velascoensis* Zone). Rare.

FRONDICULARIA cf. PARKERI Reuss
Figure **27.25** •

Frondicularia parkeri Reuss 1863, p. 91, pl. 12, fig. 7.

The specimens from Trinidad differ from the type only in having a short vertical rib on the proloculus. Length up to 0.55 mm or more (broken), breadth 0.3–0.52, maximum thickness (proloculus) 0.12 mm.

Stratigraphic range: Early Cenomanian (*Rotalipora appenninica appenninica* Zone)?, Turonian-Coniacian (*Globotruncana inornata* Zone to *G. renzi* Zone). Rare.

FRONDICULARIA cf. SEDGWICKI Reuss
Figure **27.26**

Frondicularia sedgwicki Reuss 1854, Denkschr. K. Akad. Wiss. Wien, math.- naturwiss. Cl., 7, p. 66, pl. 25, fig. 4.

The Trinidad form differs from the type in its somewhat smaller size and the presence of only one short vertical rib (not two) on the proloculus. Length 0.7 mm or more (broken), breadth 0.2–0.26, thickness 0.12–0.16 (proloculus 0.15–0.23) mm.

Stratigraphic range: Turonian (*Globotruncana inornata* Zone). Very rare.

FRONDICULARIA SURFIBRATA Harris & Jobe
Figure **27.27–28**

Frondicularia frankei (Cushman) var. *costata* Harris & Jobe 1951, Transcript Press, Norman, Oklahoma, p. 29, pl. 6, fig. 2 (non *F. costata* Kübler & Zwingli 1866).
Frondicularia surfibrata Harris & Jobe, new name, 1952, Contrib. Cushman Found. foramin. Res., 3, p. 144.

Length up to 1.7 mm or more (incomplete), width 0.46–0.6, maximum thickness 0.22–0.33 mm.

Stratigraphic range: Middle Paleocene (*Globorotalia angulata* Zone to *G. pusilla pusilla* Zone). Rare.

FRONDICULARIA TRISPICULATA Sandidge
Figure **27.29**

Frondicularia trispiculata Sandidge 1932, p. 278, pl. 42, fig. 23.

In Trinidad, the chambers may be less embracing than in Sandidge's type specimen. Length up to 1.25 mm or more (broken), breadth 0.35–0.8, thickness 0.06–0.09 (proloculus 0.12–0.18) mm.

Stratigraphic range: Coniacian-Santonian (*Globotruncana concavata* Zone). Very rare.

TRISTIX EXCAVATA (Reuss)
Figure **27.30–31**

Rhabdogonium excavatum Reuss 1863, p. 91, pl. 12, fig. 8.

Length 0.32–0.5, thickness 0.15–0.28 mm.

Stratigraphic range: Late Albian to Early Cenomanian (*Favusella washitensis* Zone to *Rotalipora appenninica appenninica* Zone). Rather rare.

Family Vaginulinidae

CRISTELLARIOPSIS PROINOPS (Israelsky)
Figure **28.3–4**

Cristellaria proinops Israelsky 1955, p. 32, pl. 12, figs. 14, 15.
Robulus oligostegius (Reuss), Cushman & Jarvis 1932, p. 22, pl. 6, figs. 8, 9. – Cushman & Renz 1946, p. 25, pl. 3, fig. 18.

Maximum diameter 0.44–0.75, thickness 0.13–0.45 mm.

Stratigraphic range: Late Paleocene (*Globorotalia pseudomenardii* Zone to *G. velascoensis* Zone). Very rare.

CRISTELLARIOPSIS WIRZI Beckmann
Figure **28.5–6, 12**

Cristallariopsis wirzi Beckmann 1991, p. 826, pl. 2, figs. 3, 4, 9.

Test ovate to thick hook-shaped; periphery broadly rounded. Initial part showing 5 planispirally coiled, subglobular and partially embracing chambers which are gradually increasing in size; this is followed by 1 to 3 similar chambers which have a tendency to uncoil and to keep an almost constant size. Sutures radial, almost straight, those of the uncoiling chambers moderately depressed. Wall smooth or slightly roughened. Aperture terminal, sitting on a small neck, not typically radiate but rounded with a beaded rim.

Apart from the variable number of chambers, the morphologic characteristics are fairly constant. The most similar species is *C. trinitatensis* (Cushman; *Marginulina t.*), which differs in having a peripheral keel and longitudinal costae. *C. inops* (Reuss; *Cristellaria i.*) has only 4 chambers in the final whorl and a more distinctly pointed final chamber. *C. proinops* (see above) differs in the absence of an uncoiled stage and the presence of a subangular and slightly thickened periphery; its sutures are distinctly curved. Length 0.45–1.25, thickness 0.32–0.53 mm.

Stratigraphic range: Middle Paleocene (*Globorotalia angulata* Zone to *G. pusilla pusilla* Zone). Rare.

DIMORPHINA (?) LAEVIUSCULA
(Cushman & Bermudez)
Figure **28.1–2**

Marginulina laeviuscula Cushman & Bermudez 1937, p. 10, pl. 1, figs. 33, 34.

In transmitted light, the lower one of the two visible chambers does not show any distinct partitions except possibly at the very bottom, where a few coiled or irregularly arranged small chambers may be present. Length 0.53–0.88, thickness 0.15–0.29 mm.

Stratigraphic range: Late Maastrichtian (*Abathomphalus mayaroensis* Zone) to Early Paleocene (*Globorotalia trinidadensis* Zone). Rare.

LENTICULINA cf. BAYROCKI Mellon & Wall
Figure **28.7–8**

Lenticulina bayrocki Mellon & Wall 1956, Rept. Alberta Res. Council, 72, p. 19, pl. 2, figs. 7, 8.

There are 8–12 chambers in the final whorl, slightly more than in the type, and the sutures are rather less distinctly curved. A *Darbyella-*

like end chamber occurs occasionally in large specimens. Diameter 0.33–0.81, thickness 0.16–0.3 mm.

Stratigraphic range: Albian? to Early Cenomanian (*Rotalipora ticinensis ticinensis* Zone?, *R. appenninica appenninica* Zone). Rare.

LENTICULINA CONVERGENS
(Bornemann)
Figure **28.9–10**

Cristellaria convergens Bornemann 1855, Zeitschr. Deut. Geol. Ges., 7, p. 327, pl. 13, figs. 16, 17.

This species has 5 to 7 chambers in the last whorl, against 8 to 9 in the otherwise similar *L. discrepans* (Reuss), and has a distinctly pointed final chamber. Maximum diameter 0.5–1.15, thickness 0.18–0.48 mm.

Stratigraphic range: Early Eocene (*Globorotalia subbotinae* Zone to *G. aragonensis* Zone). Rare.

LENTICULINA DISCREPANS (Reuss)
Figure **28.11, 13–15**

Cristellaria (Robulina) discrepans Reuss 1863, p. 78, pl. 9, fig. 7.
Robulus discrepans (Reuss), Cushman & Jarvis 1932, p. 23, pl. 7, fig. 4. – Cushman & Renz 1946, p. 25, pl. 3, figs. 15–17.

The name *discrepans* is widely used in the literature for specimens with a somewhat open spiral and gently curved sutures, which are found both in the Cretaceous and in the lower Paleogene. '*Astacolus*' *velascoensis* White (not *Lenticulina velascoensis* White) and *Lenticulina fysta* Chamney can probably also be referred to this quite variable group. Maximum diameter 0.3–1.45, thickness 0.14–0.66 mm.

Stratigraphic range: Late Albian (*Favusella washitensis* Zone) to Early Eocene (*Globorotalia aragonensis* Zone). Common. In Trinidad also found in the Eocene of the overlying Navet Formation.

LENTICULINA EXARATA (Hagenow)
Figure **28.16**

Cristellaria exarata von Hagenow 1842, N.

Jahrb. Min. Geogn. Geol. Petref.-Kunde, Stuttgart, p. 572.
Cristellaria exarata von Hagenow, Reuss 1861, Sitzungsber. K. Akad. Wiss., Vienna, math.-naturw. Cl., 44, p. 327, pl. 6, fig. 5.

The specimens from Trinidad agree well with the drawings by Reuss (1861, see synonymy) and Pozaryska (1957, textfig. 28). Diameter 0.4–0.58, thickness 0.21–0.26 mm.

Stratigraphic range: Late Albian to Early Cenomanian (*Favusella washitensis* Zone to *Rotalipora appenninica appenninica* Zone). Rare.

LENTICULINA HOWELLI (Olsson)
Figure **28.17–18**

Planularia howelli Olsson 1960, p. 12, pl. 2, figs. 12, 13.

Diameter 0.4–0.95, thickness 0.15–0.33 mm.

Stratigraphic range: Early Paleocene (*Globorotalia trinidadensis* Zone) to Early Eocene (*Globorotalia subbotinae* Zone, possibly also *G. formosa formosa* Zone). Rather rare.

LENTICULINA INSULSA (Cushman)
Figure **28.19–20**

Robulus insulsus Cushman 1947, Contrib. Cushman Lab. foramin. Res., 23, p. 83, pl. 18, figs. 2, 3.
Lenticulina whitei Tjalsma & Lohmann 1983, p. 15, pl. 7, fig. 1.

The number of chambers in the final whorl varies from 5½ to 7 (rarely 8); the end members of this range are represented by the holotypes of *L. whitei* and *L. insulsa*, respectively. Diameter 0.3–0.58, thickness 0.15–0.28 mm.

Stratigraphic range: Middle Paleocene (*Globorotalia angulata* Zone) to Early Eocene (*Globorotalia edgari* Zone). Rather rare.

LENTICULINA cf. KLAGSHAMNENSIS (Brotzen)
Figure **28.21–23**

Robulus klagshamnensis Brotzen 1948, p. 41, pl. 7, figs. 1, 2.

The Trinidad form is considerably smaller than

the types but agrees well in other characteristics. Diameter 0.43–0.77, thickness 0.12–0.21 mm.

Stratigraphic range: Coniacian-Santonian (*Globotruncana concavata* Zone). Rather rare.

LENTICULINA LEPIDA MORUGAENSIS Beckmann
Figure **28.24–27**

Lenticulina lepida morugaensis Beckmann 1991, p. 827, pl. 2, figs. 5–8.

Test compressed, in side view elongated subovoid with a pointed apex, occasionally slightly angled in outline. Periphery acute, not distinctly keeled. Mostly 6–8 chambers in the final whorl which are slightly, if at all, inflated; the final one or two may become uncoiled. Sutures slightly curved to almost straight, slightly depressed at least between the final chambers. Umbilicus closed, often slightly depressed. Wall smooth, slightly glossy. Apertural face high, convex, with an apical radiate aperture. Length 0.3–0.48 (possibly 0.64), width 0.17–0.3, thickness 0.07–0.13 mm.

The present subspecies differs from *L. lepida* (Reuss; *Cristellaria l.*) mainly in the lesser number of chambers and in the absence of an umbilical disk.

Stratigraphic range: Turonian to Coniacian (*Globotruncana inornata* Zone to *G. renzi* Zone). Rather rare.

LENTICULINA LOBATA (Reuss)
Figure **28.28–30**

Cristellaria lobata Reuss 1845, p. 34, pl. 13, fig. 59. – Reuss 1874, p. 104, pl. II/22, fig. 12; pl. II/23, fig. 1.

Reuss has later (1874) supplemented the rather poor type figures with two additional illustrations. *L. lobata* is closely similar to its possible ancestor, *L. oligostegia*, but the chambers, except for the last few, are hardly inflated; the umbilicus is normally flat to convex due to the presence of some clear shell material. Diameter 0.3–0.68, thickness 0.15–0.3 mm.

Stratigraphic range: Late Albian (*Favusella washitensis* Zone) to Santonian (*Globotruncana fornicata* Zone). Fairly common.

LENTICULINA MIDWAYENSIS CARINATA (Plummer)
Figure **29.1–2**

Cristellaria midwayensis Plummer var. *carinata* Plummer 1926, p. 97, textfig. 5 (p. 41).

Some specimens with a relatively open spiral are transitional to *L. degolyeri* (Plummer; *Cristellaria d.*). Diameter 0.38–1.2, thickness 0.24–0.55 mm.

Stratigraphic range: Early Paleocene (probably *Globorotalia pseudobulloides* Zone; Rochard facies). Rather rare.

LENTICULINA MUENSTERI (Roemer)
Figure **29.3–4**

Robulina münsteri Roemer 1839, Die Verstein. norddeut. Oolithen-Gebirges, Hannover, p. 48, pl. 20, fig. 29.
Lenticulina (L.) muensteri (Roemer 1839), Bartenstein & Bolli 1986, p. 964, pl. 4, figs. 25–26.
? *Robulus macrodiscus* (Reuss), Cushman & Jarvis 1932, p. 23, pl. 7, fig. 3. – Cushman & Renz 1946, p. 25, pl. 3, fig. 14.
? *Robulus* sp., Cushman & Renz 1942, p. 5, pl. 1, fig. 5.

This group includes forms which are rather constant in their external shape but vary considerably in test size, sutures (curved or straight) and nature of the umbilical plug (almost absent to discoid and even somewhat raised). Diameter 0.45–1.64, thickness 0.16–0.3 mm.

Stratigraphic range: Albian (*Rotalipora ticinensis ticinensis* Zone) to Early Eocene (*Globorotalia formosa formosa* Zone), possibly also *G. aragonensis* Zone). Common. Also recorded in Trinidad from the Barremian to Early Aptian Cuche Formation and from the Late Aptian to Early Albian Maridale Formation. Found also in the Eocene Navet Formation.

LENTICULINA MULTIFORMIS ROTUNDA (Franke)
Figure **29.5, 12**

Cristellaria multiformis Franke var. *rotunda* Franke 1912, Jahrb. K. Preuss. geol. Landesanst., 32 (1911), p. 111, pl. 3, figs. 3–6.

The raised sutures are often partially broken into pustules. Where the sutural ridges are more or less entire, the specimens may resemble *L. tendami* (Israelsky; *Cristellaria t.*) very much. Diameter 0.39–1.2, thickness 0.24–0.69 mm.

Stratigraphic range: Campanian (*Globotruncana stuarti* Zone) to Early Eocene (*Globorotalia subbotinae* Zone). Rare.

LENTICULINA NUDA (Reuss)
Figure **29.6–7**

Cristellaria nuda Reuss 1861, Sitzungsber. K. Akad. Wiss. Wien, 44, p. 328, pl. 6, figs. 1–3.
Lenticulina nuda (Reuss), Cushman & Jarvis 1928, p. 96, pl. 14, fig. 2. – Cushman & Jarvis 1932, p. 24, pl. 7, fig. 6. – Cushman & Renz 1946, p. 26, pl. 3, fig. 24.
Lenticulina grata (Reuss), Cushman & Jarvis 1928, p. 96, pl. 14, fig. 3.
Marginulina jarvisi Cushman, Cushman & Renz 1946, p. 27, pl. 3, figs. 27, 28.

This form shows the same variability as the types of *L. nuda* but is somewhat smaller. The test is, on the average, more elongated than in the older variety of the *L. nuda* group described below. Length 0.45–0.94, width 0.21–0.48, thickness 0.12–0.23 mm.

Stratigraphic range: Early Maastrichtian (*Globotruncana lapparenti tricarinata* Zone) to Early Eocene (*Globorotalia edgari* Zone). Rare.

LENTICULINA NUDA (Reuss), variety A
Figure **29.8–9**

Cristellaria nuda Reuss var., Reuss 1863, p. 72, pl. 8, fig. 2.

The test is compressed with a subacute periphery and a narrow convex apertural face. The umbilicus is closed. Some specimens with a rather broad final chamber resemble *L. discrepans*

(Reuss). This variety may be the same as *L. acuta* (Reuss) of Bartenstein & Bolli (1986); however, *L. acuta* is more rhombic in cross section and has a distinct umbilical plug. The Trinidad specimens are considerably smaller than the typical *L. nuda*. Maximum diameter 0.21–0.68, thickness 0.15–0.26 mm.

Stratigraphic range: Albian (*Rotalipora ticinensis ticinensis* Zone). Rather rare.

LENTICULINA OLIGOSTEGIA (Reuss)
Figure **29.10–11**

Cristellaria oligostegia Reuss 1860, p. 213, pl. 8, fig. 8.

Diameter 0.3–0.52, thickness 0.15–0.24 mm.

Stratigraphic range: Albian (*Rotalipora ticinensis ticinensis* Zone). Rather rare.

LENTICULINA REVOLUTA (Israelsky)
Figure **29.13–14**

Robulus revolutus Israelsky 1955, p. 49, pl. 15, figs. 3–6.
? *Robulus sternalis* (Berthelin), Cushman & Jarvis 1932, p. 22, pl. 6, fig. 11. – Cushman & Renz 1946, p. 25, pl. 3, fig. 19.

This form is sometimes regarded as a small variety of *L. macrodiscus* (Reuss, *Cristellaria m.*) or is assigned to different species, often new, which are difficult to distinguish. The name *L. revoluta* seems to be the best choice for the Trinidad material. Diameter 0.26–1.0, thickness 0.15–0.46 mm.

Stratigraphic range: Early Cenomanian (*Rotalipora appenninica appenninica* Zone) to Early Eocene (*Globorotalia aragonensis* Zone). Common. In Trinidad also found in the Eocene of the overlying Navet Formation.

LENTICULINA SECANS (Reuss)
Figure **29.15–16**

Cristellaria secans Reuss 1860, p. 214, pl. 9, fig. 7.
Lenticulina (L.) secans (Reuss 1860), Bartenstein & Bolli 1986, p. 966.

Diameter 0.48–1.3, thickness 0.32–0.65 mm.

Stratigraphic range: Early Cenomanian (*Rotalipora appenninica appenninica* Zone) to Santonian (*Globotruncana fornicata* Zone). Rather rare. Also recorded in Trinidad from the Late Aptian to Early Albian Maridale Formation.

LENTICULINA SORACHIENSIS
(Takayanagi)
Figure **29.17–18**

Robulus sorachiensis Takayanagi 1960, p. 116, pl. 7, fig. 10.

Diameter 0.39–1.2, thickness 0.24–0.69 mm.

Stratigraphic range: Campanian (*Globotruncana stuarti* Zone)?, Early Maastrichtian (*Globotruncana lapparenti tricarinata* Zone) to Middle Paleocene (*Globorotalia pusilla pusilla* Zone). Rare.

LENTICULINA SUBGAULTINA
Bartenstein
Figure **29.19–20**

Lenticulina subgaultina Bartenstein 1962, Senckenbergiana Lethea (Frankfurt), 43, p. 136, pl. 15, figs. 1, 2.
Lenticulina (L.) subgaultina Bartenstein 1962, Bartenstein, Bettenstaedt & Bolli 1966, p. 147, pl. 2, figs. 128–129. – Bartenstein & Bolli 1973, p. 401, pl. 4, figs. 30, 31.

Length 0.35–1.35, width 0.55–1.0, thickness 0.3–0.42 mm.

Stratigraphic range: Early Cenomanian (*Rotalipora appenninica appenninica* Zone). Rare. Also recorded in Trinidad from the Late Aptian Maridale Formation.

LENTICULINA TRUNCATA (Reuss)
Figure **29.21**

Cristellaria truncata Reuss 1851, p. 32, pl. 3, fig. 8.

Maximum diameter 0.52–0.75, thickness 0.15–0.18 mm.

Stratigraphic range: Turonian (*Globotruncana inornata* Zone). Very rare.

LENTICULINA VELASCOENSIS White
Figure **29.22–23**

Lenticulina velascoensis White 1928, p. 199, pl. 28, fig. 8.

This species is characterized by its flattened or even slightly concave umbilical area, which is filled by clear shell material, and by its broad sometimes ragged keel. Diameter 0.33–0.78, thickness 0.12–0.29 mm.

Stratigraphic range: Santonian (*Globotruncana fornicata* Zone) to Late Paleocene (*Globorotalia velascoensis* Zone). Rather rare.

LENTICULINA sp. A
Figure **29.24**

Lenticulina sp. A, Beckmann 1960, p. 62, fig. 8.

This group consists of a few characteristic specimens which have a moderately compressed ovoid test with an acute, slightly keeled periphery; the surface is sometimes ornamented with longitudinal costae which are either simple (Fig. 29.24) or branching (Beckmann 1960, fig. 8). The apertural face is either convex or truncated and curved. Length 0.54–1.05, breadth 0.39–0.78, thickness 0.3–0.51 mm.

Stratigraphic range: Late Paleocene (*Globorotalia pseudomenardii* Zone). Very rare.

MARGINULINOPSIS CEPHALOTES (Reuss)
Figure **29.25**

Cristellaria cephalotes Reuss 1863, p. 67, pl. 7, figs. 4–6.
Lenticulina (M.) cephalotes (Reuss 1863), Bartenstein, Bettenstaedt & Bolli 1966, p. 150, pl. 2, figs. 178–182. – Bartenstein & Bolli 1973, p. 403, pl. 4, figs. 41–47. – Bartenstein & Bolli 1986, p. 962, pl. 4, figs. 11–12.

The initial spiral has an acute periphery. The chambers in the uncoiled stage are variable from hardly to distinctly inflated. Length 0.32–0.6, width 0.15–0.21, thickness 0.14–0.19 mm.

Stratigraphic range: Albian (*Rotalipora ticinensis ticinensis* Zone to *Favusella washitensis* Zone). Rare. Also recorded in Trinidad from the Late Aptian to Early Albian Maridale Formation.

MARGINULINOPSIS CURVISEPTA (Cushman & Goudkoff)
Figure **29.26–27**

Marginulina curvisepta Cushman & Goudkoff 1944, Contrib. Cushman Lab. foramin. Res., 20, p. 57, pl. 9, figs. 12, 13.

Length 0.4–1.2, maximum diameter 0.18–0.3 mm.

Stratigraphic range: Coniacian-Santonian (*Globotruncana concavata* Zone) to Campanian (*Globotruncana stuarti* Zone). Rather rare.

MARGINULINOPSIS LITUOLA (Reuss)
Figure **29.28–29**

Cristellaria lituola Reuss 1846, p. 109, pl. 24, fig. 47.

The range of variation of this characteristic group probably includes *M. stephensoni* (Cushman; *Marginulina s.*), *M. amplaspira* Young and *M. frontierensis* Young. Length 0.32–0.75, maximum width 0.15–0.3, thickness 0.09–0.19 mm.

Stratigraphic range: Turonian (*Globotruncana inornata* Zone) to Santonian (*Globotruncana fornicata* Zone). Rather rare.

MARGINULINOPSIS MULTICOSTATA (Lipnik)
Figure **29.30–31**

Marginulina multicostata Lipnik 1961, Akad. Nauk Ukrain. RSR, Trudy geol. Inst., Ser. Strat. Paleontol., 35, p. 39, pl. 1, fig. 2.

This species is often slightly larger than the type of *M. multicostata*, and the test is sometimes rather more regular with distinctly inflated chambers. Length 0.55–1.25, diameter 0.25–0.4 mm.

Stratigraphic range: Campanian (*Globotruncana stuarti* Zone). Rare.

MARGINULINOPSIS PRAETSCHOPPI Trujillo
Figure **30.1–3**

Marginulinopsis praetschoppi Trujillo 1960, p. 324, pl. 46, fig. 13.

Length 0.32–1.0, width 0.2–0.35, thickness 0.11–0.26 mm.

Stratigraphic range: Coniacian-Santonian (*Globotruncana concavata* Zone) to Campanian (*Globotruncana stuarti* Zone). Rather rare.

MARGINULINOPSIS SUBRECTA (Franke)
Figure **30.4–5**

Marginulina subrecta Franke 1927, Danmarks geol. Unders., raekke 2, 46, p. 19, pl. 1, fig. 28.

A short spiral of about 5 chambers, which may be slightly compressed, is followed by 1–5 uniserial chambers. The specimens from the Paleocene are relatively shorter and have not more than 3 uniserial chambers. Length 0.54–2.4, thickness 0.33–0.7 mm.

Stratigraphic range: Campanian (*Globotruncana stuarti* Zone) to Late Paleocene (*Globorotalia pseudomenardii* Zone). Rather rare.

PERCULTAZONARIA TUBERCULATA (Plummer)
Figure **30.6**

Cristellaria subaculeata Cushman var. *tuberculata* Plummer 1926, p. 101, pl. 7, fig. 2; pl. 14, fig. 1.
Marginulina decorata (Reuss), Cushman & Jarvis 1932, p. 28, pl. 9, fig. 2.
Marginulina cf. *decorata* Reuss, Cushman & Renz 1946, p. 27, pl. 3, fig. 31.

In the present material, this species is represented only by possibly immature specimens lacking an uncoiled stage. Length 0.48–0.75, width 0.27–0.33, thickness 0.14–0.16 mm.

Stratigraphic range: Early Eocene (*Globorotalia formosa formosa* Zone). Very rare.

PRAVOSLAVLEVIA DUTROI (Tappan)
Figure **30.7**

Saracenaria dutroi Tappan 1957, p. 216, pl. 68, figs. 14–16.

Length 0.8–1.03, thickness 0.4–0.52 mm.

Stratigraphic range: Coniacian-Santonian (*Globotruncana concavata* Zone). Rare.

SARACENARIA CRETACEA Dailey
Figure **30.8–9**

Saracenaria cretacea Dailey 1970, p. 105, pl. 12, figs. 1, 2.

Length 0.36–0.75, breadth 0.22–0.42, thickness 0.21–0.33 mm.

Stratigraphic range: Late Albian (*Favusella washitensis* Zone) to Turonian (*Globotruncana inornata* Zone). Rare.

SARACENARIA NAVICULA (d'Orbigny)
Figure **30.10**

Cristellaria navicula d'Orbigny 1840, p. 27, pl. 2, figs. 19, 20.
Lenticulina navicula d'Orbigny, Cushman & Jarvis 1932, p. 24, pl. 7, fig. 5. – Cushman & Renz 1946, p. 26, pl. 3, fig. 26.

The final chamber is elongated and involute. Length 0.32–1.14, thickness 0.17–0.66 mm.

Stratigraphic range: Turonian (*Globotruncana inornata* Zone) to Maastrichtian (*Globotruncana gansseri* Zone). Rare.

SARACENARIA TRIANGULARIS (d'Orbigny)
Figure **30.11–12**

Cristellaria triangularis d'Orbigny 1840, p. 27, pl. 2, figs. 21, 22.
Saracenaria cf. *triangularis* d'Orbigny, Cushman & Renz 1946, p. 30, pl. 4, fig. 27.

Length 0.35–1.23, thickness 0.24–0.75 mm.

Stratigraphic range: Late Albian (*Favusella washitensis* Zone) to Early Eocene (*Globorotalia edgari* Zone). Common.

SARACENARIA TUNESIANA ten Dam & Sigal
Figure **30.13**

Saracenaria tunesiana ten Dam & Sigal 1950, p. 36, pl. 2, fig. 21.

S. curvabilis Harris & Jobe has a slightly less ornamented test but is otherwise hardly distinguishable and may well be a junior synonym. Length 0.48–0.63, breadth 0.27–0.39, thickness 0.27–0.36 mm. A less typical large specimen measures 0.78 × 0.63 × 0.51 mm.

Stratigraphic range: Paleocene (*Globorotalia pusilla pusilla* Zone to *G. pseudomenardii* Zone). Very rare.

NEOFLABELLINA cf. CORANICA
(Marie)
Figure **30.14–15**

Flabellina coranica Marie 1945, Bull. Soc. géol. France, sér. 5, 14 (1944), p. 402, textfigs. 11–16.

Whereas the rhombic shape of the individuals from Trinidad is typically that of *N. coranica*, their size is smaller and the pattern of the surface ornamentation is more predominantly reticulate. Length 0.6–0.85, width 0.3–0.62, thickness 0.21–0.32 mm.

Stratigraphic range: Paleocene (*Globorotalia trinidadensis* Zone to *G. angulata* Zone). Rare.

NEOFLABELLINA DELICATISSIMA
(Plummer)
Figure **30.16–17**

Frondicularia delicatissima Plummer 1926, p. 120, pl. 5, fig. 4.

Length 0.45–0.82, width 0.24–0.46, thickness 0.12–0.15 mm.

Stratigraphic range: Late Maastrichtian (*Abathomphalus mayaraoensis* Zone). Rare.

NEOFLABELLINA JARVISI (Cushman)
Figure **30.18–19, 25**

Flabellina jarvisi Cushman 1935, Contrib. Cushman Lab. foramin. Res., 11, p. 85, pl. 13, figs. 7, 8.
Flabellina interpunctata von der Marck, Cushman & Jarvis 1932, p. 38, pl. 12, fig. 1.
Palmula jarvisi (Cushman), Cushman & Renz 1946, p. 31, pl. 5, figs. 10–13 (not fig. 14).

This species is easily recognized by its large size, regular ovoid to subrhombic shape and slightly convex sides. The ornamentation varies from simple to fairly complex. The paratype figured by Cushman (fig. 7), although strikingly differ-

ent from the holotype, should be accepted as an extreme variant of the species, and the suggestion by Kristan-Tollmann & Tollmann (1976) to include it in the Cretaceous species *N. hanzlikovae* is not supported. Length 0.52–2.0, width 0.33–1.05, thickness 0.21–0.38 mm.

Stratigraphic range: Paleocene (*Globorotalia angulata* Zone to *G. velascoensis* Zone). Fairly common.

NEOFLABELLINA NUMISMALIS
(Wedekind)
Figure **30.20–21**

Flabellina numismalis Wedekind 1940, Neues Jahrb. Geol. Min. Paläontol., Stuttgart, Beilage-Bd. 84, Abt. B, 2, p. 200, pl. 9, figs. 1–3; pl. 11, figs. 8, 9.

Length 0.45–0.85, width 0.35–0.72, thickness 0.12–0.19 mm.

Stratigraphic range: Campanian (*Globotruncana stuarti* Zone). Rare.

NEOFLABELLINA PALEOCENICA
Titova
Figure **30.22–24**

Neoflabellina paleocenica Titova 1975, Paleontol. Sbornik, Lvov, 12, p. 32, pl. 2, figs. 5–8.
Flabellina reticulata Reuss, Cushman & Jarvis 1932, p. 37, pl. 11, fig. 15.
Palmula semireticulata (Cushman & Jarvis), Cushman & Renz 1946, p. 31, pl. 5, fig. 8 (not fig. 9).

The Trinidad specimens are somewhat smaller than the types but agree in other characteristics, particularly in the high rectangular pattern of the ornamentation. This species includes specimens referred to *N. semireticulata* (Cushman & Jarvis) by Cushman (1946*b*) and some subsequent authors. Length 0.48–0.87, width 0.3–0.58, thickness 0.09–0.15 mm.

Stratigraphic range: Paleocene (*Globorotalia pseudomenardii* Zone); transitional specimens from *N. semireticulata* to *N. paleocenica* occur in the *Globorotalia pusilla pusilla* Zone). Rare.

NEOFLABELLINA cf. PERMUTATA
Koch
Figure **30.26**

Neoflabellina permutata Koch 1977, Geol.
  Jahrbuch, Hannover, A 38, p. 55, pl. 17, figs.
  1–3.

The tests in Trinidad are smaller than the types
but, in spite of their rather poor preservation,
seem to show the sutural form and fine hispid
surface of the species. Length 0.53–0.94, breadth 0.3–0.52, thickness 0.14–0.17 mm.

Stratigraphic range: Probably Early Maastrichtian
(*Globotruncana lapparenti tricarinata* Zone,
Marabella facies). Very rare.

NEOFLABELLINA PRAERETICULATA
Hiltermann
Figure **30.27–28**

Neoflabellina praereticulata Hiltermann 1952,
  Geol. Jahrbuch, Hannover, 67, p. 53, fig. 3/
  57.

Length 0.45–1.5 mm (incomplete, probably
more than 2 mm), width 0.3–1.44, thickness
0.11–0.2 mm.

Stratigraphic range: Campanian (*Globotruncana
stuarti* Zone) to Early Maastrichtian
(*Globotruncana lapparenti tricarinata* Zone).
Rare.

NEOFLABELLINA PROJECTA (Carsey)
Figure **31.1–5**

Frondicularia projecta Carsey 1926, Texas
  Univ. Bull., 2612, p. 41, pl. 6, fig. 5.
Flabellina projecta (Carsey), Plummer 1931,
  p. 165, pl. 12, figs. 5–8 (5, holotype
  refigured).

This species includes the oldest representatives
of *Neoflabellina* in Trinidad. It is possible to distinguish three forms (A, B and C), which differ
in details of the surface ornamentation (pustules, sutures); the form B is the most typical
and can be regarded as the central type.

Form A (Fig. 31.1–2): The specimens are relatively coarsely built, and the sides may be
slightly convex. The sutural ridges and the apertural loops are simple, and pustules are rare or

even absent. Length 0.52–0.7 mm or more
(broken), width 0.4–0.45, thickness 0.16–
0.21 mm.

Stratigraphic range: Santonian (*Globotruncana
fornicata* Zone). Very rare.

Form B (Fig.31.3): This form is more finely
structured than the form A. The sides are flat,
and pustules are usually present. Length 0.48–
0.65 mm or more (incomplete), width 0.33–
0.45 mm (or more), thickness 0.12–0.18 mm.

Stratigraphic range: Santonian (*Globotruncana
fornicata* Zone) to Campanian (*Globotruncana
stuarti* Zone). Rare.

Form C (Figs. 31.4–5): The test may be a little
larger than in the other two forms; the sutures
have a tendency to be more complicated and
may become interrupted, wavy or doubled. Parallel to this, the apertural loops may also become
irregular or multiple. Length 0.53–0.8 or more
(incomplete), width 0.36–0.65 (or more?),
thickness 0.13–0.18 mm.

Stratigraphic range: Santonian (*Globotruncana
fornicata* Zone) to Campanian (*Globotruncana
stuarti* Zone). Rare.

NEOFLABELLINA SEMIRETICULATA
(Cushman & Jarvis)
Figure **30.29–30**

Flabellina semireticulata Cushman & Jarvis
  1928, p. 98, pl. 13, fig. 14.
Flabellina semireticulata Cushman & Jarvis,
  Cushman & Jarvis 1932, p. 38, pl. 11, fig. 14.
Palmula semireticulata (Cushman & Jarvis),
  Cushman & Renz 1946, p. 31, pl. 5, fig. 9
  (not fig. 8).
Palmula jarvisi (Cushman), Cushman & Renz
  1946, p. 31, pl. 5, fig. 14 (not figs. 10–13.)

Apart from typical specimens (Fig. 30.30), there
are a few which show a more distinct spiral which
resembles a reduced initial portion of *N. lacostei*
(Marie; *Flabellina l.*). Length 0.45–0.9, width
0.27–0.64, thickness 0.13–0.23 mm.

Stratigraphic range: Paleocene (*Globorotalia
angulata* Zone to *G. pusilla pusilla* Zone).
Fairly common.

NEOFLABELLINA THALMANNI (Finlay)
Figure 31.6–7

Palmula thalmanni Finlay 1939, Trans. Proc.
Royal Soc. New Zealand, 69, p. 315, pl. 26,
figs. 53, 54.

The reticulation, which is concentrated along the
sutures, resembles that of the paratype or, more
rarely, that of the holotype. *N. hanzlikovae* Kristan-Tollmann & Tollmann has a similar,
although weaker ornamentation, but the shape
of its test is distinctly narrower. Length 0.58–
1.35 mm or more (incomplete), width 0.28–
1.33 mm or more (incomplete), thickness 0.14–
0.26 mm.

Stratigraphic range: Campanian (*Globotruncana
stuarti* Zone) to Early Maastrichtian
(*Globotruncana lapparenti tricarinata* Zone).
Rare.

NEOFLABELLINA sp. A
Figure 31.8–9

This form resembles *N. numismalis* superficially
but is more narrowly elliptical and relatively
flatter. The sutures are less distinct and partially
replaced by pustules, and the initial spiral is
fairly large. Length 0.33–0.76, width 0.22–0.48,
thickness 0.08–0.15 mm.

Stratigraphic range: Probably Early Maastrichtian
(*Globotruncana lapparenti tricarinata* Zone;
Marabella facies). Rare.

PALMULA cf. PRIMITIVA Cushman
Figure 31.10–11

Palmula simplex Cushman 1938, Contrib.
Cushman Lab. foramin. Res., 14, p. 36, pl. 6,
fig. 1 (non *Flabellina simplex* Reuss 1851).
Palmula primitiva (new name) Cushman 1939,
Contrib. Cushman Lab. foramin. Res., 15,
p. 91.

The Trinidad specimens differ from the type in
having usually slightly inflated chambers. The
chamber height is variable and sometimes
greater than that of the type specimen. Length
0.33–1.3, width 0.21–0.3, thickness 0.11–
0.14 mm.

Stratigraphic range: Maastrichtian (*Abathomphalus
mayaroensis* Zone) to Early Paleocene
(*Globorotalia trinidadensis* Zone). Rare.

PALMULA TOULMINI ten Dam & Sigal
Figure 31.12–13

Palmula toulmini ten Dam & Sigal 1950, p. 35,
pl. 2, figs. 18, 19.

A few specimens show about 6 fine and slightly
irregular vertical costae on the spiral part.
Maximum diameter 0.9–1.3, minimum diameter
0.75–1.2 mm, thickness 0.09–0.15 mm
(thickened spiral 0.24–0.3 mm).

Stratigraphic range: Early Paleocene (probably
Globorotalia pseudobulloides Zone; Rochard
facies). Rare.

ASTACOLUS EXQUISITUS (Toulmin)
Figure 31.14

Vaginulinopsis exquisita Toulmin 1941, p. 583,
pl. 79, figs. 4–6.

There is some variability in the length and thickness of the later uniserial chambers. Length 0.6–
1.6, width 0.34–0.75, thickness 0.16–0.27 mm.

Stratigraphic range: Early Eocene (*Globorotalia
subbotinae* Zone to *G. formosa formosa* Zone).
Rare. In Trinidad also found in the Eocene of
the overlying Navet Formation.

ASTACOLUS GRATUS (Reuss)
Figure 31.15

Cristellaria (Cristellaria) grata (Reuss) 1863,
p. 70, pl. 7, fig. 14.
Lenticulina (A.) grata (Reuss 1863),
Bartenstein, Bettenstaedt & Bolli 1966,
p. 148, pl. 2, figs. 130–133. – Bartenstein &
Bolli 1973, p. 401, pl. 4, figs. 1–12. –
Bartenstein & Bolli 1986, p. 963, pl. 4,
fig. 16.

Length 0.31–0.55, width 0.18–0.25, thickness
0.07–0.11 mm.

Stratigraphic range: Late Albian (*Favusella
washitensis* Zone) to Turonian (*Globotruncana
inornata* Zone). Rare. Also recorded in
Trinidad from the Late Aptian to Early Albian
Maridale Formation.

ASTACOLUS JARVISI (Cushman)
Figure **31.16–17, 26**

Marginulina jarvisi Cushman 1938, Contrib.
    Cushman Lab. foramin. Res., 14, p. 35, pl. 5,
    figs. 17, 18.
Marginulina grata (Reuss), Cushman & Jarvis
    1932, p. 25, pl. 7, fig. 7; pl. 8, fig. 3.
Marginulina jarvisi Cushman, Cushman &
    Renz 1946, p. 27, pl. 4, figs. 5, 6.

Length 0.6–2.0, width 0.32–0.63, thickness
0.24–0.36 mm.

Stratigraphic range: Early Paleocene (*Globorotalia
    trinidadensis* Zone) to Early Eocene
    (*Globorotalia aragonensis* Zone). Rather rare.
    In Trinidad also found in the Eocene of the
    overlying Navet Formation.

ASTACOLUS NUTTALLI (Todd & Kniker)
Figure **31.18–19**

Cristellaria sublituus Nuttall 1932, J.
    Paleontol., 6, p. 11, pl. 1, figs. 13, 14 (non
    Marginulina sublituus d'Orbigny 1826).
Marginulina nuttalli (new name) Todd &
    Kniker 1952, Cushman Lab. foramin. Res.,
    Spec. Publ., 1, p. 14, pl. 2, figs. 30, 31.

Length 0.57–1.2 or more (broken), width 0.14–
0.3, thickness 0.11–0.17 mm.

Stratigraphic range: Early Eocene (*Globorotalia
    formosa formosa* Zone to *G. aragonensis* Zone).
    Rare. In Trinidad also found in the Eocene of
    the overlying Navet Formation.

ASTACOLUS PARALLELUS (Reuss)
Figure **31.20–21**

Cristellaria parallela Reuss 1863, p. 67, pl. 7,
    figs. 1, 2.

Length 0.33–1.32, width 0.12–0.27, thickness
0.09–0.22.

Stratigraphic range: Late Albian (*Favusella
    washitensis* Zone) to Santonian (*Globotruncana
    fornicata* Zone). Rather rare.

ASTACOLUS PEROBLIQUUS (Reuss)
Figure **31.22–23**

Cristellaria (Cristellaria) perobliqua Reuss
    1863, p. 67, pl. 7, fig. 3.
Lenticulina (A.) perobliqua (Reuss 1863),
    Bartenstein & Bolli 1986, p. 964, pl. 4, figs.
    27–29.

Length 0.25–0.74, width 0.12–0.25, thickness
0.07–0.11 mm.

Stratigraphic range: Late Albian (*Favusella
    washitensis* Zone) to Coniacian-Santonian
    (*Globotruncana concavata* Zone). Rather rare.
    Also recorded in Trinidad from the Late
    Aptian to Early Albian Maridale Formation.

ASTACOLUS PLANIUSCULUS (Reuss)
Figure **31.24–25**

Cristellaria (Cristellaria) planiuscula Reuss
    1863, p. 71, pl. 7, fig. 15.

Length 0.25–0.46, width 0.15–0.26, thickness
0.08–0.1 mm.

Stratigraphic range: Albian (*Rotalipora ticinensis
    ticinensis* Zone to *Favusella washitensis* Zone).
    Rare.

ASTACOLUS cf. PSEUDOMARCKI
(Cushman)
Figure **31.27**

Marginulina pseudomarcki Cushman 1937,
    Contrib. Cushman Lab. foramin. Res., 13,
    p. 94, pl. 13, figs. 19, 20.

The umbilical filling is rather less distinct and
the chambers are sometimes higher than in the
type; however, these differences may reflect the
considerable range of variation of this species.
Length 0.65–5.0, breadth 1.5–3.8, thickness
0.6–1.8 mm.

Stratigraphic range: Santonian (*Globotruncana
    fornicata* Zone) to Early Maastrichtian
    (*Globotruncana lapparenti tricarinata* Zone).
    Very rare.

ASTACOLUS RECTUS (d'Orbigny)
Figure **31.28–29**

Cristellaria recta d'Orbigny 1840, p. 28, pl. 2,
    figs. 23–25.

The test is larger and rather less compressed than in *A. perobliquus*. Length 0.42–1.8, width 0.2–0.52, thickness 0.11–0.24 mm.

Stratigraphic range: Early Cenomanian (*Rotalipora appenninica appenninica* Zone) to Early Eocene (*Globorotalia formosa formosa* Zone). Rather rare. In Trinidad also found in the Eocene of the overlying Navet Formation.

ASTACOLUS sp. A
Figure 31.30–31

The test is usually slightly curved, and the cross section is oval with subrounded peripheral edges. The morphology is intermediate between that of *A. parallelus* (Reuss; *Cristellaria p.*) and *A. scitulus* (Berthelin; *Cristellaria s.*). Length 0.42–0.9, width 0.18–0.3, thickness 0.13–0.21 mm.

Stratigraphic range: Early Cenomanian (*Rotalipora appenninica appenninica* Zone). Rather rare.

HEMIROBULINA ABBREVIATA
(Neugeboren)
Figure 31.32–33

Marginulina abbreviata Neugeboren 1851, Verhandl. Mitt. Siebenbürg. Ver. Naturwiss. Hermannstadt, 2, p. 129, pl. 5, fig. 4.
Marginulina bullata Reuss, Cushman & Jarvis 1932, p. 25, pl. 8, figs. 7, 8.
Marginulina cf. *bullata* Reuss, Cushman & Renz 1946, p. 27, pl. 3, fig. 32; pl. 4, fig. 3.

Length 0.3–0.75, thickness 0.22–0.6 mm.

Stratigraphic range: Albian (*Rotalipora ticinensis ticinensis* Zone) to Early Eocene (*Globorotalia formosa formosa* Zone). Fairly common. In Trinidad also found in the Eocene of the overlying Navet Formation.

HEMIROBULINA cf. ANCEPS
(Neugeboren)
Figure 32.1

Marginulina anceps Neugeboren 1851, Verhandl. Mitt. Siebenbürg. Ver. Naturwiss. Hermannstadt, 2, p. 125, pl. 4, fig. 10.

The type specimen is rather more slender but has the same chamber arrangement. Length 0.35–1.15, thickness 0.12–0.2 mm.

Stratigraphic range: Early Eocene (*Globorotalia formosa formosa* Zone to *G. aragonensis* Zone). Rather rare. In Trinidad also found in the Eocene of the overlying Navet Formation.

HEMIROBULINA BULLATA (Reuss)
Figure 32.2

Marginulina bullata Reuss 1845, p. 29, pl. 13, figs. 34–38.
Marginulina bullata Reuss 1845, Bartenstein & Bolli 1986, p. 967, pl. 5, figs. 17, 18.
Marginulina bullata Reuss, Cushman & Jarvis 1928, p. 96, pl. 14, figs. 7, 8.

Length 0.28–0.8, thickness 0.2–0.48 mm.

Stratigraphic range: Turonian (*Globotruncana inornata* Zone) to Early Eocene (*Globorotalia formosa formosa* Zone). Fairly common. Also recorded in Trinidad from the Late Aptian to Early Albian Maridale Formation. Found also in the Eocene Navet Formation.

HEMIROBULINA CURVATURA
(Cushman)
Figure 32.3–4

Marginulina curvatura Cushman 1938, Contrib. Cushman Lab. foramin. Res., 14, p. 34, pl. 5, figs. 13, 14.

Some less typical specimens (Fig.32.4) have have a more elongated final chamber. Length 0.33–0.7, thickness 0.14–0.26 mm.

Stratigraphic range: Albian (*Rotalipora ticinensis ticinensis* Zone) to Late Paleocene (*Globorotalia pseudomenardii* Zone). Rare.

HEMIROBULINA HAMULOIDES
(Brotzen)
Figure 32.5

Marginulina hamuloides Brotzen 1936, p. 68, pl. 4, figs. 10, 11.

The specimens are often slightly more elongated than the types and have 2 to 3 uniserial chambers. Length 0.4–0.95, diameter 0.16–0.4 mm.

Stratigraphic range: Early Maastrichtian (*Globotruncana lapparenti tricarinata* Zone) to Early Eocene (*Globorotalia formosa formosa* Zone). Rather rare. In Trinidad also found in the Eocene of the overlying Navet Formation.

HEMIROBULINA HANTKENI (Bandy)
Figure 32.6

Marginulina subbullata Hantken 1875, Magyar. kir. földt. int. Evkönyve, 4, p. 39, pl. 4, figs. 9, 10 (non *Marginulina subbullata* Guembel 1861)
Marginulina hankteni Bandy (new name) 1949, Bull. Am. Paleontol. 32/131, p. 46, pl. 6, fig. 9.

The test consists of a small spiral followed by 2–3 subglobular chambers. Length 0.53–1.2, thickness 0.25–0.52 mm.

Stratigraphic range: Late Albian (*Favusella washitensis* Zone) to Campanian (*Globotruncana stuarti* Zone). Rather rare.

HEMIROBULINA REUSSI (Takayanagi)
Figure 32.7

Marginulina inaequalis Reuss 1860, p. 207, pl. 7, fig. 3 (non *Marginulina inaequalis* Costa 1855).
Marginulina reussi Takayanagi (new name) 1960, p. 89, pl. 4, figs. 11, 12.

Length 0.35–0.67, diameter 0.12–0.2 mm.

Stratigraphic range: Turonian (*Globotruncana inornata* Zone) to Late Maastrichtian (*Abathomphalus mayaroensis* Zone). Rare.

HEMIROBULINA cf. SCHLOENBACHI
(Reuss)
Figure 32.9–10

Cristellaria schloenbachi Reuss 1863, p. 65, pl. 6, figs. 14, 15.
? *Marginulina* cf. *texasensis* Cushman, Cushman & Renz 1946, p. 27, pl. 4, figs. 8, 9.

In Trinidad, the specimens assigned here to this group may become slightly thicker than the types and sometimes have a less tapering early stage with a trace of a *Marginulinopsis*-like coil. Length 0.36–0.74, width 0.15–0.24, thickness 0.14–0.22 mm.

Stratigraphic range: Late Albian (*Favusella washitensis* Zone) to Turonian (*Globotruncana inornata* Zone). Rare.

HEMIROBULINA TRUNCATULA
(Berthelin)
Figure 32.11

Cristellaria truncatula Berthelin 1880, p. 53, pl. 3, figs. 26, 27.

Length 0.3–0.41, diameter 0.16–0.2 mm.

Stratigraphic range: Turonian (*Globotruncana inornata* Zone) to Santonian (*Globotruncana fornicata* Zone) Rare.

VAGINULINOPSIS HEMICYLINDRICA
(Noth)
Figure 32.8

Lenticulina hemicylindrica Noth 1951, Jahrbuch geol. Bundesanstalt, Vienna, 1951, Sonderbd. 3, p. 45, pl. 8, fig. 2.

Length 0.69–1.29, diameter 0.18–0.26 mm.

Stratigraphic range: Late Paleocene (*Globorotalia pseudomenardii* Zone) to Early Eocene (*Globorotalia subbotinae* Zone). Rare.

VAGINULINOPSIS cf. SILICULA
(Plummer)
Figure 32.12

Hemicristellaria silicula Plummer 1931, p. 148, pl. 10, figs. 8, 9.

In Trinidad, the tests are practically straight and more symmetrical than the types in side view; horizontal ridges are only rarely present. Length up to 1.8 mm (fragments only), width 0.45–0.65, thickness 0.3–0.46 mm.

Stratigraphic range: Campanian (*Globotruncana stuarti* Zone) to Early Maastrichtian (*Globotruncana lapparenti tricarinata* Zone). Rare.

CITHARINA ANGUSTISSIMA (Reuss)
Figure **32.13**

Vaginulina angustissima Reuss 1863, p. 45, pl. 3, fig. 3.

Length 0.8–1.8 mm or more (broken), width 0.21–0.5, thickness 0.09–0.24.

Stratigraphic range: Santonian (*Globotruncana fornicata* Zone) to Early Maastrichtian (*Globotruncana lapparenti tricarinata* Zone). Rare.

CITHARINA BARCOENSIS (Cushman & Hedberg)
Figure **32.14–15**

Vaginulina barcoensis Cushman & Hedberg 1941, p. 90, pl. 22, fig. 2.

Length 0.45–0.7, width 0.16–0.3, thickness 0.03–0.05 mm.

Stratigraphic range: Campanian (*Globotruncana stuarti* Zone) to Early Maastrichtian (*Globotruncana lapparenti tricarinata* Zone). Rare.

PLANULARIA ADVENA Cushman & Jarvis
Figure **32.17–18**

Planularia advena Cushman & Jarvis 1932, p. 24, pl. 8, figs. 1, 2.
Planularia advena Cushman & Jarvis, Cushman & Renz 1946, p. 26, pl. 3, fig. 29.

On some specimens the raised sutures and the inflated umbilical area are distinct but others are practically smooth and flat. Diameter 0.8–1.45 mm or more (incomplete), thickness 0.24–0.35 mm.

Stratigraphic range: Early Paleocene (probably *Globorotalia pseudobulloides* Zone; Rochard facies). Rare.

PLANULARIA COMPLANATA (Reuss)
Figure **32.16**

Cristellaria complanata Reuss 1845, p. 33, pl. 13, fig. 54.
Lenticulina (P.) complanata (Reuss 1845), Bartenstein & Bolli 1986, p. 962, pl. 4, fig. 13.

Length 0.2–0.75 mm or more (incomplete), width 0.15–0.37, thickness 0.07–0.11 mm.

Stratigraphic range: Late Albian (*Favusella washitensis* Zone) to Santonian (*Globotruncana fornicata* Zone). Common. Also recorded in Trinidad from the Late Aptian to Early Albian Maridale Formation.

PSILOCITHARELLA RECTA (Reuss)
Figure **32.19**

Vaginulina recta Reuss 1863, p. 48, pl. 3, figs. 14, 15.
Vaginulina recta Reuss 1863, Bartenstein, Bettenstaedt & Bolli 1966, p. 155, pl. 3, figs. 250–253. – Bartenstein & Bolli 1986, p. 971, pl. 6, figs. 7, 8.

The figured specimen is deformed near the base but otherwise clearly shows the distinctive characters of the test and of the sutures. Length 0.45–0.65 mm or more (broken), width 0.12–0.18, thickness 0.08–0.11 mm.

Stratigraphic range: Late Albian to Early Cenomanian (*Favusella washitensis* Zone to *Rotalipora appenninica appenninica* Zone). Very rare. Also recorded in Trinidad from the Barremian to Early Aptian Cuche Formation and from the Late Aptian to Early Albian Maridale Formation.

VAGINULINA ANITANA Akpati
Figure **32.20–21**

Vaginulina anitana Akpati 1966, Contrib. Cushman Found. foramin. Res, 17, p. 138, pl. 10, figs. 10, 11.
Marginulina trilobata d'Orbigny, Cushman & Jarvis 1932, p. 28, pl. 9, figs. 3, 4.
Marginulina? trilobata d'Orbigny, Cushman & Renz 1946, p. 27, pl. 4, fig. 4.

Length 0.5–1.25 mm or more (incomplete), width 0.24–0.5, thickness 0.2–0.32 mm.

Stratigraphic range: Late Maastrichtian (*Abathomphalus mayaroensis* Zone) to Late Paleocene (*Globorotalia velascoensis* Zone). Rare.

VAGINULINA PLUMMERAE (Cushman)
Figure **32.22–23**

Marginulina plummerae Cushman 1937,
    Contrib. Cushman Lab. foramin. Res., 13,
    p. 97, pl. 13, figs. 21–23.

Length 0.55–1.4 mm or more (broken), width
0.16–0.3, thickness 0.15–0.2 mm.

Stratigraphic range: Early Paleocene (probably
Globorotalia pseudobulloides Zone; Rochard
facies). Rare.

Family Lagenidae

LAGENA AMPHORA PAUCICOSTATA
Franke
Figure **32.24**

Lagena amphora Reuss forma *paucicostata*
    Franke 1928, p. 87, pl. 7, fig. 38.

Occasionally this form may have a short basal
spine. Length 0.33–0.48, thickness 0.12–
0.2 mm.

Stratigraphic range: Campanian (*Globotruncana
stuarti* Zone) to Early Eocene (*Globorotalia
formosa formosa* Zone). Very rare. In Trinidad
also found in the Eocene of the overlying Navet
Formation.

LAGENA cf. AMPHORA Reuss
Figure **32.25**

Lagena amphora Reuss 1863, Sitzungsber. K.
    Akad. Wiss. Wien, math.-naturwiss. Cl., 46/
    1 (1862), p. 330, pl. 4, fig. 57.

The apertural area is not typical and shows some
modifications, which in certain specimens
resemble those of *L. artificiosa* Buchner. Length
0.24–0.5, thickness 0.1–0.24.

Stratigraphic range: Campanian (*Globotruncana
stuarti* Zone) to Late Maastrichtian
(*Abathomphalus mayaroensis* Zone). Very
rare.

LAGENA ASPERA APICULATA White
Figure **32.26**

Lagena aspera var. *apiculata* White 1928, J.
    Paleontol., 2, p. 208, pl. 29, fig. 5.

Length 0.45–0.71, diameter 0.3–0.52 mm.

Stratigraphic range: Late Maastrichtian
(*Abathomphalus mayaroensis* Zone) to Late
Paleocene (*Globorotalia pseudomenardii*
Zone). Rare. In Trinidad also found in the
Eocene of the overlying Navet Formation.

LAGENA GRACILICOSTA Reuss
Figure **32.27**

Lagena gracilicosta Reuss 1863, Sitzungsber. K.
    Akad. Wiss. Wien, math.-naturwiss. Cl.,
    46/1 (1862), p. 327, pl. 3, figs. 42–43.

Length 0.22–0.3, thickness 0.12–0.18 mm.

Stratigraphic range: Coniacian-Santonian
(*Globotruncana concavata* Zone); Late
Paleocene (*Globorotalia velascoensis* Zone) to
Early Eocene (*Globorotalia formosa formosa*
Zone). Rare.

LAGENA ISABELLA CONSCRIPTA
Cushman & Barksdale
Figure **32.28–29**

Lagena isabella (d'Orbigny) var. *conscripta*
    Cushman & Barksdale 1930, Contrib.
    Stanford Univ., Dept. Geol., 1, p. 66, pl. 12,
    fig. 4.

This appears to be a fairly compact group,
though it shows a considerable variability in the
shape of the test and in details of the apertural
area (bifurcation of septa, thickened rim).
Length 0.2–0.41, thickness 0.12–0.25 mm.

Stratigraphic range: Campanian (*Globotruncana
stuarti* Zone) to Late Maastrichtian
(*Abathomphalus mayaroensis* Zone). Rare

LAGENA RETICULOSTRIATA Haque
Figure **32.30**

Lagena reticulostriata Haque 1956, p. 101, pl. 8,
    fig. 8; pl. 23, figs. 4, 5.

Length 0.36–0.52, thickness 0.12–0.21 mm.

Stratigraphic range: Maastrichtian (*Globotruncana lapparenti tricarinata* Zone to *Abathomphalus mayaroensis* Zone). Very rare.

LAGENA TENUISTRIATA Stache
Figure **32.31–32**

Lagena tenuistriata Stache 1865, Novara Exped. 1857–1859 (Vienna), geol. Theil, 1/2, p. 184, pl. 22, fig. 4.

Length 0.22–0.65, diameter 0.18–0.4 mm.

Stratigraphic range: Late Paleocene (*Globorotalia pseudomenardii* Zone) to Early Eocene (*Globorotalia edgari* Zone). Rare. In Trinidad also found in the Eocene of the overlying Navet Formation.

LAGENA VIRGINIANA (Nogan)
Figure **32.33**

Oolina virginiana Nogan 1964, Cushman Found. foramin. Res., Spec. Publ., 7, p. 27, pl. 1, fig. 29.

The shape of the test varies from almost spherical to thick ovoid, and a small basal spine is often present. Length 0.18–0.45, diameter 0.13–0.37 mm.

Stratigraphic range: Santonian (*Globotruncana fornicata* Zone) to Early Eocene (*Globorotalia formosa formosa* Zone). Rather rare.

REUSSOOLINA APICULATA APICULATA (Reuss)
Figure **32.34**

Oolina apiculata Reuss 1851, p. 22, pl. 2, fig. 1.
Lagena apiculata (Reuss 1851), Bartenstein & Bolli 1986, p. 960, pl. 3, figs. 44, 45.
Lagena apiculata (Reuss), Cushman & Renz 1946, p. 34, pl. 5, fig. 18

Length 0.25–1.05, thickness 0.2–0.76 mm.

Stratigraphic range: Late Maastrichtian (*Abathomphalus mayaroensis* Zone) to Late Paleocene (*Globorotalia velascoensis* Zone). Fairly common. In Trinidad also recorded from the Late Aptian to Early Albian Maridale Formation. Found also in the Eocene Navet Formation.

REUSSOOLINA APICULATA ELLIPTICA (Reuss)
Figure **32.35–36**

Lagena apiculata var. *elliptica* Reuss 1863, Sitzungsber. K. Akad. Wiss. Wien, math.-naturwiss. Cl., 46/1 (1862), p. 35, pl. 2, fig. 2.

The shape varies between that of the varieties *ovoidea* Marie and *phialaeformis* Crespin. Length 0.25–0.72, thickness 0.12–0.4 mm.

Stratigraphic range: Turonian (*Globotruncana inornata* Zone) to Late Paleocene (*Globorotalia velascoensis* Zone). Fairly common.

Family Polymorphinidae

GLOBULINA AMPULLA (Jones)
Figure **33.1–2**

Polymorphina ampulla Jones 1852, Quart. J. geol. Soc. London, 8, p. 267, pl. 16, fig. 14.
? *Pyrulina cylindroides* (Roemer), Cushman & Renz 1946, p. 35, pl. 5, fig. 24.

This group consists of ovoid to short fusiform specimens with a broadly rounded base and a subacute top. The external shape is hardly different from that of *G. obtusa* Reuss (1863; *Polymorphina o.*, non d'Orbigny 1850). Length 0.26–0.54, thickness 0.17–0.35 mm.

Stratigraphic range: Turonian (*Globotruncana inornata* Zone) to Late Paleocene (*Globorotalia pseudomenardii* Zone). Rather rare.

GLOBULINA BUCCULENTA (Berthelin)
Figure **33.3**

Polymorphina bucculenta Berthelin 1880, p. 58, pl. 4, figs. 16, 17.

Length 0.3–0.54, thickness 0.15–0.32 mm.

Stratigraphic range: Late Albian to Early Cenomanian (*Favusella washitensis* Zone to *Rotalipora appenninica appenninica* Zone). Rather rare.

GLOBULINA FUSIFORMIS (Roemer)
Figure **33.4**

Polymorphina (Globulina) fusiformis Roemer 1838, Neues Jahrb. Min. Geogn. Geol. Petref.-Kunde, Stuttgart, p. 386, pl. 3, fig. 37.

This group includes fusiform-ovoid forms with an acute apex. The base is less distinctly rounded than in *G. ampulla*, and the cross section may be slightly triangular. Length 0.35–0.97, thickness 0.23–0.45 mm.

Stratigraphic range: Campanian (*Globotruncana stuarti* Zone) to Early Eocene (*Globorotalia formosa formosa* Zone). Rather rare. In Trinidad also found in the Eocene of the overlying Navet Formation.

GLOBULINA LACRIMA HORRIDA
Reuss
Figure **33.5**

Globulina horrida Reuss 1846, p. 110, pl. 43, fig. 14.

Length 0.22–0.5, thickness 0.15–0.46 mm.

Stratigraphic range: Late Albian (*Favusella washitensis* Zone) to Middle Paleocene (*Globorotalia pusilla pusilla* Zone). Fairly common.

GLOBULINA LACRIMA LACRIMA
(Reuss)
Figure **33.6**

Polymorphina (Globulina) lacrima Reuss 1845, p. 40, pl. 12, fig. 6; pl. 13, fig. 83.
Globulina lacrima Reuss, Cushman & Renz 1946, p. 34, pl. 5, fig. 22 (not fig. 21 as indicated in text).

Length 0.22–0.65, thickness 0.2–0.55 mm.

Stratigraphic range: Albian (*Rotalipora ticinensis ticinensis* Zone) to Late Paleocene (*Globorotalia velascoensis* Zone). Common.

GLOBULINA LACRIMA SUBSPHAERICA (Berthelin)
Figure **33.7**

Polymorphina subsphaerica Berthelin 1880, p. 58, pl. 4, fig. 18.

Globulina lacrima Reuss var. *subsphaerica* (Berthelin), Cushman & Renz 1946, p. 34, pl. 5, fig. 21 (not fig. 22 as indicated in text).

Length 0.18–0.42, thickness 0.16–0.39 mm.

Stratigraphic range: Late Albian (*Favusella washitensis* Zone) to Middle Paleocene (*Globorotalia uncinata* Zone). Fairly common.

GLOBULINA cf. PRISCA (Reuss)
Figure **33.8–9**

Polymorphina (Globulina) prisca Reuss 1863, p. 79, pl. 9, fig. 8.

Most specimens in Trinidad have a fistulose end chamber; this feature is rarely described from *G. prisca* in other localities but occurs commonly in Alaska (Macbeth & Schmidt, 1973). The wall is often very finely hispid. Length 0.41–0.63, thickness 0.22–0.38 mm.

Stratigraphic range: Early Maastrichtian (*Globotruncana lapparenti tricarinata* Zone) to Early Eocene (*Globorotalia formosa formosa* Zone). Very rare.

GUTTULINA TRIGONULA (Reuss)
Figure **33.10–12**

Polymorphina (Guttulina) trigonula Reuss 1845, p. 40, pl. 13, fig. 84.
Guttulina trigonula (Reuss), Cushman & Jarvis 1946, p. 34, pl. 5, fig. 20.
? *Guttulina adhaerens* (Olszewski), Cushman & Renz 1946, p. 34, pl. 5, fig. 19.

In the early Tertiary samples, the typical form is accompanied by larger specimens which seem to add one more inflated chamber; this gives the test a more lobate outline (Fig. 33.12). Length 0.25–0.64 (lobate form 0.44–0.9) mm, thickness 0.22–0.5 (lobate form 0.4–0.75) mm.

Stratigraphic range: Campanian (*Globotruncana stuarti* Zone) to Early Eocene (*Globorotalia subbotinae* Zone). Fairly common.

PALAEOPOLYMORPHINA cf. PLEUROSTOMELLOIDES (Franke)
Figure **33.13**

Polymorphina pleurostomelloides Franke 1928, p. 121, pl. 11, fig. 11.

The specimens are rather small. In the early stages, the test is slightly compressed and the chambers are more embracing and less inflated than in the type. The final chambers are more typically inflated and pseudouniserial. Length 0.28–0.56, thickness 0.12–0.17 mm.

Stratigraphic range: Maastrichtian (*Globotruncana lapparenti tricarinata* Zone to *Abathomphalus mayaroensis* Zone). Very rare.

POLYMORPHINA BERICA (Cati)
Figure **33.14–15**

Bolivina berica Cati 1964, Giorn. Geol., Bologna, ser. 2, 32, p. 237, pl. 38, fig. 20.

As in Cati's type, the aperture is a small round terminal opening, whose irregular edges may suggest that is was originally radiate. Length 0.6–0.79, breadth 0.32–0.43, thickness 0.15–0.2 mm.

Stratigraphic range: Late Maastrichtian (*Abathomphalus mayaroensis* Zone). Very rare.

POLYMORPHINELLA (?) cf. CYLINDRACEA (Reuss)
Figure **33.16–17**

Nodosaria (Glandulina) cylindracea Reuss 1845, p. 25, pl. 13, figs. 1, 2.

The irregularly biserial-uniserial chamber arrangement observed in the available specimens may in principle fit the genus definition of *Polymorphinella*. However, the identity with the form known in the literature as '*Pseudoglandulina*' *cylindracea* and with the types of Reuss is uncertain. Length 0.65–1.95, diameter 0.2–0.56 mm.

Stratigraphic range: Late Maastrichtian (*Abathomphalus mayaroensis* Zone) to Early Eocene (*Globorotalia formosa formosa* Zone). Rare.

PSEUDOPOLYMORPHINA RUDIS (Reuss)
Figure **33.18–20**

Polymorphina rudis Reuss 1862, Paläontol. Beitr., Sitzungsber. K. Akad. Wiss. Wien, math.-naturwiss. Cl., 44/1 (1861), p. 319, pl. 3, figs. 5–8.

The test shape varies from thick fusiform to sub-cylindrical. Specimens with parallel sides can be similar to *P. digitata* (d'Orbigny) as emended by Cushman & Ozawa (1930) but are usually thicker. Length 0.6–2.9, thickness 0.26–1.15 mm.

Stratigraphic range: Campanian (*Globotruncana stuarti* Zone) to Late Paleocene (*Globorotalia velascoensis* Zone). Rare.

PYRULINA CYLINDROIDES (Roemer)
Figure **33.21–22**

Polymorphina cylindroides Roemer 1838, Neues Jahrb. Min. Geogn. Geol. Petref.-Kunde, Stuttgart, p. 385, pl. 3, fig. 26.
Pyrulina cylindroides (Roemer 1838), Bartenstein, Bettenstaedt & Bolli 1966, p. 158, pl. 3, figs. 299–302. – Bartenstein & Bolli 1973, p. 407, pl. 6, figs. 13–17. – Bartenstein & Bolli 1986, p. 973, pl. 3, figs. 24, 25.
Pyrulina cylindroides (Roemer), Cushman & Renz 1948, p. 22, pl. 5, fig. 1.
? *Globulina prisca* Reuss, Cushman & Renz 1946, p. 34, pl. 5, fig. 23.

As for other common and well-known representatives of the Polymorphinidae described in the nineteenth century, the definition and classification of Cushman & Ozawa (1930) is here followed. In Trinidad, the species includes a few specimens with a fistulose aperture. In the older samples (up to the *Globotruncana concavata* Zone) shorter tests occur which may resemble *Globulina prisca* (Reuss). Length 0.3–1.35, thickness 0.16–0.45 mm.

Stratigraphic range: Early Cenomanian (*Rotalipora appenninica appenninica* Zone) to Early Eocene (*Globorotalia aragonensis* Zone). Fairly common. Also recorded in Trinidad from the Late Aptian to Early Albian Maridale Formation. Found also in the Eocene Navet Formation.

PYRULINA VELASCOENSIS (Cushman)
Figure **33.23, 29–30**

Polymorphina velascoensis Cushman 1926, p. 604, pl. 20, fig. 16.
Pyrulina velascoensis (Cushman), Cushman & Renz 1946, p. 35, pl. 5, fig. 25.

This group differs from *P. cylindroides* in the fusiform test which may be slightly compressed and sometimes slightly asymmetric. Length 0.25–1.45, thickness 0.12–0.35 mm.

Stratigraphic range: Early Cenomanian (*Rotalipora appenninica appenninica* Zone) to Late Maastrichtian (*Abathomphalus mayaroensis* Zone). Rather rare.

PYRULINOIDES cf. KALININI Dain
Figure **33.24–25**

Pyrulinoides kalinini Dain 1958, Microfauna SSSR, Trudy VNIGRI, Leningrad, n. ser., 115/9, p. 42, pl. 7, fig. 4.

On the average, the test is slightly more compressed and the final chamber is slightly more embracing than in the type specimen. Length 0.55–0.9, breadth 0.2–0.42, thickness 0.2–0.27 mm.

Stratigraphic range: Early Eocene (*Globorotalia subbotinae* Zone to *G. aragonensis* Zone). Very rare. In Trinidad found also in the Eocene of the overlying Navet Formation.

PYRULINOIDES OBESUS Marie
Figure **33.26–27**

Pyrulinoides obesa Marie 1941, p. 172, pl. 26, fig. 253.

The degree of compression of the test is rather more variable than in Marie's types. Length 0.33–0.75, breadth 0.22–0.4, thickness 0.16–0.37 mm.

Stratigraphic range: Campanian (*Globotruncana stuarti* Zone) to Late Paleocene (*Globorotalia pseudomenardii* Zone). Rare.

RAMULINA cf. ABSCISSA Loeblich & Tappan
Figure **33.28**

Ramulina abscissa Loeblich & Tappan 1946, J. Paleontol., 20, p. 254, pl. 7, figs. 7, 8.

The fragments of branches are often thicker than those of the types. Length up to 0.75, diameter of branches 0.08–0.3 mm.

Stratigraphic range: Coniacian-Santonian (*Globotruncana concavata* Zone). Rare.

RAMULINA ACULEATA (d'Orbigny)
Figure **33.31**

Nodosaria (Dentalina) aculeata d'Orbigny 1840, p. 13, pl. 1, figs. 2, 3.
Ramulina aculeata (d'Orbigny), Cushman & Renz 1946, p. 35, pl. 5, fig. 28.
Ramulina sp., Cushman & Jarvis 1932, p. 41, pl. 12, fig. 10 only.

The shape of the fragments varies from almost cylindrical to distinctly inflated, but intermediate specimens resemble the lectotype designated by Loeblich & Tappan (1964) very closely. The wall always shows relatively coarse pustules. Length 0.8–2.5, thickness 0.35–1.0 mm.

Stratigraphic range: Late Maastrichtian (*Abathomphalus mayaroensis* Zone) to Middle Paleocene (*Globorotalia angulata* Zone). Rare.

RAMULINA MURICATINA Loeblich & Tappan
Figure **33.32**

Ramulina aculeata Wright, Tappan 1943, J. Paleontol. 17, p. 506, pl. 81, fig. 7 (designated holotype).
Ramulina muricatina Loeblich & Tappan 1949, J. Paleontol., 23, p. 261, pl. 50, figs. 5, 6.

In Trinidad, this group is represented by rather delicate tubular fragments with a finely hispid surface. Length up to 1.35, thickness 0.15–0.55 mm.

Stratigraphic range: Coniacian-Santonian (*Globotruncana concavata* Zone to *G. fornicata* Zone) Rather rare.

RAMULINA NAVARROANA Cushman
Figure **33.33**

Ramulina navarroana Cushman 1938, Contrib. Cushman Lab. foramin. Res., 14, p. 43, pl. 7, figs. 10, 11.

Total size of fragments 0.55–1.4, body only 0.35–0.6 mm

Stratigraphic range: Late Maastrichtian (*Abathomphalus mayaroensis* Zone) to Middle Paleocene (*Globorotalia pusilla pusilla* Zone). Rare.

RAMULINA ORNATA Cushman
Figure **33.34**

Ramulina ornata Cushman 1938, Contrib.
   Cushman Lab. foramin. Res., 14, p. 44, pl. 7,
   fig. 15.
Ramulina ornata Cushman, Cushman & Renz
   1946, p. 35, pl. 5, figs. 29, 30.
Ramulina sp., Cushman & Jarvis 1932, p. 41,
   pl. 12, fig. 11 only.

Size 0.4–0.75, body 0.25–0.52 mm.

Stratigraphic range: Late Paleocene (*Globorotalia pseudomenardii* Zone) to Early Eocene (*Globorotalia formosa formosa* Zone). Rather rare. In Trinidad also found in the Eocene of the overlying Navet Formation.

RAMULINA PSEUDOACULEATA
(Olsson)
Figure **33.35–36**

Dentalina pseudoaculeata Olsson 1960, p. 14,
   pl. 3, figs. 1, 2 (non *R. pseudoaculeata*
   Barnard 1972).

Length 0.35–1.7, diameter 0.23–0.95 mm.

Stratigraphic range: Probably Early Maastrichtian (*Globotruncana lapparenti tricarinata* Zone; Marabella facies), Late Maastrichtian (*Abathomphalus mayaroensis* Zone) to Late Paleocene (*Globorotalia velascoensis* Zone). Fairly common.

Family Ellipsolagenidae

BUCHNERINA CASTRENSIS (Schwager)
Figure **34.1**

Lagena castrensis Schwager 1866, Novara
   Exped. 1857–1859, Vienna, geol. Theil, 2/
   2, p. 208, pl. 5, fig. 22.

Length 0.35–0.53, width 0.27–0.38, thickness 0.18–0.26 mm.

Stratigraphic range: Late Paleocene (*Globorotalia pseudomenardii* Zone); probably also Early Eocene (*Globorotalia formosa formosa* Zone) Rare.

BUCHNERINA cf. OSLATUS (Shifflett)
Figure **34.2**

Entosolenia oslatus Shifflett 1948, Bull.
   Maryland Dept. Geol. Mines Water
   Resources, 3, p. 62, pl. 3, fig. 14.

Two thin keels meet below the aperture, which is rather more pointed than in Shifflett's types. Length 0.25–0.44, width 0.16–0.33, thickness 0.15–0.31 mm.

Stratigraphic range: Early Maastrichtian (*Globotruncana lapparenti tricarinata* Zone); Early Eocene (*Globorotalia subbotinae* Zone). Very rare.

BUCHNERINA SEGUENZIANA
(Fornasini)
Figure **34.3**

Fissurina marginata Seguenza 1862, Foram.
   monothal. Messina, p. 66, pl. 2, figs. 27, 28
   (non *Vermiculum marginatum* Montagu
   1803).
Lagena seguenziana (new name) Fornasini
   1886, Boll. Soc. geol. Ital., 5, p.351, pl. 8,
   figs. 1–8.

The thickened periphery consists of one to several (often three) keels. A basal spine is seen rarely. Length 0.3–0.44, width 0.21–0.34, thickness 0.13–0.24 mm.

Stratigraphic range: Paleocene (*Globorotalia pusilla pusilla* Zone to *G. velascoensis* Zone). Very rare.

EXSCULPTINA JACOBI (Marie)
Figure **34.4**

Lagena jacobi Marie 1941, p. 78, pl. 9, figs. 87,
   88.

Length 0.2–0.28, diameter 0.15–0.24 mm.

Stratigraphic range: Maastrichtian (*Globotruncana lapparenti tricarinata* Zone to *Abathomphalus mayaroensis* Zone). Very rare.

FAVULINA cf. GEOMETRICA (Reuss)
Figure **34.5**

Lagena geometrica Reuss 1863, Sitzungsber. K.
   Akad. Wiss. Wien, math.- naturwiss. Cl.,
   46/1 (1862), p. 334, pl. 5, fig. 74.

The shape is more spherical than in the type but the surface ornamentation is practically the same. Length 0.3–0.32, thickness 0.26–0.29 mm.

Stratigraphic range: Late Paleocene (*Globorotalia pseudomenardii* Zone). Very rare.

LAGNEA PERFECTOCOSTATA (Cushman & Renz)
Figure **34.**6

Entosolenia orbignyana Seguenza var. *perfecto-costata* Cushman & Renz, 1947*a*, p. 46, pl. 12, fig. 5.

The ornamentation of the peripheral keel is finely radial, rather than vertical as illustrated in the original drawing. Length 0.45–0.65, width 0.32–0.5 mm, thickness 0.17–0.27 mm.

Stratigraphic range: Maastrichtian (*Globotruncana lapparenti tricarinata* Zone to *Abathomphalus mayaroensis* Zone). Rare.

OOLINA GLOBOSA (Montagu)
Figure **34.**7–8

Serpula (Lagena) laevis globosa Walker & Boys 1784, Minute Shells, 3, pl. 1, fig. 8 (non-Linnean).
Vermiculum globosum Montagu 1803, Testacea Britannica. Romsey, p. 523.
Lagena globosa (Montagu 1803), Bartenstein & Bolli 1986, p. 960, pl. 3, figs. 46–48.

The earlier specimens (up to the *Globotruncana fornicata* Zone) are relatively small, but higher up there is a gradual size increase, particularly in the latest Cretaceous, where the maximum size is reached in the *Abathomphalus mayaroensis* Zone. In the Lower Lizard Springs Formation, some specimens have a more elongated shape, resembling those referred to *Oolina simplex* var. *lacrima* by White (1928). Length 0.34–1.3, thickness 0.2–1.1 mm.

Stratigraphic range: Turonian (*Globotruncana inornata* Zone) to Late Paleocene (*Globorotalia pseudomenardii* Zone). Rather rare. Also recorded in Trinidad from the Late Aptian to Early Albian Maridale Formation.

FISSURINA LAEVIGATA Reuss
Figure **34.**9–10

Fissurina laevigata Reuss 1850, Denkschr. K. Akad. Wiss. Wien, math.-naturwiss. Cl., 1, p. 366, pl. 46, fig. 1.
Entosolenia sp. C, Cushman & Renz 1946, p. 39, pl. 6, fig. 22.
? *Entosolenia marginata* (Walker & Jacob), Cushman & Renz 1946, p. 38, pl. 6, fig. 16.

Maximum diameter 0.24–0.4, thickness 0.14–0.2 mm.

Stratigraphic range: Late Maastrichtian (*Abathomphalus mayaroensis* Zone) to Early Eocene (*Globorotalia formosa formosa* Zone). Rare. In Trinidad also found in the Eocene of the overlying Navet Formation.

FISSURINA TUBATA (Matthes)
Figure **34.**11

Lagena tubata Matthes 1939, Palaeontographica, Stuttgart, A 90, p. 78, pl. 5, fig. 90.
? *Fissurina acuta* Reuss 1863, Sitzungsber. K. Akad. Wiss. Wien, math.-naturwiss. Cl., 46/1 (1862), p. 340, pl. 7, figs. 90, 91 (non *Fissurina acuta* Costa 1862).

Length 0.17–0.44, width 0.1–0.28, thickness 0.08–0.2 mm.

Stratigraphic range: Early Maastrichtian (*Globotruncana lapparenti tricarinata* Zone) to Early Eocene (*Globorotalia formosa formosa* Zone). Rare. In Trinidad also found in the Eocene of the overlying Navet Formation.

PALLIOLATELLA ORBIGNYANA (Seguenza)
Figure **34.**12

Fissurina orbignyana Seguenza 1862, Terreni Terziarii Messina, T. Capra, Messina, 2, p. 66, pl. 2, figs. 25, 26.
Lagena orbignyana (Seguenza), Cushman & Jarvis 1932, p. 40, pl. 12, fig. 6.
Entosolenia cf. *orbignyana* Seguenza, Cushman & Renz 1946, p. 38, pl. 6, fig. 17.
Entosolenia sp. A, Cushman & Renz 1946, p. 39, pl. 6, figs. 19, 20.

There is considerable variation in the length of the apertural extension; it is often somewhat longer than in Seguenza's type. Length 0.32–0.66, width 0.27–0.42, thickness 0.15–0.25 mm.

Stratigraphic range: Early Maastrichtian (*Globotruncana lapparenti tricarinata* Zone) to Early Eocene (*Globorotalia subbotinae* Zone). Rather rare. In Trinidad also found in the Eocene of the overlying Navet Formation.

Family Glandulinidae

GLANDULINA cf. BICAMERATA
Herrmann
Figure **34.14–16**

Glandulina bicamerata Herrmann 1917, Mitt. geol. Landesanst., Strassburg, 10, p. 284, pl. 2, fig. 13.
Pseudoglandulina bistegia (Olszewski), Cushman & Jarvis 1932, p. 37, pl. 11, figs. 10–12. – Cushman & Renz 1946, p. 31, pl. 4, fig. 28.

The test consists of two (rarely three) chambers which vary in shape from hemispherical to almost spherical. The specimens are usually broadly rounded at the base, but very rarely there is a very small initial chamber. Exceptionally, a large final chamber is present, and the test may then resemble that of *G. obtussissima* Reuss. 'Cristellaria' bistegia Olszewski has an asymmetric test and may not be related to *Glandulina*. Length 0.39–0.82 mm (up to 0.96 mm with three chambers), thickness 0.24–0.64 mm.

Stratigraphic range: Late Albian (*Favusella washitensis* Zone) to Campanian (*Globotruncana stuarti* Zone); probably also Late Paleocene (*Globorotalia pseudomenardii* Zone) and Early Eocene (*Globorotalia subbotinae* Zone). Rare.

GLANDULINA (?) CYLINDRACEA
CONSTRICTA Franke
Figure **34.17**

Glandulina cylindracea (Reuss) forma *constricta* Franke 1928, p. 52, pl. 4, fig. 22.
Pseudoglandulina cylindracea (Reuss), Cushman & Jarvis 1932, p. 36, pl. 11, fig. 7 (not fig. 8). – Cushman & Renz 1946, p. 31, pl. 5, fig. 5.

The early part of the test, below the constriction separating it from the elongated final chamber, suggests in transmitted light the presence of several rather indistinct and delicate partitions with an irregular, possibly polymorphinid arrangement. The final chamber may be slightly asymmetric. Length 0.4–0.95, thickness 0.1–0.3 mm.

Stratigraphic range: Late Maastrichtian (*Abathomphalus mayaroensis* Zone) to Early Eocene (*Globorotalia subbotinae* Zone). Rather rare.

GLANDULINA LAEVIGATA (d'Orbigny)
Figure **34.18**

Nodosaria (Glandulina) laevigata d'Orbigny 1826, p. 252, pl. 10, figs. 1–3.

This group includes uniserial as well as initially biserial specimens. Length 0.35–1.02, thickness 0.25–0.63 mm.

Stratigraphic range: Late Turonian/Coniacian (*Globotruncana renzi* Zone) to Early Eocene (*Globorotalia formosa formosa* Zone). Rather rare. In Trinidad also found in the Eocene of the overlying Navet Formation.

GLANDULINA SUBCONICA Alth
Figure **34.13, 19**

Glandulina subconica Alth 1850, p. 270, pl. 13, fig. 32.
Pseudoglandulina parallela (Marsson), Cushman & Jarvis 1932, p. 36, pl. 11, fig. 9. – Cushman & Renz, 1946, p. 31, pl. 5, figs. 6, 7.

The initial chambers are separated by oblique sutures and may include a trace of a biserial stage. A few practically cylindrical specimens are difficult to be distinguished from *G. parallela* Marsson. Length 0.55–1.6 (1.9?), thickness 0.26–0.85 mm.

Stratigraphic range: Campanian (*Globotruncana stuarti* Zone) to Late Paleocene (*Globorotalia velascoensis* Zone). Rare.

Family Heterohelicidae

BIFARINA ALABAMENSIS (Cushman)
Figure **34.20**

Rectoguembelina alabamensis Cushman 1940, Contrib. Cushman Lab. foramin. Res., 16, p. 65, pl. 11, fig. 16.

Length 0.18–0.27, thickness 0.06–0.09 mm.

Stratigraphic range: Early Eocene (*Globorotalia edgari* Zone to *G. subbotinae* Zone). Rather rare.

Family Bolivinidae

BRIZALINA sp. A
Figure **34.21**

Compressed, with slightly convex sides, acute periphery and subrounded base. Sutures slightly oblique, flush with the surface or, more rarely, slightly raised. *Bolivina tenuis* Marsson and *B.* sp. 2 (Sliter 1985) are both similar but more distinctly lozenge-shaped in side view. Length 0.18–0.3, breadth 0.12–0.18, thickness 0.05–0.07 mm.

Stratigraphic range: Early Cenomanian (*Rotalipora appenninica appenninica* Zone) to Late Turonian-Coniacian (*Globotruncana renzi* Zone). Very rare.

GABONITA cf. KLASZI (Roveda)
Figure **34.22–23**

Gabonella klaszi Roveda 1964, Revue Micropaléontol., Paris, 7, p. 197, pl. 1, figs. 1–6.

The Trinidad form differs from the types in the less distinctly inverted saddle-shaped sutures. Sometimes the aperture is crescentic rather than semicircular. The sutures are less deeply incised than in *G. kugleri* (see below). Length 0.24–0.45, thickness 0.11–0.19 mm.

Stratigraphic range: Coniacian-Santonian (*Globotruncana concavata* Zone). Rare.

GABONITA KUGLERI (Beckmann)
Figure **34.24–25**

Gabonella kugleri Beckmann 1974, Verhandl. Naturforsch. Ges. Basel, 84, p. 322, pl. 1, fig. 2; textfigs. 5–9.

The most similar species is *G. mammosa* deKlasz & van Hinte (1977), described from beds of the same age in Gabon. It differs from *G. kugleri* mainly in its broader and more detached chambers. Length 0.21–0.44, thickness 0.08–0.18 mm

Stratigraphic range: Coniacian-Santonian (*Globotruncana concavata* Zone). Rather rare.

LOXOSTOMOIDES APPLINAE (Plummer)
Figure **34.26–27**

Bolivina applini Plummer 1926, p. 69, pl. 4, fig. 1. .

Length 0.25–0.65, breadth 0.11–0.21, thickness 0.07–0.14 mm.

Stratigraphic range: Early Maastrichtian (probably *Globotruncana lapparenti tricarinata* Zone; Marabella facies) and Early Paleocene (probably *Globorotalia pseudobulloides* Zone; Rochard facies). Possibly also Middle Paleocene (*Globorotalia pusilla pusilla* Zone). Rather rare.

LOXOSTOMOIDES PLUMMERAE (Cushman)
Figure **34.28–29**

Loxostoma plummerae Cushman 1936, Cushman Lab. foramin. Res, Special Publ., 6, p. 59, pl. 8, fig. 13.

Length 0.21–0.35, breadth 0.07–0.11, thickness 0.07–0.09 mm.

Stratigraphic range: Middle Paleocene (*Globorotalia uncinata* Zone to *G. pusilla pusilla* Zone). Rather rare.

TAPPANINA EOUVIGERINIFORMIS
(Keller)
Figure **34.30**

Bolivinita eouvigeriniformis Keller 1935, Bull.
 Soc. Nat. Moscow, Sect. Geol., 14, p. 548,
 pl. 3. figs. 20, 21.
Tappanina eouvigeriniformis (Keller),
 Beckmann & Koch 1964, p. 53, pl. 7, figs.
 1, 2.

Length 0.15–0.25, breadth 0.12–0.21, thickness
0.07–0.16 mm.

Stratigraphic range: Turonian-Coniacian
 (*Globotruncana inornata* Zone to *G. renzi*
 Zone). Rare.

TAPPANINA SELMENSIS (Cushman),
emend. Brotzen 1948
Figure **34.31–34**

Bolivinita selmensis Cushman 1933, Contrib.
 Cushman Lab. foramin. Res., 9, p. 58, pl. 7,
 figs. 3, 4.

A few shorter and broader specimens from the
Early Eocene samples (Fig. 34.33–34) are less
typical without, however, forming a morphologically distinct group. Length 0.1–0.21, thickness
0.07–0.14 mm.

Stratigraphic range: Middle Paleocene (*Globorotalia
angulata* Zone) to Early Eocene (*Globorotalia
subbotinae* Zone). Rare.

Family Bolivinoididae

BOLIVINOIDES DECORATUS
DECORATUS (Jones)
Figure **34.35**

Bolivina decorata Jones 1886, Proc. Belfast
 Nat. Field Club, n. ser., 1, appendix 9,
 p. 330, pl. 27, figs. 7, 8.
Bolivinoides decoratus decoratus (Jones 1886),
 Beckmann & Koch 1964, p. 43, pl. 6, fig. 10.

Length 0.34–0.37, breadth 0.2–0.22, thickness
0.15 mm.

Stratigraphic range: Early Maastrichtian
 (*Globotruncana lapparenti tricarinata* Zone).
 Very rare.

BOLIVINOIDES DELICATULUS
CURTUS Reiss
Figure **34.36**

Bolivinoides curta Reiss 1954, Contrib.
 Cushman Found. foramin. Res., 5, p. 158,
 pl. 30, figs. 15, 16.
Bolivinoides delicatulus curtus (Reiss 1954),
 Beckmann & Koch 1964, p. 43, pl. 6, figs. 2,
 3.

Length 0.22–0.45, breadth 0.13–0.21, thickness
0.06–0.08 mm.

Stratigraphic range: Paleocene (*Globorotalia
trinidadensis* Zone to *G. pseudomenardii*
Zone). Rare.

BOLIVINOIDES DELICATULUS
DELICATULUS Cushman
Figure **34.37**

Bolivinoides decorata (Jones) var. *delicatula*
 Cushman 1927, Contrib. Cushman Lab.
 foramin. Res., 2, p. 90, pl. 12, fig. 8.
Bolivinoides decorata (Jones) var. *delicatula*
 Cushman, Cushman & Jarvis 1928, p. 99,
 pl. 14, fig. 9. – Cushman & Jarvis 1932, p. 42,
 pl. 13, fig. 2. – Cushman & Renz 1946,
 p. 36, pl. 6, fig. 4.
Bolivinoides delicatulus delicatulus Cushman
 1927, Beckmann & Koch 1964, p. 41, pl. 5,
 figs. 8, 9, 16, 20.

Length 0.28–0.6, breadth 0.15–0.22, thickness
0.07–0.1 mm.

Stratigraphic range: Maastrichtian (*Globotruncana
gansseri* Zone) to Middle Paleocene
(*Globorotalia angulata* Zone). Rare.

BOLIVINOIDES DRACO DRACO
(Marsson)
Figure **35.1**

Bolivina draco Marsson 1878, p. 157, pl. 3,
 fig. 25.
Bolivinoides draco draco (Marsson 1878),
 Beckmann & Koch 1964, p. 45, pl. 6, fig. 24.

Length 0.24–0.6, breadth 0.21–0.38, thickness 0.12–0.24 mm.

Stratigraphic range: Maastrichtian (top of *Globotruncana lapparenti tricarinata* Zone to *Abathomphalus mayaroensis* Zone). Rare.

BOLIVINOIDES DRACO MILIARIS
Hiltermann & Koch
Figure 35.2

Bolivinoides draco (Marsson) *miliaris* Hiltermann & Koch 1950, Geol. Jahrbuch, Hannover, 64, p. 604, figs. 2–4 (26, 32–34, 39–41, 46–48), 5 (39, holotype).
Bolivinoides draco miliaris Hiltermann & Koch 1950, Beckmann & Koch 1964, p. 44, pl. 6, fig. 21.

Length 0.26–0.51, breadth 0.24–0.36, thickness 0.15–0.24 mm.

Stratigraphic range: Early Maastrichtian (*Globotruncana lapparenti tricarinata* Zone). Very rare.

BOLIVINOIDES DRACO VERRUCOSUS
Beckmann & Koch
Figure 35.3

Bolivinoides cf. *rhomboidea* (Cushman), Beckmann 1960, p. 62, fig.10.
Bolivinoides draco verrucosus Beckmann & Koch 1964, p. 46, pl. 6, figs. 17–19.

Length 0.27–0.45, breadth 0.21–0.33, thickness 0.12–0.15 mm.

Stratigraphic range: Maastrichtian (*Globotruncana gansseri* Zone to *Abathomphalus mayaroensis* Zone). Rare.

BOLIVINOIDES GRANULATUS EMERSUS
Beckmann & Koch
Figure 35.4

Bolivinoides granulatus emersus Beckmann & Koch 1964, p. 40, pl. 5, figs. 1–3.

Length 0.33–0.42, breadth 0.15–0.19, thickness 0.07–0.09 mm.

Stratigraphic range: Campanian (*Globotruncana stuarti* Zone) to Maastrichtian (*Globotruncana gansseri* Zone). Rare.

Family Loxostomatidae

ARAGONIA ARAGONENSIS (Nuttall)
Figure 35.5

Textularia aragonensis Nuttall 1930, p. 280, pl. 23, fig. 6.
Aragonia aragonensis (Nuttall), Beckmann & Koch 1964, p. 52, pl. 7, figs. 10, 11.

Length 0.16–0.33, breadth 0.13–0.27, thickness 0.06–0.13 mm.

Stratigraphic range: Early Eocene (*Globorotalia edgari* Zone to *G. aragonensis* Zone). Rather rare.

ARAGONIA MATERNA KUGLERI
Beckmann & Koch
Figure 35.6

Aragonia materna kugleri Beckmann & Koch 1964, p. 49, pl. 7, figs. 14–17.

Length 0.22–0.33, breadth 0.15–0.22, thickness 0.08–0.1 mm.

Stratigraphic range: Turonian (*Globotruncana inornata* Zone) to Santonian (*Globotruncana fornicata* Zone). Rather rare.

ARAGONIA VELASCOENSIS (Cushman)
Figure 35.7, 15

Textularia velascoensis Cushman 1925, Contrib. Cushman Lab. foramin. Res., 1, p. 18, pl. 3, fig. 1.
Bolivinoides trinitatensis Cushman & Jarvis 1928, p. 99, pl. 14, fig. 10. – *Bolivinoides trinitatensis* Cushman & Jarvis, Cushman & Jarvis 1932, p. 43, pl. 13, fig. 3. – Cushman & Renz 1946, p. 36, pl. 6, figs. 5, 6.
Bolivinoides sp. A, Beckmann 1960, p. 62, fig. 11.
Aragonia velascoensis (Cushman), Beckmann & Koch 1964, p. 50, pl. 7, fig. 6.

The relatively compressed specimens with rather oval outline and weak ornamentation (Fig. 35.7), which predominate in the Late Cretaceous samples, were separated by Beckmann (1960) as *Bolivinoides* sp. A.; they suggest an ancestry in the *A. materna* group. Length 0.24–0.66, breadth 0.18–0.48, thickness 0.11–0.3 mm.

Stratigraphic range: Late Santonian (top of *Globotruncana fornicata* Zone) to Late Paleocene (*Globorotalia velascoensis* Zone). Fairly common.

ZEAUVIGERINA TEURIA Finlay
Figure **35.8–10**

Zeauvigerina teuria Finlay 1947, New Zealand J. Sci. Technol., B, 28, p. 276, pl. 4, figs. 49–54.

The specimens are rather poorly preserved, but nevertheless show the typical apertural characters and the finely papillate wall surface. Length 0.24–0.4, breadth 0.15–0.22, thickness 0.13–0.18 mm.

Stratigraphic range: Santonian (*Globotruncana fornicata* Zone). Rather rare.

Family Cassidulinidae

GLOBOCASSIDULINA GLOBOSA (Hantken)
Figure **35.11**

Cassidulina globosa Hantken 1875, Mitteil. Jahrbuch K. Ungar. geol. Anst., 4, p. 64, pl. 16, fig. 2.

This species, originally described from the Oligocene, is represented by minute specimens; the sutures are either flush with the surface or slightly depressed. Diameter 0.08–0.25, thickness 0.06–0.16 mm.

Stratigraphic range: Early Eocene (*Globorotalia edgari* Zone to *G. aragonensis* Zone). Rather rare. In Trinidad also found in the Eocene of the overlying Navet Formation.

Family Eouvigerinidae

EOUVIGERINA SUBSCULPTURA McNeil & Caldwell
Figure **35.12–13**

Eouvigerina aculeata Cushman 1933, Contrib. Cushman Lab. foramin. Res., 9, p. 62, pl. 7, fig. 8; non *Eouvigerina aculeata* (Ehrenberg

1854) fide McNeil & Caldwell 1981.
Eouvigerina subsculptura (new name) McNeil & Caldwell 1981, Geol. Assoc. Canada, Special P., 21, p. 231, pl. 18, figs. 20, 21.

The Trinidad specimens are probably not conspecific with *E. aculeata* Cushman but do not differ substantially from those figured as *E. subsculptura* by McNeil & Caldwell. In Trinidad, the keels of some of the chambers may be slightly weaker and the surface may sometimes be rather more regularly hispid, like in *E. pseudohispida* Nyong & Olsson (1983); the test of the latter species, however, is more slender and elongated. Length 0.3–0.44, breadth 0.12–0.18, thickness 0.09–0.12 mm.

Stratigraphic range: Late Maastrichtian (*Abathomphalus mayaroensis* Zone). Fairly common.

Family Turrilinidae

NEOBULIMINA IRREGULARIS Cushman & Parker
Figure **35.16–17**

Neobulimina irregularis Cushman & Parker 1936, Contrib. Cushman Lab. foramin. Res., 12, p. 9, pl. 2, fig. 8.

Length 0.18–0.47, thickness 0.08–0.12 mm.

Stratigraphic range: Early Cenomanian (*Rotalipora appenninica appenninica* Zone) to Coniacian-Santonian (*Globotruncana concavata* Zone). Rather rare.

NEOBULIMINA PRIMITIVA (Cushman)
Figure **35.14**

Virgulina primitiva Cushman 1936, Cushman Lab. foramin. Res., Special Publ., 6, p. 46, pl. 7, fig. 1.

Length 0.18–0.35, thickness 0.07–0.12 mm.

Stratigraphic range: Albian? to Early Cenomanian (*Rotalipora ticinensis ticinensis* Zone?, *R. appenninica appenninica* Zone), possibly also Turonian (*Globotruncana inornata* Zone). Rare.

PRAEBULIMINA KICKAPOOENSIS KICKAPOOENSIS (Cole)
Figure **35.18–19**

Bulimina kickapooensis Cole 1938, Florida
    Dept. Cons., Geol. Survey, Bull., 16, p. 45,
    pl. 3, fig. 5.

Length 0.4–0.96, thickness 0.21–0.4 mm.

Stratigraphic range: Campanian (*Globotruncana stuarti* Zone) to Late Maastrichtian (*Abathomphalus mayaroensis* Zone). Fairly common.

PRAEBULIMINA KICKAPOOENSIS PINGUA (Cushman & Parker)
Figure **35.20**

Bulimina kickapooensis Cole var. *pingua*
    Cushman & Parker 1940, Contrib.
    Cushman Lab. foramin. Res., 16, p. 44, pl. 8,
    figs. 13, 14.
Bulimina kickapooensis Cole var. *pingua*
    Cushman & Parker, Cushman & Renz 1947*a*,
    p. 45, pl. 12, fig. 3.

Length 0.3–0.74, thickness 0.22–0.4 mm

Stratigraphic range: Campanian (*Globotruncana stuarti* Zone). Fairly common.

PRAEBULIMINA NANNINA (Tappan)
Figure **35.21**

Bulimina nannina Tappan 1940, J. Paleontol.,
    14, p. 116, pl. 19, fig. 4.

Length 0.14–0.24, thickness 0.1–0.18 mm.

Stratigraphic range: Early Cenomanian (*Rotalipora appenninica appenninica* Zone). Rare.

PRAEBULIMINA PETROLEANA (Cushman & Hedberg)
Figure **35.22–23**

Bulimina petroleana Cushman & Hedberg
    1941, p. 95, pl. 22, fig. 31.

The species includes specimens which are larger than the types; the shape varies from slender and subconical to thick ovoid. Length 0.44–0.92, thickness 0.3–0.6 mm.

Stratigraphic range: Campanian (*Globotruncana stuarti* Zone) to Early Maastrichtian (*Globotruncana lapparenti tricarinata* Zone). Fairly common.

PRAEBULIMINA (?) QUADRATA (Plummer)
Figure **35.24–25**

Bulimina quadrata Plummer 1926, p. 72, pl. 4,
    figs. 4, 5.

This species is similar and possibly related to the *P. kickapooensis* group, but, at least in mature specimens, usually shows a subterminal or terminal aperture. Nevertheless, it can hardly be assigned to *Desinobulimina*, as some authors including Cushman have suggested. *Desinobulimina* is based on a Recent type with different test morphology and toothplate characteristics. The generic position of *P. quadrata* is here left open; its inclusion in *Praebulimina* appears reasonable but would require an extension of the current genus definition. Length 0.44–1.2, thickness 0.32–0.6 mm.

Stratigraphic range: Campanian (*Globotruncana stuarti* Zone) to Middle Paleocene (*Globorotalia pusilla pusilla* Zone). Rather rare.

PRAEBULIMINA REUSSI (Morrow)
Figure **35.26–28**

Bulimina ovulum Reuss 1844, Geognost.
    Skizzen aus Böhmen, 2, p. 215. – Reuss
    1845, p. 37, pl. 8, fig. 57; pl. 13, fig. 73.
Bulimina reussi Morrow 1934, J. Paleontol., 8,
    p. 195, pl. 29, fig. 12.

The species has a rather long stratigraphic range and shows a considerable variability. Length 0.16–0.45, thickness 0.12–0.35 mm.

Stratigraphic range: Coniacian-Santonian (*Globotruncana concavata* Zone) to Late Maastrichtian (*Abathomphalus mayaroensis* Zone). Common.

PRAEBULIMINA cf. SEABEENSIS Tappan
Figure **35.29–30**

Praebulimina seabeensis Tappan 1957, p. 217,
    pl. 69, figs. 14–16.

Specimens referred here to this group are not

uncommon but rather poorly preserved. The shape varies between subcylindrical and subfusiform; it may sometimes be slightly triangular in cross section but much less so than in *P. prolixa* (Cushman & Parker; *Bulimina p.*). On the average, the chambers seem to be slightly higher than in the types. Length 0.21–0.33, thickness 0.12–0.2 mm.

Stratigraphic range: Santonian (*Globotruncana fornicata* Zone) to Campanian (*Globotruncana stuarti* Zone). Rather rare.

PRAEBULIMINA cf. SPINATA (Cushman & Campbell)
Figure **35.31**

Bulimina spinata Cushman & Campbell 1935, Contrib. Cushman Lab. foramin. Res., 11, p. 72, pl. 11, fig. 11.

The specimens are rather more elongated than the type of *P. spinata* and are superficially similar to *P. arkadelphiana* Cushman; however, the arrangement of the spines is that of *P. spinata*. There are transitional specimens to *P. petroleana*, of which *P. spinata* may be a facies variant. Length 0.25–0.54, thickness 0.2–0.4 mm.

Stratigraphic range: Early Maastrichtian (probably *Globotruncana lapparenti tricarinata* Zone; Marabella facies). Rare.

PRAEBULIMINA WYOMINGENSIS (Fox)
Figure **35.32**

Bulimina wyomingensis Fox 1954, U.S. Geol. Survey, Prof. Paper, 254-E, p. 118, pl. 26, figs. 8–11.

Length 0.25–0.44, thickness 0.12–0.16 mm.

Stratigraphic range: Albian (*Rotalipora ticinensis ticinensis* Zone to *Favusella washitensis* Zone). Rare.

PSEUDOUVIGERINA NAHEOLENSIS
Cushman & Todd
Figure **35.33**

Pseudouvigerina naheolensis Cushman & Todd 1942, Contrib. Cushman Lab. foramin. Res., 18, p. 36, pl. 6, figs. 18–19.

The test is variable in thickness, and the wall may become very finely hispid in some specimens. Length 0.16–0.27, thickness 0.07–0.14 mm.

Stratigraphic range: Late Paleocene (*Globorotalia pseudomenardii* Zone) to Early Eocene (*Globorotalia subbotinae* Zone). Very rare.

PSEUDOUVIGERINA PLUMMERAE
Cushman
Figure **35.34**

Pseudouvigerina plummerae Cushman 1927, Contrib. Cushman Lab. foramin. Res., 3, p. 115, pl. 23, fig. 8.

Length 0.2–0.38, thickness 0.12–0.2 mm.

Stratigraphic range: Maastrichtian (*Globotruncana gansseri* Zone to *Abathomphalus mayaroensis* Zone). Rare.

PYRAMIDINA CUSHMANI (Brotzen)
Figure **35.35**

Reussella cushmani Brotzen 1936, p. 135, pl. 8, fig. 7; textfig. 47.

Length 0.18–0.35, thickness 0.12–0.2 mm.

Stratigraphic range: Early Maastrichtian (probably *Globotruncana lapparenti tricarinata* Zone; Marabella facies). Rather rare.

PYRAMIDINA cf. ROBUSTA Neagu
Figure **35.36**

Pyramidina minima robusta Neagu 1972, Rocznik Polsk. Towar. Geol., Krakow, 17, p. 21, pl. 9, figs. 7–12.

The Trinidad form has the shape and size of *P. robusta* but the aperture is distinctly more elongated at a right angle to the basal suture; in this respect it resembles that of *P. minima* Brotzen. Length 0.15–0.34, thickness 0.16–0.24 mm.

Stratigraphic range: Early Cenomanian (*Rotalipora appenninica appenninica* Zone). Very rare.

PYRAMIDINA RUDITA (Cushman & Parker)
Figure 35.37

Bulimina ornata Cushman & Parker 1935, p. 97, pl. 15, fig. 4 (non *B. ornata* Egger 1895).
Bulimina rudita Cushman & Parker, n. name, 1936, Contrib. Cushman Lab. foramin. Res., 12, p. 45.

Length 0.16–0.30, thickness 0.12–0.18 mm.

Stratigraphic range: Turonian-Coniacian (*Globotruncana inornata* Zone to *G. renzi* Zone). Fairly common.

PYRAMIDINA SZAJNOCHAE (Grzybowski)
Figure 36.1–2

Verneuilina szajnochae Grzybowski 1896, p. 287, pl. 9, fig. 19.
Bulimina limbata White 1929, p. 48, pl. 5, fig. 9.
Bulimina limbata White, Cushman & Renz 1947a, p. 45, pl. 12, fig. 4.
Reussella californica Cushman & Goudkoff 1944, Contrib. Cushman Lab. foramin. Res, 20, p. 59, pl. 10, figs. 3–5.

The distinctive edges of the chambers vary from strongly keeled to subrounded, the latter being predominant in the younger samples (*Globotruncana lapparenti tricarinata* Zone). Length 0.4–1.6, thickness 0.35–0.75 mm.

Stratigraphic range: Campanian (*Globotruncana stuarti* Zone; rare) to Early Maastrichtian (*Globotruncana lapparenti tricarinata* Zone). Fairly common.

SITELLA COLONENSIS (Cushman & Hedberg)
Figure 36.3

Buliminella colonensis Cushman & Hedberg 1930, Contrib. Cushman Lab. foramin. Res., 6, p. 65, pl. 9, figs. 6–7.
Buliminella colonensis Cushman & Hedberg, Cushman & Renz 1947a, p. 45, pl. 12, fig. 2.

Diameter 0.3–0.45, thickness 0.25–0.4 mm.

Stratigraphic range: Maastrichtian (*Globotruncana lapparenti tricarinata* Zone to *Abathomphalus mayaroensis* Zone). Very rare.

TURRILINA CARSEYAE (Plummer)
Figures 36.4–5

Bulimina compressa Carsey 1926, p. 29, pl. 4, fig. 14 (non *Bulimina compressa* Bailey 1851).
Buliminella carseyae Plummer, new name, 1931, p. 179, pl. 8, fig. 9.

Length 0.18–0.37, thickness 0.11–0.21 mm.

Stratigraphic range: Turonian (*Globotruncana inornata* Zone) to Coniacian-Santonian (*Globotruncana concavata* Zone, possibly also *G. fornicata* Zone). Fairly common.

TURRILINA CUSHMANI (Sandidge)
Figure 36.6

Buliminella cushmani Sandidge 1932, p. 280, pl. 42, figs. 18–19.

Length 0.26–0.5, thickness 0.2–0.32 mm.

Stratigraphic range: Maastrichtian (*Globotruncana lapparenti tricarinata* Zone to *Abathomphalus mayaroensis* Zone). Rather rare.

Family Stainforthiidae

HOPKINSINA DORREENI (Weiss)
Figure 36.7–8

Virgulina dorreeni Weiss 1955, J. Paleontol., 29, p. 16, pl. 4, figs. 6–8.

The test wall surface varies from finely hispid to almost smooth. The aperture is narrowly elliptical to slit-like; its position varies between basal and subterminal. Length 0.24–0.42, breadth 0.11–0.16, thickness 0.07–0.11 mm.

Stratigraphic range: Early Maastrichtian (probably *Globotruncana lapparenti tricarinata* Zone; Marabella facies) to Early Paleocene (probably *Globorotalia pseudobulloides* Zone; Rochard facies). Rare.

STAINFORTHIA cf. NAVARROANA
(Cushman)
Figure **36.9–10**

Virgulina navarroana Cushman 1933, Contrib.
Cushman Lab. foramin. Res., 9, p. 63, pl. 7,
figs. 9–10.

In Trinidad, the test is less distinctly pointed at
the base than that of the types; the chambers
increase regularly in size and are slightly inflated
throughout. There is an older, smaller variety
(up to *Globorotalia trinidadensis* Zone) and a
younger, somewhat bigger one (from *Globorotalia pusilla pusilla* Zone upwards); the two cannot be clearly separated otherwise. Smaller
variety: Length 0.18–0.35, thickness 0.08–
0.15 mm. Larger variety: Length 0.45–0.62,
thickness 0.16–0.24 mm.

Stratigraphic range: Smaller variety: Early
Maastrichtian (probably *Globotruncana
lapparenti tricarinata* Zone; Marabella facies)
and Early Paleocene (probably *Globorotalia
pseudobulloides* Zone; Rochard facies). Larger
variety: Middle Paleocene (*Globorotalia pusilla
pusilla* Zone) to Early Eocene (*Globorotalia
formosa formosa* Zone). Rare.

Family Siphogenerinoididae

SIPHOGENERINOIDES BRAMLETTEI
Cushman
Figure **36.11**

Siphogenerinoides bramlettei Cushman 1929*a*,
p. 56, pl. 9, figs. 5–6.

Length 0.4–0.64, thickness 0.12–0.2 mm.

Stratigraphic range: Early Maastrichtian (probably
Globotruncana lapparenti tricarinata Zone;
Marabella facies). Rare.

SIPHOGENERINOIDES LANDESI Stone
Figure **36.12**

Siphogenerinoides landesi Stone 1946, J.
Paleontol., 20, p. 476, pl. 72, figs. 13–17.

Length 0.8–1.2, thickness 0.2–0.35 mm.

Stratigraphic range: Campanian (*Globotruncana
stuarti* Zone) and Early Maastrichtian (probably

Globotruncana lapparenti tricarinata Zone;
Marabella facies). Rare.

ORTHOKARSTENIA CRETACEA
(Cushman)
Figure **36.13**

Siphogeneriniodes cretacea Cushman 1929*a*,
p. 58, pl. 9, figs. 14–15.

Length 1.2–1.28, thickness 0.36–0.48 mm.

Stratigraphic range: Early Maastrichtian (probably
Globotruncana lapparenti tricarinata Zone;
Marabella facies). Very rare.

ORTHOKARSTENIA IDKYENSIS (Colom)
Figure **36.14–15**

Siphogenerinoides cretacea Cushman subsp.
idkyensis Colom 1948, Bol. R. Soc. Espan.
Hist. Nat., 45 (1947), p. 665, pl. 48, figs. 1–
13, 15–16.

Length 0.95–2.0, thickness 0.35–0.51 mm.

Stratigraphic range: Probably Early Maastrichtian
(*Globotruncana lapparenti tricarinata* Zone;
Marabella facies); Maastrichtian
(*Globotruncana lapparenti tricarinata* Zone and
G. gansseri Zone). Very rare.

ORTHOKARSTENIA cf. PARVA
(Cushman)
Figure **36.16–17**

Siphogenerinoides parva Cushman 1929*a*, p. 58,
pl. 9, figs. 11–13.

The specimens of this rather variable group show
a mixture of characters of *O. parva* (general
shape of the test, nature of the sutures and ornamentation) and *O. cretacea* (test size, chamber
shape). Some small elongated subconical tests
resemble those of *O. castelaini* LeCalvez,
deKlasz & Brun. The main differences from the
typical *O. parva* are the somewhat larger test
size and the rather lower chambers. Length 0.9–
1.36, thickness 0.28–0.5 mm.

Stratigraphic range: Early Maastrichtian (probably
Globotruncana lapparenti tricarinata Zone;
Marabella facies). Rare.

Family Buliminidae

BULIMINA EXCAVATA Cushman & Parker
Figure **36.18**

Bulimina excavata Cushman & Parker 1936,
  Contrib. Cushman Lab. foramin. Res, 12,
  p. 41, pl. 7, fig. 4.

Length 0.27–0.41, thickness 0.15–0.3 mm.

Stratigraphic range: Paleocene (*Globorotalia
trinidadensis* Zone to *G. pseudomenardii*
Zone). Very rare.

BULIMINA MIDWAYENSIS Cushman &
Parker
Figure **36.19–21**

Bulimina arkadelphiana var. *midwayensis*
  Cushman & Parker 1936, Contrib.
  Cushman Lab. foramin. Res., 12, p. 42, pl. 7,
  figs. 9–10.
? *Bulimina petroleana* Cushman & Hedberg,
  Cushman & Renz 1946, p. 37, pl. 6, fig. 12.
? *Bulimina petroleana* Cushman & Hedberg
  var. *spinea* Cushman & Renz 1946, p. 37,
  pl. 6, fig. 13.

Length 0.15–0.38, thickness 0.14–0.3 mm.

Stratigraphic range: Campanian (*Globotruncana
stuarti* Zone) to Late Paleocene (*Globorotalia
pseudomenardii* Zone). Fairly common.

BULIMINA PLENA Brotzen
Figure **36.22**

Bulimina plena Brotzen 1940, Sverig. Geol.
  Unders., C, 435, p. 24, textfig. 6:3.

The horizontal, hardly depressed sutures are
typical for this species. *B. tarda* Parker & Bermudez is very similar but somewhat larger and
relatively fatter, with slightly depressed sutures.
Length 0.19–0.27, thickness 0.12–0.15 mm.

Stratigraphic range: Late Paleocene (*Globorotalia
pseudomenardii* Zone). Very rare.

BULIMINA SEMICOSTATA Nuttall
Figure **36.23**

Bulimina semicostata Nuttall 1930, p. 285,
  pl. 23, figs. 15–16.

The shape of the test is weakly triangular. The
sutures are not well visible but in the later chambers, which are less distinctly costate, they are
occasionally slightly indented. Some of the fatter
specimens are close to the variety *lacrima* Mallory which probably does not deserve a formal
status. Length 0.25–0.56, thickness 0.15–
0.28 mm.

Stratigraphic range: Early Eocene (*Globorotalia
formosa formosa* Zone). Fairly common. In
Trinidad also found in the Eocene of the
overlying Navet Formation.

BULIMINA STOKESI Cushman & Renz
Figure **36.24–26**

Bulimina stokesi Cushman & Renz 1946, p. 37,
  pl. 6, fig. 14.

This species is distinguished from the similar *B.
midwayensis* in its coarser ornamentation and
somewhat larger size. Length (0.14 juvenile?)
0.27–0.46, thickness (0.1 juvenile?) 0.18–
0.31 mm.

Stratigraphic range: Early Eocene (*Globorotalia
edgari* Zone to *G. formosa formosa* Zone).
Rather rare. In Trinidad also found in the
Eocene of the overlying Navet Formation.

BULIMINA TAYLORENSIS Cushman &
Parker
Figure **36.27**

Bulimina taylorensis Cushman & Parker 1935,
  p. 96, pl. 15, fig. 3.

Length 0.27–0.5, thickness 0.18–0.34 mm.

Stratigraphic range: Maastrichtian (*Globotruncana
lapparenti tricarinata* Zone to *Abathomphalus
mayaroensis* Zone). Rather rare.

BULIMINA TRINITATENSIS Cushman &
Jarvis
Figure **36.28–29**

Bulimina trinitatensis Cushman & Jarvis 1928,
  p. 102, pl. 14, fig. 12. – Cushman & Jarvis

1932, p. 44, pl. 13, fig. 4. – Cushman & Renz 1946, p. 37, pl. 6, figs. 8, 9.

B. trinitatensis, apparently a successor of *B. taylorensis*, is distinguished from this species in the slightly larger test size, more inflated chambers, and in a gradual loss of the costae; in the final stage the wall develops a hispid surface or even a finely reticulate pattern (Fig. 36.29). Transitional specimens between the two species occur up to the *Globorotalia pseudomenardii* Zone. Length 0.26–0.6, thickness 0.18–0.38 mm.

Stratigraphic range: Early Paleocene (*Globorotalia trinidadensis* Zone) to Early Eocene (*Globorotalia formosa formosa* Zone). Rather rare.

BULIMINA TRUNCANELLA Finlay
Figure 36.30–31

Bulimina truncanella Finlay 1940, p. 455, pl. 64, figs. 89–91.

The specimens are mostly subtriangular in cross section. There are about 20 costae which are slightly irregular to almost straight and show occasional branching. They may represent an ancestral form to *B. corrugata* Cushman & Siegfus, which is rather larger and has a smooth, inflated final chamber. Length 0.21–0.29, thickness 0.13–0.22 mm.

Stratigraphic range: Early Eocene (*Globorotalia subbotinae* Zone to *G. aragonensis* Zone). Rare.

BULIMINA TUXPAMENSIS Cole
Figure 36.32–33

Bulimina tuxpamensis Cole 1928, Bull. Am. Paleontol., 14/53, p. 212, pl. 32, fig. 23.

The Trinidad form is slightly smaller than the type but otherwise identical. Length 0.22–0.64, thickness 0.2–0.4 mm.

Stratigraphic range: Late Paleocene (*Globorotalia pseudomenardii* Zone) to Early Eocene (*Globorotalia aragonensis* Zone). Fairly common. In Trinidad also found in the Eocene of the overlying Navet Formation.

GLOBOBULIMINA (?) SUTERI (Cushman & Renz)
Figure 36.34–36

Bulimina (Desinobulimina) suteri Cushman & Renz, 1946, p. 38, pl. 6, fig. 15.

This highly variable species presents an unusual combination of features of the Buliminidae (coiling mode) and the Pleurostomellidae (apertural characteristics). Length 0.4–1.55, thickness 0.17–0.72 mm.

Stratigraphic range: Campanian (*Globotruncana stuarti* Zone) to Late Paleocene (*Globorotalia velascoensis* Zone). Rare.

PRAEGLOBOBULIMINA OVATA (d'Orbigny)
Figure 37.1–2

Bulimina ovata d'Orbigny 1846, Foram. foss. bassin tert. Vienne, Paris, p. 185, pl. 11, figs. 13–14.
Bulimina pupoides d'Orbigny, Cushman & Renz 1948, p. 25, pl. 5, fig. 16.

The present specimens correspond rather well to *P. ovata* as emended by Haynes (1954). The most similar American species is probably *P. guayabalensis ampla* (Cushman & Parker; *Bulimina guayabalensis* var. *ampla*), which is rather fatter and has more horizontal sutures. Length 0.5–0.75, thickness 0.3–0.54 mm.

Stratigraphic range: Early Eocene (*Globorotalia edgari* Zone to *G. aragonensis* Zone). Rare. In Trinidad also recorded in the Eocene of the overlying Navet Formation.

Family Buliminellidae

BULIMINELLA GRATA Parker & Bermudez
Figure 37.3

Buliminella grata Parker & Bermudez 1937, J. Paleontol., 11, p. 515, pl. 59, fig. 6.
Buliminella beaumonti Cushman & Renz 1946, p. 36, pl. 6, fig. 7.
Buliminella grata Parker & Bermudez, Cushman & Renz 1948, p. 24, pl. 5, fig. 12.

Length 0.33–0.5, thickness 0.27–0.33 mm.

Stratigraphic range: Early Eocene (*Globorotalia formosa formosa* Zone to *G. aragonensis* Zone). Rare. In Trinidad also found in the Eocene of the overlying Navet Formation.

Family Uvigerinidae

EUUVIGERINA ELONGATA (Cole)
Figure 37.4

Uvigerina elongata Cole 1927, Bull. Am. Paleontol., 14/51, p. 26, pl. 4, figs. 2–3.

Length 0.15–0.28, thickness 0.08–0.14 mm.

Stratigraphic range: Paleocene (*Globorotalia trinidadensis* Zone to *G. angulata* Zone). Rare.

Family Fursenkoinidae

CORYPHOSTOMA CRASSUM (Vasilenko & Myatlyuk)
Figure 37.5–7

Bolivina incrassata Reuss, var. *crassa* Vasilenko & Myatlyuk 1947, Microfauna oil fields Caucasus, Emba and Central Asia, VNIGRI, Leningrad, p. 203, pl. 2, figs. 3–5.
Bolivina incrassata Reuss, forma *gigantea* Wicher 1949, Bull. Mus. Hist. Natur. Pays Serbe, Belgrad, A, 2, p. 57, pl. 5, figs. 2–3.
Bolivina incrassata Reuss, thick variety, Beckmann 1960, p. 62, fig. 14.

Length 0.8–1.2, breadth 0.34–0.45, thickness 0.2–0.31 mm.

Stratigraphic range: Late Maastrichtian (*Abathomphalus mayaroensis* Zone). Rare.

CORYPHOSTOMA cf. GEMMA (Cushman)
Figure 37.8–9

Bolivina gemma Cushman 1927, Contrib. Cushman Lab. foramin. Res., 2, p. 87, pl. 12, fig. 3.

This may be a facies variant of the *C. crassum/ incrassatum* group; it consists of relatively large and broad specimens which have a more angular

periphery than *C. crassum* and show distinct thickenings of the sutures near their axial ends. Length 0.75–1.15, breadth 0.25–0.55, thickness 0.16–0.21 mm.

Stratigraphic range: Early Maastrichtian (probably *Globotruncana lapparenti tricarinata* Zone; Marabella facies). Rare.

CORYPHOSTOMA INCRASSATUM (Reuss)
Figure 37.10–12

Bolivina incrassata Reuss 1851, p. 45, pl. 5, fig. 13.
Bolivina incrassata Reuss, Beckmann 1960, p. 62, fig. 13.

Length 0.6–1.25, breadth 0.2–0.35, thickness 0.08–0.14 mm.

Stratigraphic range: Campanian (*Globotruncana stuarti* Zone) to Maastrichtian (*Globotruncana gansseri* Zone). Fairly common.

CORYPHOSTOMA MIDWAYENSE (Cushman)
Figure 37.13–15

Bolivina midwayensis Cushman 1936, Cushman Lab. foramin. Res, Special Publ., 6, p. 50, pl. 7, fig. 12.
Coryphostoma paleocenica Akpati 1966, Contrib. Cushman Found. foramin. Res., 17, p. 138, pl. 10, fig. 12.
? *Bolivina primatumida* White 1929, J. Paleontol., 3, p. 44, pl. 4, fig. 20.
Loxostomum limonense (Cushman), Cushman & Renz 1946, p. 39, pl. 6, fig. 23.

Length 0.32–0.8, breadth 0.12–0.25, thickness 0.06–0.1 mm.

Stratigraphic range: Paleocene (*Globorotalia trinidadensis* Zone to *G. velascoensis* Zone). Common.

CORYPHOSTOMA sp. A
Figure 37.16

Test compressed, wedge-shaped in outline, with relatively high biserially arranged chambers. Sutures oblique and curved, slightly depressed. Aperture a narrow slit at the base of the chamber. *Bolivina subcretacea* Khan is somewhat similar but differs in its more distinctly lozenge-shaped side view and straighter sutures. Length 0.21–0.29, breadth 0.12–0.16, thickness 0.03–0.05 mm.

Stratigraphic range: Turonian-Coniacian (*Globotruncana inornata* Zone to *G. renzi* Zone). Very rare.

Family Pleurostomellidae

This family contains some species which show similar or even practically identical morphological characteristics, except for differences in the arrangement of the early chambers (uniserial, biserial, occasionally triserial). A strict adherence to typological generic definitions (e.g. in Loeblich & Tappan, 1987) is therefore not always possible. Various authors have felt that such a separation of genera based on these early chambers disrupts natural units (see also Belford 1960). In the present text, the genera are used in this sense and are named according to which character is predominant, taking account of the rules of nomenclature as far as possible.

ELLIPSODIMORPHINA COALINGENSIS (Cushman & Church)
Figure **37.17–19**

Nodosarella coalingensis Cushman & Church 1929, p. 514, pl. 39, figs. 20–22.
Nodosarella coalingensis Cushman & Church, Cushman & Renz 1946, p. 43, pl. 7, figs. 2, 3.

The species includes specimens with biserial early chambers and entirely uniserial specimens. Length 0.35–1.3, thickness 0.22–0.53 mm.

Stratigraphic range: Middle Paleocene (*Globorotalia pusilla pusilla* Zone) to Early Eocene (*Globorotalia formosa formosa* Zone). Rather rare.

ELLIPSODIMORPHINA MORROWI (van Wessem)
Figure **37.20**

Nodosarella nov. sp., Morrow 1934, p. 197, pl. 29, figs. 2, 3.
Nodosarella morrowi van Wessem 1943, Geogr. Geol. Meded., Physiogr.-geol. Reeks, Utrecht, ser. 2, no. 5, p. 47, pl. 1, fig. 48.

This species is distinguished from the similar *Nodosarella paleocenica* Cushman & Todd in its short but distinct biserial early stage and the rather longer and usually slightly more inflated adult chambers. Length 0.75–1.8, thickness 0.18–0.32 mm.

Stratigraphic range: Early Eocene (*Globorotalia subbotinae* Zone to *G. aragonensis* Zone). Very rare.

ELLIPSODIMORPHINA SOLIDA (Brotzen)
Figure **37.21–22**

Nodosarella solida Brotzen 1936, p. 140, pl. 9, fig. 11.

The early chambers are biserially to loosely biserially or pseudo-uniserially arranged. Length 0.24–1.0, thickness 0.14–0.25 mm.

Stratigraphic range: Early Cenomanian (*Rotalipora appenninica appenninica* Zone) to Maastrichtian (*Abathomphalus mayaroensis* Zone). Very rare.

ELLIPSOGLANDULINA CHILOSTOMA (Rzehak)
Figure **37.24–25**

Glandulina laevigata d'Orbigny var. *chilostoma* Rzehak 1895, Ann. Naturhist. Mus. Wien, 10, p. 19, pl. 7, fig. 6.
Ellipsopleurostomella curta Cushman, Cushman & Jarvis 1928, p. 102, pl. 14, fig. 19 only.
Ellipsoglandulina exponens (H. B. Brady), Cushman & Jarvis 1932, p. 43, pl. 13, fig. 14 only.

Together with typical specimens (Fig. 37.25), there are some with a pointed initial part, which resemble *E. laevigata* Silvestri (Fig. 37.24). The two types cannot be clearly separated and are here combined. It is possible, but not clearly visible, that some of the pointed specimens contain a tiny biserial initial portion. Length 0.45–1.25, thickness 0.25–0.72 mm.

Stratigraphic range: Campanian (*Globotruncana stuarti* Zone) to Early Eocene (*Globorotalia aragonensis* Zone). Rare. In Trinidad also found in the Eocene of the overlying Navet Formation.

ELLIPSOGLANDULINA LABIATA
(Schwager)
Figure **37.23**

Glandulina labiata Schwager 1866, Novara
    Exped. 1857–1859, Vienna, Geol. Theil, 2,
    p. 237, pl. 6, fig. 77.
Ellipsopleurostomella curta Cushman,
    Cushman & Jarvis 1928, p. 102, pl. 14, fig. 18
    only.
Ellipsoglandulina exponens (H.B. Brady),
    Cushman & Jarvis 1932, p. 45, pl. 13, fig. 16
    only.
? *Ellipsoglandulina labiata* (Schwager),
    Cushman & Renz 1948, p. 33, pl. 6, fig. 17.

Length 0.24–0.8, thickness 0.2–0.62 mm.

Stratigraphic range: Late Maastrichtian
(*Abathomphalus mayaroensis* Zone) to Early
Eocene (*Globorotalia aragonensis* Zone).
Rare. In Trinidad also found in the Eocene of
the overlying Navet Formation.

ELLIPSOIDELLA ATTENUATA
(Plummer)
Figure **37.26**

Ellipsopleurostomella attenuata Plummer 1926,
    p. 131, pl. 8, fig. 6.

The Trinidad specimens are sometimes not quite
as slender as the types but do not differ otherwise. The long biserial stage distinguishes them
from *Ellipsodimorphina morrowi*. Length 0.3–
1.15, thickness 0.11–0.26 mm.

Stratigraphic range: Early Paleocene (*Globorotalia
trinidadensis* Zone) to Early Eocene
(*Globorotalia subbotinae* Zone). Rare.

ELLIPSOIDELLA BINARIA Belford
Figure **37.27–28, 37**

Ellipsoidella binaria Belford 1960, p. 71, pl. 19,
    figs. 14–22.

The arrangement of the early chambers varies
mostly between irregularly biserial to almost
uniserial. Very rare specimens with a minute
proloculus (microspheric?) have a triserial initial
stage and could be placed in *Bandyella*. Length
0.35–1.15, thickness 0.16–0.32 mm.

Stratigraphic range: Santonian (*Globotruncana
fornicata* Zone) to Early Maastrichtian

(*Globotruncana lapparenti tricarinata* Zone).
Rare.

ELLIPSOIDELLA KUGLERI (Cushman &
Renz)
Figure **37.29–33**

Nodosarella kugleri Cushman & Renz 1946,
    p. 42, pl. 6, figs. 30, 33.
? *Ellipsonodosaria* sp., Cushman & Renz 1946,
    p. 43, pl. 7, fig. 8.

This group is represented in Trinidad by specimens of variable shape and size. The chamber
arrangement of most of the specimens is biserial
to irregularly or loosely biserial; uniserial chambers may be added in the adult or may even
constitute the entire test. In the Upper Lizard
Springs Formation (from the *Globorotalia subbotinae* Zone upwards), we find occasional
specimens with a subacute base showing a somewhat irregular triserial chamber arrangement.
Length 0.33–1.5, thickness 0.19–0.55 mm.

Stratigraphic range: Late Maastrichtian
(*Abathomphalus mayaroensis* Zone) to Early
Eocene (*Globorotalia aragonensis* Zone).
Rather rare. In Trinidad also found in the
Eocene of the overlying Navet Formation.

ELLIPSOIDELLA cf. ROBUSTA
(Cushman)
Figure **37.34**

Nodosarella robusta Cushman 1943, Contrib.
    Cushman Lab. foramin. Res., 19, p. 92, pl. 16,
    fig. 8.
Nodosarella robusta Cushman, Cushman &
    Renz 1948, p. 31, pl. 6, fig. 5.

The chambers are slightly more inflated than in
typical specimens. The Trinidad form may be
the ancestor of this common Eocene to Miocene
species. Length 0.95–1.4, thickness 0.25–
0.32 mm.

Stratigraphic range: Early Eocene (*Globorotalia
formosa formosa* Zone). Very rare.

ELLIPSOIDINA ABBREVIATA Seguenza
Figure **37.35**

Ellipsoidina abbreviata Seguenza 1859, Eco
  Peloritano, Messina, ser. 2, 5, p. 14, fig. 5.
Ellipsoidina cf. *ellipsoides* Seguenza var.
  abbreviata Seguenza, Cushman & Renz
  1948, p. 34, pl. 6, fig. 19.

Length 0.2–0.8, thickness 0.18–0.7 mm.

Stratigraphic range: Late Maastrichtian
  (*Abathomphalus mayaroensis* Zone) to Early
  Eocene (*Globorotalia aragonensis* Zone).
  Fairly common. In Trinidad also found in the
  Eocene of the overlying Navet Formation.

ELLIPSOIDINA ELLIPSOIDES Seguenza
Figure **37.36**

Ellipsoidina ellipsoides Seguenza 1859, Eco
  Peloritano, Messina, ser. 2, 5, p. 12, figs.
  1–3.

Length 0.48–0.95, thickness 0.3–0.55 mm.

Stratigraphic range: Late Paleocene (*Globorotalia
  pseudomenardii* Zone) to Early Eocene
  (*Globorotalia aragonensis* Zone). Rare. In
  Trinidad also found in the Eocene of the
  overlying Navet Formation.

ELLIPSOPOLYMORPHINA
PRINCIPIENSIS (Cushman & Bermudez)
Figure **37.38**

Ellipsoglandulina principiensis Cushman &
  Bermudez 1937, p. 18, pl. 2, figs. 1–3.
Ellipsoglandulina principiensis Cushman &
  Bermudez, Cushman & Renz 1948, p. 33,
  pl. 6, fig. 16.

The species includes specimens with biserial and
with uniserial early chambers. Length 0.4–1.16,
thickness 0.28–0.65 mm.

Stratigraphic range: Coniacian-Santonian
  (*Globotruncana concavata* Zone) to Early
  Eocene (*Globorotalia formosa formosa* Zone).
  Rare. In Trinidad also found in the Eocene of
  the overlying Navet Formation.

ELLIPSOPOLYMORPHINA
VELASCOENSIS (Cushman)
Figure **37.39**

Ellipsoglandulina velascoensis Cushman 1926,
  p. 590, pl. 16, fig. 7.

The early chambers are biserial to pseudouniserial, possibly rarely uniserial. Length 0.25–0.9,
thickness 0.22–0.66 mm.

Stratigraphic range: Campanian (*Globotruncana
  stuarti* Zone) to Early Eocene (*Globorotalia
  formosa formosa* Zone). Rare. In Trinidad also
  found in the Eocene of the overlying Navet
  Formation.

NODOSARELLA CONSTRICTA Cushman
& Bermudez
Figure **37.40**

Nodosarella constricta Cushman & Bermudez
  1937, p. 18, pl. 2, figs. 4–7.

Length 0.52–1.4, thickness 0.23–0.5 mm.

Stratigraphic range: Late Maastrichtian
  (*Abathomphalus mayaroensis* Zone) to Early
  Eocene (*Globorotalia formosa formosa* Zone).
  Very rare. In Trinidad also found in the Eocene
  of the overlying Navet Formation.

NODOSARELLA HEDBERGI Cushman &
Renz
Figure **37.41–42**

Nodosarella hedbergi Cushman & Renz 1946,
  p. 42, pl. 7, fig. 1.
Ellipsonodosaria subnodosa (Guppy),
  Cushman & Jarvis 1928, p. 102, pl. 14,
  fig. 15 only. – Cushman & Jarvis 1932, p. 45,
  pl. 13, fig. 12 only.

Length 0.85–1.8, thickness 0.25–0.56 mm.

Stratigraphic range: Late Maastrichtian
  (*Abathomphalus mayaroensis* Zone) to Late
  Paleocene (*Globorotalia pseudomenardii*
  Zone). Rare.

NODOSARELLA PALEOCENICA
Cushman & Todd
Figure 37.43–44

Nodosarella paleocenica Cushman & Todd
  1946, Contrib. Cushman Lab. foramin. Res.,
  22, p. 60, pl. 10, fig. 23.
Nodosarella paleocenica Cushman & Todd,
  Cushman & Renz 1948, p. 31, pl. 6, fig. 7.

The chamber arrangement is uniserial, but
sometimes the earliest chamber sutures are
alternately oblique. Length 0.44–1.3, thickness
0.17–0.28 mm.

Stratigraphic range: Early Paleocene (*Globorotalia
trinidadensis* Zone) to Early Eocene
(*Globorotalia aragonensis* Zone). Rare. In
Trinidad also found in the Eocene of the
overlying Navet Formation.

NODOSARELLA SUBNODOSA (Guppy)
Figure 37.45–46

Ellipsoidina subnodosa Guppy 1894, Proc.
  Zool. Soc. London, 1894, p. 650, pl. 41,
  fig. 13.
Nodosarella subnodosa (Guppy), Cushman &
  Renz 1948, p. 31, pl. 6, fig. 6.

Specimens representing this group in the Tertiary of Trinidad and other areas are mostly
referred to *N. subnodosa* in the literature,
although the test is more compact with less
depressed sutures than the illustrated type.
Loosely or irregularly biserial early chambers
are occasionally present; they are also mentioned by Guppy. Length 0.1–1.75, thickness
0.23–0.5 mm.

Stratigraphic range: Late Maastrichtian
(*Abathomphalus mayaroensis* Zone) to Early
Eocene (*Globorotalia aragonensis* Zone).
Rare. In Trinidad also found in the Eocene of
the overlying Navet Formation.

NODOSARELLA WINTERERI Trujillo
Figure 38.1

Nodosarella wintereri Trujillo 1960, p. 345,
  pl. 50, fig. 8.

The test consists of 2–4 chambers and may be
slightly curved. The frequent association with
Bandyella greatvalleyensis and the similar

chamber shape in the uniserial/pseudouniserial
portion in both species could suggest a close
relationship of the two species, even one of
belonging to two generations (megalospheric
and microspheric). However, their stratigraphic
ranges are not exactly identical. Length 0.3–1.0,
thickness 0.23–0.5 mm.

Stratigraphic range: Late Albian (*Favusella
washitensis* Zone) to Santonian (*Globotruncana
fornicata* Zone). Fairly common.

PLEUROSTOMELLA CLAVATA Cushman
Figure 38.2

Pleurostomella clavata Cushman 1926, p. 590,
  pl. 16, fig. 5.
Pleurostomella clavata Cushman, Cushman &
  Jarvis 1932, p. 44, pl. 13, fig. 8. – Cushman
  & Renz 1946, p. 42, pl. 6, figs. 28, 29.

Length 0.57–0.9, thickness 0.21–0.33 mm.

Stratigraphic range: Late Paleocene (*Globorotalia
pseudomenardii* Zone) to Early Eocene
(*Globorotalia formosa formosa* Zone). Rare.

PLEUROSTOMELLA COPIOSA Bulakova
Figure 38.3

Pleurostomella copiosa Bulakova 1960, Trudy
  VNIGRI, Leningrad, vyp. 6, Paleontol.
  Sbornik, 3, p. 229, pl. 1, fig. 8.

Length 0.62–1.2, thickness 0.2 mm.

Stratigraphic range: Santonian (*Globotruncana
fornicata* Zone). Very rare.

PLEUROSTOMELLA CUBENSIS Cushman
& Bermudez
Figure 38.4–5

Pleurostomella alazanensis Cushman, var.
  cubensis Cushman & Bermudez 1937, p. 17,
  pl. 1, figs. 64, 65.
Pleurostomella cubensis Cushman &
  Bermudez, Cushman & Renz 1948, p. 30,
  pl. 6, figs. 1, 3.

The average of the specimens is somewhat
smaller and more slender than the types. Length
0.32–0.9, thickness 0.13–0.24 mm.

Stratigraphic range: Early Eocene (*Globorotalia subbotinae* Zone to *Globorotalia aragonensis* Zone). Rare. In Trinidad also found in the Eocene of the overlying Navet Formation.

PLEUROSTOMELLA NARANJOENSIS
Cushman & Bermudez
Figure **38.6–7**

Pleurostomella naranjoensis Cushmen & Bermudez 1937, p. 16, pl. 1, figs. 59, 60.
Pleurostomella naranjoensis Cushman & Bermudez, Cushman & Renz 1948, p. 30, pl. 5, fig. 21.

Length 0.32–1.25, thickness 0.2–0.45 mm.

Stratigraphic range: Late Paleocene (*Globorotalia pseudomenardii* Zone) to Early Eocene (*Globorotalia aragonensis* Zone). Rather rare. In Trinidad also found in the Eocene of the overlying Navet Formation.

PLEUROSTOMELLA OBTUSA Berthelin
Figure **38.8–10**

Pleurostomella obtusa Berthelin 1880, p. 29, pl. 1, fig. 9.

The specimens vary considerably in size and in the relative thickness of the test and the chambers. The larger specimens (Fig. 38.8) are confined to the Early Cenomanian. Length 0.33–1.2, thickness 0.08–0.3 mm.

Stratigraphic range: Early Cenomanian (*Rotalipora appenninica appenninica* Zone) to Turonian/Coniacian (*Globotruncana renzi* Zone). Fairly common (rare only in the *Globotruncana inornata* Zone and *G. renzi* Zone).

PLEUROSTOMELLA PALEOCENICA
Cushman
Figure **38.11–12**

Pleurostomella paleocenica Cushman 1947, Contrib. Cushman Lab. foramin. Res., 23, p. 86, pl. 18, figs. 14, 15.

Length 0.2–0.63, thickness 0.06–0.15 mm.

Stratigraphic range: Middle Paleocene (*Globorotalia angulata* Zone) to Early Eocene (*Globorotalia subbotinae* Zone). Rather rare.

PLEUROSTOMELLA cf. REUSSI Berthelin
Figure **38.13**

Pleurostomella reussi Berthelin 1880, p. 28, pl. 1, figs. 10–12.

The specimens have more distinctly inflated chambers than the types. The initial chambers are arranged in an irregularly triserial manner, though not as distinctly as in *Bandyella*. It is possible that this is a microspheric generation of *P. obtusa*. Length 0.52–1.1, thickness 0.14–0.3 mm.

Stratigraphic range: Early Cenomanian (*Rotalipora appenninica appenninica* Zone). Rare. Very rare smaller, less typical specimens occur in the Turonian (*Globotruncana inornata* Zone).

PLEUROSTOMELLA VELASCOENSIS
Cushman
Figure **38.14–15**

Pleurostomella velascoensis Cushman 1926, p. 590, pl. 16, fig. 4.
? *Pleurostomella torta* Cushman, Cushman & Renz 1946, p. 39, pl. 6, figs. 26, 27.

Specimens with a pointed base (microspheric?) appear to have a very small triserial early stage. Length 0.35–1.05, thickness 0.17–0.3 mm.

Stratigraphic range: Paleocene (*Globorotalia pusilla pusilla* Zone to *G. velascoensis* Zone). Rare.

BANDYELLA GREATVALLEYENSIS
(Trujillo)
Figure **38.17–18**

Pleurostomella greatvalleyensis Trujillo 1960, p. 345, pl. 50, figs. 5, 6.

The triserial early stage, which is apparently always present, is often followed by an irregularly arranged portion which may become pseudouniserial. Length 0.18–1.1, thickness 0.14–0.5 mm.

Stratigraphic range: Turonian (*Globotruncana inornata* Zone) to Santonian (*Globotruncana fornicata* Zone). Fairly common.

CZARKOWYELLA TORTA (Cushman)
Figure **38.16**

Pleurostomella torta Cushman 1926, Contrib.
   Cushman Lab. foramin. Res., 2, p. 18, pl. 2,
   fig. 7.
? *Pleurostomella subnodosa* Reuss var. *gigantia*
   White 1929, p. 53, pl. 5, fig. 16.

As already suggested by Cushman's type figure,
the earliest chambers are irregularly triserial
rather than biserial. In this, as in the apertural
characteristics, the species fits well into the
genus definition of *Czarkowyella* by Gawor-
Biedova (1987). *C. wadowicensis* (Grzybowski;
Pleurostomella w.) is sometimes cited as a senior
synonym of *C. torta*; however, the revision work
of Liszka & Liszkova (1981) seems to indicate
that *'Pleurostomella' wadowicensis* is rather
more slender with slightly longer chambers.
Length 0.65–3.0, thickness 0.27–0.85 mm.

Stratigraphic range: Campanian (*Globotruncana
   stuarti* Zone) to Late Maastrichtian
   (*Abathomphalus mayaroensis* Zone). Rare.

Family Stilostomellidae

ORTHOMORPHINA ROHRI (Cushman &
Stainforth)
Figure **38.21–22**

Nodogenerina rohri Cushman & Stainforth
   1945, Cushman Lab. foramin. Res., Special
   Publ., 14, p. 39, pl. 5, fig. 26.
Nodosaria monile Hagenow, Cushman & Jarvis
   1932, p. 33, pl. 10, fig. 9. – Cushman &
   Renz 1946, p. 30, pl. 4, fig. 26.
Nodogenerina rohri Cushman & Stainforth,
   Cushman & Renz 1948, p. 24, pl. 5, figs. 9–
   11.

Length 0.3–1.75, thickness 0.12–0.3 mm.

Stratigraphic range: Early Maastrichtian
   (*Globotruncana lapparenti tricarinata* Zone)
   to Early Eocene (*Globorotalia aragonensis*
   Zone). Rather rare. In Trinidad also found in
   the Eocene of the overlying Navet Formation.

SIPHONODOSARIA
DENTATAGLABRATA (Cushman)
Figure **38.23**

Ellipsonodosaria dentata-glabrata Cushman
   1936, Contrib. Cushman Lab. foramin.
   Res., 12, p. 54, pl. 9, figs. 22, 23.

The Trinidad specimens are slightly bigger than
the types from the Late Cretaceous. Length
1.27–1.4, thickness 0.24–0.28 mm.

Stratigraphic range: Early Eocene (*Globorotalia
   subbotinae* Zone). Very rare. In Trinidad also
   found in the Eocene of the overlying Navet
   Formation.

SIPHONODOSARIA LONGAE McLean
Figure **38.24**

Siphonodosaria longae McLean 1952, Acad.
   Nat. Sci. Philadelphia, Notulae Naturae, 242,
   p. 10, pl. 2, figs. 17–20.
Ellipsonodosaria recta Palmer & Bermudez,
   Cushman & Renz 1948, p. 32, pl. 6, fig. 14.

Length 0.45–1.43, thickness 0.13–0.22 mm.

Stratigraphic range: Late Paleocene (*Globorotalia
   velascoensis* Zone) to Early Eocene
   (*Globorotalia aragonensis* Zone). Rare. In
   Trinidad also found in the Eocene of the
   overlying Navet Formation.

SIPHONODOSARIA NUTTALLI
NUTTALLI (Cushman & Jarvis)
Figure **38.25**

Ellipsonodosaria nuttalli Cushman & Jarvis
   1934, p. 72, pl. 10, fig. 6.

The early Tertiary specimens are rather smaller
than the type from the Miocene but are otherwise morphologically the same. Length 0.39–
1.35, thickness 0.13–0.24 mm.

Stratigraphic range: Late Paleocene (*Globorotalia
   pseudomenardii* Zone) to Early Eocene
   (*Globorotalia aragonensis* Zone). Rare. In
   Trinidad also found in the Eocene of the
   overlying Navet Formation.

SIPHONODOSARIA NUTTALLI GRACILLIMA (Cushman & Jarvis)
Figure **38.26**

Ellipsonodosaria nuttalli Cushman & Jarvis var. *gracillima* Cushman & Jarvis 1934, Contrib. Cushman Lab. foramin. Res., 10, p. 72, pl. 10, fig. 7.
? *Ellipsonodosaria nuttalli* Cushman & Jarvis var. *gracillima* Cushman & Jarvis, Cushman & Renz 1948, p. 32, pl. 6, fig. 9.

Length up to 1.25, thickness 0.15–0.20 mm.

Stratigraphic range: Early Eocene (*Globorotalia subbotinae* Zone to *G. aragonensis* Zone). Rare. In Trinidad also found in the Eocene of the overlying Navet Formation.

STILOSTOMELLA ALEXANDERI (Cushman)
Figure **38.27–28**

Ellipsonodosaria alexanderi Cushman 1936, Contrib. Cushman Lab. foramin. Res., 12, p. 52, pl. 9, figs. 6–9.
Ellipsonodosaria alexanderi var. *impensia* Cushman 1938, Contrib. Cushman Lab. foramin. Res., 14, p. 48, pl. 8, figs. 4, 5. – Cushman & Renz 1947a, p. 48, pl. 12, figs. 9, 10.

The long chambers carry spines which may be directed backwards. Length up to 1.4 mm or more (always broken), thickness 0.15–0.4 mm.

Stratigraphic range: Santonian (*Globotruncana fornicata* Zone) to Early Maastrichtian (*Globotruncana lapparenti tricarinata* Zone). Rare.

STILOSTOMELLA PALEOCENICA (Cushman & Todd)
Figure **38.29–30**

Ellipsonodosaria paleocenica Cushman & Todd 1946, Contrib. Cushman Lab. foramin. Res., 22, p. 61, pl. 10, fig. 26.

In Trinidad, the specimens are somewhat smaller than the holotype but otherwise identical. Length up to more than 1.0 mm (broken), thickness 0.06–0.18 mm.

Stratigraphic range: Santonian (*Globotruncana fornicata* Zone) to Early Eocene (*Globorotalia*

formosa formosa Zone). Fairly common. In Trinidad also found in the Eocene of the overlying Navet Formation.

STILOSTOMELLA SPINEA (Cushman)
Figure **38.31–33**

Ellipsonodosaria curvatura Cushman var. *spinea* Cushman 1939, Contrib. Cushman Lab. foramin. Res., 15, p. 71, pl. 12, figs. 7–11.

The surface ornamentation is variable and consists of pustules or short spines projected backwards. Length up to more than 3.2 mm (specimens mostly broken), thickness 0.24–0.55 mm.

Stratigraphic range: Campanian (*Globotruncana stuarti* Zone) to Middle Paleocene (*Globorotalia pusilla pusilla* Zone). Rare.

Family Conorbinidae

CONORBINA cf. HAIDINGERI (d'Orbigny)
Figure **38.34–36, 39–41**

Rotalina haidingerii d'Orbigny 1846, Foram. Foss. Bassin Tert. Vienne, p. 154, pl. 8, figs. 7–9.

The Trinidad specimens from the Late Cretaceaus vary in shape from distinctly planoconvex to rounded biconvex; intermediate stages are figured by Cushman (1931) from Tennessee. In the basal Tertiary they are more poorly preserved and differ in usually having a small open umbilicus (Fig. 38.41) and occasionally slightly raised sutures in the umbilical area; these represent possibly a separate species or subspecies. The basal Tertiary form may be the same as *Rotalia*-6 described by Fuenmayor (1989) from the Paleocene Guasare Formation of western Venezuela. Diameter 0.25–0.48, thickness 0.16–0.29 mm.

Stratigraphic range: Late Maastrichtian (*Abathomphalus mayaroensis* Zone) to Early Paleocene (probably *Globorotalia pseudobulloides* Zone; Rochard facies). Rather rare.

Family Bagginidae

ROTAMORPHINA CUSHMANI Finlay
Figure 38.37–38, 42–44

Rotamorphina cushmani Finlay 1939, Trans.
  Proc. Royal Soc. New Zealand, 69, p. 325,
  pl. 28, figs. 130–133.
Valvulineria jarvisi Martin 1964, p. 103, pl. 15,
  fig. 4.
Valvulineria teuriensis n. name Loeblich &
  Tappan 1964, Treatise Invert. Paleontol.,
  C, p. C587 (for *V. cushmani* (Finlay 1939),
  non *V. cushmani* Coryell & Embich 1937).
Valvulineria allomorphinoides (Reuss),
  Cushman & Jarvis 1932, p. 46, pl. 13, fig. 17.

The number of chambers in Trinidad varies from
4½ to 7. Except for some specimens in the Paleocene (*Globorotalia pusilla pusilla* Zone to *G. velascoensis* Zone), the periphery is rather less
rounded than figured by Cushman & Jarvis.
Relatively compressed specimens, like those
described by Martin (1964) as *V. jarvisi*, can be
found in some samples of Maastrichtian and
Early Paleocene age. Diameter 0.24–0.85,
thickness 0.14–0.51 mm.

Stratigraphic range: Maastrichtian (*Globotruncana
gansseri* Zone) to Late Paleocene (*Globorotalia
velascoensis* Zone). Fairly common.

VALVULINERIA PARVA Khan
Figure 39.1–2

Valvulineria parva Khan 1950, J. Royal Micr.
  Soc. London, ser. 3, 70, p. 275, pl. 2, figs.12–
  14, 19.

V. gracillima ten Dam is very similar but is
rather larger and has only 7 chambers in the last
whorl (against 7–9 in *V. parva*). Diameter 0.16–
0.24, thickness 0.1–0.14 mm.

Stratigraphic range: Late Albian (*Rotalipora
ticinensis ticinensis* Zone to *Favusella
washitensis* Zone). Rare.

VALVULINERIA SUBINFREQUENS
Dailey
Figure 39.3–5

Valvulineria subinfrequens Dailey 1970, p. 107,
  pl. 12, fig. 11.

Diameter 0.2–0.32, thickness 0.08–0.15 mm.

Stratigraphic range: Late Albian (*Favusella
washitensis* Zone). Very rare.

Family Eponididae

A few species which are often referred to the genus
Eponides are listed here in quotation marks, as they
differ from this genus as defined by the type species
E. repandus.

'EPONIDES' cf. HILLEBRANDTI (Fisher)
Figure 39.7–9

Eponides whitei Hillebrandt 1962, p. 106, pl. 8,
  fig. 11.
Neoeponides hillebrandti, new name, Fisher
  1969, Palaeontology, London, 12, p. 196
  (for *Eponides whitei* Hillebrandt 1962, non
  Brotzen 1936).

This species lacks the stellate pattern of *Neoeponides* on the umbilical side and is therefore
tentatively assigned to *Eponides*. The available
specimens are slightly smaller than Hillebrandt's
types and have rather less chambers (mostly 6–
7) in the last whorl. Diameter 0.18–0.29, thickness 0.13–0.18 mm.

Stratigraphic range: Maastrichtian (*Globotruncana
gansseri* Zone) to Middle Paleocene
(*Globorotalia angulata* Zone). Rare.

'EPONIDES' cf. HUAYNAI Frizzell
Figure 39.10–11

Eponides huaynai Frizzell 1943, J. Paleontol.,
  17, p. 351, pl. 57, fig. 7.

The test has 8–10 chambers in the final whorl.
The periphery is usually slightly rounded and
not quite as acute as indicated by Frizzell. The
sutures in the early part of the last whorl may
be slightly raised and limbate. Diameter 0.34–
0.52, thickness 0.13–0.25 mm.

Stratigraphic range: Probably Early Maastrichtian (*Globotruncana lapparenti tricarinata* Zone; Marabella facies). Rare.

'EPONIDES' PLUMMERAE Cushman
Figure **39.13–15**

Truncatulina tenera Plummer 1926 (non
  Brady), p. 146, pl. 9, fig. 5.
Eponides plummerae Cushman 1948, Contrib.
  Cushman Lab. foramin. Res., 24, p. 44, pl. 8,
  fig. 9.
Gyroidina beisseli White, Cushman & Renz
  1946, p. 44, pl. 7, figs. 21, 22.

The sutures on the spiral side are usually oblique. *E. praemegastomus* Myatlyuk may be a junior synonym. Diameter 0.25–0.72, thickness 0.2–0.48 mm.

Stratigraphic range: Late Maastrichtian (*Abathomphalus mayaroensis* Zone) to Late Paleocene (*Globorotalia velascoensis* Zone). Rather rare.

'EPONIDES' SUBUMBONATUS Myatlyuk
Figure **39.16–18**

Eponides subumbonatus Myatlyuk 1953,
  p. 109, pl. 15, figs. 2, 3.
? *Eponides simplex* (White), Cushman & Renz
  1946, p. 46, pl. 8, fig. 1.
? *Eponides umbonatus* (Reuss), Cushman &
  Renz 1948, p. 35, pl. 7, fig. 7 (not fig. 6).

The test is similar to that of the genus *Oridorsalis*, but lacks the dorsal supplementary apertures. Diameter 0.25–0.6, thickness 0.12–0.33 mm.

Stratigraphic range: Early Eocene (*Globorotalia edgari* Zone to *G. aragonensis* Zone). Fairly common. In Trinidad also found in the Eocene of the overlying Navet Formation.

Family Mississippinidae

STOMATORBINA (?) sp. A
Figure **39.6, 12**

Test flat to slightly planoconvex; spiral side slightly convex, umbilical side flat or slightly concave; periphery keeled. Sutures slightly depressed, curved backwards towards the periphery. Aperture peripheral to umbilical, forming grooves along the sutures on the umbilical side.

Because of the poor preservation of the specimens, important morphological features such as the sutures and apertures are not well seen, and the generic position must be left open. The tests differ from similar large forms in the Late Cretaceous like *Mississippina (?) binkhorsti* (Reuss; *Rosalina b.*), *Stomatorbina ranikotensis* Haque and *Discorbis midwayensis* var. *trinitatensis* Cushman & Renz mainly in the absence of distinctly raised and thickend sutures on the spiral side. '*Cibicides*' *hedbergi* Petters is of about the same size and shape, but has a flat spiral side and a peripheral aperture. Diameter: 0.55–1.2, thickness 0.15–0.3 mm.

Stratigraphic range: Probably Early Maastrichtian (*Globotruncana lapparenti tricarinata* Zone; Marabella facies). Rare.

Family Parrelloididae

CIBICIDOIDES MERUS (Cushman & Renz)
Figure **39.19–21**

Eponides bronnimanni Cushman & Renz var.
  mera Cushman & Renz 1946, p. 45, pl. 7,
  fig. 25.

The type in the Cushman Collection (Washington, D.C.) is not an *Eponides* but appears to be conspecific with the present material. *C. subspiratus* (Nuttall; *Cibicides s.*) is a probable successor; it has more (10–11) chambers and more strongly curved sutures. Diameter 0.35–0.65, thickness 0.16–0.28 mm.

Stratigraphic range: Late Paleocene (*Globorotalia velascoensis* Zone) to Early Eocene (*Globorotalia subbotinae* Zone). Rare.

CIBICIDOIDES PADELLA (Jennings)
Figure **39.22–24**

Cibicides padella Jennings 1936, p. 40, pl. 5,
  fig. 6.

Diameter 0.22–0.42, thickness 0.11–0.2 mm.

Stratigraphic range: Maastrichtian (*Globotruncana*

gansseri Zone) to Early Paleocene (*Globorotalia trinidadensis* Zone). Rather rare.

CIBICIDOIDES PRAEMUNDULUS
Berggren & Miller
Figure 39.25–27

Cibicidoides praemundulus Berggren & Miller 1986, in: Morkhoven *et al.* 1986, p. 164, pl. 87, figs. 1–3; textfigs. 5, 6.

Diameter 0.27–0.58, thickness 0.13–0.2 mm.

Stratigraphic range: Early Eocene (*Globorotalia edgari* Zone to *G. aragonensis* Zone). Rare. In Trinidad also found in the Eocene of the overlying Navet Formation.

CIBICIDOIDES PROPRIUS (Brotzen)
Figure 39.28–30

Cibicides (Cibicidoides) proprius Brotzen 1948, p. 78, pl. 12, figs. 3, 4.

The species shows some variability in the shape of the test (biconvex to planoconvex) and the width of the spiral. It appears to be the ancestor of the group around *C. pseudoungerianus* (Cushman; *Cibicides p.*). Generally, the periphery is more lobate than in *C. praemundulus*, and there are less than 10 chambers in the final whorl. Diameter 0.22–0.8, thickness 0.12–0.3 mm.

Stratigraphic range: Early Paleocene (*Globorotalia trinidadensis* Zone) to Early Eocene (*Globorotalia aragonensis* Zone). Common. In Trinidad also found in the Eocene of the overlying Navet Formation.

CIBICIDOIDES SUCCEDENS (Brotzen)
Figure 40.7–9

Cibicides succedens Brotzen 1948, p. 80, pl. 12, figs. 1, 2; textfig. 21.

Diameter 0.25–0.62, thickness 0.15–0.26 mm.

Stratigraphic range: Late Maastrichtian (*Abathomphalus mayaroensis* Zone), possibly also Early Paleocene (*Globorotalia trinidadensis* Zone). Rare.

CIBICIDOIDES TUXPAMENSIS LAXISPIRALIS Beckmann
Figure 40.4–6

Cibicidoides tuxpamensis laxispiralis Beckmann 1991, p. 827, pl. 2, figs. 10, 14, 15.

Test relatively large, biconvex to planoconvex with a more or less flattened dorsal (spiral) side; periphery subacute, sometimes with a keel. 7–10 chambers in the last whorl. Dorsal side moderately to nearly involute, with oblique and slightly curved sutures and a distinct, sometimes spirally shaped central glassy thickening. Ventral side convex with a prominent clear plug. Aperture peripheral, extending to both sides of the test.

This subspecies may be an ancestral form of *C. tuxpamensis tuxpamensis*; it shows a wider spiral and semi-involute coiling on the dorsal side. There are also specimens which show transitional characters between the two subspecies. *C. tuxpamensis* var. *jabacoensis* Bermudez is quite similar in dorsal view but has more (12–13) chambers in the final whorl. Diameter 0.4–1.1, thickness 0.24–0.55 mm.

Stratigraphic range: Early Eocene (*Globorotalia subbotinae* Zone to *G. formosa formosa* Zone). Fairly common

CIBICIDOIDES TUXPAMENSIS TUXPAMENSIS (Cole)
Figure 40.1–3

Cibicides tuxpamensis Cole 1928, Bull. Am. Paleontol., 14/53, p. 219, pl. 1, figs. 2, 3; pl. 3, figs. 5, 6.

Diameter 0.36–0.7, thickness 0.18–0.38 mm.

Stratigraphic range: Early Eocene (*Globorotalia subbotinae* Zone to *G. aragonensis* Zone). Fairly common. In Trinidad also found in the Eocene of the overlying Navet Formation.

CIBICIDOIDES UMBONIFER (Schwager)
Figure 40.10–11

Discorbina umbonifera Schwager 1883, Palaeontographica, Cassel, 30, part 2, p. 126, pl. 27, fig. 14.

This species differs from *C. padella* in the lesser number of chambers (8–10 against 9–12) and the rather more lobate periphery. *C. harperi* (Sandidge; *Anomalina h.*) may be closely related but has a more biconvex test. Diameter 0.3–0.46, thickness 0.12–0.18 mm.

Stratigraphic range: Maastrichtian (*Globotruncana gansseri* Zone to *Abathomphalus mayaroensis* Zone). Rare.

Family Cibicididae

CIBICIDINA NEWMANAE (Plummer)
Figure 40.12–14

Discorbis newmanae Plummer 1926, p. 138, pl. 9, fig. 4.
Cibicides cf. *semiplectus* (Schwager), Cushman & Renz 1942, p. 14, pl. 3, fig. 11.

Diameter 0.21–0.36, thickness 0.06–0.09 mm.

Stratigraphic range: Early Paleocene (probably *Globorotalia pseudobulloides* Zone; Rochard facies). Rather scarce.

CIBICIDINA cf. SIMPLEX (Brotzen)
Figure 40.15–17

Cibicides simplex Brotzen, 1948, p. 83, pl. 13, figs. 4, 5.

Test bi-involute, with 8–9 chambers in the final whorl. Similar to *C. simplex* but rather thinner. Umbilicus small, slightly deepened, without the small umbilical disk sometimes present in the types. Diameter 0.16–0.35 (rarely up to 0.45), thickness 0.08–0.2 mm.

Stratigraphic range: Early Paleocene (probably *Globorotalia pseudobulloides* Zone; Rochard facies). Rather scarce.

FALSOPLANULINA ACUTA (Plummer)
Figure 40.18–20

Anomalina ammonoides (Reuss) var. *acuta* Plummer 1926, p. 149, pl. 10, fig. 2.
Anomalina acuta Plummer, Cushman & Renz 1942, p. 12, pl. 3, fig. 6.
? *Planulina schloenbachi* (Reuss), Cushman & Jarvis 1932, p. 52, pl. 16, fig. 7.

Diameter 0.24–0.5, thickness 0.09–0.19 mm.

Stratigraphic range: Paleocene (probably *Globorotalia pseudobulloides* Zone, Rochard facies; *Globorotalia trinidadensis* Zone to *G. pusilla pusilla* Zone). Fairly common.

FALSOPLANULINA WALTONENSIS (Applin & Jordan)
Figure 40.21–23

Planulina waltonensis Applin & Jordan 1945, J. Paleontol., 19, p. 147, pl. 20, fig. 5.

This is probably the successor of *F. acuta*; transitional specimens exist in the Middle Paleocene (*Globorotalia pusilla pusilla* Zone) in particular, where they start to develop an evolute dorsal side. Diameter 0.3–0.64, thickness 0.1–0.19 mm.

Stratigraphic range: Paleocene (*Globorotalia pusilla pusilla* Zone?, *G. pseudomenardii* Zone) to Early Eocene (*Globorotalia formosa formosa* Zone). Fairly common.

Family Epistomariidae

NUTTALLIDES TRUEMPYI CRASSAFORMIS (Cushman & Siegfus)
Figure 40.24–26

Asterigerina crassaformis Cushman & Siegfus 1935, Contrib. Cushman Lab. foramin. Res., 11, p. 94, pl. 14, fig. 10.

This subspecies includes fairly highly conical specimens without an umbilical plug. Diameter 0.2–0.33, thickness 0.12–0.21 mm.

Stratigraphic range: Middle Paleocene (*Globorotalia uncinata* Zone) to Early Eocene (*Globorotalia formosa formosa* Zone). Rare. In Trinidad also found in the Eocene of the overlying Navet Formation.

NUTTALLIDES TRUEMPYI TRUEMPYI (Nuttall)
Figure 40.27–28

Eponides trümpyi Nuttall 1930, p. 287, pl. 24, figs. 9, 13, 14.
Eponides bronnimanni Cushman & Renz 1946, p. 45, pl. 7, fig. 24.
Eponides trumpyi Nuttall, Cushman & Renz

1948, p. 35, pl. 7, figs. 6, 8 (not fig. 7, which should be labelled *Eponides umbonatus*).

Nuttallides carinotuempyi Finlay and *Asterigerina crassaformis* var. *umbilicatula* Mallory are probably synonyms. Diameter 0.22–0.7, thickness 0.13–0.4 mm.

Stratigraphic range: Early Maastrichtian (*Globotruncana lapparenti tricarinata* Zone) to Early Eocene (*Globorotalia aragonensis* Zone). Common. In Trinidad also found in the Eocene of the overlying Navet Formation.

NUTTALLINELLA CORONULA (Belford)
Figure **41.1–2**

Nuttallina coronula Belford 1958, Contrib. Cushman Found. foramin. Res., 9, p. 97, pl. 19, figs. 1–14; textfig. 4.

The specimens are flatter and larger than those of *N. florealis*, and the broad, usually ragged keel lies in a plane at a right angle to the axis of coiling. Diameter 0.45–1.25, thickness 0.17–0.3 mm.

Stratigraphic range: Campanian (*Globotruncana stuarti* Zone) to Early Maastrichtian (*Globotruncana lapparenti tricarinata* Zone). Rare.

NUTTALLINELLA FLOREALIS (White)
Figure **41.3–5**

Gyroidina florealis White 1928, J. Paleontol., 2, p. 293, pl. 40, fig. 3.
Pulvinulinella alata (Marsson), Cushman & Jarvis 1932, p. 48, pl. 15, figs. 1, 2.
Pulvinulinella? florealis (White), Cushman & Renz 1946, p. 46, pl. 8, figs. 4, 5.

Typical representatives are fairly high conical, usually with a narrow keel (Fig. 41.3); a broader keel (Fig. 41.5) is sometimes found in parts of the Lower Lizard Springs Formation (*Globorotalia pusilla pusilla* Zone to *G. velascoensis* Zone). There is a noticeable size increase in the Maastrichtian between the *Globotruncana lapparenti tricarinata* Zone and the *G. gansseri* Zone. Diameter 0.21–0.9, thickness 0.16–0.54 mm.

Stratigraphic range: Santonian (*Globotruncana fornicata* Zone) to Late Paleocene (*Globorotalia velasoensis* Zone).

NUTTALLINELLA cf. FLOREALIS
(White)
Figure **41.6–7**

Slightly smaller specimens of the *N. florealis* group with a faint or indistinct keel are found predominantly in the lowermost part of the Lizard Springs Formation, where they sometimes replace the typical form almost completely. Diameter 0.22–0.5, thickness 0.17–0.33 mm.

Stratigraphic range: Paleocene (*Globorotalia trinidadensis* Zone to *G. pusilla pusilla* Zone). Rather rare.

NUTTALLINELLA LUSITANICA Fisher
Figure **41.8–10**

Nuttallinella lusitanica Fisher 1969, Palaeontology, London, 12, p. 195, textfig. 2 (p. 196).

In some specimens, the umbilicus is not completely closed but is occupied by a small clear plug. Diameter 0.3–0.42, thickness 0.16–0.21 mm.

Stratigraphic range: Middle Paleocene (*Globorotalia pusilla pusilla* Zone). Rare.

Family Nonionidae

SPIROTECTA (?) OVATA (Balakhmatova)
Figure **41.11–13**

Nonion ovatus Balakhmatova 1955, Materialy VSEGEI, Moscow, n. ser., 2, p. 33, pl. 3, fig. 1.

The test is not as compressed as in the type species of *Spirotecta*, *S. pellicula*, and is only slightly asymmetric, sometimes practically symmetric. The periphery is subrounded, rarely subangular, and there are 4–5½ chambers in the final whorl. Rounded, symmetric specimens are close to *Pullenia eggeri* Cushman & Todd, and compressed, fairly symmetric specimens resemble '*Allomorphinella*' *nonioides* Dain. Diameter 0.25–0.45, thickness 0.16–0.3 mm.

Stratigraphic range: Late Maastrichtian (*Globotruncana gansseri* Zone?, *Abathomphalus mayaroensis* Zone). Rare.

NONION HAVANENSE Cushman &
Bermudez
Figure **41.14–15**

Nonion havanense Cushman & Bermudez 1937,
　p. 19, pl. 2, figs. 13, 14.
Nonion havanense Cushman & Bermudez,
　Cushman & Renz 1948, p. 22, pl. 5, fig. 4.

7–10 chambers in the final whorl. Diameter
0.23–0.55, thickness 0.1–0.28 mm.

Stratigraphic range: Late Paleocene (*Globorotalia
pseudomenardii* Zone) to Early Eocene
(*Globorotalia aragonensis* Zone). Rather rare.
In Trinidad also found in the Eocene of the
overlying Navet Formation.

NONIONELLA AUSTINANA Cushman
Figure **41.16–18**

Nonionella austinana Cushman 1933, Contrib.
　Cushman Lab. foramin. Res., 9, p. 57, pl. 7,
　fig. 2.

Diameter 0.2–0.32, thickness 0.07–0.15 mm.

Stratigraphic range: Santonian to Campanian
(*Globotruncana fornicata* Zone to *G. stuarti*
Zone). Rare.

NONIONELLA ROBUSTA Plummer
Figure **41.19, 25**

Nonionella robusta Plummer 1931, p. 175,
　pl. 14, fig. 12.
Nonoinella robusta Plummer, Cushman &
　Renz 1946, p. 36, pl. 5, figs. 32, 33.

The test is sometimes rather less pointed than
in the type specimen. Diameter 0.25–0.42,
thickness 0.15–0.2 mm. A few small but otherwise practically identical specimens have a diameter of 0.12–0.16 mm.

Stratigraphic range: Middle Paleocene (*Globorotalia
pusilla pusilla* Zone) to Early Eocene
(*Globorotalia aragonensis* Zone). Very rare.

NONIONELLA cf. SOLDADOENSIS
Cushman & Renz
Figure **41.20, 26**

Nonionella soldadoensis Cushman & Renz
　1942, p. 7, pl. 2, fig. 7.

The types of *N. soldadoensis* in the Cushman
collection are badly preserved and probably secondarily compressed. The specimens in the present material are quite variable in shape and
size; some resemble *N. jacksonensis* Cushman
but are considerably smaller; small specimens
are very close to *N. africana* LeRoy or *N. excavata* var. *thalmanni* Haque. Diameter 0.18–0.48,
thickness 0.05–0.2 mm.

Stratigraphic range: Middle Paleocene (*Globorotalia
pusilla pusilla* Zone) to Early Eocene
(*Globorotalia formosa formosa* Zone).
Specimens larger than 0.3 mm are confined to
the Early Eocene. Rare.

PULLENIA ANGUSTA Cushman & Todd
Figure **41.21–22**

Pullenia quinqueloba (Reuss) var. *angusta*
　Cushman & Todd 1943, Contrib. Cushman
　Lab. foramin. Res., 19, p. 10, pl. 2, fig. 3,
　4.
? *Pullenia minuta* Cushman, Cushman & Renz
　1947a, p. 49, pl. 12, fig. 8.

P. minuta Cushman is very similar in shape but
is distinctly smaller. Diameter 0.22–0.38, thickness 0.12–0.27 mm.

Stratigraphic range: Probably Early Maastrichtian
(*Globotruncana lapparenti tricarinata* Zone;
Marabella facies); Maastrichtian
(*Globotruncana gansseri* Zone) to Early Eocene
(*Globorotalia aragonensis* Zone). Rare.

PULLENIA CORYELLI White
Figure **41.23–24**

Pullenia coryelli White 1929, J. Paleontol., 3,
　p. 56, pl. 5, fig. 22.
Pullenia coryelli White, Cushman & Jarvis
　1932, p. 50, pl. 15, fig. 5. – Cushman & Renz
　1946, p. 47, pl. 7, fig. 9.

Diameter 0.2–0.55, thickness 0.18–0.48 mm.

Stratigraphic range: Early Maastrichtian
(*Globotruncana lapparenti tricarinata* Zone)
to Late Paleocene (*Globorotalia velascoensis*
Zone). Common.

PULLENIA CRETACEA Cushman
Figure **41.27–28**

Pullenia cretacea Cushman 1936, Contrib.
Cushman Lab. foramin. Res., 12, p. 75, pl. 13, fig. 8.

Diameter 0.2–0.33, thickness 0.18–0.25 mm.

Stratigraphic range: Santonian (*Globotruncana fornicata* Zone) to Late Maastrichtian (*Abathomphalus mayaroensis* Zone). Rather rare.

PULLENIA EOCENICA Cushman & Siegfus
Figure **41.29–30**

Pullenia eocenica Cushman & Siegfus 1939, Contrib. Cushman Lab. foramin. Res., 15, p. 31, pl. 7, fig. 1.
Pullenia compressiuscula Reuss, Cushman & Renz 1948, p. 38, pl. 7, fig. 16.

Diameter 0.28–0.72, thickness 0.24–0.55 mm.

Stratigraphic range: Late Maastrichtian (*Abathomphalus mayaroensis* Zone) to Early Eocene (*Globorotalia formosa formosa* Zone). Rare. In Trinidad also found in the Eocene of the overlying Navet Formation.

PULLENIA JARVISI Cushman
Figure **42.1–3**

Pullenia jarvisi Cushman 1936, Contrib. Cushman Lab. foramin. Res., 12, p. 77, pl. 13, fig. 6.
Pullenia jarvisi Cushman, Cushman & Renz 1946, p. 47, pl. 8, fig. 10.

The test is variable in thickness, and the number of chambers is mostly 5 but can vary between 4 and 6. Diameter 0.21–0.67, thickness 0.15–0.44 mm.

Stratigraphic range: Santonian (*Globotruncana fornicata* Zone) to Early Eocene (*Globorotalia aragonensis* Zone). Fairly common. In Trinidad also found in the Eocene of the overlying Navet Formation.

PULLENIA QUINQUELOBA (Reuss)
Figure **41.31–32**

Nonionina quinqueloba Reuss 1851, Zeitschr. Deut. Geol. Ges., Berlin, 3, p. 71, pl. 5, fig. 31.
Pullenia quinqueloba (Reuss), Cushman & Jarvis 1932, p. 49, pl. 15, fig. 4.

In general, the test is smaller and more compressed than in *P. jarvisi*, but in the zones where the two species overlap, a clear separation is sometimes difficult. Diameter 0.26–0.41, thickness 0.21–0.28 mm.

Stratigraphic range: Middle Paleocene (*Globorotalia angulata* Zone) to Early Eocene (*Globorotalia aragonensis* Zone). Fairly common. In Trinidad also found in the Eocene of the overlying Navet Formation.

Family Chilostomellidae

ALLOMORPHINA CRETACEA Reuss
Figure **42.4**

Allomorphina cretacea Reuss 1851, p. 42, pl. 5, fig. 6.

Length 0.24–0.5, width 0.21–0.39, thickness 0.15–0.3 mm.

Stratigraphic range: Early Paleocene (probably *Globorotalia pseudobulloides* Zone, Rochard facies). Very rare.

CHILOSTOMELLA CZIZEKI Reuss
Figure **42.5**

Chilostomella czizeki Reuss 1850, Denkschr. K. Akad. Wiss. Wien, math.-naturwiss. Cl., 1, p. 380, pl. 48, fig. 13.
Chilostomella cf. *ovoidea* Reuss, Cushman & Renz 1946, p. 47, pl. 8, fig. 8.

Length 0.8–0.9, thickness 0.4–0.52 mm.

Stratigraphic range: Early Eocene (*Globorotalia edgari* Zone). Very rare. In Trinidad also found in the Eocene of the overlying Navet Formation.

GLOBIMORPHINA TROCHOIDES
(Reuss)
Figure **42.6–7**

Globigerina trochoides Reuss 1845, p. 36,
   pl. 12, fig. 22.
Allomorphina conica Cushman & Todd 1949,
   p. 62, pl. 11, fig. 8.
Allomorphina trochoides (Reuss), Cushman &
   Jarvis 1932, p. 49, pl. 15, fig. 3. – Cushman
   & Renz 1946, p. 46, pl. 8, fig. 6.
Eggerella? trochoides (Reuss), Cushman &
   Renz 1946, p. 22, pl. 2, fig. 20.
Eggerella trochoides (Reuss), Kaminski et al.
   1988, p. 195, pl. 9, figs. 12, 13.

The specimens from the Upper Lizard Springs
Formation (Fig. 42.7) have a tendency to
become more elongated conical with slightly less
inflated chambers. Length 0.3–0.52, thickness
0.22–0.5 mm.

Stratigraphic range: Possibly Campanian
   (*Globotruncana stuarti* Zone); Early
   Maastrichtian (*Globotruncana lapparenti
   tricarinata* Zone) to Early Eocene (*Globorotalia
   formosa formosa* Zone). Common.

PALLAIMORPHINA cf. YAMAGUCHII
Takayanagi
Figure **42.8–9**

Pallaimorphina yamaguchii Takayanagi 1960,
   p. 130, pl. 9, fig. 9.

The Trinidad specimens have a rather less
rounded periphery than the type but agree in
other characters. Diameter 0.24–0.42, thickness
0.12–0.18 mm.

Stratigraphic range: Turonian (*Globotruncana
   inornata* Zone). Very rare.

QUADRIMORPHINELLA sp. A
Figure **42.10–11**

Subspherical. Spiral side slightly evolute, with 3
to 3½ visible chambers. Umbilical side showing
3 chambers and a low semi-elliptical aperture.
Diameter 0.22–0.3 mm.

Stratigraphic range: Late Maastrichtian
   (*Abathomphalus mayaroensis* Zone). Very rare.

Family Quadrimorphinidae

QUADRIMORPHINA ADVENA (Cushman
& Siegfus)
Figure **42.12–13**

Valvulineria advena Cushman & Siegfus 1939,
   Contrib. Cushman Lab. foramin. Res., 15,
   p. 31, pl. 6, fig. 22.
Valvulineria allomorphinoides (Reuss),
   Cushman & Renz 1946, p. 44, pl. 7, figs. 13,
   14.

Diameter 0.3–0.68, thickness 0.18–0.42 mm.

Stratigraphic range: Early Eocene (*Globorotalia
   edgari* Zone to *G. aragonensis* Zone). Rare. In
   Trinidad also found in the Eocene of the
   overlying Navet Formation.

QUADRIMORPHINA
ALLOMORPHINOIDES (Reuss)
Figure **42.14–15**

Valvulina allomorphinoides Reuss 1860, p. 223,
   pl. 11, fig. 6.

Following Cushman & Todd (1949) and
common usage, specimens with a relatively large
final chamber and with only slightly depressed
sutures and a fairly large umbilical lip are
included here. Diameter 0.25–0.62, thickness
0.12–0.3 mm.

Stratigraphic range: Turonian to Campanian
   (*Globotruncana inornata* Zone to *G. stuarti*
   Zone). Fairly common.

QUADRIMORPHINA CAMERATA
CAMERATA (Brotzen)
Figure **42.16–17, 23**

Valvulineria camerata Brotzen 1936, p. 155,
   pl. 10, fig. 2; textfig. 57 (1, 2).

The umbilical area is variable, and it is mostly
the number and shape of the chambers which
allows a distinction of the typical 5-chambered
form from the subspecies *umbilicata* (see
below). Diameter 0.3–0.54, thickness 0.16–
0.34 mm.

Stratigraphic range: Santonian to Campanian
   (*Globotruncana fornicata* Zone to *G. stuarti*
   Zone). Rather rare.

QUADRIMORPHINA CAMERATA UMBILICATA (Brotzen)
Figure **42.18–20**

Valvulineria camerata Brotzen var. *umbilicata* Brotzen 1936, p. 156, pl. 10, fig. 1, textfig. 57 (3,4).

Diameter 0.2–0.42, thickness 0.15–0.26 mm.

Stratigraphic range: Early Cenomanian (*Rotalipora appenninica appenninica* Zone) to Campanian (*Globotruncana stuarti* Zone). Rather rare.

QUADRIMORPHINA HALLI (Jennings)
Figure **42.21–22**

Allomorphina halli Jennings 1936, Bull. Am. Paleontol., 23/78, p. 34, pl. 4, fig. 5.

Diameter 0.2–0.32, width 0.08–0.27 mm.

Stratigraphic range: Early Paleocene (*Globorotalia trinidadensis* Zone). Rare.

QUADRIMORPHINA MINUTA (Cushman)
Figure **42.24**

Allomorphina minuta Cushman 1936, Contrib. Cushman Lab. foramin. Res., 12, p. 72, pl. 13, fig. 3.

Diameter 0.25–0.35, thickness 0.08–0.14 mm.

Stratigraphic range: Maastrichtian (*Globotruncana gansseri* Zone to *Abathomphalus mayaroensis* Zone). Very rare.

QUADRIMORPHINA PROFUNDA
Schnitker & Tjalsma
Figure **42.25–26**

Quadrimorphina profunda Schnitker & Tjalsma 1980, J. foramin. Res., 10, p. 239, pl. 1, figs. 16–21.

Diameter 0.14–0.27, thickness 0.06–0.16 mm.

Stratigraphic range: Early Eocene (*Globorotalia subbotinae* Zone). Very rare.

QUADRIMORPHINA PYRIFORMIS (Taylor)
Figure **42.27–29**

Allomorphina pyriformis Taylor 1964, Proc. Royal Soc. Victoria, Melbourne, n. ser., 77, p. 582, pl. 82, fig. 5.

The last whorl shows 3 to 3½ chambers. The apertural flap has a fairly straight border which differs from the more distinctly curved flap of the otherwise similar *Allomorphina* (?) *wanghaia* Finlay. A short internal partition, similar to that of *Qu. halli*, is visible in broken specimens. Maximum diameter 0.21–0.48, thickness 0.14–0.3 mm.

Stratigraphic range: Late Turonian to Coniacian (*Globotruncana renzi* Zone). Rare.

QUADRIMORPHINA sp. A
Figure **42.31–32**

Test thick ovoid to subspherical; early chambers usually raised to a small point. Chambers increasing rapidly in size, four forming the last whorl. Sutures slightly depressed. Umbilicus small, partially covered by a plate-like chamber extension. Diameter 0.25–0.6 mm.

Stratigraphic range: Santonian (*Globotruncana fornicata* Zone) to Early Maastrichtian (*Globotruncana lapparenti tricarinata* Zone). Very rare.

Family Alabaminidae

ALABAMINA DISSONATA (Cushman & Renz)
Figure **42.30**

Pulvinulinella atlantisae Cushman var. *dissonata* Cushman & Renz 1948, p. 35, pl. 7, figs. 11, 12.

Diameter 0.51–0.55, thickness 0.21–0.27 mm.

Stratigraphic range: Early Eocene (*Globorotalia aragonensis* Zone). Very rare. In Trinidad also found in the Eocene of the overlying Navet Formation.

ALABAMINA MIDWAYENSIS Brotzen
Figure 42.33–34

Pulvinulina exigua H. B. Brady var. *obtusa*
Plummer 1926, p. 151, pl. 11, fig. 2 (non *P.
exigua* var. *obtusa* Burrows & Holland 1897).
Alabamina midwayensis Brotzen 1948, p. 99,
pl. 16, figs. 1, 2; textfigs. 25, 26.

Diameter 0.2–0.36, thickness 0.1–0.18 mm.

Stratigraphic range: Early Paleocene (probably
Globorotalia pseudobulloides Zone; Rochard
facies). Rare.

ALABAMINA cf. WILCOXENSIS Toulmin
Figure 42.35–37

Alabamina wilcoxensis Toulmin 1941, p. 603,
pl. 81, figs. 10–14; textfigs. 4A–C.
? *Pulvinulinella obtusa* (Burrows & Holland),
Cushman & Renz 1942, p. 11, pl. 2, fig. 16.

The Trinidad specimens differ from Toulmin's
types in the slightly bigger test size and in the
greater number of chambers (6 to 8, usually 7);
the periphery may be somewhat less angular. *A.
haitensis* Bermudez is quite similar in general
morphology but is considerably larger. Diameter
0.45–0.62, thickness 0.27–0.34 mm.

Stratigraphic range: Early Eocene (*Globorotalia
subbotinae* Zone). Rare.

Family Globorotalitidae

Goel (1962, p.113 ff.) has studied material
belonging to the genus *Globorotalites* from the
Paris Basin and has erected 27 new species, in
addition to the previously known species
michelinianus, *conicus* and *alabamensis*. The
material from France includes mostly specimens
with a keel and with raised sutures, and usually
with rather many chambers in the last whorl.
None of the 30 species can be recognized with
certainty in the Trinidad material, with the possible exception of the *G. michelinianus* group.

CONOROTALITES APTIENSIS
(Bettenstaedt)
Figure 43.1–3

Globorotalites bartensteini subsp. *aptiensis*
Bettenstaedt 1952, Senckenbergiana, 33,

p. 282, pl. 2, fig. 32; pl. 4, figs. 59–72.
Conorotalites aptiensis (Bettenstaedt 1952),
Bartenstein, Bettenstaedt & Bolli 1966,
p. 162, pl. 4, figs. 357–359. – Bartenstein &
Bolli 1973, p. 411, pl. 6, figs. 62–65. –
Bartenstein & Bolli 1986, p. 974, pl. 6, figs.
11–13.

The test shape is usually high conical, with a flat
to slightly concave spiral side and a narrow but
distinct keel. Diameter 0.26–0.55, thickness
0.18–0.42 mm.

Stratigraphic range: Late Albian to Early
Cenomanian (*Favusella washitensis* Zone to
Rotalipora appenninica appenninica Zone).
Rare. Also recorded in Trinidad from the
Early Aptian Cuche Formation and from the
Late Aptian to Early Albian Maridale
Formation.

CONOROTALITES SUBCONICUS
(Morrow)
Figure 43.10–13

Globorotalia subconica Morrow 1934, p. 200,
pl. 30, figs 11, 18.

In the upper part of their stratigraphic range
(from the *Globotruncana concavata* Zone
upwards), the specimens have a tendency to
become slightly larger and flatter (Fig. 43.13).
Diameter 0.15–0.4, thickness 0.11–0.26 mm.

Stratigraphic range: Turonian (*Globotruncana
inornata* Zone) to Campanian (*Globotruncana
stuarti* Zone). Common.

GLOBOROTALITES cf. CONICUS
(Carsey)
Figure 43.4–6

Truncatulina refulgens Montfort var. *conica*
Carsey 1926, p. 46, pl. 4, fig. 15.

The Trinidad form is slightly older than Carsey's
type and also rather smaller. There are 5½ to
6½ chambers in the final whorl. Because these
differences are relatively minor and the original
description of *G. conicus* is not sufficiently accurate, no formal name is here given to this group.
Diameter 0.2–0.45, thickness 0.12–0.34 mm.

Stratigraphic range: Turonian (*Globotruncana inornata* Zone) to Santonian (*Globotruncana fornicata* Zone). Fairly common.

GLOBOROTALITES MICHELINIANUS
(d'Orbigny)
Figure **43.7–9**

Rotalina micheliniana d'Orbigny 1840, p. 31, pl. 3, figs. 1–3.
Gyroidina sp. B., Beckmann 1960, p. 62, fig. 17.

The discussions by Goel (1962) reflect the uncertainties concerning this species and other species of *Globorotalites*. The specimens from Trinidad correspond to the concept of Cushman (1946), Marie (1941), Kaever (1961) and ten Dam & Magné (1948); they also seem to agree fairly well with d'Orbigny's type description. *G. michelinianus* in Goel (1962) is more distinctly conical and has more chambers in the final whorl (9 against 6–8 in Trinidad). Diameter 0.34–0.65, thickness 0.2–0.44 mm.

Stratigraphic range: Late Maastrichtian (*Abathomphalus mayaroensis* Zone). Rare.

Family Osangulariidae

CHARLTONINA AUSTRALIS Scheibnerova
Figure **43.14–16**

Charltonina australis Scheibnerova 1978, Bureau Min. Res. Geol. Geophys, Canberra, 192, p. 140, pl. 5, figs. 2–5.

Diameter 0.15–0.4, thickness 0.06–0.12 mm.

Stratigraphic range: Early Cenomanian (*Rotalipora appenninica appenninica* Zone); possibly also Turonian (*Globotruncana inornata* Zone), where transitional forms to *Osangularia* cf. *alata* (see below) seem to occur. Rather scarce.

OSANGULARIA cf. ALATA (Marsson)
Figure **43.17–19**

Discorbina alata Marsson 1878, p. 165, pl. 4, fig. 33.

The specimens from Trinidad are smaller than the type and have less extremely oblique sutures on the spiral side. The umbilicus is small, nearly closed and usually slightly depressed. Diameter 0.2–0.46, thickness 0.1–0.19 mm.

Stratigraphic range: Late Turonian/Coniacian (*Globotruncana renzi* Zone) to Campanian (*Globotruncana stuarti* Zone). Rare.

OSANGULARIA CALIFORNICA Dailey
Figure **43.21–24**

Osangularia californica Dailey 1970, p. 108, pl. 13, figs. 3, 4.

The sutures on the ventral side may be either straight or curved, raised or flush with the surface. This species is here not regarded as identical, perhaps not even generically, with '*Rotalia*' *schloenbachi* Reuss, as proposed by Crittenden (1983; neotype designated by Crittenden & Price, 1991). Because of its raised sutures on the spiral side, it appears not to be conspecific with '*Eponides*' *utaturensis* Sastry & Sastri (1966) either, as several authors have suggested. Diameter 0.28–0.5, thickness 0.14–0.24 mm.

Stratigraphic range: Early Cenomanian (*Rotalipora appenninica appenninica* Zone). Fairly common.

OSANGULARIA EXPANSA (Toulmin)
Figure **43.20, 25–26**

Parrella expansa Toulmin 1941, p. 604, textfig. 3; p. 605, textfigs. 4F, 4G.

Diameter 0.33–0.65, thickness 0.12–0.28 mm.

Stratigraphic range: Middle Paleocene (*Globorotalia pusilla pusilla* Zone) to Early Eocene (*Globorotalia formosa formosa* Zone). Rather rare.

OSANGULARIA INSIGNA SECUNDA
Dailey
Figure **43.27–29**

Osangularia insigna secunda Dailey 1970, p. 109, pl. 14, fig. 2.

Diameter 0.3–0.42, thickness 0.11–0.14 mm.

Stratigraphic range: Early Cenomanian (*Rotalipora appenninica appenninica* Zone). Rather rare.

OSANGULARIA MEXICANA (Cole)
Figure **43.30–32**

Pulvinulinella culter (Parker & Jones) var. *mexicana* Cole 1927, Bull. Am. Paleontol., 14/51, p. 31, pl. 1, figs. 15, 16.
Parrella mexicana (Cole), Cushman & Renz 1948, p. 35, pl. 7, figs. 9, 10.

O. mexicana differs from *O. expansa* in its slightly larger size and greater number of chambers (8½–11 against 7–9). Diameter 0.34–0.78, thickness 0.21–0.34 mm.

Stratigraphic range: Early Eocene (*Globorotalia formosa formosa* Zone to *G. aragonensis* Zone). Rare. In Trinidad also recorded in the Eocene of the overlying Navet Formation.

OSANGULARIA TEXANA (Cushman)
Figure **43.33–35**

Pulvinulinella texana Cushman 1938, Contrib. Cushman Lab. foramin. Res., 14, p. 49, pl. 8, fig. 8.

This species differs from its probable ancestor *O.* cf. *alata* in its less lobate periphery and less depressed ventral sutures. The umbilicus is closed and may have a small plug or clear area. Diameter 0.23–0.52, thickness 0.12–0.24 mm.

Stratigraphic range: Santonian to Maastrichtian (*Globotruncana fornicata* Zone to *Abathomphalus mayaroensis* Zone). Rather rare.

OSANGULARIA VELASCOENSIS (Cushman)
Figure **43.36–38**

Truncatulina velascoensis Cushman 1925, Contrib. Cushman Lab. foramin. Res., 1, p. 20, pl. 3, fig. 2.
Pulvinulinella velascoensis (Cushman), Cushman & Jarvis 1932, p. 48, pl. 14, fig. 6. – Cushman & Renz 1946, p. 46, pl. 8, figs. 2, 3.

Diameter 0.3–0.95, thickness 0.15–0.42 mm.

Stratigraphic range: Paleocene (*Globorotalia trinidadenis* Zone to *G. velascoensis* Zone). Fairly common.

Family Heterolepidae

ANOMALINOIDES cf. CHIRANUS (Cushman & Stone)
Figure **44.1–2**

Anomalina chirana Cushman & Stone 1947, Contrib. Cushman Lab. foramin. Res., 20, p. 25, pl. 4, fig. 1.

The specimens in Trinidad are slightly smaller than the figured holotype but have about the same number of chambers, 7 to 7½, in the final whorl (the type description mentions 10 or more). Diameter 0.24–0.44, thickness 0.15–0.27 mm.

Stratigraphic range: Early Eocene (*Globorotalia formosa formosa* Zone). Very rare.

ANOMALINOIDES DORRI ARAGONENSIS (Nuttall)
Figure **44.3–5**

Anomalina dorri Cole var. *aragonensis* Nuttall 1930, p. 219, pl. 24, fig. 18; pl. 25, fig. 1.
Anomalina visenda Finlay 1940, p. 458, pl. 65, figs. 116–119.
Anomalina rubiginosa Cushman, Cushman & Jarvis 1932, p. 52, pl. 16, figs. 3–5.
Anomalina dorri Cole var. *aragonensis* Nuttall, Cushman & Renz 1948, p. 41, pl. 8, figs. 13, 14.

Diameter 0.38–0.96, thickness 0.18–0.5 mm.

Stratigraphic range: Middle Paleocene (*Globorotalia pusilla pusilla* Zone) to Early Eocene (*Globorotalia aragonensis* Zone). Fairly common. In Trinidad also found in the Eocene of the overlying Navet Formation.

ANOMALINOIDES GUATEMALENSIS (Bermudez)
Figure **44.7–9**

Anomalina guatemalensis Bermudez 1963, p. 16, pl. 6, figs. 1–3.

The test is more planispiral/involute and somewhat larger than that of *A. welleri* (see below) but otherwise similar. Diameter 0.45–0.75, thickness 0.24–0.33 mm.

Stratigraphic range: Middle Paleocene (*Globorotalia pusilla pusilla* Zone). Rare.

ANOMALINOIDES cf.
MADRUGAENSIS (Cushman & Bermudez)
Figure **44.10–12**

Anomalina madrugaensis Cushman &
  Bermudez 1948, Contrib. Cushman Lab.
  foramin. Res., 24, p. 86, pl. 15, figs. 4–6.

The specimens look like a variant of *A. dorri
aragonensis* which is less trochoid, and the early
part of the test is rather thinner. The number of
chambers in the last whorl is the same, but the
sutures of the early chambers on the spiral side
in the final whorl are often distinctly curved,
similar to those of *Gavelinella lellingensis*
Brotzen. On the whole, the test is rather less
symmetrical in peripheral view than in the typical *A. madrugaensis*. Diameter 0.28–0.6, thickness 0.18–0.33 mm.

Stratigraphic range: Early Eocene (*Globorotalia
  subbotinae* Zone to *G. aragonensis* Zone). Rare.

ANOMALINOIDES NOBILIS Brotzen
Figure **44.14–17**

Anomalinoides nobilis Brotzen 1948, p. 89,
  pl. 14, fig. 5.

Traces of apertural flaps are often noticed on
the ventral (umbilical) side of the final chambers, but they are rather less distinct than in
Brotzen's type. On the dorsal side, the spiral is
usually fairly open, but occasionally the coiling
may become almost involute. *A. tokachiensis*
Yoshida may be closely related or even a synonym. Diameter 0.18–0.42, thickness 0.06–
0.24 mm.

Stratigraphic range: Late Maastrichtian
  (*Abathomphalus mayaroensis* Zone) to Early
  Eocene (*Globorotalia subbotinae* Zone). Rare.

ANOMALINOIDES RUBIGINOSUS
(Cushman)
Figure **44.6, 13, 18–19**

Anomalina rubiginosa Cushman 1926, p. 607,
  pl. 21, fig. 6.
Anomalina rubiginosa Cushman, Cushman &
  Jarvis 1932, p. 52, pl. 16, figs. 3–5. – Cushman
  & Renz 1946, p. 48, pl. 8, figs. 17, 18.

7–10 chambers in the final whorl. The shape varies from moderately planoconvex (predominant

in the Late Cretaceous) to thick biconvex and
pseudoplanispiral (mostly in the Paleocene), and
the coiling is usually nearly involute. The species
concept of Cushman and his co-authors also
includes some more evolute tests with a strongly
pustulose ventral side which may indicate a
relationship to the *A. dorri aragonensis* group.
Diameter 0.26–1.05, thickness 0.16–0.53 mm.

Stratigraphic range: Campanian (*Globotruncana
  stuarti* Zone) to Late Paleocene (*Globorotalia
  pseudomenardii* Zone). Fairly common.

ANOMALINOIDES UMBONATUS
(Cushman)
Figure **44.20–22**

Anomalina umbonata Cushman 1925, Bull.
  Am. Assoc. Petrol. Geologists, 9, p. 300,
  pl. 7, figs. 5, 6.

Diameter 0.33–0.5, thickness 0.16–0.24 mm.

Stratigraphic range: Early Eocene (*Globorotalia
  subbotinae* Zone to *G. formosa formosa* Zone).
  Very rare. In Trinidad also found in the Eocene
  of the overlying Navet Formation.

ANOMALINOIDES WELLERI (Plummer)
Figure **44.23–25**

Truncatulina welleri Plummer 1926, p. 143,
  pl. 9, fig. 6.
Anomalina ammonoides (Reuss), Cushman &
  Jarvis 1932, p. 51, pl. 16, fig. 1.

This is a very widespread group in the Late Cretaceous and Early Paleogene and is also known
under various species names such as *brotzeni*
Said & Kenawy, *loweryi* Mallory, *praespissi-
formis* Cushman & Bermudez, *pseudowelleri*
Olsson and *regina* Martin. The distinctly porous
wall separates it from the superficially similar
groups around *Valvalabamina lenticula* (Reuss)
and *Gyroidinoides depressus* (Alth). Diameter
0.2–0.6, thickness 0.11–0.24 mm.

Stratigraphic range: Santonian (*Globotruncana
  fornicata* Zone) to Late Paleocene (*Globorotalia
  pseudomenardii* Zone). Fairly common.

HETEROLEPA cf. HISPANIOLAE
(Bermudez)
Figure **44.26–28**

Cibicides hispaniolae Bermudez 1949, p. 300, pl. 25, figs. 34–36.
? *Cibicides praecursorius* (Schwager), Cushman & Renz 1942, p. 13, pl. 3, fig. 9.

11–16 chambers in the last whorl. The dorsal side is more involute than in *H. hispaniolae*, and the septa of the earlier chambers are thickened. In this respect, the Trinidad specimens are intermediate between this species and *Cibicides suzakensis* Bykova. Diameter 0.36–0.55, thickness 0.15–0.24 mm.

Stratigraphic range: Middle Paleocene (*Globorotalia pusilla pusilla* Zone). Rare.

Family Gavelinellidae

GYROIDINOIDES
ANGUSTIUMBILICATUS (ten Dam)
Figure **44.31, 35**

Gyroidina angustiumbilicata ten Dam 1944, Meded. Geol. Stichting, Haarlem, ser. C, 5, p. 117, pl. 4, fig. 7.

G. octocameratus Cushman & Hanna (*Gyroidina soldanii* var. *octocamerata*), a probable successor, is distinguished mainly by its somewhat larger size and more depressed sutures on the spiral side. Diameter 0.2–0.48, thickness 0.13–0.33 mm.

Stratigraphic range: Early Paleocene (probably *Globorotalia pseudobulloides* Zone, Rochard facies; *Globorotalia trinidadensis* Zone) to Early Eocene (*Globorotalia aragonensis* Zone). Rather scarce. In Trinidad also found in the Eocene of the overlying Navet Formation.

GYROIDINOIDES DEPRESSUS (Alth)
Figure **44.29–30, 32–34**

Rotalina depressa Alth 1850, p. 266, pl. 13, fig. 21.
Gyroidina depressa (Alth), Cushman & Jarvis 1932, p. 46, pl. 14, fig. 1. – Cushman & Renz 1946, p. 44, pl. 7, figs. 16, 17.
? *Gyroidina depressa* (Alth) var. *colombiana*

Cushman & Hedberg, Cushman & Renz 1946, p. 44, pl. 7, figs. 18, 19.

The specimens from Trinidad have a small open umbilicus; there is no apertural flap, or rarely a trace of it, and in this respect they differ from the otherwise very similar '*Rotalina*' *lenticula* Reuss (type species of *Valvalabamina* Reiss). The species has normally 8–11 chambers in the last whorl but includes, from the Middle Paleocene onwards, increasing proportions of a variant with 9–13 slightly compressed chambers (Fig. 44.33–34). Diameter 0.25–0.72, thickness 0.11–0.32 mm.

Stratigraphic range: Santonian (*Globotruncana fornicata* Zone) to Early Eocene (*Globorotalia aragonensis* Zone). Common. In Trinidad also found in the Eocene of the overlying Navet Formation.

GYROIDINOIDES GLOBOSUS (Hagenow)
Figure **45.1–3**

Nonionina globosa Hagenow 1842, Neues Jahrb. Min. Geogn. Geol. Petref.-Kunde, Stuttgart, p. 574.
Rotalia globosa v. Hagenow, Reuss 1862, Sitzungsber. Akad. Wiss. Wien, math.-naturwiss. Kl., 44/1 (1861), p. 330, pl. 7, fig. 2.
Gyroidina globosa (Hagenow), Cushman & Jarvis 1932, p. 47, pl. 14, figs. 3, 4. – Cushman & Renz 1946, p. 44, pl. 7, fig. 15.

The name *G. globosus*, as used here, corresponds to the usage of Cushman (see Cushman & Jarvis 1932, Cushman 1946b and others). According to Mello (1969), who had access to topotypic material, this concept probably is different from that of the original author. As proposed by Mello, Cushman's interpretation is here retained until the species is redefined. Diameter 0.2–0.65, thickness 0.18–0.55 mm.

Stratigraphic range: Campanian (*Globotruncana stuarti* Zone) to Late Paleocene (*Globorotalia velascoensis* Zone). Common.

GYROIDINOIDES GRAHAMI Martin
Figure **45.4–6**

Gyroidinoides grahami Martin 1964, p. 95,
pl. 13, fig. 1.

Diameter 0.2–0.55, thickness 0.15–0.36 mm.

Stratigraphic range: Santonian (*Globotruncana
fornicata* Zone) to Maastrichtian
(*Globotruncana gansseri* Zone). Rather rare.

GYROIDINOIDES KAMINSKII Beckmann
Figure **45.7–12**

Gyroidinoides kaminskii Beckmann 1991,
p. 828, pl. 2, figs. 11–13, 16–18.

Test thick biconvex, sometimes slightly flattened
on the dorsal side; periphery broadly rounded. 6
to 8 chambers in the final whorl, which increase
slowly in size as added and are moderately to
distinctly inflated. Sutures on both sides
depressed and slightly curved, usually slightly
oblique on the dorsal side. Umbilicus open but
rather small, often partially covered by broad
but short chamber extensions. Wall smooth,
glossy. Aperture interiomarginal-umbilical.
There is a gradual but distinct change in morphology from the older to the younger specimens: The chambers decrease in number from
7–8 to 6–7 and become higher; sometimes the
spiral becomes slightly wider (compare Fig.
45.7–9 with Fig. 45.10–12).

The most similar species, *G. primitivus*
Hofker and *G. subglobosus* Dailey, both have a
wider spiral and straighter and more radial
sutures. Diameter: 0.18–0.35, thickness 0.12–
0.26 mm.

Stratigraphic range: Turonian (*Globotruncana
inornata* Zone) to Santonian (*Globotruncana
fornicata* Zone). Fairly common.

GYROIDINOIDES NITIDUS (Reuss)
Figure **45.13–16**

Rotalina nitida Reuss 1844, Geognost. Skizzen
Böhmen, Prague, 2, p. 214.
Rotalia nitida Reuss, 1845, p. 35, pl. 8, fig. 52;
pl. 12, figs. 8, 20.

The Trinidad form shows a considerable variability, both in shape and in size. Older specimens
(particularly in the Santonian, see Fig. 45.13–
14) correspond to the one figured in Loeblich

& Tappan (1987, plate 713, figs. 7–9). In the
Campanian-Maastrichtian, the umbilicus
becomes more closed and the shape is more distinctly planoconvex (Fig. 45.15–16). Such specimens are very similar to *Gyroidina megastoma*
(see below) which, however, is more distinctly
conical in peripheral view. Diameter 0.25–0.78,
thickness 0.18–0.6 mm.

Stratigraphic range: Santonian (*Globotruncana
fornicata* Zone) to Maastrichtian
(*Globotruncana gansseri* Zone). Rather rare.

GYROIDINOIDES cf. PLUMMERAE
(Cushman & Bermudez)
Figure **45.17–18, 24**

Gyroidina plummerae Cushman & Bermudez
1937, p. 22, pl. 2, figs. 42–44.

The Trinidad form has more chambers than the
type (7–9 against 5) and is slightly smaller. It is
close to *G. octocameratus* (Cushman & Hanna;
G. soldanii var. *o.*) but has a more involute
spiral on the dorsal side. Diameter 0.2–0.46,
thickness 0.14–0.22 mm.

Stratigraphic range: Early Eocene (*Globorotalia
aragonensis* Zone). Rare.

GYROIDINOIDES QUADRATUS
(Cushman & Church)
Figure **45.19–21**

Gyroidina quadrata Cushman & Church 1929,
p. 516, pl. 41, figs. 7–9.
Gyroidina sp. A, Beckmann 1960, p. 62, fig. 16.

Diameter 0.18–0.35, thickness 0.12–0.3 mm.

Stratigraphic range: Turonian (*Globotruncana
inornata* Zone) to Late Maastrichtian
(*Abathomphalus mayaroensis* Zone). Rather
rare.

GYROIDINOIDES (?) cf. QUADRATUS
(Cushman & Church)
Figure **45.22–23**

A few specimens have the general shape of *G.
quadratus* but differ in their more angular to
weakly keeled spiral suture which may carry fine
pustules, reminiscent of those of the genus *Stensioeina*. Where the pustules are weak or absent,
the distinction from *G. quadratus* may become

difficult. There are rather coarse wall perforations on the umbilical side. Diameter 0.18–0.28, thickness 0.12–0.20 m.

Stratigraphic range: Santonian to Campanian (*Globotruncana fornicata* Zone to *G. stuarti* Zone). Very rare.

GYROIDINOIDES SUBANGULATUS (Plummer)
Figure **45.25–27**

Rotalia soldanii d'Orbigny var. *subangulata* Plummer 1926, p. 154, pl. 12, fig. 1.

Relatively large and robust specimens are found in the Campanian and Early Maastrichtian. Later, the size becomes normal (diameter 0.4 mm). Diameter 0.28–0.63, thickness 0.24–0.52 mm.

Stratigraphic range: Campanian (*Globotruncana stuarti* Zone) to Early Eocene (*Globorotalia aragonensis* Zone). Rather rare. In Trinidad also found in the Eocene of the overlying Navet Formation.

GYROIDINOIDES SUBGLOBOSUS Dailey
Figure **45.28–30**

Gyroidinoides subglobosa Dailey 1970, p. 108, pl. 13, fig. 2.

Some specimens have less inflated chambers than the type and resemble *G. infracretacea* Morozova (*Gyroidina nitida* var. *i.*). Diameter 0.18–0.33, thickness 0.12–0.23 mm.

Stratigraphic range: Early Cenomanian (*Rotalipora appenninica appenninica* Zone). Rather rare.

STENSIOEINA EXCOLATA (Cushman)
Figure **45.31–33**

Truncatulina excolata Cushman 1926, Contrib. Cushman Lab. foramin. Res., 2, p. 22, pl. 3, fig. 2.

Diameter 0.36–0.72, thickness 0.15–0.35 mm.

Stratigraphic range: Maastrichtian (*Globotruncana gansseri* Zone to *Abathomphalus mayaroensis* Zone). Rare.

ANGULOGAVELINELLA AVNIMELECHI (Reiss)
Figure **45.34–36**

Pseudovalvulineria avnimelechi Reiss 1952, Bull. Res. Council., Israel, 2, p. 269, textfig. 2.
Pseudovalvulineria sp. A, Beckmann 1960, p. 62, fig. 15.

Diameter 0.27–0.65, thickness 0.15–0.38 mm.

Stratigraphic range: Maastrichtian (*Globotruncana lapparenti tricarinata* Zone?, *Abathomphalus mayaroensis* Zone) to Late Paleocene (*Globorotalia velascoensis* Zone). Fairly common.

GAVELINELLA ARACAJUENSIS (Petri)
Figure **46.1–3**

Eponides aracajuensis Petri 1962, Bol. Fac. Filosofia, Ciencias, Letras, Univ. Sao Paulo, 256 (Geol. no. 20), p. 115, pl. 15, figs. 1–3.

Gavelinopsis (?) proelevata Koch (1968) is very similar but rather larger and has more distinct calcifications on the spiral side. Diameter 0.28–0.42, thickness 0.13–0.22 mm.

Stratigraphic range: Maastrichtian (*Globotruncana gansseri* Zone to *Abathomphalus mayaroensis* Zone). Rare.

GAVELINELLA BECCARIIFORMIS (White)
Figure **46.4–6**

Rotalia beccariiformis (& varieties) White 1928, p. 287, pl. 39, figs. 2–4.
Anomalina beccariiformis (White), Cushman & Renz 1946, p. 48, pl. 8, figs. 21, 22. ?Cushman & Renz 1947a, p. 50, pl. 12, fig. 14.
Anomalina whitei Martin 1964, p. 106, pl. 16, fig. 4.

The very rare specimens in the Campanian (*Globotruncana stuarti* Zone), with a slightly angular periphery and a relatively strong ventral ornamentation, resemble '*Gyroidina*' *infrafossa* Finlay. Apart from this, the variability in the Guayaguayare and Lower Lizard Springs formations is such that the three varieties of White cannot be clearly separated. Diameter 0.2–0.52, thickness 0.09–0.28 mm.

Stratigraphic range: Campanian (*Globotruncana stuarti* Zone) to Late Paleocene (*Globorotalia velascoensis* Zone). Common.

GAVELINELLA CENOMANICA (Brotzen)
Figure **46.7–10**

Cibicidoides (Cibicides) cenomanica Brotzen 1945, Sveriges Geol. Unders., Stockholm, ser. C, 465 (arsb. 38, 1944, no. 7), p. 54, pl. 2, fig. 2.

The Trinidad specimens are identical with the hypotypes of *G. cenomanica* deposited in the collection of the Geologische Bundesanstalt, Hannover. They differ from Brotzen's holotype in being slightly more asymmetric in peripheral view, with a flatter spiral side. The affinities with very similar species like *Anomalina ammonoides* var. *crassisepta* Vasilenko & Myatlyuk and *Planulina andersoni* Church will have to be investigated. In Trinidad, the test gradually develops a flatter spiral side and a more angular periphery in its later occurrences (*Rotalipora appenninica appenninica* Zone; see Fig. 46.10). Diameter 0.26–0.6, thickness 0.12–0.23 mm.

Stratigraphic range: Late Albian to Early Cenomanian (*Favusella washitensis* Zone to *Rotalipora appenninica appenninica* Zone). Fairly common.

GAVELINELLA DAKOTENSIS (Fox)
Figure **46.14–16**

Planulina dakotensis Fox 1954, Prof. Pap. U.S. Geol. Survey, 254-E, p. 119, pl. 26, figs. 19–21.

Apart from the typical form, a smaller variety with a more compressed test is found in the Early Cenomanian samples.
Diameter 0.18–0.38, thickness 0.05–0.09 mm. The smaller variety has a diameter of 0.15–0.25 and a thickness of 0.04 mm.

Stratigraphic range: Early Cenomanian (*Rotalipora appenninica appenninica* Zone), Coniacian/Santonian (*Globotruncana concavata* Zone). Rather rare.

GAVELINELLA DAYI (White)
Figure **46.11–13**

Planulina dayi White 1928, J. Paleontol., 2, p. 300, pl. 41, fig. 3.
Planulina constricta (v. Hagenow), Cushman & Jarvis 1932, p. 52, pl. 16, fig. 6.
Cibicides stephensoni Cushman 1938, Contrib. Cushman Lab. foramin. Res., 14, p. 70, pl. 12, fig. 5.
Cibicides stephensoni Cushman, Cushman & Renz 1946, p. 48, pl. 8, fig. 25. – Cushman & Renz 1947a, p. 51, pl. 12, fig. 15.
? *Trochammina trinitatensis* Cushman & Jarvis 1928, p. 95, pl. 13, fig. 13 (a collapsed and corroded specimen, which should be regarded as a nomen dubium). – Cushman & Jarvis 1932, p. 21, pl. 6, fig. 6. – Cushman & Renz 1946, p. 24, pl. 3, fig. 12.

Diameter 0.33–0.85, thickness 0.15–0.52 mm.

Stratigraphic range: Probably Early Maastrichtian (*Globotruncana lapparenti tricarinata* Zone, Marabella facies); Early Maastrichtian (*Globotruncana lapparenti tricarinata* Zone) to Late Paleocene (*Globorotalia velascoensis* Zone). Fairly common.

GAVELINELLA DRYCREEKENSIS
Dailey
Figure **46.17–19**

Gavelinella drycreekensis Dailey 1970, p. 107, pl. 13, fig. 1.

Diameter 0.24–0.55, thickness 0.09–0.17 mm.

Stratigraphic range: Early Cenomanian (*Rotalipora appenninica appenninica* Zone). Fairly common.

GAVELINELLA ERIKSDALENSIS
(Brotzen)
Figure **46.21–25**

Cibicides (Cibicidoides) eriksdalensis Brotzen 1936, p. 193, pl. 14, fig. 5; textfig. 69.

Biconvex to planoconvex, with an angular periphery which may have a faint keel. The more convex side, called the spiral side by Brotzen, usually bears a central elevated knob. The opposite side has a tendency to become evolute. Planoconvex specimens with an angular periphery (Figs. 46.24–25) are found mostly in the *Glo-*

botruncana renzi and *G. concavata* zones. Diameter 0.21–0.67, thickness 0.11–0.23 mm.

Stratigraphic range: Turonian (*Globotruncana inornata* Zone) to Coniacian/Early Santonian (*Globotruncana concavata* Zone). Fairly common.

GAVELINELLA MIDWAYENSIS TROCHOIDEA (Plummer)
Figure **46.20, 26–27**

Truncatulina midwayensis Plummer var. *trochoidea* Plummer 1926, p. 142, pl. 9, fig. 8.

Diameter 0.3–0.46, thickness 0.1–0.17 mm.

Stratigraphic range: Early Paleocene (probably *Globorotalia pseudobulloides* Zone; Rochard facies). Rather scarce.

GAVELINELLA cf. MONTERELENSIS (Marie)
Figure **46.28–30**

Anomalina monterelensis Marie 1941, p. 243, pl. 37, fig. 342.

Some specimens differ from the types in having an almost involute spiral side, and their tests are often rather more compressed, like the specimens of *G. monterelensis* figured by Jenkins & Murray (1989). This species may be an ancestor of *G. dayi*, from which it differs in the flatter test and the often more distinctly inflated chambers. Diameter 0.36–0.66, thickness 0.12–0.25 mm.

Stratigraphic range: Campanian (*Globotruncana stuarti* Zone) to Early Maastrichtian (*Globotruncana lapparenti tricarinata* Zone). Rather rare.

GAVELINELLA PETITA (Carsey)
Figure **46.31–32**

Anomalina petita Carsey nom nud., 1926, p. 48, pl. 7, fig. 3 (figured specimen is a *Cristellaria*).
Anomalina falcata Plummer 1931 (non Reuss 1869), p. 202, pl. 14, figs. 7–8 (holotype fig. 7).
Anomalina plummerae Tappan, n. name, 1940, J. Paleontol., 14, p. 124.

The final whorl has 8–10 relativey high cham-

bers; the sutures are slightly curved to almost radial. An elevated central boss may or may not be present on the spiral side. *Anomalina plummerae* Tappan, originally created as a substitute for *A. petita*, is rather bigger and has more distinctly curved sutures; it may therefore not be conspecific. Diameter 0.18–0.33, thickness 0.07–0.12 mm.

Stratigraphic range: Albian (*Rotalipora ticinensis ticinensis* Zone). Very rare.

GAVELINELLA ROCHARDENSIS Beckmann
Figure **46.35–39**

Gavelinella rochardensis Beckmann 1991, p. 828, pl. 2, figs. 19–24.

Test planoconvex to nearly biconvex, normally with a flattened dorsal side but exceptionally more convex dorsally; periphery subrounded to subangular, occasionally very slightly lobate. 7 to 9 chambers in the final whorl. Sutures on both sides somewhat curved, usually more so on the spiral side but sometimes almost radial, hardly depressed except in the final chambers. Umbilicus open, sometimes fairly large and deep. Wall smooth, glossy. Aperture extending from the periphery to the umbilicus, often with a distinct umbilical flap.
'*Eponides*' *huaynai* Frizzell, which occurs in a few Late Cretaceous samples of the present material, may be an ancestral form; is has a rather more acute periphery, hardly depressed sutures and an interiomarginal to umbilical aperture. Diameter 0.3–0.66, thickness 0.16–0.27 mm.

Stratigraphic range: Early Paleocene (probably *Globorotalia pseudobulloides* Zone, Rochard facies). Rather rare.

GAVELINELLA cf. SANDIDGEI (Brotzen)
Figure **47.1–5**

Cibicides sandidgei Brotzen 1936, p. 191, pl. 14, figs. 2–4.

As in Brotzen's type material, this species is represented by two morphotypes. Both are planoconvex and nearly bi-involute; the flatter side is called the umbilical side by Brotzen. Type A (Fig. 47.1–3) is stratigraphically older; it is not quite typical as it has a subrounded periphery and only 6½–7 moderately inflated chambers in the final whorl. Type B (Fig. 47.4–5) is more distinctly planoconvex with a more acute periphery; it has 7 to 9 chambers which are hardly if at all inflated. Except for the smaller average number of chambers in the final whorl and a difference in age, the Trinidad form cannot be clearly separated from Brotzen's very variable species. Diameter: 0.22–0.44, thickness 0.12–0.16 mm.

Stratigraphic range: Late Albian (*Favusella washitensis* Zone; mostly type A) to Early Cenomanian (*Rotalipora appenninica appenninica* Zone; mostly type B). Rather rare.

GAVELINELLA SCHLOENBACHI
(Reuss)
Figure **46.33–34, 40–41**

Rotalia schlönbachi Reuss 1863, p. 84, pl. 10, fig. 5.

The younger specimens (*Globotruncana concavata* and *G. fornicata* zones; Fig. 46.41) are often slightly thicker and more distinctly planoconvex and resemble '*Anomalina*' *popenoei* Trujillo. Diameter 0.2–0.52, thickness 0.07–0.18 mm.

Stratigraphic range: Turonian (*Globotruncana inornata* Zone) to Santonian (*Globotruncana fornicata* Zone). Fairly common.

GAVELINELLA cf. SIBIRICA (Dain)
Figure **47.6–7, 14–15**

Anomalina (Pseudovalvulineria) sibirica Dain 1954, Trudy VNIGRI, Leningrad, n. s., vyp. 80, p. 103, pl. 14, fig. 1.

In the general shape of the test and the arrangement of the chambers, the Trinidad specimens are similar to the type; however, the umbilical flaps are usually less distinct, and in some specimens the later chambers may become more inflated and have almost radial sutures. Diameter 0.18–0.33, thickness 0.09–0.2 mm.

Stratigraphic range: Early Cenomanian (*Rotalipora appenninica appenninica* Zone) to Coniacian-Santonian (*Globotruncana concavata* Zone). Rather rare.

GAVELINELLA SPISSOCOSTATA
(Cushman)
Figure **47.8–10**

Planulina spissocostata Cushman 1938, Contrib. Cushman Lab. foramin. Res., 14, p. 69, pl. 12, fig. 4.

G. compressa Sliter is very similar but has slightly less chambers (11–13 against 12–15) in the final whorl which increase more rapidly in size. Diameter 0.32–0.55, thickness 0.1–0.15 mm.

Stratigraphic range: Probably Early Maastrichtian (*Globotruncana lapparenti tricarinata* Zone, Marabella facies); Early Maastrichtian (*Globotruncana lapparenti tricarinata* Zone). Rather rare.

GAVELINELLA cf. VESCA (Bykova)
Figure **47.11–13**

Discorbis vescus Bykova 1939, Trudy Neft. Geol.-Razved. Inst., Leningrad, ser. A, 121, p. 28, pl. 3, figs. 1–6.

The Trinidad specimens differ from *G. vesca* in often having slightly raised and limbate chamber sutures, particularly in the early chambers of the last whorl. Diameter 0.21–0.24, thickness 0.07–0.09 mm.

Stratigraphic range: Early Cenomanian (*Rotalipora appenninica appenninica* Zone). Rare.

GYROIDINA BANDYI (Trujillo)
Figure **47.16–19**

Eponides bandyi Trujillo 1960, p. 332, pl. 48, fig. 3.

The typical shape in axial section is lenticular, but occasionally it may be more planoconvex (with a flatter spiral side) or subspherical and then become very similar to that of *G. mauretanica*. The aperture is interiomarginal, usually reaching neither the umbilicus nor the periphery. The Trinidad specimens are somewhat smaller than the types. Diameter 0.16–0.47, thickness 0.1–0.32 mm.

Stratigraphic range: Albian (*Rotalipora ticinensis ticinensis* Zone) to Santonian (*Globotruncana fornicata* Zone). Common.

GYROIDINA BOLLII (Cushman & Renz)
Figure 47.20–22

Eponides bollii Cushman & Renz 1946, p. 44,
   pl. 7, fig 23.
? *Eponides haidingerii* (d'Orbigny), Cushman
   & Jarvis 1932, p. 47, pl. 14, fig. 5.

This species may be a phylogenetic successor of
G. bandyi, becoming larger and more distinctly
planoconvex. It is similar to the somewhat
smaller and more acutely lenticular *'Eponides'
plummerae* (see p. 147). *'Eponides' sigali* Said &
Kenawy is possibly a junior synonym. Diameter
0.32–1.1, thickness 0.26–0.48 mm.

Stratigraphic range: Campanian (*Globotruncana
stuarti* Zone) to Late Paleocene (*Globorotalia
velascoensis* Zone). Fairly common.

GYROIDINA MAURETANICA Carbonnier
Figure 47.23–24

Gyroidina mauretanica Carbonnier 1952, Bull.
   Soc. géol. France, ser. 6, 2, p. 113, pl. 5, fig. 5.

The aperture is interiomarginal and does not
quite reach the umbilicus in most specimens.
Diameter 0.16–0.32, thickness 0.15–0.24 mm.

Stratigraphic range: Late Albian (*Rotalipora
ticinensis ticinensis* Zone) to Santonian
(*Globotruncana fornicata* Zone). Common.

GYROIDINA MEGASTOMA (Grzybowski)
Figure 47.25–27

Pulvinulina megastoma Grzybowski 1896,
   p. 303, pl. 11, fig. 9.

Test planoconvex; the periphery of the last 2 or
3 chambers may become subacute. 5–6 chambers in the final whorl. Aperture interiomarginal, not extending into the umbilicus.
   Liszka & Liszkova (1981) have redescribed
and refigured Grzybowski's material. These tests
are planoconvex with a subrounded to acute periphery and with sutures that are oblique on the
dorsal side and slightly curved on the ventral
side. *G. beisseli* White and possibly *G. mendez-
ensis* White, which are reported to differ in
details of the periphery and the sutures, may be
junior synonyms. Diameter 0.18–0.75, thickness
0.14–0.38 mm.

Stratigraphic range: Santonian (*Globotruncana
fornicata* Zone) to Late Paleocene (*Globorotalia
velascoensis* Zone). Common.

GYROIDINA cf. NODA Belford
Figure 47.28, 33–34

Gyroidina noda Belford 1960, p. 79, pl. 21, figs.
   16–27.

Except for the lesser number of chambers in the
final whorl (5½–6½) and the somewhat flatter
dorsal side, the Trinidad form is closely similar
to the types. The periphery may be slightly
keeled. Diameter 0.22–0.5, thickness 0.16–
0.34 mm.

Stratigraphic range: Coniacian-Santonian
(*Globotruncana concavata* Zone) to
Campanian (*Globotruncana stuarti* Zone).
Rare.

GYROIDINA PRAEGLOBOSA Brotzen
Figure 47.29–32

Gyroidina praeglobosa Brotzen 1936, p. 159,
   pl. 11, fig. 4.

The sutures on the dorsal side vary from slightly
to distinctly oblique and curved. In the youngest
samples (*Globotruncana stuarti* Zone) the spiral
often becomes narrower and the periphery is
then less lobate. Diameter 0.18–0.42, thickness
0.12–0.25 mm.

Stratigraphic range: Turonian (*Globotruncana
inornata* Zone) to Campanian (*Globotruncana
stuarti* Zone). Fairly common.

LINARESIA SEMICRIBRATA (Beckmann)
Figure 47.35–38

Anomalina pompilioides Galloway &
   Heminway var. *semicribrata* Beckmann
   1954, Eclog. geol. Helv., 46 (1953), p. 400,
   pl. 27, fig. 3; textfigs. 24, 25.

The coarse perforations of the ventral (umbilical) side seem to develop gradually. In the
Santonian, some relatively small specimens are
found which usually lack these perforations but
in the higher Cretaceous samples, the coarse
perforations are often concentrated on the axial
part of the umbilical side; in the Early Paleogene
of the Lizard Springs Formation they normally

extend to the periphery. Also in the Lizard Springs Formation (*Globorotalia pseudomenardii* Zone to *G. subbotinae* Zone in particular), we can observe the temporary appearance of a subgroup which shows a tendency to become more symmetrical (Fig. 47.38), with a deepened umbilicus on both sides (a parallel development to that observed in the *Anomalinoides dorri aragonensis/A. madrugaensis* group). Diameter 0.24–0.6, thickness 0.15–0.33 mm.

Stratigraphic range: Santonian (*Globotruncana fornicata* Zone) to Early Eocene (*Globorotalia aragonensis* Zone). Fairly common. In Trinidad also found in the Eocene of the overlying Navet Formation.

Family Rotaliidae

ROTALIA (?) HERMI Hillebrandt
Figure **47.39–40**

Rotalia hermi Hillebrandt 1962, p. 116, pl. 10, figs. 13–17; textfig. 10.

The aperture is poorly visible, possibly interiomarginal. Diameter 0.42–0.67, thickness 0.22–0.28 mm.

Stratigraphic range: Maastrichtian (*Globotruncana gansseri* Zone or *Abathomphalus mayaroensis* Zone). Very rare.

Figs. 18–47

Fig. 18. Bathysiphonidae, Rhabdamminidae, Saccamminidae, Hippocrepinidae, Ammodiscidae

1–2 *Bathysiphon* cf. *discretus* (Brady). × 30
1 Guayaguayare well no. 159, core at 3986–4000 feet, *Globorotalia pseudomenardii* Zone, Late Paleocene. **2** Sample K9415, *Globorotalia velascoensis* Zone, Late Paleocene.

3 *Bathysiphon eocenicus* Cushman & Hanna. × 30
Guayaguayare well no. 159, core at 4778–4790 feet, *Globorotalia pusilla pusilla* Zone, Middle Paleocene.

4 *Bathysiphon robustus* (Grzybowski). × 25
Morne Diablo well no. 34, core at 13774–13821 feet, *Globotruncana inornata* Zone, Turonian.

5–6 *Bathysiphon* sp. A
5 Moruga well no. 3, core at 10258–10264 feet, *Globorotalia trinidadensis* Zone, Early Paleocene. × 45. **6** Moruga well no. 3, core at 11428–11448 feet, *Globorotalia trinidadensis* Zone, Early Paleocene. × 40.

7 *Rhizammina* (?) *indivisa* Brady. × 50
Moruga well no. 3, core at 10258–10264 feet, *Globorotalia trinidadensis* Zone, Early Paleocene.

8 *Dendrophrya* (?) *excelsa* Grzybowski. × 50
Sample K9415, *Globorotalia velascoensis* Zone, Late Paleocene.

9–10 *Lagenammina grzybowskii* (Schubert). × 40
9 Guayaguayare well no. 159, core at 3986–4000 feet, *Globorotalia pseudomenardii* Zone, Late Paleocene. **10** Morne Diablo well no. 34, core at 12720–12750 feet, *Globotruncana fornicata* Zone, Santonian.

11 *Saccammina complanata* (Franke). × 40
Sample Hk1831, *Globorotalia formosa formosa* Zone, Early Eocene.

12–13 *Saccammina globosa* Crespin
12 Morne Diablo well no. 34, core at 13965–13984 feet, *Rotalipora appenninica appenninica* Zone, Early Cenomanian. × 50. **13** Same locality. × 40.

14–15 *Saccammina placenta* (Grzybowski).
14 Guayaguayare well no. 163, core at 5588–5598 feet, *Abathomphalus mayaroensis* Zone, Late Maastrichtian. × 35. **15** A subspherical specimen. Sample K9415, *Globorotalia velascoensis* Zone, Late Paleocene. × 40.

16–17 *Thurammina* (?) spp. × 50
16 Moruga well no. 15, core at 4617–4637 feet, *Rzehakina epigona* Zonule, Early Paleocene. **17** Sample KR23575, *Globorotalia angulata* Zone, Middle Paleocene.

18–19 *Hyperammina elongata* Brady. × 22
18 Sample K9415, *Globorotalia velascoensis* Zone, Late Paleocene. **19** Guayaguayare well no. 163, core at 5882–5902 feet, *Globotruncana lapparenti tricarinata* Zone, Early Maastrichtian.

20 *Hyperammina erugata* Sliter. × 25

Moruga well no. 15, core at 3796–3816 feet, *Globorotalia pseudomenardii* Zone, Late Paleocene.

21–22 *Hyperammina* cf. *subnodosiformis* Grzybowski. × 25
Guayaguayare well no. 159, core at 3986–4000 feet, *Globorotalia pseudomenardii* Zone, Late Paleocene.

23–24 *Ammodiscus glabratus* Cushman & Jarvis. × 25
23 Guayaguayare well no. 163, core at 5588–5598 feet, *Abathomphalus mayaroensis* Zone, Late Maastrichtian. **24** Guayaguayare well no. 159, core at 4524–4536 feet, *Globorotalia pusilla pusilla* Zone, Middle Paleocene.

25 *Ammodiscus latus* Grzybowski. × 35
Sample Hk1831, *Globorotalia subbotinae* Zone, Early Eocene.

26–27 *Ammodiscus macilentus* Myatlyuk. × 80
26 Moruga well no. 15, core at 3796–3816 feet, *Globorotalia pseudomenardii* Zone, Late Paleocene. **27** Guayaguayare well no. 159, core at 3986–4000 feet, *Globorotalia pseudomenardii* Zone, Late Paleocene.

28, 36 *Ammodiscus pennyi* Cushman & Jarvis
28 Guayaguayare well no. 163, screen sample at 5672–5682, *Globotruncana gansseri* Zone, Maastrichtian. × 28. **36** Moruga well no. 15, core at 4296–4316 feet, *Globorotalia uncinata* Zone, Middle Paleocene. × 25.

29 *Ammodiscus planus* Loeblich. × 100
Guayaguayare well no. 163, core at 5588–5598 feet, *Abathomphalus mayaroensis* Zone, Late Maastrichtian.

30 *Glomospira gordialis diffundens* Cushman & Renz. × 50
Guayaguayare well no. 163, core at 5588–5598 feet, *Abathomphalus mayaroensis* Zone, Late Maastrichtian.

31 *Glomospira serpens* (Grzybowski). × 35
Oropouche well no. 4, core at 7829–7849 feet, *Globorotalia subbotinae* Zone, Early Eocene.

32 *Ammolagena clavata* (Jones & Parker). × 25
Specimen attached to *Ammodiscus* sp.; Guayaguayare well no. 159, core at 3986–4000 feet, *Globorotalia pseudomenardii* Zone, Late Paleocene.

33 *Glomospira irregularis* (Grzybowski). × 25
Sample K10832, *Globorotalia pseudomenardii* Zone, Late Paleocene.

34–35 *Glomospirella gaultina gaultina* (Berthelin), emend. Bartenstein 1954. × 50
34 Guayaguayare well no. 163, core at 5588–5598 feet, *Abathomphalus mayaroensis* Zone, Late Maastrichtian. **35** Sample Hk1831, *Globorotalia formosa formosa* Zone, Early Eocene.

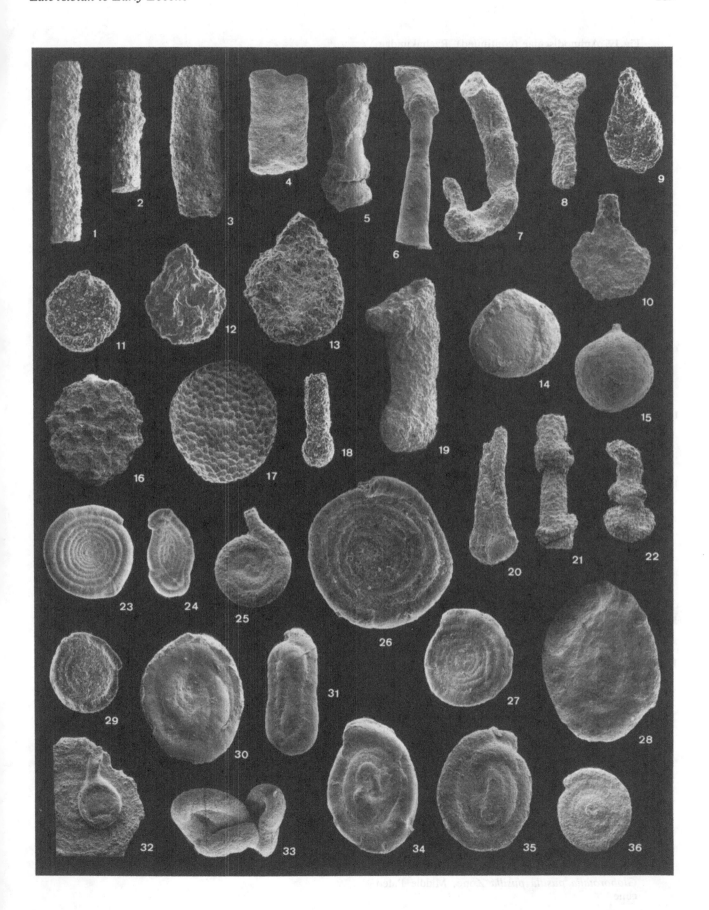

Fig.19. Ammodiscidae (continued), Rzehakinidae, Aschemocellidae, Hormosinidae, Haplophragmoididae

1–2 *Glomospirella gaultina confusa* (Zaspelova)
1 Moruga well no. 3, core at 10258–10264 feet, *Globorotalia trinidadensis* Zone, Early Paleocene. × 50. 2 Guayaguayare well no. 163, core at 5588–5598 feet, *Abathomphalus mayaroensis* Zone, Late Maastrichtian. × 35.

3 *Repmanina charoides corona* (Cushman & Jarvis). × 70
Guayaguayare well no. 159, core at 3986–4000 feet, *Globorotalia pseudomenardii* Zone, Late Paleocene.

4–6 *Repmanina charoides favilla* (Emiliani). × 50
Guayaguayare well no. 159, core at 3986–4000 feet, *Globorotalia pseudomenardii* Zone, Late Paleocene.

7–8 *Rzehakina epigona* (Rzehak). × 50
7 Sample KR23575, *Globorotalia angulata* Zone, Middle Paleocene. 8 Sample K9415, *Globorotalia velascoensis* Zone, Late Paleocene.

9 *Rzehakina minima* Cushman & Renz. × 50
Guayaguayare well no. 163, core at 5882–5902 feet, *Globotruncana lapparenti tricarinata* Zone, Early Maastrichtian.

10–11 *Silicosigmoilina akkeshiensis* Yoshida
10 Side view. Moruga well no. 15, core at 4831–4849 feet, *Globotruncana lapparenti tricarinata* Zone, Early Maastrichtian, × 50. 11 Apertural view. Same locality. × 60.

12–14 *Silicosigmoilina angusta* Turenko
12 Side view. Moruga well no. 15, core at 4831–4849 feet, *Globotruncana lapparenti tricarinata* Zone, Early Maastrichtian, × 60. 13 Side view. Rock Dome well no. 1, core at 6032–6087 feet, sample at 6061 feet, *Globorotalia pseudobulloides* Zone or *G. trinidadensis* Zone, Early Paleocene. × 60. 14 Apertural view. Rochard well no. 1, core at 8556–8571 feet, probably *Globorotalia pseudobulloides* Zone, Early Paleocene. × 100.

15–16 *Aschemocella moniliformis* (Neagu). × 25
Moruga well no. 15, core at 4296–4316 feet, *Globorotalia uncinata* Zone, Middle Paleocene.

17–19 *Hormosinella ovulum* (Grzybowski)
17 Sample KR23575, *Globorotalia angulata* Zone, Middle Paleocene. × 55. 18 Guayaguayare well no. 163, core at 5588–5598 feet, *Abathomphalus mayaroensis* Zone, Late Maastrichtian. × 55. 19 Sample Hk1831, *Globorotalia subbotinae* Zone, Early Eocene. × 65.

20 *Hormosinella* sp. A. × 20
Guayaguayare well no. 159, core at 4524–4536 feet, *Globorotalia pusilla pusilla* Zone, Middle Paleocene.

21 *Reophax elongatus* Grzybowski. × 25
Sample Hk1831, *Globorotalia subbotinae* Zone, Early Eocene.

22 *Reophax* cf. *folkestonensis* (Khan). × 28
Guayaguayare well no. 159, core at 4524–4536 feet, *Globorotalia pusilla pusilla* Zone, Middle Paleocene.

23–24 *Reophax globosus* Sliter. × 30
23 Morne Diablo well no. 34, core at 12680–12700 feet, *Globotruncana fornicata* Zone, Santonian. 24 Marabella well no. 1, core at 3512–3540 feet, probably *Globotruncana lapparenti tricarinata* Zone, Early Maastrichtian (Marabella facies).

25 *Reophax prolatus* Sliter. × 50
Guayaguayare well no. 163, core at 5588–5598 feet, *Abathomphalus mayaroensis* Zone, Late Maastrichtian.

26 *Reophax* cf. *troyeri* Tappan. × 50
Moruga well no. 3, core at 10258–10262 feet, *Globorotalia trinidadensis* Zone, Early Paleocene.

27 *Subreophax pseudoscalarius* (Samuel). × 20
Sample K10832, *Globorotalia pseudomenardii* Zone, Late Paleocene.

28–29 *Subreophax scalarius* (Grzybowski)
28 Sample K9415, *Globorotalia velascoensis* Zone, Late Paleocene. × 25. 29 Guayaguayare well no. 159, core at 4524–4536 feet, *Globorotalia pusilla pusilla* Zone, Middle Paleocene. × 50.

30–31, 36–37 *Hormosina trinitatensis* Cushman & Renz
30 Guayaguayare well no. 163, core at 5882–5902 feet, *Globotruncana lapparenti tricarinata* Zone, Early Maastrichtian. × 20. 31 Guayaguayare well no. 163, core at 5588–5598 feet, *Abathomphalus mayaroensis* Zone, Late Maastrichtian. × 30. 36 Sample KR23575, *Globorotalia angulata* Zone, Middle Paleocene. × 25. 37 Single chamber. Sample K10832, *Globorotalia pseudomenardii* Zone, Late Paleocene. × 25.

32, 39 *Reticulophragmoides jarvisi* (Thalmann). × 50
32 Peripheral view. Guayaguayare well no. 159, core at 3986–4000 feet, *Globorotalia pseudomenardii* Zone, Late Paleocene. 39 Side view. Same locality.

33–34, 40 *Asanospira walteri* (Grzybowski). × 50
33 Peripheral view. Moruga well no. 15, core at 3796–3816 feet, *Globorotalia pseudomenardii* Zone, Late Paleocene. 34 Peripheral view. Sample K10832, *Globorotalia pseudomenardii* Zone, Late Paleocene. 40 Side view. Same locality.

35, 41 *Cribrostomoides trinitatensis* Cushman & Jarvis. × 50
35 Side view. Guayaguayare well no. 163, core at 5588–5598 feet, *Abathomphalus mayaroensis* Zone, Late Maastrichtian. 41 Peripheral view. Sample RM287, *Globorotalia pseudomenardii* Zone, Late Paleocene.

38 *Subreophax velascoensis* (Cushman). × 25
Moruga well no. 15, core at 4296–4316 feet, *Globorotalia uncinata* Zone, Middle Paleocene.

42–43 *Evolutinella flagleri* (Cushman & Hedberg). × 50
42 Side view. Morne Diablo well no. 34, core at 13842–13889 feet, *Globotruncana inornata* Zone, Turonian. 43 Peripheral view. Same locality.

Fig. 20. Haplophragmoididae (continued), Lituotubidae, Cyclamminidae (part), Lituolidae

1–3 *Evolutinella renzi* Beckmann
1 Side view. Marac well no. 1, core at 9853–9891 feet, sample at 9883 feet, *Favusella washitensis* Zone, Late Albian. × 50. **2** Peripheral view. Same locality. × 50. **3** Side view of holotype. Morne Diablo well no. 34, core at 13889–13936 feet, sample at 13928 feet, *Rotalipora appenninica appenninica* Zone, Early Cenomanian. × 45.

4–5 *Haplophragmoides caucasicus* Shutskaya
4 Side view. Guayaguayare well no. 163, screen sample at 5744–5754 feet, *Globotruncana gansseri* Zone, Maastrichtian. × 55. **5** Peripheral view. Sample KB 6972, *Globorotalia formosa formosa* Zone, Early Eocene. × 50.

6, 12 *Haplophragmoides concavus* (Chapman). × 50
6 Side view. Marac well no. 1, core at 9853–9891 feet, *Favusella washitensis* Zone, Late Albian. **12** Oblique side view. Morne Diablo well no. 34, core at 12641–12660 feet, *Globotruncana fornicata* Zone, Santonian.

7 *Haplophragmoides formosus* Takayanagi. × 40
Marac well no. 1, core at 9853–9891 feet, *Favusella washitensis* Zone, Late Albian.

8–9 *Haplophragmoides horridus* (Grzybowski)
8 Moruga well no. 15, core at 4831–4849 feet, *Globotruncana lapparenti tricarinata* Zone, Early Maastrichtian. × 25. **9** Same locality. × 30.

10 *Haplophragmoides kirki* Wickenden. × 65
Sample Rz413, *Globorotalia aragonensis* Zone, Early Eocene.

11 *Haplophragmoides mjatliukae* Maslakova. × 100
Sample K10832, *Globorotalia pseudomenardii* Zone, Late Paleocene.

13–14 *Haplophragmoides neochapmani* Beckmann. × 50
Morne Diablo well no. 34, core at 13260–12300 feet, *Globotruncana renzi* Zone, Late Turonian-Coniacian.

15–16 *Haplophragmoides porrectus* Maslakova. × 50
15 Side view. Guayaguayare well no. 159, core at 4524–4536 feet, *Globorotalia pusilla pusilla* Zone, Middle Paleocene. **16** Oblique side view of a compressed specimen. Moruga well no. 15, core at 4617–4637 feet, *Rzehakina epigona* Zonule, Early Paleocene.

17 *Haplophragmoides retroseptus* (Grzybowski). × 25
Guayaguayare well no. 163, core at 5882–5902 feet, *Globotruncana lapparenti tricarinata* Zone, Early Maastrichtian.

18 *Haplophragmoides* cf. *suborbicularis* (Grzybowski). × 35

Sample KR23575, *Globorotalia angulata* Zone, Middle Paleocene.

19–20 *Haplophragmoides trinitatensis* Cushman & Renz. × 50
19 Side view. Moruga well no. 3, core at 10258–10264 feet, *Globorotalia trinidadensis* Zone, Early Paleocene. **20** Oblique side view. Guayaguayare well no. 159, core at 4778–4790 feet, *Globorotalia pusilla pusilla* Zone, Middle Paleocene.

21–22 *Lituotuba lituiformis* (Brady). × 25
21 Oropouche well no. 4, core at 7829–7849 feet, sample at 7849 feet, *Globorotalia subbotinae* Zone, Early Eocene. **22** Sample Hk1831, *Globorotalia subbotinae* Zone, Early Eocene.

23–24 *Paratrochamminoides irregularis* (White)
23 Sample K9415, *Globorotalia velascoensis* Zone, Late Paleocene. × 25. **24** Guayaguayare well no. 163, core at 5588–5598 feet, *Abathomphalus mayaroensis* Zone, Late Maastrichtian. × 40.

25 *Trochamminoides proteus* (Karrer). × 33
Guayaguayare well no. 163, core at 5588–5598 feet, *Abathomphalus mayaroensis* Zone, Late Maastrichtian.

26 *Trochamminoides subcoronatus* (Grzybowski). × 40
Guayaguayare well no. 163, core at 5588–5598 feet, *Abathomphalus mayaroensis* Zone, Late Maastrichtian.

27–28 *Trochamminoides dubius* (Grzybowski)
27 Sample K10832, *Globorotalia pseudomenardii* Zone, Late Paleocene. × 40. **28** Guayaguayare well no. 163, core at 5588–5598 feet, *Abathomphalus mayaroensis* Zone, Late Maastrichtian. × 50.

29–30 *Popovia beckmanni* (Kaminski & Geroch). × 50
Moruga well no. 15, core at 4617–4637 feet, *Rzehakina epigona* Zonule, Early Paleocene.

31–33 *Ammobaculites fragmentarius* Cushman
31 Morne Diablo well no. 34, core 12578–12608 feet, *Globotruncana fornicata* Zone, Santonian. × 35. **32** Sample KR23575, *Globorotalia angulata* Zone, Middle Paleocene. × 35. **33** Guayaguayare well no. 159, core at 4524–4536 feet, *Globorotalia pusilla pusilla* Zone, Middle Paleocene. × 25.

34 *Ammobaculites subcretaceus* Cushman & Alexander. × 55
Morne Diablo well no. 34, core at 13889–13936 feet, *Rotalipora appenninica appenninica* Zone, Early Cenomanian.

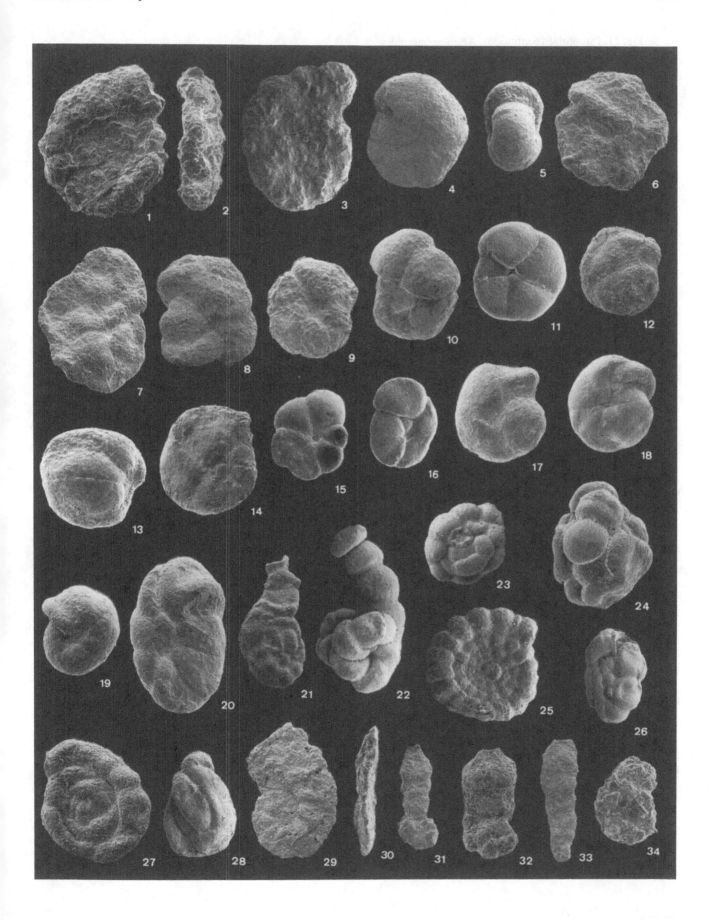

Fig. 21. Lituolidae (continued), Ammosphaeroidinidae, Cyclamminidae (continued), Spiroplectamminidae

1–4 *Ammobaculites lacertae* Beckmann
  1 Holotype, side view. Sample Hk1831, *Globorotalia subbotinae* Zone, Early Eocene. × 40. **2** Peripheral view. Sample KB 6972, *Globorotalia formosa formosa* Zone, Early Eocene. × 35. **3** Side view of a long specimen. Boodoosingh well no.1, 6892–6909 feet, *Globorotalia subbotinae* Zone, Early Eocene. × 25. **4** Juvenile? spiral stage, side view. Same locality as for Fig. 21.2. × 40.
  5 *Ammomarginulina* cf. *expansa* (Plummer). × 25 Sample Hk1832, *Globorotalia edgari* Zone, Early Eocene.
6–8, *Sculptobaculines barri* Beckmann. × 25
15 **6** Side view of holotype. Moruga well no. 15, core at 4296–4316 feet, *Globorotalia uncinata* Zone, Middle Paleocene. **7** Peripheral view. Guayaguayare well no. 159, core at 3986–4000 feet, *Globorotalia pseudomenardii* Zone, Late Paleocene. **8** Specimen with coarse sand grains. Sample Hk1831, *Globorotalia subbotinae* Zone, Early Eocene. **15** Variant with thick fine grained wall. Same locality as for Fig. 21.7.
9–10 *Triplasia fundibularis* (Harris & Jobe)
  9 Moruga well no. 15, core at 4296–4316 feet, sample at 4316 feet, *Globorotalia uncinata* Zone, Middle Paleocene. × 20. **10** Same locality. × 25.
11 *Triplasia inaequalis* Hagn. × 25 Sample Hk1831, *Globorotalia subbotinae* Zone, Early Eocene.
12–13 *Ammosphaeroidina pseudopauciloculata* (Myatlyuk). × 50
  12 Spiral view. Guayaguayare well no. 163, core at 5588–5598 feet, *Abathomphalus mayaroensis* Zone, Late Maastrichtian. **13** Umbilical view. Sample Hk1831, *Globorotalia subbotinae* Zone, Early Eocene.
14 *Recurvoides deflexiformis* (Noth). × 50 Guayaguayare well no. 159, core at 3986–4000 feet, *Globorotalia pseudomenardii* Zone, Late Paleocene.
16 *Recurvoides* cf. *loczyi* (Majzon). × 100 Apertural view. Morne Diablo well no. 34, core at 13842–13889 feet, *Globotruncana inornata* Zone, Turonian.
17–18 *Recurvoides* cf. *walteri* (Grzybowski). × 50
  17 Apertural view. Sample Hk1831, *Globorotalia subbotinae* Zone, Early Eocene. **18** Oblique side view. Sample KB6972, *Globorotalia formosa formosa* Zone, Early Eocene.
19–20 *Thalmannammina subturbinata* (Grzybowski)
  19 Side view. Sample Rz413, *Globorotalia ara-*

gonensis Zone, Early Eocene. × 60. **20** Apertural view. Guayaguayare well no. 159, core at 3986–4000 feet, *Globorotalia pseudomenardii* Zone, Late Paleocene. × 80.
21–22 *Quasicyclammina*? sp. A
  21 Peripheral view. Sample KB6972, *Globorotalia formosa formosa* Zone, Early Eocene. × 65. **22** Side view. Same locality. × 50.
23–24, *Reticulophragmium* cf. *garcilassoi* (Frizzell). × 50
30–31 **23** Side view. Guayaguayare well no. 159, core at 3986–4000 feet, *Globorotalia pseudomenardii* Zone, Late Paleocene. **24** Peripheral view. Same locality. **30** Side view. Guayaguayare well no. 159, core at 3707–3713 feet, *Globorotalia subbotinae* Zone, Early Eocene. **31** Peripheral view. Same locality.
25 *Bulbobaculites jarvisi* (Cushman & Renz). × 25 Guayaguayare well no. 159, core at 3986–4000 feet, *Globorotalia pseudomenardii* Zone, Late Paleocene.
26–27 *Bulbobaculites lueckei* (Cushman & Hedberg). × 50
  26 Guayaguayare well no. 159, core at 3986–4000 feet, *Globorotalia pseudomenardii* Zone, Late Paleocene. **27** Moruga well no. 15, core at 4401–4421 feet, *Globorotalia trinidadensis* Zone, Early Paleocene.
28–29 *Quasispiroplectammina chicoana* (Lalicker). × 50
  28 Morne Diablo well no. 34, core at 12680–12700 feet, *Globotruncana fornicata* Zone, Santonian. **29** Moruga well no. 10, core sample at 7987 feet, *Globotruncana stuarti* Zone, Campanian.
32–33 *Bolivinopsis spectabilis* (Grzybowski)
  32 Megalospheric specimen. Guayaguayare well no. 163, core at 5882–5902 feet, *Globotruncana lapparenti tricarinata* Zone, Early Maastrichtian. × 50. **33** Microspheric specimen. Same locality. × 40.
34–35 *Bolivinopsis trinitatensis* (Cushman & Renz)
  34 Megalospheric specimen. Sample RM278, *Globorotalia pseudomenardii* Zone, Late Paleocene. × 40. **35** Microspheric specimen. Same locality. × 50.
36–38 *Quasispiroplectammina* cf. *embaensis* (Myatlyuk)
  36 Side view of a specimen with spinose periphery. Moruga well no. 15, core at 6519–6544 feet, *Globotruncana renzi* Zone, Late Turonian-Santonian. × 60. **37** Peripheral view. Same locality. × 60. **38** Side view. Same locality. × 45.
39 *Quasispiroplectammina* cf. *gandolfii* (Carbonnier). × 50 Sample KR8385, *Rotalipora appenninica appenninica* Zone, Early Cenomanian.

Fig. 22. Spiroplectamminidae (continued), Textulariopsidae, Trochamminidae, Conotrochamminidae

1–2 *Quasispiroplectammina jarvisi* (Cushman). × 50
Guayaguayare well no. 159, core at 3986–4000, *Globorotalia pseudomenardii* Zone, Late Paleocene.

3 *Quasispiroplectammina nuda* (Lalicker). × 60
Moruga well no. 15, core at 6980–7005 feet, *Globotruncana inornata* Zone, Turonian.

4–7 *Spiroplectinella dentata* (Alth). × 50.
4 Peripheral view. Guayaguayare well no. 163, screen sample at 5707–5719 feet, *Globotruncana gansseri* Zone, Maastrichtian. **5** Side view. Same locality. **6** Side view. Sample Bn61, Pickering Quarry, *Globotruncana stuarti* Zone or younger, Campanian-Maastrichtian. **7** Side view. Marabella well no. 1, core at 3512–3540 feet, probably *Globotruncana lapparenti tricarinata* Zone, Early Maastrichtian (Marabella facies).

8, 17 *Spiroplectinella excolata* (Cushman)
8 Side view. Sample K9415, *Globorotalia velascoensis* Zone, Late Paleocene. × 30. **17** Peripheral view. Same locality. × 35.

9, 18 *Textulariopsis losangica* (Loeblich & Tappan). × 50
9 Oblique side view. KR8385, *Rotalipora appenninica appenninica* Zone, Early Cenomanian. **18** Oblique side view. Marac well no. 1, core at 9578–9621 feet, sample at 9602 feet, *Rotalipora appenninica appenninica* Zone, Early Cenomanian.

10–12 *Spiroplectammina* (?) *navarroana* (Cushman)
10 Guayaguayare well no. 163, core at 5588–5598 feet, *Abathomphalus mayaroensis* Zone, Late Maastrichtian. × 45. **11–12** Guayaguayare well no 159, core at 3986–4000 feet, *Globorotalia pseudomenardii* Zone, Late Paleocene. × 50.

13 *Vulvulina colei* Cushman. × 50
Sample Rz281, *Globorotalia formosa formosa* Zone, Early Eocene.

14–16 *Textulariopsis hikagezawensis* (Takayanagi). × 50
14–15 Side views. Sample KR8385, *Rotalipora appenninica appenninica* Zone, Early Cenomanian. **16** Peripheral view. Same locality.

19 *Ammoglobigerina boehmi* (Franke). × 32
Morne Diablo well no. 34, core at 13889–13936 feet, sample at 13936 feet, *Rotalipora appenninica appenninica* Zone, Early Cenomanian.

20–22 *Ammoglobigerina altiformis* (Cushman & Renz)
20 Oblique spiral view. Moruga well no. 15, core at 4831–4849 feet, *Globotruncana lapparenti tricarinata* Zone, Early Maastrichtian. × 25. **21** Spiral view.

Sample KR23575, *Globorotalia angulata* Zone, Middle Paleocene. × 25. **22** Umbilical view. Sample K10832, *Globorotalia pseudomenardii* Zone, Late Paleocene. × 32.

23 *Ammoglobigerina* cf. *altiformis* (Cushman & Renz). × 55
Sample Rz281, *Globorotalia formosa formosa* Zone, Early Eocene.

24–25 *Ammoglobigerina globigeriniformis* (Cushman & Renz), forma A. × 50
24 Spiral view. Sample Hk1831, *Globorotalia subbotinae* Zone, Early Eocene. **25** Umbilical view. Same locality.

26–27 *Trochammina* (?) cf. *deformis* Grzybowski. × 50
Oblique views of compressed specimens. Morne Diablo well no. 34, core at 12578–12608 feet, sample at 12608 feet, *Globotruncana fornicata* Zone, Santonian.

28–29 *Trochammina ruthvenmurrayi* Cushman & Renz. × 50
28 Spiral view. Guayaguayare well no. 159, core at 3986–4000 feet, *Globorotalia pseudomenardii* Zone, Late Paleocene. **29** Side view. Same locality.

30–34 *Trochammina globolaevigata* Beckmann
30 Spiral view of holotype. Marac well no. 1, core at 9853–9891 feet, *Favusella washitensis* Zone, Late Albian. × 50. **31** Apertural view of a juvenile? specimen. Sample KR8385, *Rotalipora appenninica appenninica* Zone, Early Cenomanian. × 50. **32** Large spherical specimen with collapsed final chamber. Same locality as for Fig. 22.30. × 40. **33** Apertural view. Same locality as for Fig. 22.30. × 40. **34** Side view of a slightly conical specimen. Sample JS1019, *Rotalipora appenninica appenninica* Zone, Early Cenomanian. × 50.

35 *Trochammina wickendeni* Loeblich. × 100
Morne Diablo well no. 34, core at 12660–12680 feet, *Globotruncana fornicata* Zone, Santonian.

36–37 *Conotrochammina whangaia* Finlay. × 50
36 Umbilical view. Sample K10832, *Globorotalia pseudomenardii* Zone, Late Paleocene. **37** Spiral view. Same locality.

38–39 *Conotrochammina* (?) cf. *depressa* Finlay. × 50
38 Spiral view. Guayaguayare well no. 159, core at 3707–3713 feet, *Globorotalia subbotinae* Zone, Early Eocene. **39** Side view. Same locality.

Fig. 23. Prolixoplectidae, Verneuilinidae

1–2 *Karrerulina bortonica* Finlay. × 50
1 Side view. Sample Hk1831, *Globorotalia subbotinae* Zone, Early Eocene. 2 Edge view. Same locality.

3–4 *Karrerulina coniformis* (Grzybowski). × 50
3 Moruga well no. 15, core at 4617–4637 feet, *Rzehakina epigona* Zonule, Early Paleocene. 4 Sample Rz281, *Globorotalia formosa formosa* Zone, Early Eocene.

5–8 *Karrerulina conversa* (Grzybowski). × 50
5 Morne Diablo well no. 34, core at 12578–12608 feet, *Globotruncana fornicata* Zone, Santonian. 6 Specimen with relatively fine grained wall. Guayaguayare well no. 163, core at 5588–5598 feet, *Abathomphalus mayaroensis* Zone, Late Maastrichtian. 7–8 Two specimens with distinctly terminal final chamber. Guayaguayare well no. 159, core at 3986–4000 feet, *Globorotalia pseudomenardii* Zone, Late Paleocene.

9–10 *Karrerulina smokyensis* (Wall). × 50
9 Guayaguayare well no. 163, core at 5882–5902 feet, *Globotruncana lapparenti tricarinata* Zone, Early Maastrichtian. 10 Morne Diablo well no. 34, core at 12700–12720 feet, *Globotruncana fornicata* Zone, Santonian.

11–12 *Plectina kugleri* Beckmann. × 40
11 Side view of holotype. Guayaguayare well no. 163, core at 5588–5598 feet, *Abathomphalus mayaroensis* Zone, Late Maastrichtian. 12 Edge view. Guayaguayare well no. 163, screen sample at 5744–5754 feet, *Globotruncana gansseri* Zone, Maastrichtian.

13 *Verneuilinella* sp. A. × 100
Sample KB6972, *Globorotalia formosa formosa* Zone, Early Eocene.

14–15 *Gaudryinopsis bosquensis* Loeblich & Tappan. × 50
14 Side view. Moruga well no. 15, core at 8506–8518 feet, *Rotalipora appenninica appenninica* Zone, Early Cenomanian. 15 Side view. Same locality.

16 *Gaudryinopsis canadensis* (Cushman). × 50
Marac well no. 1, core at 9578–9621 feet, sample at 9602 feet, *Rotalipora appenninica appenninica* Zone, Early Cenomanian.

17 *Gaudryinopsis gradata* (Berthelin). × 40
Morne Diablo well no. 34, core at 13889–13936 feet, *Rotalipora appenninica appenninica* Zone, Early Cenomanian.

18–19 *Verneuilinoides polystropha* (Reuss). × 50
Moruga well no. 3, core at 10258–10264 feet, *Globorotalia trinidadensis* Zone, Early Paleocene.

20–21 *Verneuilinoides tailleuri* Tappan. × 50
20 Morne Diablo well no. 34, core at 12660–12680 feet, *Globotruncana fornicata* Zone, Santonian. 21 Guayaguayare well no. 163, core at 5588–5598 feet, *Abathomphalus mayaroensis* Zone, Late Maastrichtian.

22–23 *Verneuilinoides triformis* (Sliter). × 25
22 Moruge well no. 3, core at 10258–10264 feet, *Globorotalia trinidadensis* Zone, Early Paleocene. 23 Sample K9415, *Globorotalia velascoensis* Zone, Late Paleocene.

24–25 *Belorussiella* cf. *ossipovae* (Bykova)
24 Moruga well no. 3, core at 10258–10264 feet, *Globorotalia trinidadensis* Zone, Early Paleocene. × 100. 25 Guayaguayare well no. 159, core at 4524–4536 feet, *Globorotalia pusilla pusilla* Zone, Middle Paleocene. × 110.

26 *Gaudryina cretacea* (Karrer). × 25
Guayaguayare well no. 163, core at 5588–5598 feet, *Abathomphalus mayaroensis* Zone, Late Maastrichtian.

27–28 *Gaudryina expansa* Israelsky. × 40
27 Side view. Boodoosingh well no. 1, core at 6892–6909, *Globorotalia subbotinae* Zone, Early Eocene. 28 Oblique apertural view. Same locality.

29–30 *Gaudryina frankei* Brotzen
29 Side view. Moruga well no. 15, core at 5195–5210 feet, *Globotruncana stuarti* Zone, Campanian. × 40. 30 Edge view. Moruga well no. 15, core at 5427–5452 feet, *Globotruncana fornicata* Zone, Santonian. × 50.

31–32 *Gaudryina* sp. aff. *G. inflata* Israelsky. × 40
31 Side view. Sample K9415, *Globorotalia velascoensis* Zone, Late Paleocene. 32 Edge view. Oropouche well no. 4, core at 7829–7849 feet, *Globorotalia subbotinae* Zone, Early Eocene.

33–34 *Gaudryina ingens* Voloshinova. × 25
33 Side view. Moruga well no. 15, core at 4831–4849 feet, *Globotruncana lapparenti tricarinata* Zone, Early Maastrichtian. 34 Edge view. Same locality.

35–37 *Gaudryina laevigata* Franke. × 50
35 Edge view. Lizard well no. 1x, core at 7198–7200 feet, *Globotruncana inornata* Zone, Turonian. 36 Oblique edge view. Lizard well no. 1x, core at 7378–7390 feet, *Globotruncana inornata* Zone, Turonian. 37 Side view. Marac well no. 1, core at 8657–8698 feet, sample at 8657 feet, *Globotruncana concavata* Zone, Coniacian/Santonian.

38–40 *Gaudryina* cf. *modica* Bermudez. × 25
38 Side view. Guayaguayare well no. 159, core at 4524–4536 feet, *Globorotalia pusilla pusilla* Zone, Middle Paleocene. 39 Edge view. Same locality. 40 Side view of an immature specimen. Sample Hk1832, *Globorotalia edgari* Zone, Early Eocene.

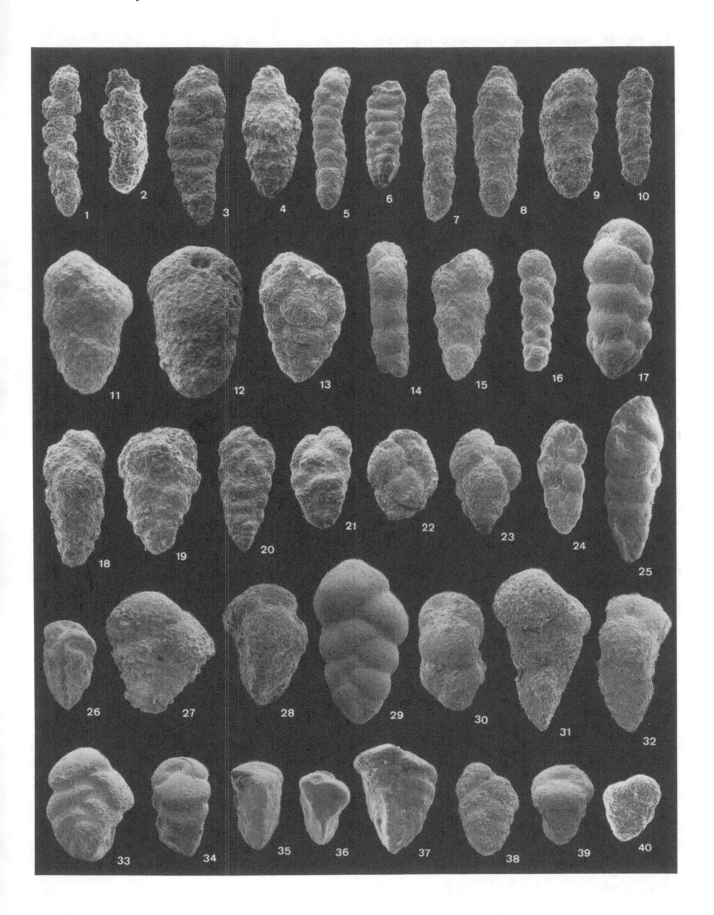

Fig. 24. Verneuilinidae (continued), Tritaxiidae, Ataxophragmiidae, Globotextulariidae, Eggerellidae

1–3 *Gaudryina* sp. aff. *G. nitida* Haque. × 50
1 Oblique side view. Oropouche no. 4, core at 7829–7849 feet, *Globorotalia subbotinae* Zone, Early Eocene. **2** Side view. Sample KB6972, *Globorotalia formosa formosa* Zone, Early Eocene. **3** Edge view. Same locality.

4–6 *Gaudryina pyramidata* Cushman. × 50
4 Oblique apertural view of a large flaring specimen. Guayaguayare well no. 163, core at 5588–5598 feet, *Abathomphalus mayaroensis* Zone, Late Maastrichtian. **5** Edge view of the large variant. Sample KR23575, *Globorotalia angulata* Zone, Middle Paleocene. **6** Edge view of the small type. Guayaguayare well no. 159, core at 3896–4000 feet, *Globorotalia pseudomenardii* Zone, Late Paleocene.

7–8 *Gaudryina* cf. *rugosa* d'Orbigny. × 25
7 Guayaguayare well no.159, core at 4524–4536 feet, *Globorotalia pusilla pusilla* Zone, Middle Paleocene. **8** Moruga well no. 3, core at 10258–10264 feet, *Globorotalia trinidadensis* Zone, Early Paleocene.

9–10 *Gaudryina rutteni* Cushman & Bermudez
Oblique edge views from opposite sides. **9** Sample Rz413, *Globorotalia aragonensis* Zone, Early Eocene. × 35. **10** Same locality. × 25.

11–13 *Gaudryina* sp. A. × 50
11 Side view of megalospheric specimen. Sample KR8385, *Rotalipora appenninica appenninica* Zone, Early Cenomanian. **12** Side view of microspheric specimen. Same locality. **13** Edge view. Same locality.

14 *Gaudryinella pusilla* Magniez-Jannin. × 100
Side view of a compressed specimen. Morne Diablo well no. 34, core at 12700–12720 feet, *Globotruncana fornicata* Zone, Santonian.

15–16 *Pseudogaudryinella compacta* ten Dam & Sigal. × 40
15 Guayaguayare well no. 163, core at 5588–5598 feet, *Abathomphalus mayaroensis* Zone, Late Maastrichtian. **16** Marabella well no. 1, core at 3512–3540 feet, probably *Globotruncana lapparenti tricarinata* Zone, Early Maastrichtian (Marabella facies).

17 *Tritaxia tricarinata* (Reuss). × 50
Morne Diablo well no. 34, core at 13889–13936 feet, *Rotalipora appenninica appenninica* Zone, Early Cenomanian.

18 *Arenobulimina puschi* (Reuss). × 50
Marabella well no. 1, core at 3512–3540 feet, probably *Globotruncana lapparenti tricarinata* Zone, Early Maastrichtian (Marabella facies).

19–20 *Arenobulimina truncata* Reuss. × 80
Moruga well no. 15, core at 4296–4316 feet, *Globorotalia uncinata* Zone, Middle Paleocene.

21, 31 *Dorothia beloides* Hillebrandt. × 50
21 A cylindrical variant. Guayaguayare well no. 159, core at 4524–4536 feet, *Globorotalia pusilla pusilla* Zone, Middle Paleocene. **31** A typical specimen. Sample K10832, *Globorotalia pseudomenardii* Zone, Late Paleocene.

22–24 *Remesella varians* (Glaessner)
22 Guayaguayare well no. 163, core at 5588–5598 feet, *Abathomphalus mayaroensis* Zone, Late Maastrichtian. × 40. **23** Same locality. × 50. **24** Cylindrical, megalospheric specimen. Sample KR23575, *Globorotalia angulata* Zone, Middle Paleocene. × 50.

25–27 *Bannerella biformis* (Finlay). × 25
25 Microspheric specimen. Marabella well no. 1, core at 3512–3540 feet, probably *Globotruncana lapparenti tricarinata* Zone, Early Maastrichtian (Marabella facies). **26–27** Megalospheric specimens. Same locality.

28 *Dorothia bulbosa* Israelsky. × 25
Sample KR23575, *Globorotalia angulata* Zone, Middle Paleocene.

29 *Dorothia pupa* (Reuss). × 25
Sample K9415, *Globorotalia velascoensis* Zone, Late Paleocene.

30 *Dorothia subretusa* Israelsky. × 25
Sample KR23575, *Globorotalia angulata* Zone, Middle Paleocene.

32–34 *Dorothia cylindracea* Bermudez. × 50
32 Relatively short specimen with flattened top. Sample K10832, *Globorotalia pseudomenardii* Zone, Late Paleocene. **33** Oblique side view. Boodoosingh well no. 1, core at 6892–6909 feet, *Globorotalia subbotinae* Zone, Early Eocene. **34** A more elongated variant. Guayaguayare well no. 159, core at 3986–4000 feet, *Globorotalia pseudomenardii* Zone, Late Paleocene.

35–36 *Dorothia hyperconica* Risch. × 50
35 Side view. Sample KR8385, *Rotalipora appenninica appenninica* Zone, Early Cenomanian. **36** Edge view. Same locality.

37 *Marssonella indentata* (Cushman & Jarvis). × 25
Moruga well no. 15, core at 3796–4816 feet, *Globorotalia pseudomenardii* Zone, Late Paleocene.

38 *Marssonella* cf. *nammalensis* Haque. × 50
Guayaguayare well no. 159, core at 3896–4000 feet, *Globorotalia pseudomenardii* Zone, Late Paleocene.

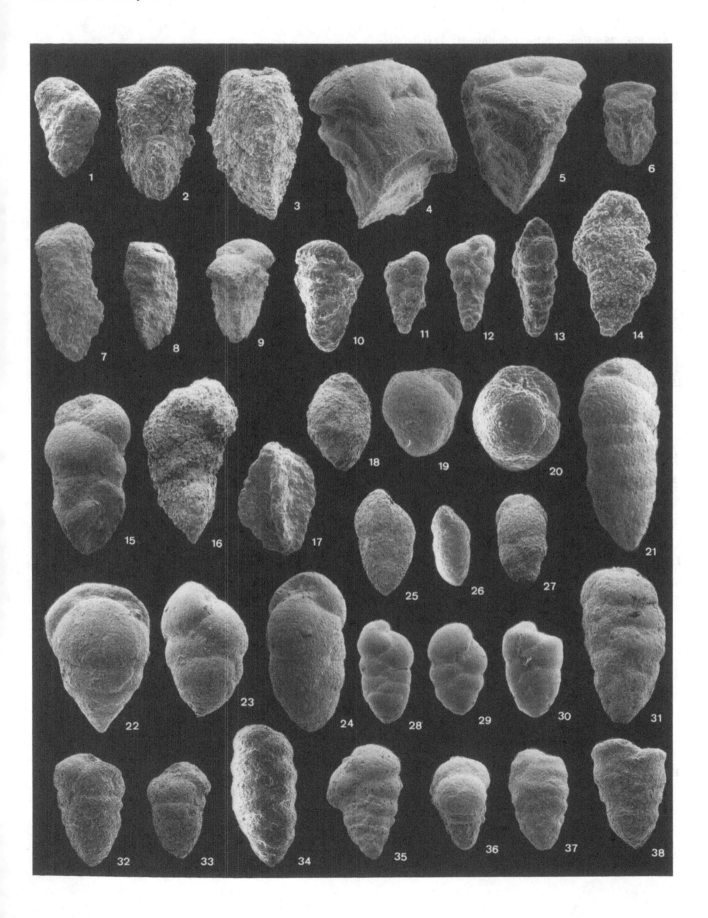

Fig. 25. Eggerellidae (continued), Textulariidae, Pseudogaudryinidae, Patellinidae, Hauerinidae

1–2 *Marssonella oxycona floridana* Applin & Jordan
1 Small type. Guayaguayare well no. 159, core at 4778–4790 feet, *Globorotalia pusilla pusilla* Zone, Middle Paleocene. × 50. **2** Larger type. Guayaguayare well no, 159, core at 3986–4000 feet, *Globorotalia pseudomenardii* Zone, Late Paleocene. × 35.

3–4 *Marssonella oxycona trinitatensis* Cushman & Renz. × 25
3 Apertural side view. Guayaguayare well no. 163, core at 5588–5598 feet, *Abathomphalus mayaroensis* Zone, Late Maastrichtian. **4** Side view. Guayaguayare well no. 159, core at 4778–4790 feet, *Globorotalia pusilla pusilla* Zone, Middle Paleocene.

5–6 *Marssonella oxycona oxycona* (Reuss). × 50
5 Sample KR8385, *Rotalipora appenninica appenninica* Zone, Early Cenomanian. **6** Moruga well no. 15, core at 5427–5452 feet, *Globorotalia fornicata* Zone, Santonian.

7 *Marssonella trochoides* (Marsson). × 40
Sample PJ253, probably *Globotruncana stuarti* Zone, Campanian.

8, *Clavulinoides asper* (Cushman). × 25
15–16 **8** Typical specimen. Sample KR23575, *Globorotalia angulata* Zone, Middle Paleocene. **15** Cylindrical specimen. Sample K9415, *Globorotalia velascoensis* Zone, Late Paleocene. **16** Large broad form of the 'whitei' type. Guayaguayare well no. 159, core at 4524–4536 feet, *Globorotalia pusilla pusilla* Zone, Middle Paleocene.

9–10 *Eggerina subovata* Beckmann. × 65
9 Holotype, side view showing aperture. Guayaguayare well no. 163, core at 5588–5598 feet, *Abathomphalus mayaroensis* Zone, Late Maastrichtian. **10** Side view at right angle to apertural view. Same locality.

11–12 *Karreriella lodoensis* Israelsky
11 Side view. Sample Hk1832, *Globorotalia edgari* Zone, Early Eocene. × 50. **12** Sample Hk1831, *Globorotalia subbotinae* Zone, Early Eocene. × 70.

13 *Martinottiella?* sp. A. × 50
Sample Rz281, *Globorotalia formosa formosa* Zone, Early Eocene.

14 *Textularia midwayana* Lalicker. × 50
Rochard well no. 1, core at 8556–8565 feet, probably *Globorotalia pseudobulloides* Zone, Early Paleocene.

17–18 *Clavulinoides disjunctus* (Cushman). × 50
Marabella well no. 1, core at 3512–3540 feet, probably *Globotruncana lapparenti tricarinata* Zone, Early Maastrichtian (Marabella facies).

19–20 *Clavulinoides plummerae* (Sandidge). × 25
Guayaguayare well no. 163, core at 5588–5598 feet, *Abathomphalus mayaroensis* Zone, Late Maastrichtian.

21 *Clavulinoides trilaterus* (Cushman). × 25
Guayaguayare well no. 159, core at 4524–4536 feet, *Globorotalia pusilla pusilla* Zone, Middle Paleocene.

22 *Pseudoclavulina amorpha* (Cushman). × 25
Sample K10832, *Globorotalia pseudomenardii* Zone, Late Paleocene.

23–24 *Pseudoclavulina anglica* Cushman. × 25
23 Typical specimen. Sample KB6972, *Globorotalia formosa formosa* Zone, Early Eocene. **24** Subcylindrical specimen. Sample Rz281, *Globorotalia formosa formosa* Zone, Early Eocene.

25–26 *Pseudoclavulina californica* Cushman & Todd. × 50
25 Sample KR8385, *Rotalipora appenninica appenninica* Zone, Early Cenomanian. **26** Sample JS1019, *Rotalipora appenninica appenninica* Zone, Early Cenomanian.

27 *Pseudoclavulina carinata* (Neagu). × 25
Sample KR8385, *Rotalipora appenninica appenninica* Zone, Early Cenomanian.

28 *Pseudoclavulina chitinosa* (Cushman & Jarvis). × 50
Guayaguayare well no. 163, core at 5588–5598 feet, *Abathomphalus mayaroensis* Zone, Late Maastrichtian.

29 *Pseudoclavulina trinitatensis* Cushman & Renz. × 50
Sample Hk 1832, *Globorotalia edgari* Zone, Early Eocene.

30–32 *Patellina subcretacea* Cushman & Alexander. × 80
Spiral, lateral and umbilical views. Sample KR8385, *Rotalipora appenninica appenninica* Zone, Early Cenomanian.

33 *Quinqueloculina antiqua* (Franke). × 100
Marac well no. 1, core at 9853–9891 feet, *Favusella washitensis* Zone, Late Albian.

34–35 *Quinqueloculina eocenica* Cushman. × 80
34 Side view. Sample Bo521, *Globorotalia aragonensis* Zone, Early Eocene. **35** Peripheral view. Sample Rz413, *Globorotalia aragonensis* Zone, Early Eocene.

36–37 *Quinqueloculina sandiegoensis* Sliter
36 Apertural view. Moruga well no. 15, core at 6980–7005 feet, *Globotruncana inornata* Zone, Turonian. × 100. **37** Side view. Marac well no. 1, core at 9578–9621 feet, *Rotalipora appenninica appenninica* Zone, Early Cenomanian. × 80.

38–39 *Quinqueloculina* sp. A
38 Apertural view. Marac well no. 1, core at 9853–9891 feet, *Favusella washitensis* Zone, Late Albian. × 100. **39** Side view. Same locality. × 80.

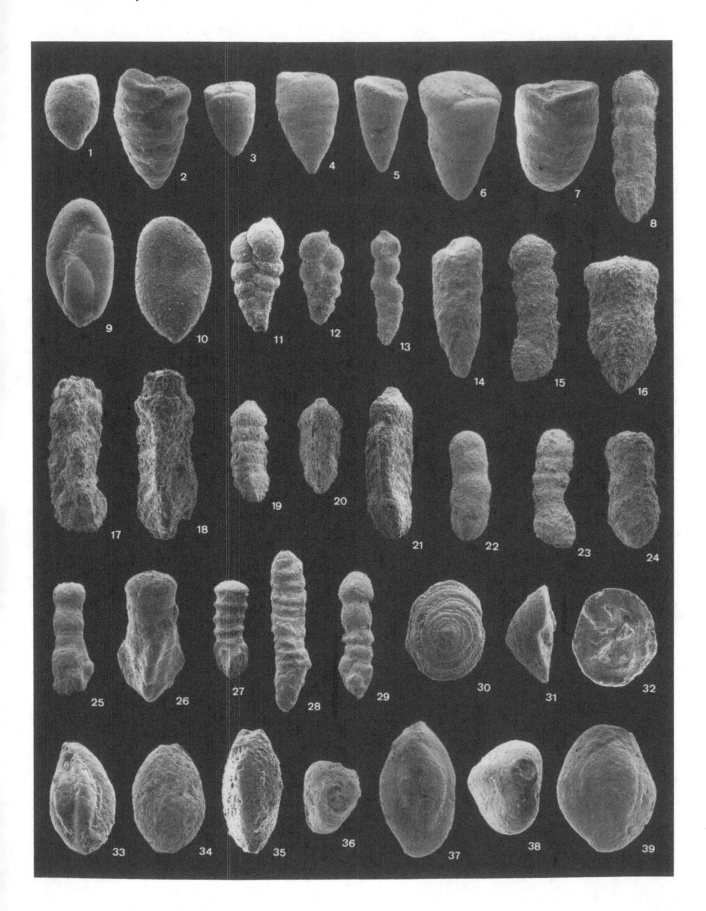

Fig. 26. Nodosariidae

1–2 *Chrysalogonium cretaceum* Cushman & Church
1 Sample Hk1832, *Globorotalia edgari* Zone, Early Eocene. × 35. **2** Final chamber. Same locality. × 50.

3–4 *Chrysalogonium elongatum* Cushman & Jarvis. × 25 Guayaguayare well no. 163, core at 5882–5902 feet, *Globotruncana lapparenti tricarinata* Zone, Early Maastrichtian.

5 *Chrysalogonium eocenicum* Cushman & Todd. × 35 Sample Hk1831, *Globorotalia subbotinae* Zone, Early Eocene.

6–7 *Chrysalogonium lanceolum* Cushman & Jarvis. × 25 Sample Rz413, *Globorotalia aragonensis* Zone. Early Eocene.

8–9 *Chrysalogonium velascoense* (Cushman). × 25 **8** Megalospheric specimen. Sample K9415, *Globorotalia velascoensis* Zone, Late Paleocene. **9** Microspheric specimen. Sample K10832, *Globorotalia pseudomenardii* Zone, Late Paleocene.

10 *Dentalina confluens* (Reuss). × 25 Sample K9415, *Globorotalia velascoensis* Zone, Late Paleocene.

11 *Dentalina frontierensis* Petersen. × 50 Marac well no. 1, core at 9578–9621 feeet, sample at 9602 feet, *Rotalipora appenninica appenninica* Zone, Early Cenomanian.

12–13 *Laevidentalina catenula* (Reuss)
12 Guayaguayare well no. 163, core at 7358–7393 feet, *Globotruncana fornicata* Zone, Santonian. × 50. **13** Guayaguayare well no. 163, core at 5588–5598 feet, *Abathomphalus mayaroensis* Zone, Late Maastrichtian. × 40.

14 *Laevidentalina gracilis* (d'Orbigny). × 50 Sample KB6972, *Globorotalia formosa formosa* Zone, Early Eocene.

15 *Laevidentalina cylindroides* (Reuss). × 25 Guayaguayare well no. 163, core at 5588–5598 feet, *Abathomphalus mayaroensis* Zone, Late Maastrichtian.

16 *Laevidentalina haeggi* (Brotzen). × 50 Moruga well no. 15, core at 3796–3816 feet, *Globorotalia pseudomenardii* Zone, Late Paleocene.

17 *Laevidentalina hammensis* (Franke). × 40 Guayaguayare well no. 159, core at 4524–4536 feet, *Globorotalia pusilla pusilla* Zone, Middle Paleocene.

18 *Laevidentalina havanensis* (Cushman & Bermudez). × 50 Guayaguayare well no. 163, core at 5588–5598 feet, *Abathomphalus mayaroensis* Zone, Late Maastrichtian.

19–20 *Laevidentalina legumen* (Reuss)
19 Moruga well no. 15, core at 5427–5452 feet, *Globotruncana fornicata* Zone, Santonian. × 50. **20** Sample KB6972, *Globorotalia formosa formosa* Zone, Early Eocene. × 35.

21 *Dentalina rancocasensis* Olsson. × 50 Sample KB6972, *Globorotalia formosa formosa* Zone, Early Eocene.

22–23 *Laevidentalina luma* (Belford). × 30 Sample K9415, *Globorotalia velascoensis* Zone, Late Paleocene.

24–25 *Laevidentalina megalopolitana* (Reuss). × 35 **24** Sample KR23575, *Globorotalia angulata* Zone, Middle Paleocene. **25** Sample KB6972, *Globorotalia formosa formosa* Zone, Early Eocene.

26 *Laevidentalina distincta* (Reuss). × 50 Marac well no. 1, core at 9853–9891 feet, *Favusella washitensis* Zone, Late Albian.

27 *Laevidentalina paradoxa* (Hussey). × 50 Guayaguayare well no. 159, core at 3986–4000 feet, *Globorotalia pseudomenardii* Zone, Late Paleocene.

28 *Laevidentalina reussi* (Neugeboren). × 50 Sample Bo521, *Globorotalia aragonensis* Zone, Early Eocene.

29 *Laevidentalina communis* (d'Orbigny). × 50 Guayaguayare well no. 163, core at 5588–5598 feet, *Abathomphalus mayaroensis* Zone, Late Maastrichtian.

30–31 *Nodosaria aspera aspera* Reuss. × 50 **30** Moruga well no. 15, core at 5427–5452 feet, *Globotruncana fornicata* Zone, Santonian. **31** Sample KB6972, *Globorotalia formosa formosa* Zone, Early Eocene.

32 *Nodosaria aspera parvavermiculata* Beckmann. × 40.
Holotype. Moruga well no. 15, core at 4296–4316 feet, *Globorotalia uncinata* Zone, Middle Paleocene.

33–34 *Nodosaria limbata* d'Orbigny
33 Guayaguayare well no. 163, core at 5882–5902 feet, *Globotruncana lapparenti tricarinata* Zone, Early Maastrichtian. × 15. **34** Guayaguayare well no. 163, core at 5588–5598 feet, *Abathomphalus mayaroensis* Zone, Late Maastrichtian. × 25.

35–36 *Nodosaria* (?) *longiscata* d'Orbigny. × 50 **35** Sample Rz389, *Globotruncana stuarti* Zone, Campanian. **36** Guayaguayare well no. 163, core at 6248–6260 feet, *Globotruncana stuarti* Zone, Campanian.

37 *Pseudonodosaria clearwaterensis* Mellon & Wall. × 80 Marac well no. 1, core at 9853–9891 feet, *Favusella washitensis* Zone, Late Albian.

38 *Pseudonodosaria* (?) *cylindrica* (Neugeboren). × 50 Guayaguayare well no. 163, core at 5588–5598 feet, *Abathomphalus mayaroensis* Zone, Late Maastrichtian.

39 *Pseudonodosaria humilis* (Roemer). × 80 Marac well no. 1, core at 9853–9891 feet, *Favusella washitensis* Zone, Late Albian.

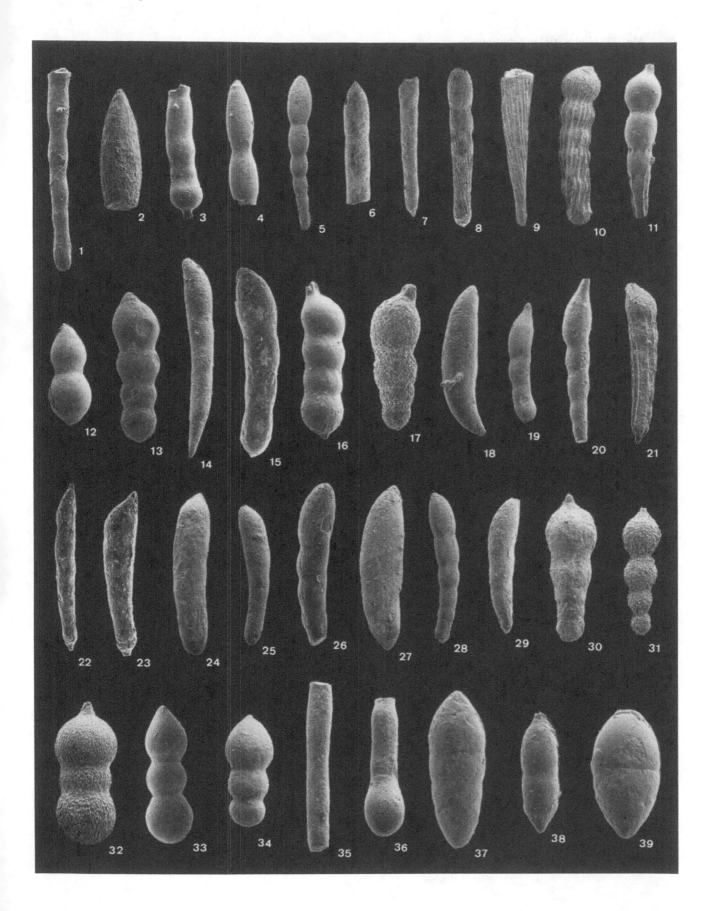

Fig. 27. Nodosariidae (continued)

1–2 *Pyramidulina obscura* (Reuss)
1 Guayaguayare well no. 163, core at 5882–5902 feet, *Globotruncana lapparenti tricarinata* Zone, Early Maastrichtian. × 25. **2** Guayaguayare well no. 159, core at 4524–4536 feet, *Globorotalia pusilla pusilla* Zone, Middle Paleocene. × 40.

3–4 *Pyramidulina latejugata* (Guembel)
3 Rochard well no. 1, core at 8556–8571 feet, probably *Globorotalia pseudobulloides* Zone, Early Paleocene. × 25. **4** Guayaguayare well no. 159, core at 3986–4000 feet, *Globorotalia pseudomenardii* Zone, Late Paleocene. × 40.

5–6 *Pyramidulina multicostata* (d'Orbigny)
5 Moruga well no. 15, core at 4831–4849 feet, *Globotruncana lapparenti tricarinata* Zone, Early Maastrichtian. × 35. **6** Guayaguayare well no. 159, core at 4524–4536 feet, *Globorotalia pusilla pusilla* Zone, Middle Paleocene. × 50.

   7 *Pyramidulina obsolescens* (Reuss). × 25
Sample K9415, *Globorotalia velascoensis* Zone, Late Paleocene.

8–9 *Pyramidulina orthopleura* (Reuss)
8 Guayaguayare well no. 163, core at 5882–5902 feet, *Globotruncana lapparenti tricarinata* Zone, Early Maastrichtian. × 50. **9** Sample G3644, *Abathomphalus mayaroensis* Zone, Late Maastrichtian. × 25.

10–11 *Pyramidulina paupercula* (Reuss). × 25
10 Marabella well no. 1, core at 3512–3540 feet, probably *Globotruncana lapparenti tricarinata* Zone, Early Maastrichtian (Marabella facies). **11** Guayaguayare well no. 163, core at 5588–5598 feet, *Abathomphalus mayaroensis* Zone, Late Maastrichtian.

12–13 *Pyramidulina prismatica* (Reuss). × 50
Guayaguayare well no. 163, core at 7641–7656 feet, *Globotruncana fornicata* Zone, Santonian.

   14 *Pyramidulina tetragona* (Reuss). × 50
Moruga well no. 15, core at 5427–5452 feet, *Globotruncana fornicata* Zone, Santonian.

   15 *Gonatosphaera* cf. *principiensis* Cushman & Bermudez. × 50
Sample KB9672, *Globorotalia formosa formosa* Zone, Early Eocene.

16–17, *Lingulina taylorana* Cushman
   24 **16** Moruga well no. 15, core at 6121–6140 feet, *Globotruncana concavata* Zone, Coniacian-Santonian. × 70. **17** Lizard well no. 1x, core at 7089–7099 feet, *Globotruncana fornicata* Zone, Santonian. × 70. **24** Guayaguayare well no. 163, core at 5588–5598 feet, *Abathomphalus mayaroensis* Zone, Late Maastrichtian. × 75.

18–19 *Frondicularia clarki* Bagg. × 40
Marabella well no. 1, core at 3489–3512, sample at 3510 feet, probably *Globotruncana lapparenti tricarinata* Zone, Early Maastrichtian (Marabella facies?).

20–21 *Frondicularia jarvisi* Cushman. × 50
20 Guayaguayare well no. 163, core at 7358–7393 feet, *Globotruncana fornicata* Zone, Santonian. **21** Guayaguayare no. 159, core at 4524–4536 feet, *Globorotalia pusilla pusilla* Zone, Middle Paleocene.

22–23 *Frondicularia mucronata* Reuss
22 Guayaguayare well no. 163, core at 5882–5902 feet, *Globotruncana lapparenti tricarinata* Zone, Early Maastrichtian. × 25. **23** Guayaguayare well no. 163, core at 5588–5598 feet, *Abathomphalus mayaroensis* Zone, Late Maastrichtian. × 50.

   25 *Frondicularia* cf. *parkeri* Reuss. × 70
Lizard well no. 1x, core at 7378–7390 feet, *Globotruncana inornata* Zone, Turonian.

   26 *Frondicularia* cf. *sedgwicki* Reuss. × 50
Lizard well no. 1x, core at 7186–7198 feet, *Globotruncana inornata* Zone, Turonian.

27–28 *Frondicularia surfibrata* Harris & Jobe. × 30
Sample KR23575, *Globorotalia angulata* Zone, Middle Paleocene.

   29 *Frondicularia trispiculata* Sandidge. × 40
Moruga well no. 15, core at 6330–6355 feet, *Globotruncana concavata* Zone, Coniacian-Santonian.

30–31 *Tristix excavata* (Reuss)
30 Marac well no. 1, core at 9853–9891 feet, *Favusella washitensis* Zone, Late Albian. × 75. **31** Sample KR8385, *Rotalipora appenninica appenninica* Zone, Early Cenomanian. × 100.

Fig. 28. Vaginulinidae

1–2 *Dimorphina* (?) *laeviuscula* (Cushman & Bermudez)
1 Sample RM287, *Globorotalia pseudomenardii* Zone, Late Paleocene. × 40. **2** Sample Hk1831, *Globorotalia subbotinae* Zone, Early Eocene. × 50.

3–4 *Cristellariopsis proinops* (Israelsky). × 50
3 Side view. Guayaguayare well no. 159, core at 3986–4000 feet, *Globorotalia pseudomenardii* Zone, Late Paleocene. **4** Peripheral view. Same locality.

5–6, 12 *Cristellariopsis wirzi* Beckmann. × 50
5 Holotype, side view. Sample KR23575, *Globorotalia angulata* Zone, Middle Paleocene. **6** Peripheral view. Guayaguayare well no. 159, core at 4524–4536 feet, *Globorotalia pusilla pusilla* Zone, Middle Paleocene. **12** Side view of a small (immature?) specimen. Same locality.

7–8 *Lenticulina* cf. *bayrocki* Mellon & Wall
7 Side view. Sample KR8385, *Rotalipora appenninica appenninica* Zone, Early Cenomanian. × 80. **8** Oblique peripheral view of a slightly asymmetrical specimen. Same locality. × 40.

9–10 *Lenticulina convergens* (Bornemann). × 50
Sample KB6972, *Globorotalia formosa formosa* Zone, Early Eocene.

11, 13–15 *Lenticulina discrepans* (Reuss)
11 Guayaguayare well no. 163, core at 5588–5598 feet, *Abathomphalus mayaroensis* Zone, Late Maastrichtian. × 50. **13** Lizard well no. 1x, core at 7390–7402 feet, *Globotruncana inornata* Zone, Turonian. × 80. **14–15** Sample K9415, *Globorotalia velascoensis* Zone, Late Paleocene. × 50.

16 *Lenticulina exarata* (Hagenow). × 50
Marac well no. 1, core at 9853–9891, *Favusella washitensis* Zone, Late Albian.

17–18 *Lenticulina howelli* (Olsson)
17 Side view. Sample Hk1832, *Globorotalia edgari* Zone, Early Eocene. × 50. **18** Peripheral view. Same locality. × 65.

19–20 *Lenticulina insulsa* (Cushman). × 70
Sample K10832, *Globorotalia pseudomenardii* Zone, Late Paleocene.

21–23 *Lenticulina* cf. *klagshamnensis* (Brotzen)
21 Side view. Marac well no. 1, core at 8180–8237 feet, sample at 8214 feet, *Globotruncana concavata* Zone, Coniacian-Santonian. × 70. **22** Side view. Same locality. × 50. **23** Peripheral view. Same locality. × 50.

24–27 *Lenticulina lepida morugaensis* Beckmann. × 75
24 Holotype, side view. Moruga well no. 15, core at 6802–6827 feet, *Globotruncana renzi* Zone, Late Turonian-Coniacian. **25** Peripheral view. Same locality. **26–27** Side views. Same localitiy.

28–30 *Lenticulina lobata* (Reuss)
28–29 Two specimens (side and peripheral views) with distinctly inflated final chambers. Marac well no. 1, core at 9347–9403 feet, sample at 9376 feet, *Globotruncana renzi* Zone, Late Turonian-Coniacian. × 80. **30** Side view of a less typical specimen with hardly inflated chambers. Morne Diablo well no. 1, core at 13984–14018 feet, sample at 14014 feet, *Rotalipora appenninica appenninica* Zone, Early Cenomanian. × 75.

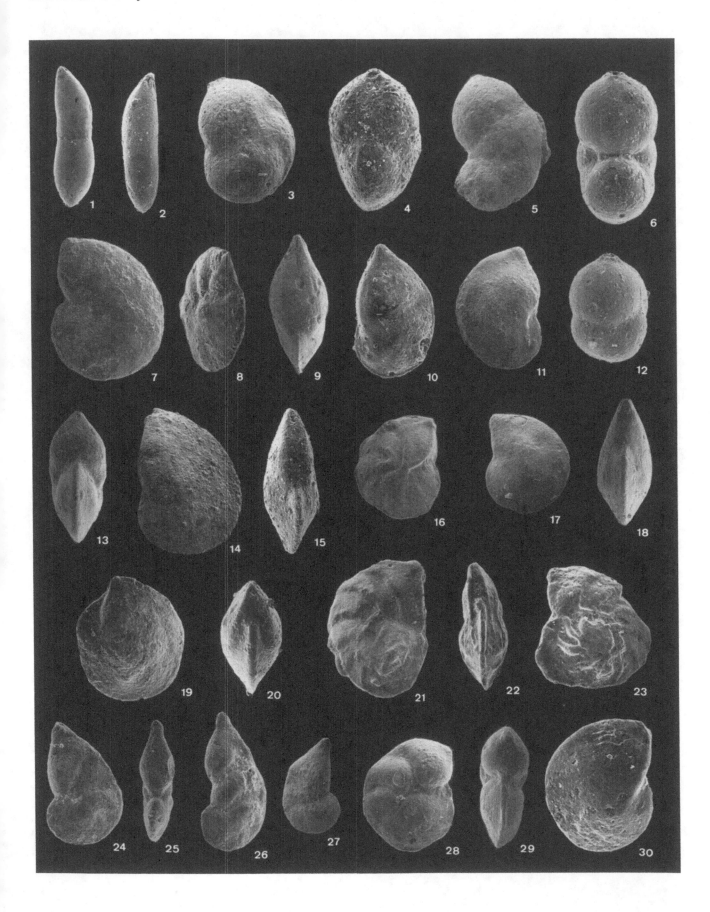

Fig. 29. Vaginulinidae (continued)

1–2 *Lenticulina midwayensis carinata* (Plummer)
1 Side view. Rochard well no. 1, core at 8556–8571 feet, sample at 8568 feet, probably *Globorotalia pseudobulloides* Zone, Early Paleocene. × 40. **2** Peripheral view. Same locality. × 50.

3–4 *Lenticulina muensteri* (Roemer). × 25
3 Guayaguayare well no. 163, core at 7358–7393 feet, *Globotruncana fornicata* Zone, Santonian. **4** Sample KB6972, *Globorotalia formosa formosa* Zone, Early Eocene.

5, 12 *Lenticulina multiformis rotunda* (Franke). × 25
5 Moruga well no. 15, core at 4831–4849 feet, *Globotruncana lapparenti tricarinata* Zone, Early Maastrichtian. **12** Sample Hk1831, *Globorotalia subbotinae* Zone, Early Eocene.

6–7 *Lenticulina nuda* (Reuss)
6 Side view. Moruga well no. 3, core at 10258–10264 feet, *Globorotalia trinidadensis* Zone, Early Paleocene. × 40. **7** Peripheral view. Same locality. × 50.

8–9 *Lenticulina nuda* (Reuss), variety A
8 Side view. Marac well no. 1, screen sample at 11979 feet, *Rotalipora ticinensis ticinensis* Zone, Albian, × 50. **9** Peripheral view. Same locality. × 65.

10–11 *Lenticulina oligostegia* (Reuss)
10 Side view. Marac well no. 1, screen sample at 11959–11989 feet, *Rotalipora ticinensis ticinensis* Zone, Albian. × 80. **11** Peripheral view. Same locality. × 100.

13–14 *Lenticulina revoluta* Israelsky. × 50
13 Side view. KR23575, *Globorotalia angulata* Zone, Middle Paleocene. **14** Peripheral view. Moruga well no. 3, core at 10258–10264 feet, *Globorotalia trinidadensis* Zone, Early Paleocene.

15–16 *Lenticulina secans* (Reuss)
15 Side view. Lizard well no. 1x, core at 7378–7390 feet, *Globotruncana inornata* Zone, Turonian. × 25. **16** Peripheral view. Same locality. × 40.

17–18 *Lenticulina sorachiensis* (Takayanagi)

17 Side view. Guayaguayare well no. 163, core at 5882–5902 feet, *Globotruncana lapparenti tricarinata* Zone, Early Maastrichtian. × 25. **18** Peripheral view. Guayaguayare well no. 159, core at 4778–4790 feet, *Globorotalia pusilla pusilla* Zone, Middle Paleocene. × 50.

19–20 *Lenticulina subgaultina* Bartenstein
19 Side view. Sample JS1019, *Rotalipora appenninica appenninica* Zone, Early Cenomanian. × 25. **20** Peripheral view. Same locality. × 40.

21 *Lenticulina truncata* (Reuss). × 50
Lizard well no. 1x, core at 7186–7198 feet, *Globotruncana inornata* Zone, Turonian.

22–23 *Lenticulina velascoensis* White
22 Side view. Sample K10832, *Globorotalia pseudomenardii* Zone, Late Paleocene. × 50. **23** Peripheral view. Sample G3644, *Abathomphalus mayaroensis* Zone, Late Maastrichtian. × 65.

24 *Lenticulina* sp. A. × 40
Guayaguayare well no. 159, core at 4778–4790 feet, *Globorotalia pusilla pusilla* Zone, Middle Paleocene.

25 *Marginulinopsis cephalotes* (Reuss). × 50
Oblique peripheral view. Marac well no. 1, core at 9853–9891 feet, *Favusella washitensis* Zone, Late Albian.

26–27 *Marginulinopsis curvisepta* (Cushman & Goudkoff). × 70
Moruga well no. 15, core at 6121–6140 feet, transition *Globotruncana concavata* Zone to *G. fornicata* Zone, Santonian.

28–29 *Marginulinopsis lituola* (Reuss)
28 Side view. Marac well no. 1, core at 8965–9017 feet, sample at 9002 feet, *Globotruncana concavata* Zone, Coniacian-Santonian. × 50. **29** Peripheral view. Same locality. × 80.

30–31 *Marginulinopsis multicostata* (Lipnik). × 50
Moruga well no. 10, core sample at 7987 feet, *Globotruncana stuarti* Zone, Campanian.

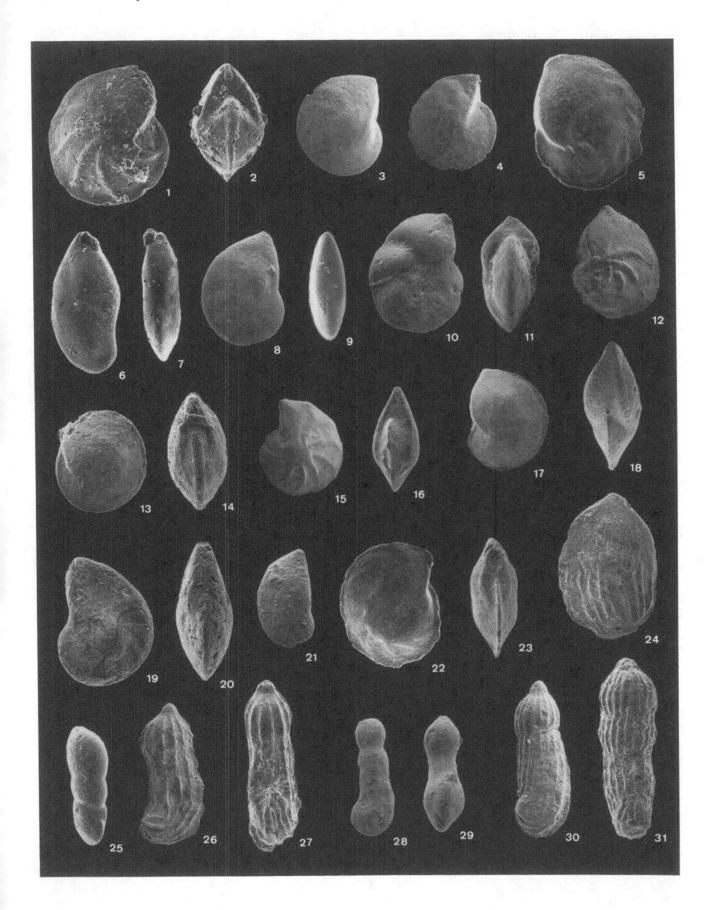

Fig. 30. Vaginulinidae (continued)

1–3 *Marginulinopsis praetschoppi* Trujillo
1 Side view. Moruga well no. 15, core at 6121–6140 feet, transition *Globotruncana concavata* Zone to *G. fornicata* Zone, Santonian. × 70. **2** Peripheral view of a broken specimen. Same locality. × 65. **3** Peripheral view. Guayaguayare well no. 163, core at 7641–7656 feet, *Globotruncana fornicata* Zone, Santonian. × 40.

4–5 *Marginulinopsis subrecta* (Franke). × 25
4 Moruga well no. 15, core at 4831–4849 feet, *Globotruncana lapparenti tricarinata* Zone, Early Maastrichtian. **5** Guayaguayare well no. 159, core at 4778–4790 feet, *Globorotalia pusilla pusilla* Zone, Middle Paleocene.

6 *Percultazonaria tuberculata* (Plummer). × 50
Sample KB6972, *Globorotalia formosa formosa* Zone, Early Eocene.

7 *Pravoslavlevia dutroi* (Tappan). × 40
Moruga well no.15, core at 6121–6140 feet, transition *Globotruncana concavata* Zone to *G. fornicata* Zone, Santonian.

8–9 *Saracenaria cretacea* Dailey. × 50
8 Marac well no. 1, core at 9578–9621, sample at 9602 feet, *Rotalipora appenninica appenninica* Zone, Early Cenomanian. **9** Marac well no. 1, core at 9347–9403 feet, sample at 9403 feet, *Globotruncana inornata* Zone, Turonian.

10 *Saracenaria navicula* (d'Orbigny). × 100
Moruga well no. 15, core at 6980–7005 feet, *Globotruncana inornata* Zone, Turonian.

11–12 *Saracenaria triangularis* (d'Orbigny)
11 Marac well no. 1, core at 8260–8305 feet, sample at 8297 feet, *Globotruncana concavata* Zone, Coniacian-Santonian. × 50. **12** Guayaguayare well no. 163, core at 5882–5902 feet, *Globotruncana lapparenti tricarinata* Zone, Early Maastrichtian. × 40.

13 *Saracenaria tunesiana* ten Dam & Sigal. × 50
Guayaguayare well no. 159, core at 4524–4536 feet, *Globorotalia pusilla pusilla* Zone, Middle Paleocene.

14–15 *Neoflabellina* cf. *coranica* (Marie). × 50
14 Sample KR23575, *Globorotalia angulata* Zone, Middle Paleocene. **15** Moruga well no. 3, core at 10258–10264 feet, *Globorotalia trinidadensis* Zone, Early Paleocene.

16–17 *Neoflabellina delicatissima* (Plummer). × 50
Guayaguayare well no. 163, core at 5588–5598 feet, *Abathomphalus mayaroensis* Zone, Late Maastrichtian.

18–19, *Neoflabellina jarvisi* (Cushman). × 25
25 **18–19** Sample K10832, *Globorotalia pseudomenardii* Zone, Late Paleocene. **25** Sample K9415, *Globorotalia velascoensis* Zone, Late Paleocene.

20–21 *Neoflabellina numismalis* (Wedekind). × 50
20 Moruga well no. 10, core sample at 7987 feet, *Globotruncana stuarti* Zone, Campanian. **21** Guayaguayare well no. 163, core at 6046–6052 feet, sample at 6047 feet, *Globotruncana stuarti* Zone, Campanian.

22–24 *Neoflabellina paleocenica* Titova. × 50
22 Guayaguayare well no. 159, core at 4524–4536 feet, *Globorotalia pusilla pusilla* Zone, Middle Paleocene. **23** Sample K10832, *Globorotalia pseudomenardii* Zone, Late Paleocene. **24** Guayaguayare well no. 159, core at 3986–4000 feet, *Globorotalia pseudomenardii* Zone, Late Paleocene.

26 *Neoflabellina* cf. *permutata* Koch. × 50
Marabella well no. 1, core at 3512–3540, probably *Globotruncana lapparenti tricarinata* Zone, Early Maastrichtian (Marabella facies).

27–28 *Neoflabellina praereticulata* Hiltermann. × 40
Guayaguayare well no. 163, core at 6052–6079 feet, sample at 6067–6068 feet, *Globotruncana stuarti* Zone, Campanian.

29–30 *Neoflabellina semireticulata* (Cushman & Jarvis)
29 Sample KR23575, *Globorotalia angulata* Zone, Middle Paleocene. × 40. **30** Guayaguayare well no. 159, core at 4524–4536 feet, *Globorotalia pusilla pusilla* Zone, Middle Paleocene. × 50.

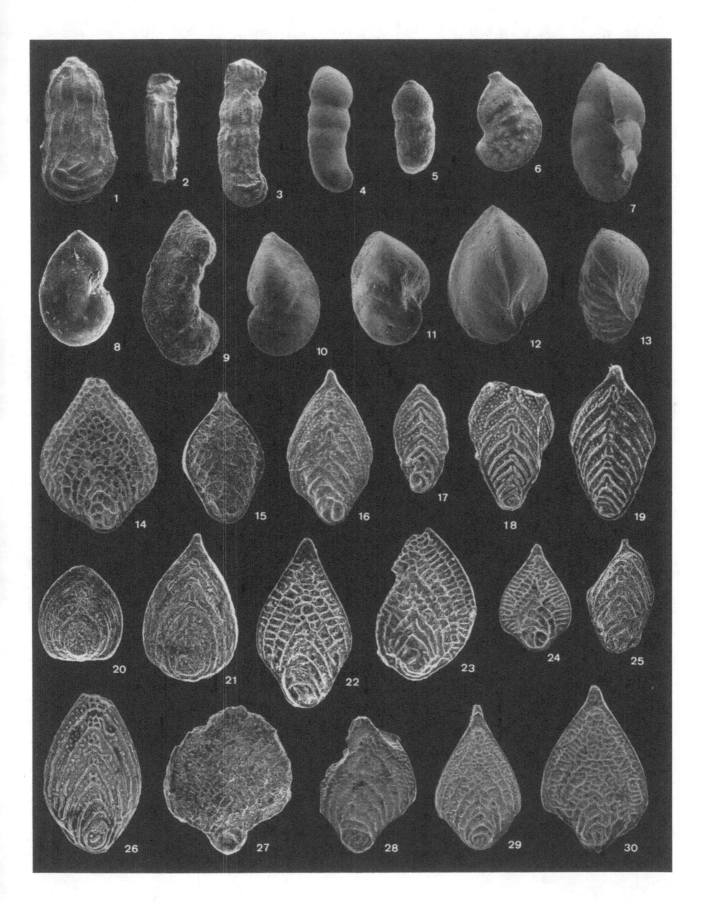

Fig. 31. Vaginulinidae (continued)

1–5 *Neoflabellina projecta* (Carsey)
1–2 Marac well no. 1, core at 7947–7967 feet, *Globotruncana fornicata* Zone, Santonian. × 50. **3** Moruga well no. 15, core at 5647–5671 feet, *Globotruncana fornicata* Zone, Santonian. × 50. **4** Moruga well no. 15, core at 5195–5210 feet, *Globotruncana stuarti* Zone, Campanian. × 70. **5** Moruga well no. 10, core sample at 7987 feet, *Globotruncana stuarti* Zone, Campanian. × 50.

6–7 *Neoflabellina thalmanni* (Finlay)
6 Guayaguayare well no. 163, core at 6052–6079 feet, sample at 6067–6068 feet, *Globotruncana stuarti* Zone, Campanian. × 35. **7** Same locality. × 50.

8–9 *Neoflabellina* sp. A
8 Marabella well no. 1, core at 3512–3540 feet, probably *Globotruncana lapparenti tricarinata* Zone, Early Maastrichtian (Marabella facies). × 50. **9** Same locality. × 75.

10–11 *Palmula* cf. *primitiva* Cushman. × 50
10 Guayaguayare well no. 163, core at 5588–5598 feet, *Abathomphalus mayaroensis* Zone, Late Maastrichtian. **11** Moruga well no. 3, core at 11428–11448 feet, *Globorotalia trinidadensis* Zone, Early Paleocene.

12–13 *Palmula toulmini* ten Dam & Sigal
12 Rochard well no. 1, core at 8565–8571 feet, sample at 8568 feet, probably *Globorotalia pseudobulloides* Zone, Early Paleocene. × 25. **13** Same locality. × 30.

14 *Astacolus exquisitus* (Toulmin). × 25
Oropouche well no. 4, core at 7829–7849 feet, sample at 7849 feet, *Globorotalia subbotinae* Zone, Early Eocene.

15 *Astacolus gratus* (Reuss). × 100
Sample KR8385, *Rotalipora appenninica appenninica* Zone, Early Cenomanian.

16–17, *Astacolus jarvisi* (Cushman)
26 **16–17** Sample K10832, *Globorotalia pseudomenardii* Zone, Late Paleocene. × 35. **26** Moruga well no.

15, core at 4296–4316 feet, *Globorotalia uncinata* Zone, Middle Paleocene. × 50.

18–19 *Astacolus nuttalli* (Todd & Kniker). × 40
KB6972, *Globorotalia formosa formosa* Zone, Early Eocene.

20–21 *Astacolus parallelus* (Reuss). × 50
20 Moruga well no 15, core at 6980–7005 feet, *Globotruncana inornata* Zone, Turonian. **21** Marac well no. 1, core at 9347–9403, feet, sample at 9376 feet, *Globotruncana renzi* Zone, Late Turonian-Coniacian.

22–23 *Astacolus perobliquus* (Reuss)
22 Peripheral view. Sample KR8385, *Rotalipora appenninica appenninica* Zone, Early Cenomanian. × 75. **23** Side view. Same locality. × 100.

24–25 *Astacolus planiusculus* (Reuss)
24 Side view. Marac well no. 1, screen sample at 11979 feet, *Rotalipora ticinensis ticinensis* Zone, Albian. × 75. **25** Peripheral view. Sample KR8385, *Rotalipora appenninica appenninica* Zone, Early Cenomanian. × 100.

27 *Astacolus* cf. *pseudomarcki* (Cushman). × 8
Guayaguayare well no. 163, core at 5882–5902 feet, *Globotruncana lapparenti tricarinata* Zone, Early Maastrichtian.

28–29 *Astacolus rectus* (d'Orbigny). × 50
28 Side view. Guayaguayare well no. 163, core at 5588–5598 feet, *Abathomphalus mayaroensis* Zone, Late Maastrichtian. **29** Peripheral view. Sample Hk1832, *Globorotalia edgari* Zone, Early Eocene.

30–31 *Astacolus* sp. A. × 50
Sample KR8385, *Rotalipora appenninica appenninica* Zone, Early Cenomanian.

32–33 *Hemirobulina abbreviata* (Neugeboren)
32 Marac well no. 1, core at 9347–9403 feet, *Globotruncana renzi* Zone, Late Turonian-Coniacian. × 50. **33** Guayaguayare well no. 163, core at 7358–7393 feet, *Globotruncana fornicata* Zone, Santonian. × 65.

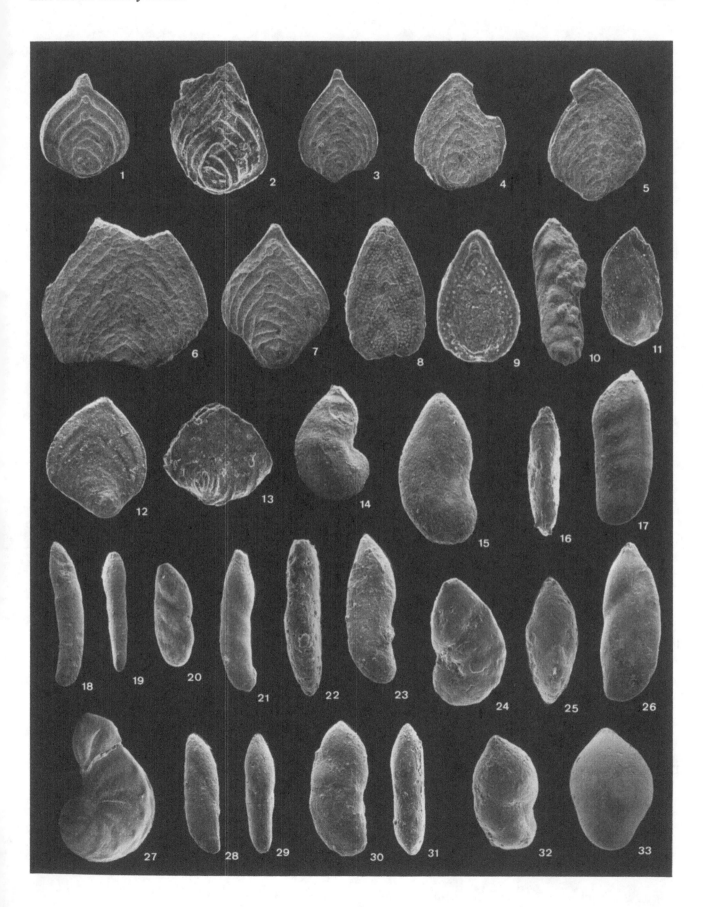

Fig. 32. Vaginulinidae (continued), Lagenidae

1 *Hemirobulina* cf. *anceps* (Neugeboren). × 50
Sample Hk1831, *Globorotalia subbotinae* Zone,
Early Eocene.

2 *Hemirobulina bullata* (Reuss). × 80
Moruga well no. 15, core at 6330–6355, *Globotruncana concavata* Zone, Coniacian-Santonian.

3–4 *Hemirobulina curvatura* (Cushman)
3 Sample KR8385, *Rotalipora appenninica appenninica* Zone, Early Cenomanian. × 65. 4 Moruga well 15, core at 5195–5210 feet, *Globotruncana stuarti* Zone, Campanian. × 50.

5 *Hemirobulina hamuloides* (Brotzen). × 50
Guayaguayare well no. 159, core at 4524–4536 feet, *Globorotalia pusilla pusilla* Zone, Middle Paleocene.

6 *Hemirobulina hantkeni* (Bandy). × 40
Sample KR8385, *Rotalipora appenninica appenninica* Zone, Early Cenomanian.

7 *Hemirobulina reussi* (Takayanagi). × 100
Marac well no. 1, core at 8332–8362 feet, sample at 8339 feet, *Globotruncana concavata* Zone, Coniacian-Santonian.

8 *Vaginulinopsis hemicylindrica* (Noth). × 40
Sample K9415, *Globorotalia velascoensis* Zone,
Late Paleocene.

9–10 *Hemirobulina* cf. *schloenbachi* (Reuss)
9 Side view. Marac well no. 1, core at 9853–9891 feet, *Favusella washitensis* Zone, Late Albian. × 75.
10 Peripheral view. Sample KR8385, *Rotalipora appenninica appenninica* Zone, Early Cenomanian. × 50.

11 *Hemirobulina truncatula* (Berthelin). × 100
Moruga well no. 15, core at 6980–7005 feet, *Globotruncana inornata* Zone, Turonian.

12 *Vaginulinopsis* cf. *silicula* (Plummer). × 25
Marabella well no. 1, core at 3512–3540 feet, probably *Globotruncana lapparenti tricarinata* Zone, Early Maastrichtian (Marabella facies).

13 *Citharina angustissima* (Reuss). × 25
Moruga well no. 15, core at 5427–5452 feet, *Globotruncana fornicata* Zone, Santonian.

14–15 *Citharina barcoensis* (Cushman & Hedberg)
14 Guayaguayare well no. 163, core at 6052–6079 feet, *Globotruncana stuarti* Zone, Campanian. × 75.
15 Marabella well no. 1, core at 3512–3540 feet, probably *Globotruncana lapparenti tricarinata* Zone, Early Maastrichtian (Marabella facies). × 80.

16 *Planularia complanata* (Reuss). × 75
Moruga well no. 15, core at 6330–6355 feet, *Globotruncana concavata* Zone, Coniacian-Santonian.

17–18 *Planularia advena* Cushman & Jarvis. × 25
Rochard well no. 1, core at 8556–8565 feet, sample at 8561 feet, probably *Globorotalia pseudobulloides* Zone, Early Paleocene.

19 *Psilocitharella recta* (Reuss). × 50
Sample KR8385, *Rotalipora appenninica appenninica* Zone, Early Cenomanian.

20–21 *Vaginulina anitana* Akpati. × 40
20 Side view. Guayaguayare well no. 159, core at 3986–4000 feet, *Globorotalia pseudomenardii* Zone, Late Paleocene. 21 Peripheral view. Sample K9415, *Globorotalia velascoensis* Zone, Late Paleocene.

22–23 *Vaginulina plummerae* (Cushman)
22 Rochard well no. 1, core at 8556–8565 feet, sample at 8561 feet, probably *Globorotalia pseudobulloides* Zone, Early Paleocene. × 50. 23 Same locality. × 25.

24 *Lagena amphora paucicostata* Franke. × 100
Guayaguare well no. 163, core at 6052–6079 feet, *Globotruncana stuarti* Zone, Campanian.

25 *Lagena* cf. *amphora* Reuss. × 100
Guayaguayare well no. 163, core at 5882–5902 feet, *Globotruncana lapparenti tricarinata* Zone, Early Maastrichtian.

26 *Lagena aspera apiculata* White. × 50
Guayaguayare well no. 159, core at 3986–4000 feet, *Globorotalia pseudomenardii* Zone, Late Paleocene.

27 *Lagena gracilicosta* Reuss. × 100
Sample Hk1832, *Globorotalia edgari* Zone, Early Eocene.

28–29 *Lagena isabella conscripta* Cushman & Barksdale. × 120
28 Guayaguayare well no. 163, core at 6052–6079 feet, *Globotruncana stuarti* Zone, Campanian. 29 Sample Hk1831, *Globorotalia subbotinae* Zone, Early Eocene.

30 *Lagena reticulostriata* Haque. × 50
Guayaguayare well no. 163, core at 5588–5598 feet, *Abathomphalus mayaroensis* Zone, Late Maastrichtian.

31–32 *Lagena tenuistriata* Stache
31 Sample K9415, *Globorotalia velascoensis* Zone, Late Paleocene. × 50. 32 Sample Hk1832, *Globorotalia edgari* Zone, Early Eocene. × 80.

33 *Lagena virginiana* (Nogan). × 100
Guayaguayare well no. 159, core at 3986–4000 feet, *Globorotalia pseudomenardii* Zone, Late Paleocene.

34 *Reussoolina apiculata apiculata* (Reuss). × 80
Guayaguayare well no. 163, core at 5882–5902 feet, *Globotruncana lapparenti tricarinata* Zone, Early Maastrichtian.

35–36 *Reussoolina apiculata elliptica* (Reuss)
35 Sample Rz389, *Globotruncana stuarti* Zone, Campanian. × 50. 36 Same locality. × 80.

Fig. 33. Polymorphinidae

1–2 *Globulina ampulla* (Jones). × 50
1 Guayaguayare well no. 163, core at 5588–5598 feet, *Abathomphalus mayaroensis* Zone, Late Maastrichtian. **2** Sample RM287, *Globorotalia pseudomenardii* Zone, Late Paleocene.

3 *Globulina bucculenta* (Berthelin). × 100
Marac well no. 1, core at 9578–9621 feet, sample at 9602 feet, *Rotalipora appenninica appenninica* Zone, Early Cenomanian.

4 *Globulina fusiformis* (Roemer). ×60
Guayaguayare well no. 159, core at 4524–4536 feet, *Globorotalia pusilla pusilla* Zone, Middle Paleocene.

5 *Globulina lacrima horrida* Reuss. × 80
Lizard well no. 1x, core at 7186–7198 feet, *Globotruncana inornata* Zone, Turonian.

6 *Globulina lacrima lacrima* (Reuss). × 80
Sample Rz389, *Globotruncana stuarti* Zone, Campanian.

7 *Globulina lacrima subsphaerica* (Berthelin). × 80
Marac well no. 1, core at 8332–8362 feet, *Globotruncana concavata* Zone, Coniacian-Santonian.

8–9 *Globulina* cf. *prisca* (Reuss). × 50
8 Guayaguayare well no. 159, core at 4524–4536 feet, *Globorotalia pusilla pusilla* Zone, Middle Paleocene. **9** Sample K9415, *Globorotalia velascoensis* Zone, Late Paleocene.

10–12 *Guttulina trigonula* (Reuss)
10 Side view. Moruga well no. 15, core at 4831–4849 feet, *Globotruncana lapparenti tricarinata* Zone, Campanian. × 80. **11** Apertural view. Same locality. × 80. **12** Large lobate specimen. Guayaguayare well no. 159, core at 3986–4000 feet, *Globorotalia pseudomenardii* Zone, Late Paleocene. × 40.

13 *Palaeopolymorphina* cf. *pleurostomelloides* (Franke). × 60
Guayaguayare well no. 163, core at 5588–5598 feet, *Abathomphalus mayaroensis* Zone, Late Maastrichtian.

14–15 *Polymorphina berica* (Cati). × 50
Sample G3644, *Abathomphalus mayaroensis* Zone, Late Maastrichtian.

16–17 *Polymorphinella* (?) cf. *cylindracea* (Reuss)
16 Guayaguayare well no. 159, core at 3986–4000 feet, *Globorotalia pseudomenardii* Zone, Late Paleocene. × 40. **17** Sample KB6972, *Globorotalia formosa formosa* Zone, Early Eocene. × 30.

18–20 *Pseudopolymorphina rudis* (Reuss)

18–19 Guayaguayare well no. 163, core at 5882–5902 feet, *Globotruncana lapparenti tricarinata* Zone, Early Maastrichtian. × 25. **20** Moruga well no. 15, core at 4831–4849 feet, *Globotruncana lapparenti tricarinata* Zone, Early Maastrichtian. × 20.

21–22 *Pyrulina cylindroides* (Roemer)
21 Sample KR8385, *Rotalipora appenninica appenninica* Zone, Early Cenomanian. × 80. **22** Guayaguayare well no. 159, core at 4524–4536 feet, *Globorotalia pusilla pusilla* Zone, Middle Paleocene. × 40.

23, *Pyrulina velascoensis* (Cushman)
29–30 **23** Marac well no. 1, core at 8260–8305 feet, *Globotruncana concavata* Zone, Coniacian-Santonian. × 80. **29** Moruga well no. 15, core at 6330–6355 feet, *Globotruncana concavata* Zone, Coniacian-Santonian. × 65. **30** Sample G3644, *Abathomphalus mayaroensis* Zone, Late Maastrichtian. × 40.

24–25 *Pyrulinoides* cf. *kalinini* Dain
24 Side view. Sample KB6972, *Globorotalia formosa formosa* Zone, Early Eocene. × 65. **25** Peripheral view. Same locality. × 50.

26–27 *Pyrulinoides obesus* Marie
26 Side view. Guayaguayare well no. 159, core at 3986–4000 feet, *Globorotalia pseudomenardii* Zone, Late Paleocene. × 50. **27** Peripheral view. Moruga well no. 15, core at 5195–5210 feet, *Globotruncana stuarti* Zone, Campanian. × 75.

28 *Ramulina* cf. *abscissa* Loeblich & Tappan. × 50
Marac well no. 1, core at 8332–8362 feet, *Globotruncana concavata* Zone, Coniacian-Santonian.

31 *Ramulina aculeata* (d'Orbigny). × 25
Guayaguayare well no. 163, core at 5588–5598 feet, *Abathomphalus mayaroensis* Zone, Late Maastrichtian.

32 *Ramulina muricatina* Loeblich & Tappan. × 50
Moruga well no. 15, core at 5427–5452 feet, *Globotruncana fornicata* Zone, Santonian.

33 *Ramulina navarroana* Cushman. × 25
Moruga well no. 15, core at 4296–4316 feet, *Globorotalia uncinata* Zone, Middle Paleocene.

34 *Ramulina ornata* Cushman. × 50
Sample KB6972, *Globorotalia formosa formosa* Zone, Early Eocene.

35–36 *Ramulina pseudoaculeata* (Olsson)
35 Sample KR23575, *Globorotalia angulata* Zone, Middle Paleocene. × 50. **36** Sample K9415, *Globorotalia velascoensis* Zone, Late Paleocene. × 25.

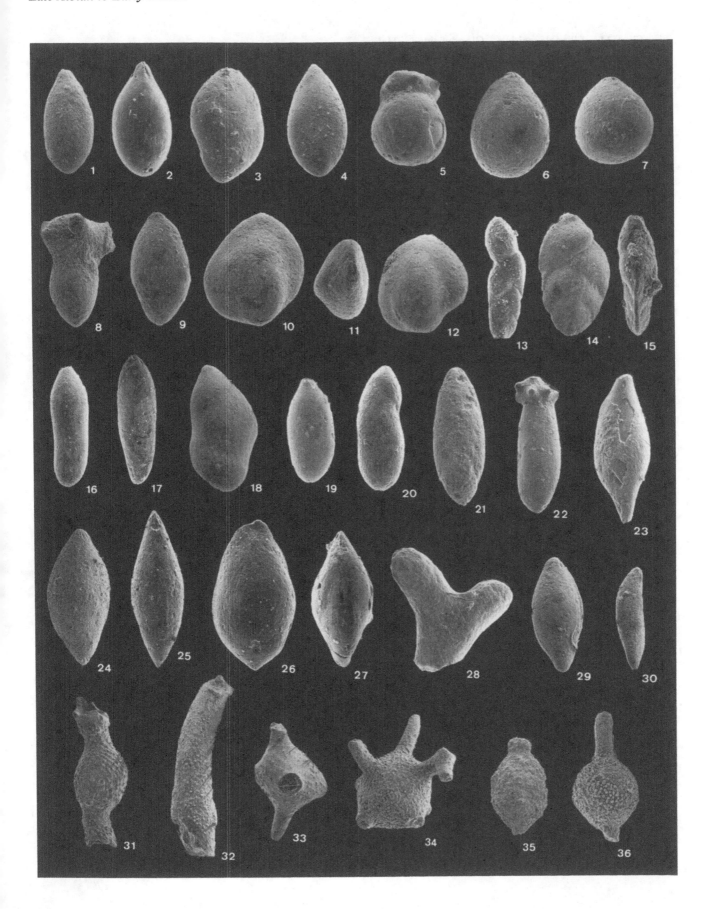

Fig. 34. Ellipsolagenidae, Glandulinidae, Heterohelicidae, Bolivinidae, Bolivinoididae

1 *Buchnerina castrensis* (Schwager). × 80
Guayaguayare well no. 159, core at 3986–4000 feet, *Globorotalia pseudomenardii* Zone, Late Paleocene.

2 *Buchnerina* cf. *oslatus* (Shifflett). × 100
Sample Hk1831, *Globorotalia subbotinae* Zone, Early Eocene.

3 *Buchnerina seguenziana* (Fornasini). × 100
Guayaguayare well no. 159, core at 3986–4000 feet, *Globorotalia pseudomenardii* Zone, Late Paleocene.

4 *Exsculptina jacobi* (Marie). × 140
Guayaguayare well no. 163, core at 5882–5902 feet, *Globotruncana lapparenti tricarinata* Zone, Early Maastrichtian.

5 *Favulina* cf. *geometrica* (Reuss). × 100
Guayaguayare well no. 159, core at 3986–4000 feet, *Globorotalia pseudomenardii* Zone, Late Paleocene.

6 *Lagnea perfectocostata* (Cushman & Renz). × 70
Sample G3644, *Abathomphalus mayaroensis* Zone, Late Maastrichtian.

7–8 *Oolina globosa* (Montagu). × 50
7 Guayaguayare well no. 163, core at 5882–5902 feet, *Globotruncana lapparenti tricarinata* Zone, Early Maastrichtian. 8 Sample K10832, *Globorotalia pseudomenardii* Zone, Late Paleocene.

9–10 *Fissurina laevigata* Reuss
9 Side view. Guayaguayare well no. 159, core at 3986–4000 feet, *Globorotalia pseudomenardii* Zone, Late Paleocene. × 80. 10 Peripheral view. Sample Hk1832, *Globorotalia edgari* Zone, Early Eocene. × 100.

11 *Fissurina tubata* (Matthes). × 100
Guayaguayare well no. 163, core at 5588–5598 feet, *Abathomphalus mayaroensis* Zone, Late Maastrichtian.

12 *Palliolatella orbignyana* (Seguenza). × 70
Guayaguayare well no. 163, core at 5882–5902 feet, *Globotruncana lapparenti tricarinata* Zone, Early Maastrichtian.

13, 19 *Glandulina subconica* Alth
13 Guayaguayare well no. 163, core at 5882–5902 feet, *Globotruncana lapparenti tricarinata* Zone, Early Maastrichtian. × 25. 19 Same locality. × 35.

14–16 *Glandulina* cf. *bicamerata* Herrmann. × 50
14–15 Sample KR8385, *Rotalipora appenninica appenninica* Zone, Early Cenomanian. 16 Moruga well no. 10, core sample at 7987 feet, *Globotruncana stuarti* Zone, Campanian.

17 *Glandulina* (?) *cylindracea constricta* Franke. × 70
Moruga well no. 15, core at 5427–5452 feet, *Globotruncana fornicata* Zone, Santonian.

18 *Glandulina laevigata* (d'Orbigny). × 50
Guayaguayare well no. 159, core at 3986–4000 feet, *Globorotalia pseudomenardii* Zone, Late Paleocene.

20 *Bifarina alabamensis* (Cushman). × 150
Sample Hk1831, *Globorotalia subbotinae* Zone, Early Eocene.

21 *Brizalina* sp. A. × 100
Lizard well no. 1x, core at 7186–7198 feet, *Globotruncana inornata* Zone, Turonian.

22–23 *Gabonita* cf. *klaszi* (Roveda)
22 Side view. Esmeralda well no. 1, core at 10678–10694 feet, *Globotruncana concavata* Zone, Coniacian-Santonian. × 100. 23 Oblique side view showing aperture. Same locality. × 80.

24–25 *Gabonita kugleri* (Beckmann). × 100
Two side views. Marabella well no. 1, core at 5018–5038 feet, *Globotruncana concavata* Zone, Coniacian-Santonian.

26–27 *Loxostomoides applinae* (Plummer). × 50
26 Side view. Marabella well no. 1, core at 3512–3540 feet, probably *Globotruncana lapparenti tricarinata* Zone, Early Maastrichtian (Marabella facies). 27 Edge view. Same locality.

28–29 *Loxostomoides plummerae* (Cushman). × 100
28 Side view. Guayaguayare well no. 159, core at 4778–4790 feet, *Globorotalia pusilla pusilla* Zone, Middle Paleocene. 29 Peripheral view. Same locality.

30 *Tappanina eouvigeriniformis* (Keller). × 150
Marac well no. 1, core at 9347–9403 feet, sample at 9347 feet, *Globotruncana renzi* Zone, Late Turonian-Coniacian.

31–34 *Tappanina selmensis* (Cushman). × 150
31–32 Sample KR23575, *Globorotalia angulata* Zone, Middle Paleocene. 33–34 Sample Hk1832, *Globorotalia edgari* Zone, Early Eocene.

35 *Bolivinoides decoratus decoratus* (Jones). × 100
Guayaguayare well no. 163, core at 5882–5902 feet, *Globotruncana lapparenti tricarinata* Zone, Early Maastrichtian.

36 *Bolivinoides delicatulus curtus* Reiss. × 100
Guayaguayare well no. 159, core at 4778–4790 feet, *Globorotalia pusilla pusilla* Zone, Middle Paleocene.

37 *Bolivinoides delicatulus delicatulus* Cushman. × 100
Guayaguayare well no. 163, core at 6052–6079 feet, sample at 6067 feet, *Globotruncana stuarti* Zone, Campanian.

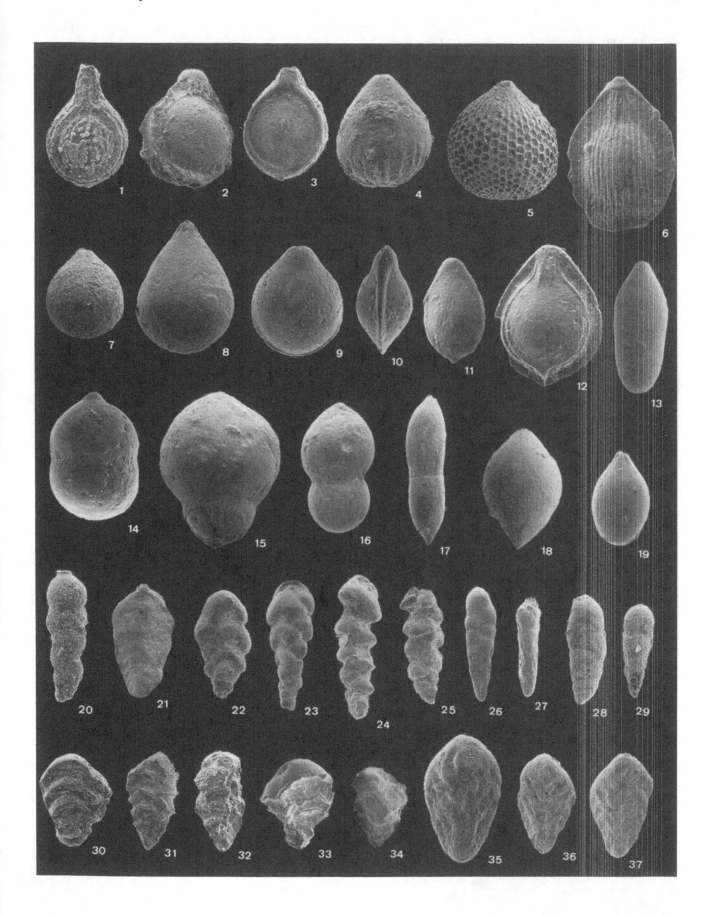

Fig. 35. Bolivinoididae (continued), Loxostomatidae, Cassidulinidae, Eouvigerinidae, Turrilinidae

1 *Bolivinoides draco draco* (Marsson). × 70
Guayaguayare well no. 163, core at 5588–98 feet, *Abathomphalus mayaroensis* Zone, Late Maastrichtian.

2 *Bolivinoides draco miliaris* Hiltermann & Koch. × 70
Guayaguayare well no. 163, core at 5882–5902 feet, *Globotruncana lapparenti tricarinata* Zone, Early Maastrichtian.

3 *Bolivinoides draco verrucosus* Beckmann & Koch. × 100
Guayaguayare well no. 163, core at 5588–5598 feet, *Abathomphalus mayaroensis* Zone, Late Maastrichtian.

4 *Bolivinoides granulatus emersus* Beckmann & Koch. × 70
Guayaguayare well no. 163, screen sample at 5729 feet, *Globotruncana gansseri* Zone, Maastrichtian.

5 *Aragonia aragonensis* (Nuttall). × 100
Sample Rz281, *Globorotalia formosa formosa* Zone, Early Eocene.

6 *Aragonia materna kugleri* Beckmann & Koch. × 100
Moruga well no. 15, core at 6519–6544 feet, *Globotruncana renzi* Zone, Late Turonian-Coniacian.

7, 15 *Aragonia velascoensis* (Cushman). × 70
7 Guayaguayare well no. 163, core at 5588–5598 feet, *Abathomphalus mayaroensis* Zone, Late Maastrichtian. 15 Sample KR23575, *Globorotalia angulata* Zone, Middle Paleocene.

8–10 *Zeauvigerina teuria* Finlay
8–9 Side views of two specimens with uniserial final chamber. Marabella well no. 1, core at 4465–4476 feet, *Globotruncana fornicata* Zone, Santonian. × 80. 10 Oblique edge view. Same locality. × 100.

11 *Globocassidulina globosa* (Hantken). × 200
Sample Hk1832, *Globorotalia edgari* Zone, Early Eocene.

12–13 *Eouvigerina subsculptura* McNeil & Caldwell. × 100
Guayaguayare well no. 163, core at 5588–5598 feet, *Abathomphalus mayaroensis* Zone, Late Maastrichtian.

14 *Neobulimina primitiva* (Cushman). × 150
Sample KR8385, *Rotalipora appenninica appenninica* Zone, Early Cenomanian.

16–17 *Neobulimina irregularis* Cushman & Parker. × 100
Sample KR8385, *Rotalipora appenninica appenninica* Zone, Early Cenomanian.

18–19 *Praebulimina kickapooensis kickapooensis* (Cole). × 50
Moruga well no. 15, core at 4831–4849 feet, *Globotruncana lapparenti tricarinata* Zone, Early Maastrichtian.

20 *Praebulimina kickapooensis pingua* (Cushman & Parker). × 50

Sample Rz389, *Globotruncana stuarti* Zone, Campanian.

21 *Praebulimina nannina* (Tappan). × 100
Sample KR8385, *Rotalipora appenninica appenninica* Zone, Early Cenomanian.

22–23 *Praebulimina petroleana* (Cushman & Hedberg). × 50
22 Guayaguayare well no. 163, core at 6052–6079 feet, *Globotruncana stuarti* Zone, Campanian. 23 Guayaguayare well no. 163, core at 5882–5902 feet, *Globotruncana lapparenti tricarinata* Zone, Early Maastrichtian.

24–25 *Praebulimina* (?) *quadrata* (Plummer). × 50
Moruga well no. 3, core at 10258–10264 feet, *Globorotalia trinidadensis* Zone, Early Paleocene.

26–28 *Praebulimina reussi* (Morrow). × 100
26 Lizard well no. 1x, core at 7089–7099 feet, *Globotruncana fornicata* Zone, Santonian. 27 Guayaguayare well no. 163, screen sample at 5744–5754 feet, *Globotruncana gansseri* Zone, Maastrichtian. 28 Guayaguayare well no. 163, core at 5588–5598 feet, *Abathomphalus mayaroensis* Zone, Late Maastrichtian.

29–30 *Praebulimina* cf. *seabeensis* Tappan. × 100
29 Marabella well no. 1, core at 4465–4476 feet, *Globotruncana fornicata* Zone, Santonian. 30 Marabella well no. 1, core at 4210–4232 feet, *Globotruncana stuarti* Zone, Campanian.

31 *Praebulimina* cf. *spinata* (Cushman & Campbell). × 60
Marabella well no. 1, core at 3512–3540 feet, probably *Globotruncana lapparenti tricarinata* Zone, Early Maastrichtian (Marabella facies).

32 *Praebulimina wyomingensis* (Fox). × 100
Marac well no. 1, core at 9853–9891 feet, *Favusella washitensis* Zone, Late Albian.

33 *Pseudouvigerina naheolensis* Cushman & Todd. × 150
Sample Hk1832, *Globorotalia edgari* Zone, Early Eocene.

34 *Pseudouvigerina plummerae* Cushman. × 100
Guayaguayare well no. 163, core at 5588–5598 feet, *Abathomphalus mayaroensis* Zone, Late Maastrichtian.

35 *Pyramidina cushmani* (Brotzen). × 100
Marabella well no. 1, core at 3512–3540 feet, probably *Globotruncana lapparenti tricarinata* Zone, Early Maastrichtian (Marabella facies).

36 *Pyramidina* cf. *robusta* Neagu. × 100
Sample KR8385, *Rotalipora appenninica appenninica* Zone, Early Cenomanian.

37 *Pyramidina rudita* (Cushman & Parker). × 150
Moruga well no. 15, core at 6980–7005 feet, *Globotruncana inornata* Zone, Turonian.

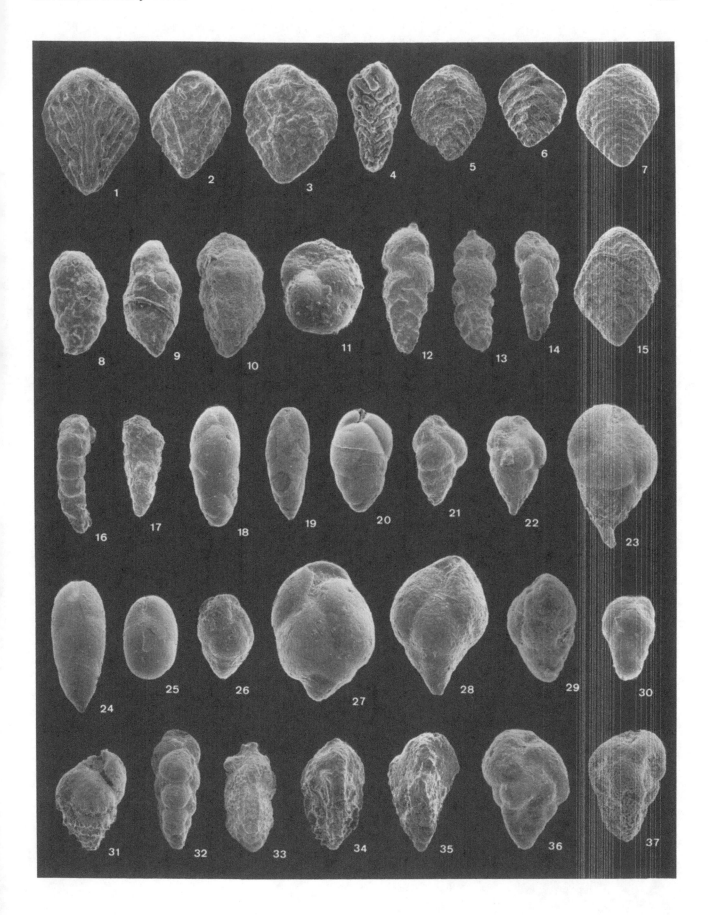

Fig. 36. Turrilinidae (continued), Stainforthiidae, Siphogenerinoididae, Buliminidae

1–2 *Pyramidina szajnochae* (Grzybowski). × 25
1 Rock Dome well no. 1, core at 6087–6146 feet, sample at 6131 feet, *Globotruncana lapparenti tricarinata* Zone, Early Maastrichtian. **2** Moruga well no. 15, core at 4831–49 feet, *Globotruncana lapparenti tricarinata* Zone, Early Maastrichtian.

3 *Sitella colonensis* (Cushman & Hedberg). × 75
Guayaguayare well no. 163, core at 5882–5902 feet, *Globotruncana lapparenti tricarinata* Zone, Early Maastrichtian.

4–5 *Turrilina carseyae* (Plummer). × 100
4 Morne Diablo well no. 34, core at 13 774–821, sample at 13 782 feet, *Globotruncana inornata* Zone, Turonian. **5** Marac well no. 1, core at 9347–9403 feet, sample at 9376 feet, *Globotruncana renzi* Zone, Late Turonian-Coniacian.

6 *Turrilina cushmani* (Sandidge). × 75
Guayaguayare well no. 163, core at 5588–98 feet, *Abathomphalus mayaroensis* Zone, Late Maastrichtian.

7–8 *Hopkinsina dorreeni* (Weiss). × 100
7 Side view. Rochard well no. 1, core at 8556–65 feet, probably *Globorotalia pseudobulloides* Zone, Early Paleocene. **8** Edge view. Same locality.

9–10 *Stainforthia* cf. *navarroana* (Cushman). × 80
9 Rochard well no. 1, core at 8556–65 feet, probably *Globorotalia pseudobulloides* Zone, Early Paleocene. **10** Guayaguayare well no. 159, core at 4524–36 feet, *Globorotalia pusilla pusilla* Zone, Late Paleocene.

11 *Siphogenerinoides bramlettei* Cushman. × 50
Marabella well no. 1, core at 3512–40 feet, probably *Globotruncana lapparenti tricarinata* Zone, Early Maastrichtian (Marabella facies).

12 *Siphogenerinoides landesi* Stone. × 50
Sample KB7907 (Lower Tarouba formation), probably *Globotruncana stuarti* Zone, Campanian.

13 *Orthokarstenia cretacea* (Cushman). × 25
Marabella well no. 1, core at 3512–40 feet, probably *Globotruncana lapparenti tricarinata* Zone, Early Maastrichtian (Marabella facies).

14–15 *Orthokarstenia idkyensis* (Colom)
14 Megalospheric form. FW well no. 214, core at 10 290–310 feet, *Globotruncana gansseri* Zone or *Abathomphalus mayaroensis* Zone, Maastrichtian. × 20. **15** Microspheric(?) form. Same locality. × 30.

16–17 *Orthokarstenia* cf. *parva* (Cushman)
16 Microspheric form. Marabella well no. 1, core at 3512–40 feet, probably *Globotruncana lapparenti tricarinata* Zone, Early Maastrichtian (Marabella

facies). × 40. **17** Megalospheric form. Same locality. × 25.

18 *Bulimina excavata* Cushman & Parker. × 100
Moruga well no. 15, core at 4401–21 feet, *Globorotalia trinidadensis* Zone, Early Paleocene.

19–21 *Bulimina midwayensis* Cushman & Parker. × 100
19 Guayaguayare well no. 163, core at 5882–5902 feet, *Globotruncana lapparenti tricarinata* Zone, Early Maastrichtian. **20** Guayaguayare well no. 163, core at 5588–98 feet, *Abathomphalus mayaroensis* Zone, Late Maastrichtian. **21** Moruga well no. 15, core at 4296–4316 feet, *Globorotalia uncinata* Zone, Middle Paleocene.

22 *Bulimina plena* Brotzen. × 100
Moruga well no. 15, core at 3796–3816 feet, *Globorotalia pseudomenardii* Zone, Late Paleocene.

23 *Bulimina semicostata* Nuttall. × 100
Sample KB6972, *Globorotalia formosa formosa* Zone, Early Eocene.

24–26 *Bulimina stokesi* Cushman & Renz. × 100
24 Guayaguayare well no. 159, core at 3986–4000 feet, *Globorotalia pseudomenardii* Zone, Late Paleocene. **25** Sample Hk1832, *Globorotalia edgari* Zone, Early Eocene. **26** Sample KB6972, *Globorotalia formosa formosa* Zone, Early Eocene.

27 *Bulimina taylorensis* Cushman & Parker. × 80
Guayaguayare well no. 163, core at 5588–98 feet, *Abathomphalus mayaroensis* Zone, Late Maastrichtian.

28–29 *Bulimina trinitatensis* Cushman & Jarvis. × 80
28 Sample K10832, *Globorotalia pseudomenardii* Zone, Late Paleocene. **29** Sample Hk1831, *Globorotalia subbotinae* Zone, Early Eocene.

30–31 *Bulimina truncanella* Finlay. × 100
30 Sample Hk1831, *Globorotalia subbotinae* Zone, Early Eocene. **31** Sample Rz413, *Globorotalia aragonensis* Zone, Early Eocene.

32–33 *Bulimina tuxpamensis* Cole
32 Guayaguayare well no. 159, core at 3986–4000 feet, *Globorotalia pseudomenardii* Zone, Late Paleocene. × 80. **33** Sample Hk1831, *Globorotalia subbotinae* Zone, Early Eocene. × 65.

34–36 *Globobulimina* (?) *suteri* (Cushman & Renz)
34 Sample KB7907 (Lower Taruba Formation), probably *Globotruncana stuarti* Zone, Campanian, × 50. **35** Sample G3644, *Abathomphalus mayaroensis* Zone, Late Maastrichtian. × 40. **36** Sample K10832, *Globorotalia pseudomenardii* Zone, Late Paleocene. × 50.

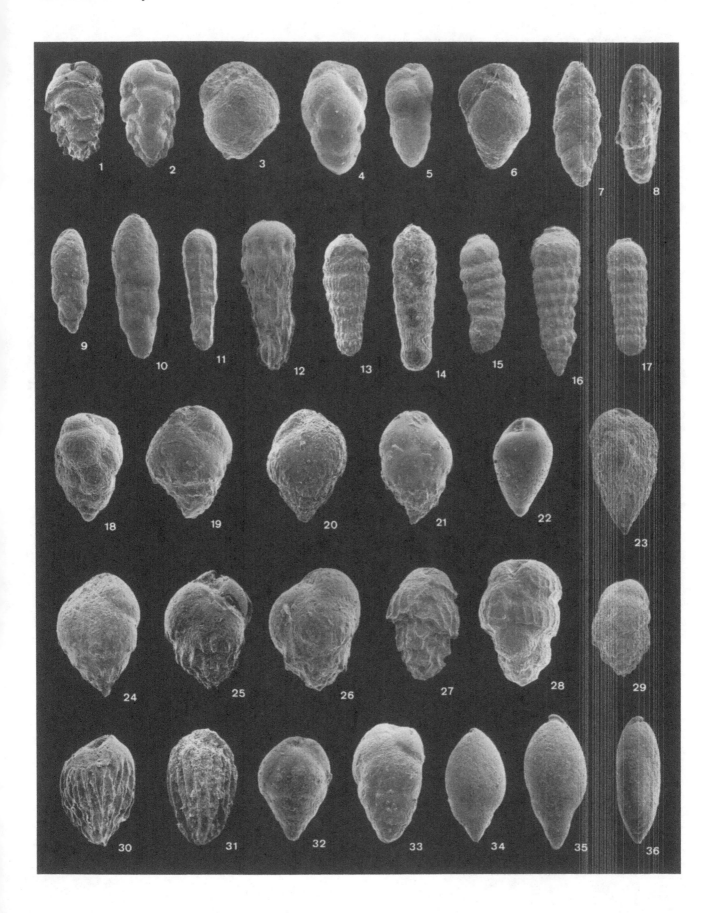

Fig. 37. Buliminidae (continued), Buliminellidae, Uvigerinidae, Fursenkoinidae, Pleurostomellidae

1–2 *Praeglobobulimina ovata* (d'Orbigny). × 40
Sample Rz413, *Globorotalia aragonensis* Zone, Early Eocene.

3 *Buliminella grata* Parker & Bermudez. × 80
Sample Rz281, *Globorotalia formosa formosa* Zone, Early Eocene.

4 *Euuvigerina elongata* (Cole). × 100
Sample KR23575, *Globorotalia angulata* Zone, Middle Paleocene.

5–7 *Coryphostoma crassum* (Vasilenko & Myatlyuk). × 30
5 Megalospheric form, side view. Guayaguayare well no. 163, core at 5588–98 feet, *Abathomphalus mayaroensis* Zone, Late Maastrichtian. **6** Edge view. Same locality. **7** Microspheric form, side view. Same locality.

8–9 *Coryphostoma* cf. *gemma* (Cushman). × 40
Marabella well no. 1, core at 3512–40 feet, probably *Globotruncana lapparenti tricarinata* Zone, Early Maastrichtian.

10–12 *Coryphostoma incrassatum* (Reuss). × 50
10 Side view, megalospheric specimen. Moruga well no. 15, core at 4831–49 feet, *Globotruncana lapparenti tricarinata* Zone, Early Maastrichtian. **11** Edge view. Same locality. **12** Side view, microspheric specimen. Same locality.

13–15 *Coryphostoma midwayense* (Cushman). × 50
13 Side view, megalospheric form. Sample K10832, *Globorotalia pseudomenardii* Zone, Late Paleocene. **14** Oblique edge view. Same locality. **15** Side view, microspheric specimen. Same locality.

16 *Coryphostoma* sp. A. × 100
Morne Diablo well no. 34, core at 12911–13 feet, *Globotruncana renzi* Zone, Late Turonian-Coniacian.

17–19 *Ellipsodimorphina coalingensis* (Cushman & Church)
17 Guayaguayare well no. 159, core at 4524–36 feet, *Globorotalia pusilla pusilla* Zone, Middle Paleocene. × 40. **18** Sample Hk1832, *Globorotalia edgari* Zone, Early Eocene. × 30. **19** Sample Hk1831, *Globorotalia subbotinae* Zone, Early Eocene. × 40.

20 *Ellipsodimorphina morrowi* (Van Wessem). × 40
Sample Hk1831, *Globorotalia subbotinae* Zone, Early Eocene.

21–22 *Ellipsodimorphina solida* (Brotzen). × 50
21 Sample KR8385, *Rotalipora appenninica apppenninica* Zone, Early Cenomanian. **22** Sample Rz389, *Globotruncana stuarti* Zone, Campanian.

23 *Ellipsoglandulina labiata* (Schwager). × 50
Guayaguayare well no. 159, core at 4524–36 feet, *Globorotalia pusilla pusilla* Zone, Middle Paleocene.

24–25 *Ellipsoglandulina chilostoma* (Rzehak). × 50
24 Sample Rz389, *Globotruncana stuarti* Zone, Campanian. **25** Sample Hk1832, *Globorotalia edgari* Zone, Early Eocene.

26 *Ellipsoidella attenuata* (Plummer). × 50
Guayaguayare well no. 159, core at 4524–38 feet, *Globorotalia pusilla pusilla* Zone, Middle Paleocene.

27–28, *Ellipsoidella binaria* Belford. × 40
37 **27** Specimen with triserial early stage. Moruga well no. 15, core at 5427–52 feet, *Globotruncana fornicata* Zone, Santonian. **28** Specimen with loosely biserial early stage. Same locality. **37** Pseudouniserial specimen. Guayaguayare well no. 163, core at 5882–5902 feet, *Globotruncana lapparenti tricarinata* Zone, Early Maastrichtian.

29–33 *Ellipsoidella kugleri* (Cushman & Renz)
29–30 Guayaguayare well no. 163, core at 5588–98 feet, *Abathomphalus mayaroensis* Zone, Late Maastrichtian. × 50. **31** Sample Hk1832, *Globorotalia edgari* Zone, Early Eocene. × 40. **32** Oropouche well no. 4, core at 7829–49 feet, *Globorotalia subbotinae* Zone, Early Eocene. × 25. **33** Sample Hk1831, *Globorotalia subbotinae* Zone, Early Eocene. × 40.

34 *Ellipsoidella* cf. *robusta* (Cushman). × 20
Sample KB6972, *Globorotalia formosa formosa* Zone, Early Eocene

35 *Ellipsoidina abbreviata* Seguenza. × 50
Marac well no. 1, core at 8651–98 feet, *Globotruncana concavata* Zone, Coniacian-Santonian.

36 *Ellipsoidina ellipsoides* Seguenza. × 50
Sample Hk1831, *Globorotalia subbotinae* Zone, Early Eocene.

38 *Ellipsopolymorphina principiensis* (Cushman & Bermudez). × 40
Guayaguayare well no. 163, core at 5588–98 feet, *Abathomphalus mayaroensis* Zone, Late Maastrichtian.

39 *Ellipsopolymorphina velascoensis* (Cushman). × 65
Sample K9415, *Globorotalia velascoensis* Zone, Late Paleocene.

40 *Nodosarella constricta* Cushman & Bermudez. × 40
Sample Hk1831, *Globorotalia subbotinae* Zone, Early Eocene.

41–42 *Nodosarella hedbergi* Cushman & Renz. × 25
41 Guayaguayare well no. 163, core at 5588–98 feet, *Abathomphalus mayaroensis* Zone, Late Maastrichtian. **42** Sample KR23575, *Globorotalia angulata* Zone, Middle Paleocene.

43–44 *Nodosarella paleocenica* Cushman & Todd. × 40
43 Moruga well no. 3, core at 11428–48 feet, *Globorotalia trinidadensis* Zone, Early Paleocene. **44** Guayaguayare well no. 159, core at 3986–4000 feet, *Globorotalia pseudomenardii* Zone, Late Paleocene.

45–46 *Nodosarella subnodosa* (Guppy). × 25
45 Uniserial specimen. Guayaguayare well no. 163, core at 5588–98 feet, *Abathomphalus mayaroensis* Zone, Late Maastrichtian. **46** Specimen with irregularly biserial early stage. Sample K10832, *Globorotalia pseudomenardii* Zone, Late Paleocene.

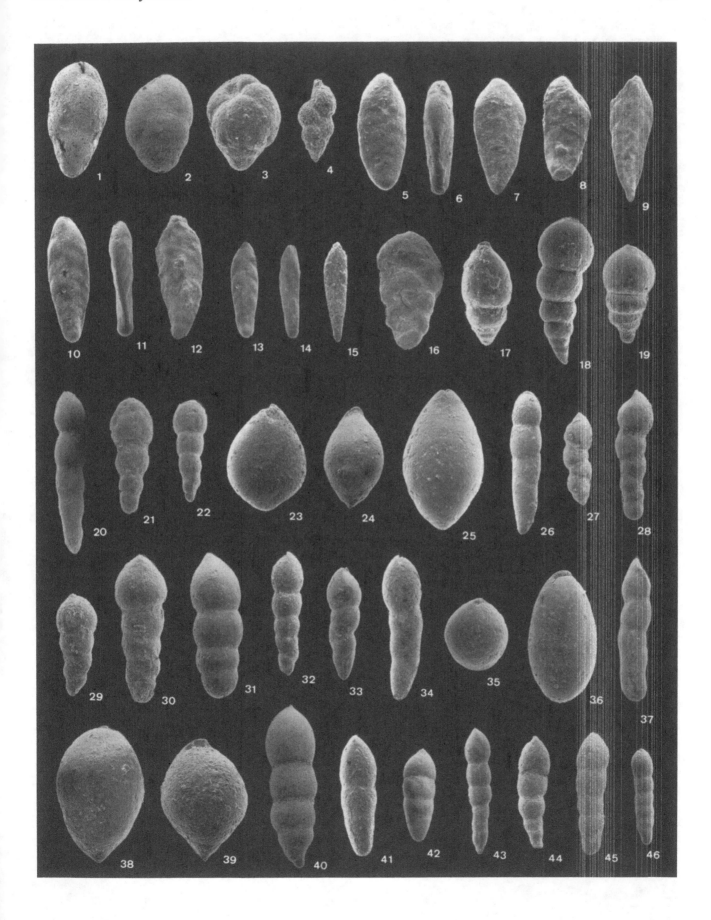

Fig. 38. Nodosariidae (continued), Pleurostomellidae (continued), Stilostomellidae, Conorbinidae, Bagginidae

1 *Nodosarella wintereri* Trujillo. × 50
Lizard well no. 1x, core at 7378–90 feet, *Globotruncana inornata* Zone, Turonian.

2 *Pleurostomella clavata* Cushman. × 50
Sample Hk1831, *Globorotalia subbotinae* Zone, Early Eocene.

3 *Pleurostomella copiosa* Bulakova. × 25
Moruga well no. 15, core at 5427–52 feet, *Globotruncana fornicata* Zone, Santonian.

4–5 *Pleurostomella cubensis* Cushman & Bermudez. × 50
4 Sample Hk1832, *Globorotalia edgari* Zone, Early Eocene. 5 Sample Bo521, *Globorotalia aragonensis* Zone, Early Eocene.

6–7 *Pleurostomella naranjoensis* Cushman & Bermudez. × 50
Sample Hk1831, *Globorotalia subbotinae* Zone, Early Eocene.

8–10 *Pleurostomella obtusa* Berthelin
8 Sample KR8385, *Rotalipora appenninica appenninica* Zone, Early Cenomanian. × 40. 9 Same locality. × 50. 10 Marac well no. 1, core at 9347–9403 feet, *Globorotalia renzi* Zone, Late Turonian-Coniacian. × 50.

11–12 *Pleurostomella paleocenica* Cushman. × 100
Sample Hk1831, *Globorotalia subbotinae* Zone, Early Eocene.

13 *Pleurostomella* cf. *reussi* Berthelin. × 40
Sample KR8385, *Rotalipora appenninica appenninica* Zone, Early Cenomanian.

14–15 *Pleurostomella velascoensis* Cushman. × 50
14 Guayaguayare well no. 159, core at 4524–36 feet, *Globorotalia pusilla pusilla* Zone, Middle Paleocene. 15 Guayaguayare well no. 159, core at 3986–4000 feet, *Globorotalia pseudomenardii* Zone, Late Paleocene.

16 *Czarkowyella torta* (Cushman). × 25
Moruga well no. 15, core at 4831–49 feet, *Globotruncana lapparenti tricarinata* Zone, Early Maastrichtian.

17–18 *Bandyella greatvalleyensis* (Trujillo)
17 Marac well no. 1, core at 8332–62 feet, *Globotruncana concavata* Zone, Coniacian-Santonian. × 50. 18 Marac well no. 1, core at 8965–9017 feet, *Globotruncana concavata* Zone, Coniacian-Santonian. × 65.

19–20 *Laevidentalina spinescens* (Reuss). × 30
Sample Hk1831, *Globorotalia subbotinae* Zone, Early Eocene.

21–22 *Orthomorphina rohri* (Cushman & Stainforth)
21 Guayaguayare well no. 163, core at 5588–98 feet, *Abathomphalus mayaroensis* Zone, Late Maastrichtian. × 50. 22 Sample KB6972, *Globorotalia formosa formosa* Zone, Early Eocene. × 35.

23 *Siphonodosaria dentataglabrata* (Cushman), × 25
Sample Hk1831, *Globorotalia subbotinae* Zone, Early Eocene.

24 *Siphonodosaria longae* McLean. × 25
Sample KB6972, *Globorotalia formosa formosa* Zone, Early Eocene.

25 *Siphonodosaria nuttalli nuttalli* (Cushman & Jarvis). × 40
Sample K9415, *Globorotalia velascoensis* Zone, Late Paleocene.

26 *Siphonodosaria nuttalli gracillima* (Cushman & Jarvis). × 50
Sample Hk1831, *Globorotalia subbotinae* Zone, Early Eocene.

27–28 *Stilostomella alexanderi* (Cushman). × 50
27 Moruga well no. 15, core at 5647–71 feet, *Globotruncana fornicata* Zone, Santonian. 28 Moruga well no. 15, core at 5427–52 feet, *Globorotalia fornicata* Zone, Santonian.

29–30 *Stilostomella paleocenica* (Cushman & Todd). × 80
29 Moruga well no. 15, core at 5427–52 feet, *Globotruncana fornicata* Zone, Santonian. 30 Moruga well no. 15, core at 4296–4316 feet, *Globorotalia uncinata* Zone, Middle Paleocene.

31–33 *Stilostomella spinea* (Cushman). × 25
31 Guayaguayare well no. 163, core at 5882–5902 feet, *Globotruncana lapparenti tricarinata* Zone, Early Maastrichtian. 32 Moruga well no. 15, core at 4401–21 feet, *Globorotalia trinidadensis* Zone, Early Paleocene. 33 Moruga well no. 15, core at 4296–4316 feet, *Globorotalia uncinata* Zone, Middle Paleocene.

34–36, 39–41 *Conorbina* cf. *haidingeri* (d'Orbigny). × 65
34–36 Spiral, side and umbilical views. Sample G3644, *Abathomphalus mayaroensis* Zone, Late Maastrichtian. 39 Oblique side view. Same locality. 40–41 Spiral and umbilical views. Rochard well no. 1, core at 8556–71 feet, probably *Globorotalia pseudobulloides* Zone, Early Paleocene.

37–38, 42–44 *Rotamorphina cushmani* Finlay. × 50
37–38 Umbilical and side views. Guayaguayare well no. 163, core at 5588–98 feet, *Abathomphalus mayaroensis* Zone, Late Maastrichtian. 42–44 Umbilical, side and spiral views. Guayaguayare well no. 159, core at 4778–90 feet, *Globorotalia pusilla pusilla* Zone, Middle Paleocene.

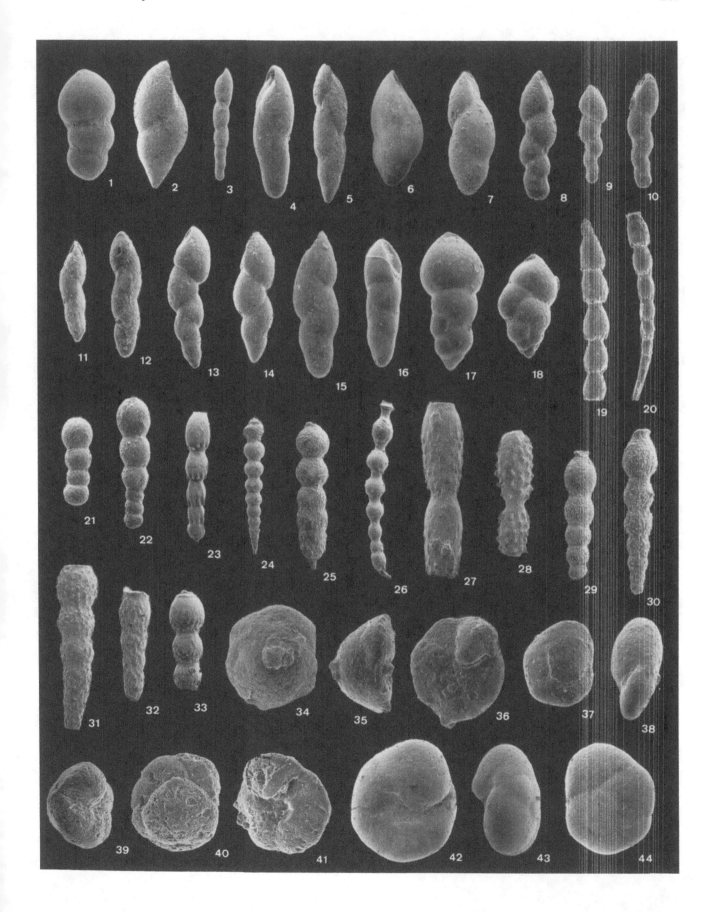

Fig. 39. Bagginidae (continued), Mississippinidae, Eponididae, Parrelloididae

1–2 *Valvulineria parva* Khan. × 150
Marac well no. 1, core at 9853–91 feet, *Favusella washitensis* Zone, Late Albian.

3–5 *Valvulineria subinfrequens* Dailey. × 150
Marac well no. 1, core at 9853–91 feet, *Favusella washitensis* Zone, Late Albian.

6, 12 *Stomatorbina* (?) sp. A. × 40
6 Peripheral view. Marabella well no. 1, core at 3512–40 feet, probably *Globotruncana lapparenti tricarinata* Zone, Early Maastrichtian (Marabella facies). **12** Ventral side. Same locality.

7–9 *"Eponides"* cf. *hillebrandti* (Fisher). × 100
Ventral, lateral and dorsal (spiral) views. Guayaguayare well no. 163, core at 5588–98 feet, *Abathomphalus mayaroensis* Zone, Late Maastrichtian.

10–11 *'Eponides'* cf. *huaynai* Frizzell. × 70
Dorsal and lateral views. Marabella well no. 1, core at 3512–40 feet, probably *Globotruncana lapparenti tricarinata* Zone, Early Maastrichtian (Marabella facies).

13–15 *"Eponides" plummerae* Cushman. × 80
Dorsal, lateral and ventral views. Guayaguayare

well no. 159, core at 3986–4000 feet, *Globorotalia pseudomenardii* Zone, Late Paleocene.

16–18 *'Eponides' subumbonatus* Myatlyuk. × 60
Dorsal, lateral and ventral views. Sample Hk1832, *Globorotalia edgari* Zone, Early Eocene.

19–21 *Cibicidoides merus* (Cushman & Renz). × 50
Ventral, lateral and dorsal views. Sample K9415, *Globorotalia velascoensis* Zone, Late Paleocene.

22–24 *Cibicidoides padella* (Jennings). × 100
Ventral, lateral and dorsal views. Guayaguayare well no. 163, core at 5588–98 feet, *Abathomphalus mayaroensis* Zone. Late Maastrichtian.

25–27 *Cibicidoides praemundulus* Berggren & Miller. × 80
25–26 Ventral and lateral views. Sample Hk1832, *Globorotalia edgari* Zone, Early Eocene. **27** Dorsal view. Sample Rz413, *Globorotalia aragonensis* Zone, Early Eocene.

28–30 *Cibicidoides proprius* (Brotzen)
28–29 Dorsal and lateral views. Sample K10832, *Globorotalia pseudomenardii* Zone, Late Paleocene. × 65. **30** Ventral view. Same locality. × 50.

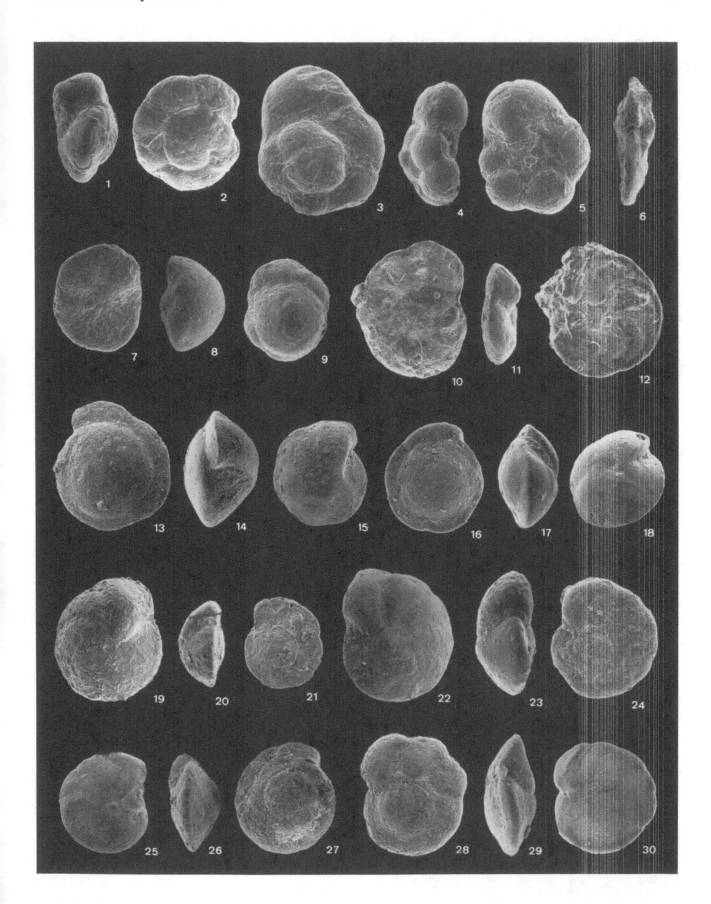

Fig. 40. Parrelloididae (continued), Cibicididae, Epistomariidae

1–3 *Cibicidoides tuxpamensis tuxpamensis* (Cole). × 50
Dorsal, peripheral and ventral views. Sample Bo521, *Globorotalia aragonensis* Zone, Early Eocene.

4–6 *Cibicidoides tuxpamensis laxispiralis* Beckmann. × 40
4 Dorsal view of holotype. Sample KB6972, *Globorotalia formosa formosa* Zone, Early Eocene. **5–6** Peripheral and ventral views. Same locality.

7–9 *Cibicidoides succedens* (Brotzen). × 80
Dorsal, peripheral and ventral views. Moruga well no. 15, core at 4401–21 feet, *Globorotalia trinidadensis* Zone, Early Paleocene.

10–11 *Cibicidoides umbonifer* (Schwager). × 100
Peripheral and dorsal views. Guayaguayare well no. 163, core at 5588–98 feet, *Abathomphalus mayaroensis* Zone, Late Maastrichtian.

12–14 *Cibicidina newmanae* (Plummer). × 125
Dorsal, peripheral and ventral views. Rochard well no. 1, core at 8556–65 feet, probably *Globorotalia pseudobulloides* Zone, Early Paleocene.

15–17 *Cibicidina* cf. *simplex* (Brotzen). × 125

Dorsal, peripheral and ventral views. Rochard well no 1, core at 8556–65 feet, probably *Globorotalia pseudobulloides* Zone, Early Paleocene.

18–20 *Falsoplanulina acuta* (Plummer). × 70
Dorsal, peripheral and ventral views. Guayaguayare well no. 159, core at 4778–90 feet, *Globorotalia pusilla pusilla* Zone, Middle Paleocene.

21–23 *Falsoplanulina waltonensis* (Applin & Jordan). × 70
Ventral, peripheral and dorsal views. Guayaguayare well no. 159, core at 3986–4000 feet, *Globorotalia pseudomenardii* Zone, Late Paleocene.

24–26 *Nuttallides truempyi crassaformis* (Cushman & Siegfus). × 100.
24 Dorsal view. Sample KR23575, *Globorotalia angulata* Zone, Middle Paleocene. **25** Peripheral view. Sample KB6972, *Globorotalia formosa formosa* Zone, Early Eocene. **26** Ventral view. Locality as for fig. 40.24.

27–28 *Nuttallides truempyi truempyi* (Nuttall). × 80
Ventral and lateral views. Sample RM287, *Globorotalia pseudomenardii* Zone, Late Paleocene.

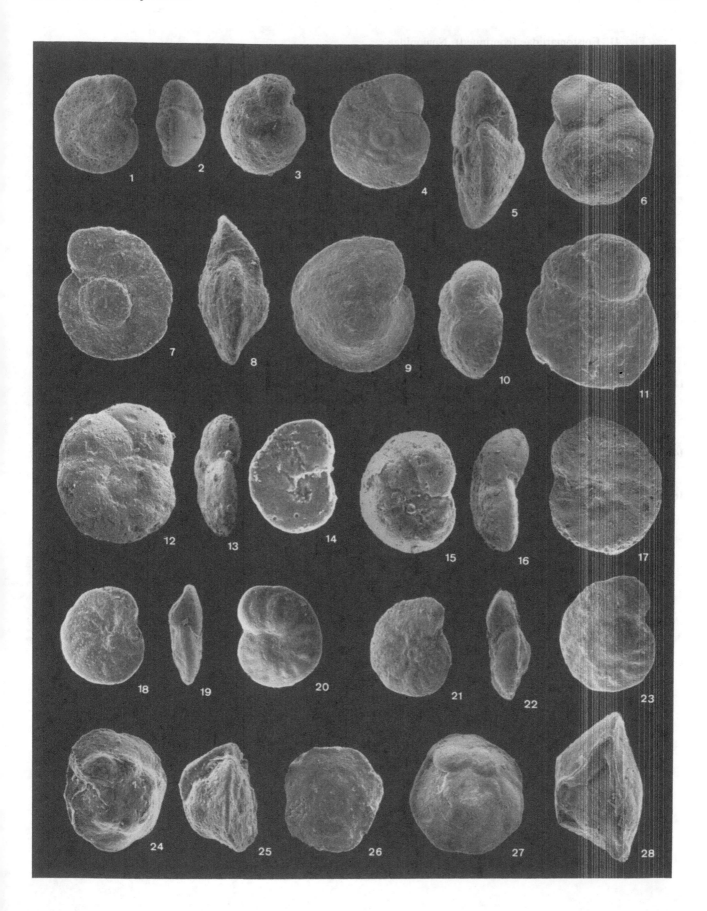

Fig. 41. Epistomariidae (continued), Nonionidae

1–2 *Nuttallinella coronula* (Belford). × 40
Ventral and peripheral views. Guayaguayare well no. 163, core at 5882–5902 feet, *Globotruncana lapparenti tricarinata* Zone, Early Maastrichtian.

3–5 *Nuttallinella florealis* (White)
3 Peripheral view of a typical specimen. Guayaguayare well no. 163, core at 5588–98 feet, *Abathomphalus mayaroensis* Zone. × 50. **4** Dorsal view. Moruga well no. 15, core at 5195–5210 feet, *Globotruncana stuarti* Zone, Campanian. × 80. **5** Peripheral view of a specimen with a broad keel. Moruga well no. 15, core at 4831–49 feet, *Globotruncana lapparenti tricarinata* Zone, Early Maastrichtian. × 40.

6–7 *Nuttallinella* cf. *florealis* (White). × 50
Dorsal and peripheral views. Guayaguayare well no. 159, core at 4778–90 feet, *Globorotalia pusilla pusilla* Zone, Middle Paleocene.

8–10 *Nuttallinella lusitanica* Fisher
8 Ventral view. Guayaguayare well no. 159, core at 4524–36 feet, *Globorotalia pusilla pusilla* Zone, Middle Paleocene. × 90. **9–10** Peripheral and dorsal views. Same locality. × 100.

11–13 *Spirotecta* (?) *ovata* (Balakhmatova). × 75
11–12 Two side views. Guayaguayare well no. 163, core at 5588–98 feet, *Abathomphalus mayaroensis* Zone, Late Maastrichtian. **13** Peripheral view of a relatively thick specimen. Same locality.

14–15 *Nonion havanense* Cushman & Bermudez. × 80
Side and peripheral views. Sample K9415, *Globorotalia velascoensis* Zone, Late Paleocene.

16–18 *Nonionella austinana* Cushman. × 120
16 Ventral view. Marabella well no. 1, core at 4465–76 feet, *Globotruncana fornicata* Zone. Santonian. **17–18** Peripheral and dorsal views. Marabella well no. 1, core at 4210–32 feet, *Globotruncana fornicata* Zone, Santonian.

19, 25 *Nonionella robusta* Plummer. × 100
Dorsal and peripheral views. Guayaguayare well no. 159, core at 3986–4000 feet, *Globorotalia pseudomenardii* Zone, Late Paleocene.

20, 26 *Nonionella* cf. *soldadoensis* Cushman & Renz. × 80
Ventral and peripheral views. Sample Hk1831, *Globorotalia subbotinae* Zone. Early Eocene.

21–22 *Pullenia angusta* Cushman & Todd. × 100
21 Side view. Marabella well no. 1, core at 3512–40 feet, probably *Globotruncana lapparenti tricarinata* Zone, Early Maastrichtian (Marabella facies). **22** Peripheral view. Guayaguayare well no. 159, core at 4778–90 feet, *Globorotalia pusilla pusilla* Zone, Middle Paleocene.

23–24 *Pullenia coryelli* White
23 Side view. Guayaguayare well no. 163, core at 5588–98 feet, *Abathomphalus mayaroensis* Zone, Late Maastrichtian. × 80. **24** Peripheral view. Same locality. × 65.

27–28 *Pullenia cretacea* Cushman. × 80
27 Peripheral view. Marabella well no. 1, core at 4465–76 feet, *Globotruncana fornicata* Zone, Santonian. **28** Side view. Moruga well no. 15, core at 5195–5210 feet, *Globotruncana stuarti* Zone, Campanian.

29–30 *Pullenia eocenica* Cushman & Siegfus. × 50
29 Oblique peripheral view. Sample K9415, *Globorotalia velascoensis* Zone. Late Paleocene. **30** Side view. Sample Hk1832, *Globorotalia edgari* Zone, Early Eocene.

31–32 *Pullenia quinqueloba* (Reuss). × 50
Side and peripheral views. Sample Hk1832, *Globorotalia edgari* Zone, Early Eocene.

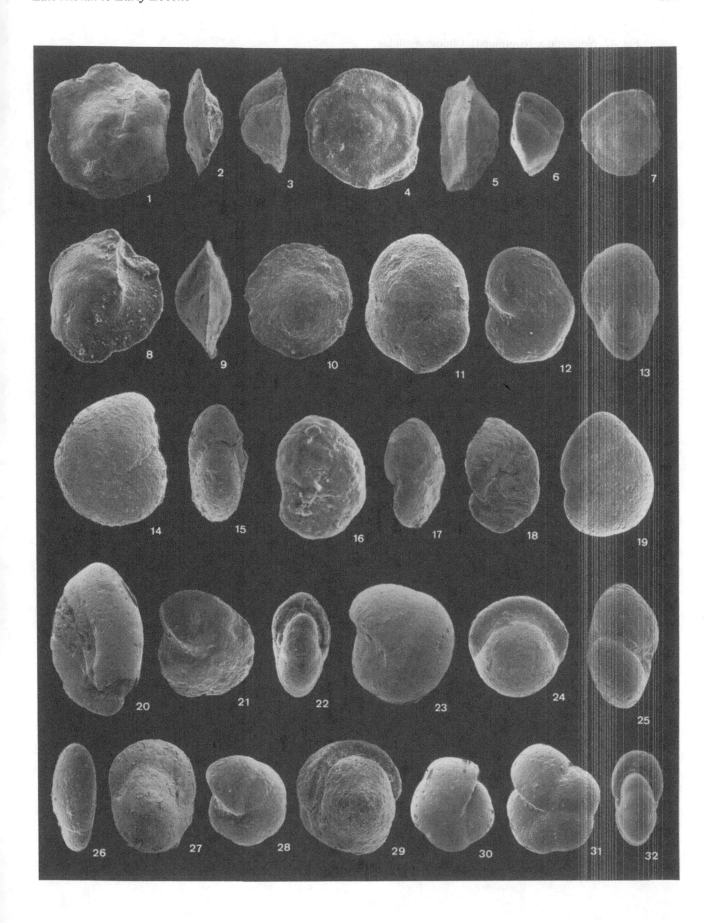

Fig. 42. Nonionidae (continued), Chilostomellidae, Quadrimorphinidae, Alabaminidae

1–3 *Pullenia jarvisi* Cushman
  1 Side view. Guayaguayare well no. 163, core at 7358–93 feet, *Globotruncana fornicata* Zone, Santonian. × 80. **2** Peripheral view. Guayaguayare well no. 163, core at 5588–98 feet, *Abathomphalus mayaroensis* Zone, Late Maastrichtian. × 60. **3** Side view. Moruga well no. 3, core at 10258–64 feet, *Globorotalia trinidadensis* Zone, Early Paleocene. × 80.

  4 *Allomorphina cretacea* Reuss. × 100
  Side view. Rochard well no. 1, core at 8565–71 feet, probably *Globorotalia pseudobulloides* Zone, Early Paleocene.

  5 *Chilostomella czizeki* Reuss. × 40
  Sample Hk1832, *Globorotalia edgari* Zone, Early Eocene.

6–7 *Globimorphina trochoides* (Reuss). × 70
  6 Guayaguayare well no. 163, core at 5588–98 feet, *Abathomphalus mayaroensis* Zone, Late Maastrichtian. **7** Sample Hk1831, *Globorotalia subbotinae* Zone, Early Eocene.

8–9 *Pallaimorphina* cf. *yamaguchii* Takayanagi. × 120
  Side view and ventral view. Lizard well no. 1x, core at 7198–7200 feet, *Globotruncana inornata* Zone, Turonian.

10–11 *Quadrimorphinella* sp. A. × 100
  Dorsal (spiral) and ventral views. Guayaguayare well no. 163, core at 5588–98 feet, *Abathomphalus mayaroensis* Zone, Late Maastrichtian.

12–13 *Quadrimorphina advena* (Cushman & Siegfus). × 50
  Ventral and peripheral views. Sample Hk1832, *Globorotalia edgari* Zone, Early Eocene.

14–15 *Quadrimorphina allomorphinoides* (Reuss). × 80
  14 Ventral view. Sample Rz389, *Globotruncana stuarti* Zone, Campanian. **15** Peripheral view. Moruga well no. 15, core at 5195–5210 feet, *Globotruncana stuarti* Zone, Campanian.

16–17, 23 *Quadrimorphina camerata camerata* (Brotzen). × 65
  16–17 Dorsal and ventral views. Moruga well no. 15, core at 5329–36 feet, *Globotruncana fornicata* Zone, Santonian. **23** Peripheral view. Marabella well no. 1, core at 4465–76 feet, *Globotruncana fornicata* Zone, Santonian.

18–20 *Quadrimorphina camerata umbilicata* (Brotzen)
  18–19 Dorsal and peripheral views. Marac well no. 1, core at 8332–62 feet, *Globotruncana concavata* Zone, Coniacian-Santonian. × 100. **20** Ventral view. Same locality. × 80.

21–22 *Quadrimorphina halli* (Jennings). × 100
  Dorsal and vertral views. Moruga well no. 3, core at 10258–64 feet, *Globorotalia trinidadensis* Zone, Early Paleocene.

  24 *Quadrimorphina minuta* (Cushman). × 120
  Guayaguayare well no. 163, screen sample at 5672–82 feet, *Globotruncana gansseri* Zone, Maastrichtian.

25–26 *Quadrimorphina profunda* Schnitker & Tjalsma
  25 Dorsal view. Sample Hk1831, *Globorotalia subbotinae* Zone, Early Eocene. × 100. **26** Peripheral view. Same locality. × 150.

27–29 *Quadrimorphina pyriformis* (Taylor). × 50
  Ventral, peripheral and dorsal views. Moruga well no. 15, core at 6802–27 feet, *Globotruncana renzi* Zone, Late Turonian-Coniacian.

  30 *Alabamina dissonata* (Cushman & Renz). × 50
  Oblique ventral view. Sample Bo521, *Globorotalia aragonensis* Zone, Early Eocene.

31–32 *Quadrimorphina* sp. A
  31 Dorsal view. Guayaguayare well no. 163, core at 5882–5902 feet, *Globotruncana lapparenti tricarinata* Zone, Early Maastrichtian. × 40. **32** Peripheral view. Same locality. × 50.

33–34 *Alabamina midwayensis* Brotzen
  33 Ventral view. Rochard well no. 1, core at 8556–65 feet, probably *Globorotalia pseudobulloides* Zone, Early Paleocene. × 100. **34** Peripheral view. Same locality. × 70.

35–37 *Alabamina* cf. *wilcoxensis* Toulmin. × 50
  Ventral, peripheral and dorsal views. Sample Hk1831, *Globorotalia subbotinae* Zone. Early Eocene.

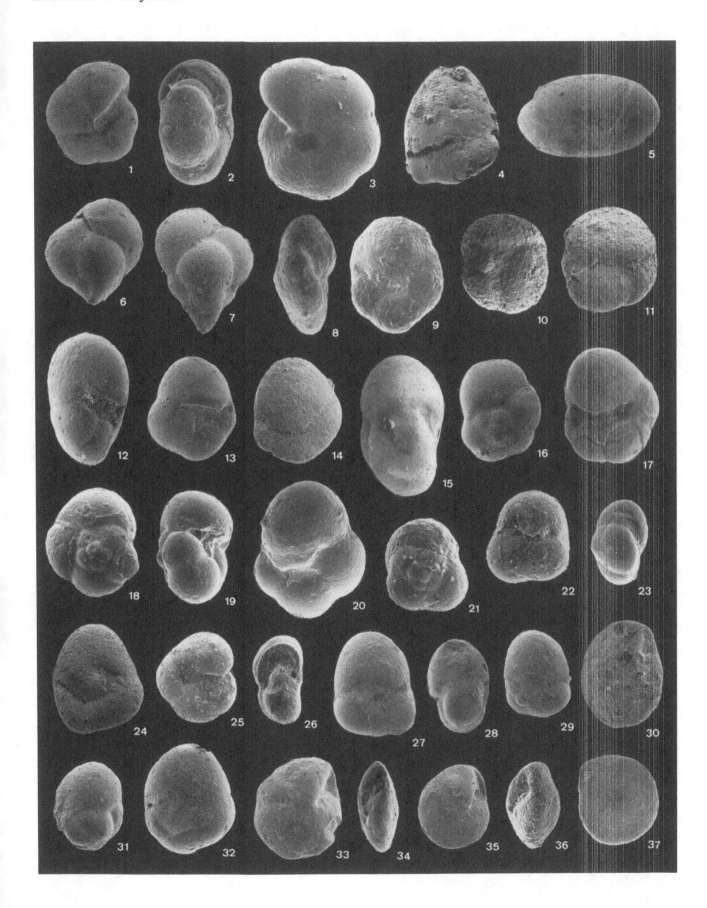

Fig. 43. Globorotalitidae, Osangulariidae

1–3 *Conorotalites aptiensis* (Bettenstaedt)
1 Ventral view. Marac well no. 1, core at 9853–91 feet, *Favusella washitensis* Zone, Late Albian. × 50. **2–3** Peripheral and dorsal views. Same locality. × 80.

4–6 *Conorotalites* cf. *conicus* (Carsey). × 100
Ventral, peripheral and dorsal views. Moruga well no. 15, core at 6980–7005 feet, *Globotruncana inornata* Zone, Turonian.

7–9 *Globorotalites michelinianus* (d'Orbigny). × 50
Dorsal, peripheral and ventral views. Guayaguayare well no. 163, core at 5588–98 feet, *Abathomphalus mayaroensis* Zone, Late Maastrichtian.

10–13 *Globorotalites subconicus* (Morrow). × 80
10–12 Dorsal, lateral and ventral views. Marac well no. 1, core at 8332–62 feet, sample at 8339 feet, *Globotruncana concavata* Zone, Coniacian-Santonian. **13** Peripheral view. Moruga well no. 15, core at 6121–40 feet, *Globotruncana concavata* Zone, Coniacian-Santonian.

14–16 *Charltonina australis* Scheibnerova. × 100
14 Dorsal view. Marac well no. 1, core at 9578–9621 feet, *Rotalipora appenninica appenninica* Zone, Early Cenomanian. **15** Peripheral view. Sample KR8385, *Rotalipora appenninica appenninica* Zone, Early Cenomanian. **16** Ventral view. Same locality as for Fig. 43.14.

17–19 *Osangularia* cf. *alata* (Marsson)
17 Dorsal view. Marac well no. 1, core at 9347–9403 feet, sample at 9403 feet, *Globotruncana renzi* Zone, Late Turonian-Coniacian. × 90. **18** Peripheral view. Moruga well no. 15, core at 5329–36 feet, *Globotruncana fornicata* Zone, Santonian. × 90. **19** Ventral view. Same locality. × 70

20, 25–26 *Osangularia expansa* (Toulmin). × 50
20 Peripheral view. Sample KB6972, *Globorotalia formosa formosa* Zone, Early Eocene. **25–26** Dorsal and ventral views. Same locality.

21–24 *Osangularia californica* Dailey
21 Ventral view. Sample KR8385, *Rotalipora appenninica appenninica* Zone, Early Cenomanian. × 70. **22–24** Ventral, peripheral and dorsal views. Same locality. × 80.

27–29 *Osangularia insigna secunda* Dailey. x 80
Dorsal, peripheral and ventral views. Sample KR8385, *Rotalipora appenninica appenninica* Zone, Early Cenomanian.

30–32 *Osangularia mexicana* (Cole). × 50
Dorsal, peripheral and ventral views. Sample Bo521, *Globorotalia aragonensis* Zone, Early Eocene.

33–35 *Osangularia texana* (Cushman). × 70
33–34 Peripheral and ventral views. Moruga well no. 15, core at 5427–52 feet, *Globotruncana fornicata* Zone, Santonian. **35** Dorsal view. Sample G3644, *Abathomphalus mayaroensis* Zone, Late Maastrichtian.

36–38 *Osangularia velascoensis* (Cushman). × 40
Dorsal, peripheral and ventral views. Guayaguayare well no. 159, core at 3986–4000 feet, *Globorotalia pseudomenardii* Zone, Late Paleocene.

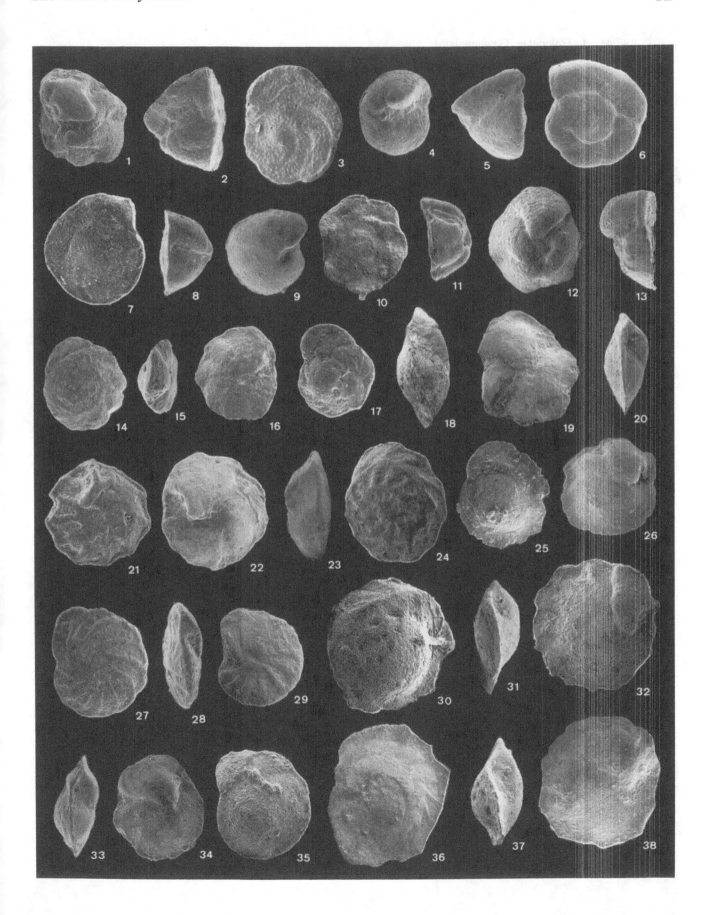

Fig. 44. Heterolepidae, Gavelinellidae

1–2   *Anomalinoides* cf. *chiranus* (Cushman & Stone). ×
80
Peripheral and ventral views. Sample KB6972, *Globorotalia formosa formosa* Zone, Early Eocene.

3–5   *Anomalinoides dorri aragonensis* (Nuttall). × 50
Ventral, peripheral and dorsal views. Sample
RM287, *Globorotalia pseudomenardii* Zone, Late
Paleocene.

6, 13, *Anomalinoides rubiginosus* (Cushman). × 50
18–19 **6, 13** Ventral and peripheral views. Guayaguayare
well no. 163, core at 5588–98 feet, *Abathomphalus mayaroensis* Zone, Late Maastrichtian. **18–19** Dorsal and peripheral views. Sample KR23575, *Globorotalia angulata* Zone, Middle Paleocene.

7–9   *Anomalinoides guatemalensis* (Bermudez). × 50
7–8 Ventral and peripheral views. Guayaguayare
well no. 159, core at 4524–36 feet, *Globorotalia pusilla pusilla* Zone, Middle Paleocene. **9** Dorsal
view. Guayaguayare well no. 159, core at 4778–
90 feet, *Globorotalia pusilla pusilla* Zone, Middle
Paleocene.

10–12 *Anomalinoides* cf. *madrugaensis* (Cushman & Bermudez). × 50
Dorsal, peripheral and ventral views. Sample
KB6972, *Globorotalia formosa formosa* Zone,
Early Eocene.

14–17 *Anomalinoides nobilis* Brotzen. × 100
14 Dorsal view. Guayaguayare well no. 163, core at
5588–98 feet, *Abathomphalus mayaroensis* Zone,
Late Maastrichtian. **15–16** Dorsal and peripheral
views. Sample Hk1831, *Globorotalia subbotinae*

Zone, Early Eocene. **17** Ventral view. Sample
Hk1832, *Globorotalia edgari* Zone, Early Eocene.

20–22 *Anomalinoides umbonatus* (Cushman). × 70
Ventral, peripheral and dorsal views. Sample
KB6972, *Globorotalia formosa formosa* Zone,
Early Eocene.

23–25 *Anomalinoides welleri* (Plummer). × 70
Dorsal, peripheral and ventral views. Moruga well
no. 3, core at 10258–64 feet, *Globorotalia trinidadensis* Zone, Early Paleocene.

26–28 *Heterolepa* cf. *hispaniolae* (Bermudez). × 80
26–27 Dorsal and peripheral views. Guayaguayare
well no. 159, core at 4524–36 feet, *Globorotalia pusilla pusilla* Zone, Middle Paleocene. **28** Ventral
view. Guayaguayare well no. 159, core at 4478–
90 feet, *Globorotalia pusilla pusilla* Zone, Middle
Paleocene.

29–30, *Gyroidinoides depressus* (Alth). × 50
32–34 **29** Dorsal view, Guayaguayare well no. 163, core
at 6052–79 feet, *Globotruncana stuarti* Zone, Campanian. **30, 32** Peripheral and ventral views. Guayaguayare well no. 159, core at 4778–90 feet,
Globorotalia pusilla pusilla Zone, Middle Paleocene. **33–34** Dorsal and peripheral views. Sample
Hk1831, *Globorotalia subbotinae* Zone, Early
Eocene.

31, 35 *Gyroidinoides angustiumbilicatus* (ten Dam). × 80
Dorsal and peripheral views. Guayaguayare well
no. 159, core at 3986–4000 feet, *Globorotalia pseudomenardii* Zone, Late Paleocene.

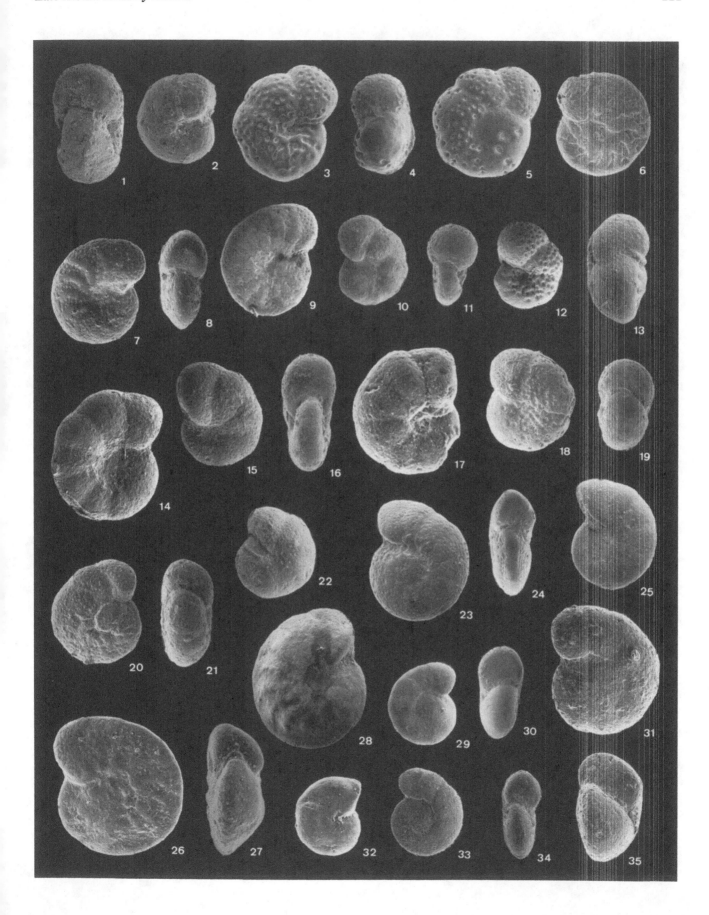

Fig. 45. Gavelinellidae (continued)

1–3 *Gyroidinoides globosus* (Hagenow). × 60
1 Ventral view. Sample KR23575, *Globorotalia angulata* Zone, Middle Paleocene. **2–3** Peripheral and dorsal views. Sample Rz389, *Globotruncana stuarti* Zone, Campanian.

4–6 *Gyroidinoides grahami* Martin. × 70
Dorsal, peripheral and ventral views. Moruga well no. 15, core at 5195–5210 feet, *Globotruncana stuarti* Zone, Campanian.

7–12 *Gyroidinoides kaminskii* Beckmann. × 100
7–9 Ventral, peripheral and dorsal views. Moruga well no. 15, core at 6980–7005 feet, *Globotruncana inornata* Zone, Turonian. **10** Holotype, dorsal view. Marac well no. 1, core at 8657–98 feet, sample at 8657 feet, *Globotruncana concavata* Zone, Coniacian-Santonian. **11–12** Peripheral and ventral views. Same locality as Fig. 45.10.

13–16 *Gyroidinoides nitidus* (Reuss). × 50
13–14 Dorsal and peripheral views. Moruga well no. 15, core at 5410–15 feet, *Globotruncana fornicata* Zone, Santonian. **15–16** Dorsal and ventral views. Sample Rz389, *Globotruncana stuarti* Zone, Campanian.

17–18, *Gyroidinoides* cf. *plummerae* (Cushman & Ber-
24 mudez), × 70
Ventral, peripheral and dorsal views. Sample Rz413, *Globorotalia aragonensis* Zone, Early Eocene.

19–21 *Gyroidinoides quadratus* (Cushman & Church). × 80
Ventral, peripheral and dorsal views. Moruga well no. 15, core at 5195–5210 feet, *Globotruncana stuarti* Zone, Campanian.

22–23 *Gyroidinoides* (?) cf. *quadratus* (Cushman & Church). × 100
Dorsal and peripheral views. Moruga well no. 15, core at 5427–52, *Globotruncana fornicata* Zone, Santonian.

25–27 *Gyroidinoides subangulatus* (Plummer). × 80
Ventral, peripheral and dorsal views. Guayaguayare well no. 159, core at 3986–4000 feet, *Globorotalia pseudomenardii* Zone, Late Paleocene.

28–30 *Gyroidinoides subglobosus* Dailey. × 100
Dorsal, peripheral and ventral views. Sample KR8385, *Rotalipora appenninica appenninica* Zone, Early Cenomanian.

31–33 *Stensioeina excolata* (Cushman). × 50
Ventral, peripheral and dorsal views. Guayaguayare well no. 163, core at 5588–98 feet, *Abathomphalus mayaroensis* Zone, Late Maastrichtian.

34–36 *Angulogavelinella avnimelechi* (Reiss). × 80
Ventral, peripheral and dorsal views. Moruga well no. 3, core at 10258–64 feet, *Globorotalia trinidadensis* Zone, Early Paleocene.

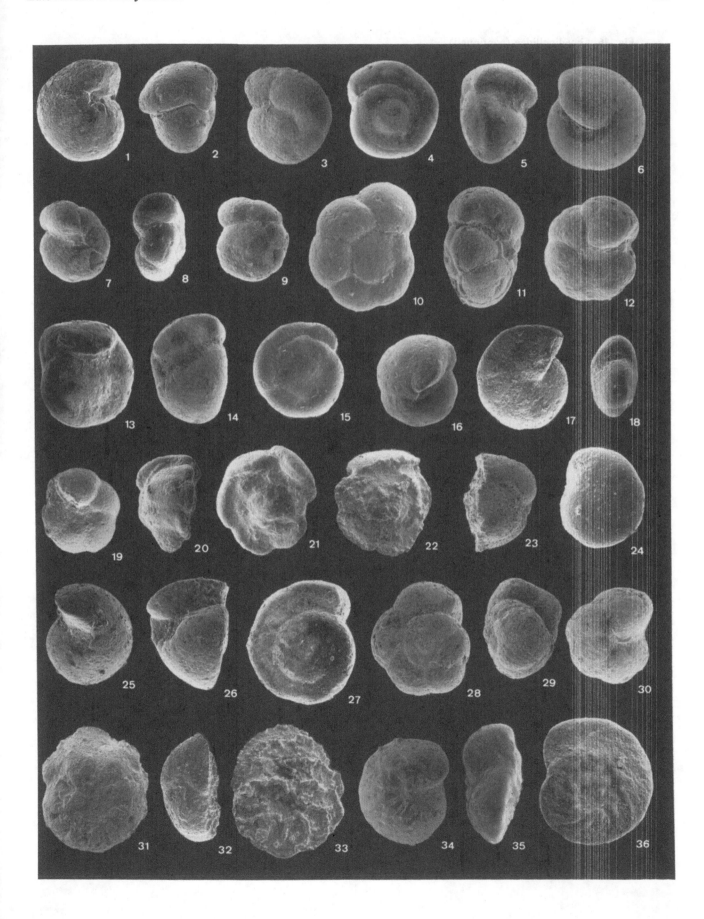

Fig. 46. Gavelinellidae (continued)

1–3 *Gavelinella aracajuensis* (Petri). × 80
Dorsal, peripheral and ventral view. Guayaguayare well no. 163, screen sample at 5672–82 feet, *Globotruncana gansseri* Zone, Maastrichtian.

4–6 *Gavelinella beccariiformis* (White). × 70
4 Dorsal view. Moruga well no. 15, core at 4831–49 feet, *Globotruncana lapparenti tricarinata* Zone, Early Maastrichtian. **5–6** Peripheral and ventral views. Sample K10832, *Globorotalia pseudomenardii* Zone, Late Paleocene.

7–10 *Gavelinella cenomanica* (Brotzen). × 50
7–9 Ventral, peripheral and dorsal views. Marac well no. 1, core at 9853–91 feet, *Favusella washitensis* Zone, Late Albian. **10** Dorsal view. Marac well no. 1, core at 9578–9621 feet, sample at 9602 feet, *Rotalipora appenninica appenninica* Zone, Early Cenomanian.

11–13 *Gavelinella dayi* (White). × 50
11, 13 dorsal and ventral views. Guayaguayare well no. 159, core at 4524–36 feet, *Globorotalia pusilla pusilla* Zone, Middle Paleocene. **12** Peripheral view. Guayaguayare well no. 163, core at 5588–98 feet, *Abathomphalus mayaroensis* Zone, Late Maastrichtian.

14–16 *Gavelinella dakotensis* (Fox). × 100
Ventral, peripheral and dorsal views. Marabella well no. 1, core at 5018–38 feet, sample at 5020 feet, *Globotruncana concavata* Zone, Coniacian-Santonian.

17–19 *Gavelinella drycreekensis* Dailey. × 50
Dorsal, peripheral and ventral views. Sample KR8385, *Rotalipora appenninica appenninica* Zone, Late Albian/Early Cenomanian.

20, *Gavelinella midwayensis trochoidea* (Plummer). ×
26–27 80
Dorsal, peripheral and ventral views. Rochard well no. 1, core at 8565–71 feet, probably *Globorotalia pseudobulloides* Zone, Early Paleocene.

21–25 *Gavelinella eriksdalensis* (Brotzen)
21–22 Dorsal and ventral views. Lizard well no. 1x, core at 7390–7402 feet, *Globotruncana inornata* Zone, Turonian. × 70. **23** Peripheral view. Marac well no. 1, core at 9347–9403 feet, sample at 9347 feet, *Globotruncana renzi* Zone, Late Turonian-Coniacian. × 70. **24** Peripheral view. Moruga well no. 15, core at 6519–44 feet, *Globotruncana renzi* Zone, Late Turonian-Coniacian. × 70. **25** Ventral view. Same locality. × 60.

28–30 *Gavelinella* cf. *monterelensis* (Marie). × 50
Dorsal, peripheral and ventral views. Moruga well no. 15, core at 4831–49 feet, *Globotruncana lapparenti tricarinata* Zone, Early Maastrichtian.

31–32 *Gavelinella petita* (Carsey). × 100
Ventral and peripheral views. Marac well no. 1, screen sample at 11959–89 feet, *Rotalipora ticinensis ticinensis* Zone, Albian.

33–34, *Gavelinella schloenbachi* (Reuss)
40–41 **33** Ventral view. Moruga well no.15, core at 6980–7005 feet, *Globotruncana inornata* Zone, Turonian. × 70. **34** Peripheral view. Same locality. × 100. **40** Dorsal view. Same locality. × 70. **41** Peripheral view. Marac well no. 1, core at 8332–62 feet, *Globotruncana concavata* Zone, Coniacian-Santonian. × 70.

35–39 *Gavelinella rochardensis* Beckmann. × 50
35 Holotype, dorsal view. Rochard well no. 1, core at 8565–71 feet, probably *Globorotalia pseudobulloides* Zone, Early Paleocene. **36–37** peripheral and ventral views. Same locality. **38** Peripheral view of a planoconvex specimen. Same locality. **39** Ventral view showing apertural flaps. Same locality.

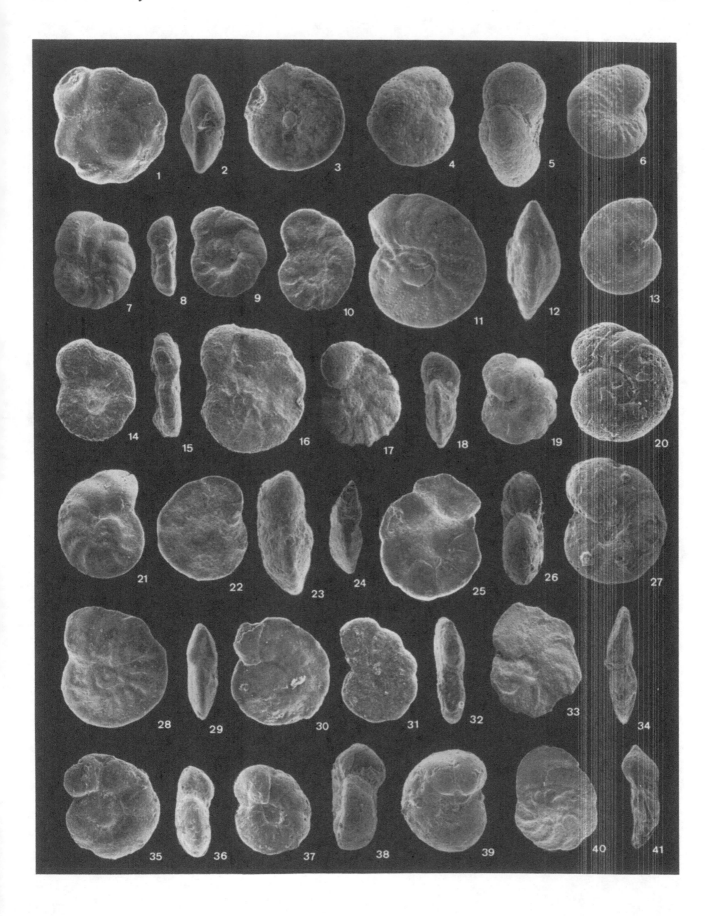

Fig. 47. Gavelinellidae (continued), Rotaliidae

1–5 *Gavelinella* cf. *sandidgei* (Brotzen). × 70
1–3 Ventral, peripheral and dorsal views. Marac well no. 1, core at 9853–91 feet, *Favusella washitensis* Zone, Late Albian. **4–5** Ventral and peripheral views. Morne Diablo well no. 34, core at 13965–84 feet, *Rotalipora appenninica appenninica* Zone, Early Cenomanian.

6–7, *Gavelinella* cf. *sibirica* (Dain). × 100
14–15 **6–7** Dorsal and ventral views. Lizard well no. 1x, core at 7186–98 feet, *Globotruncana inornata* Zone, Turonian. **14** Peripheral view. Same locality. **15** Ventral view. Morne Diablo well no. 34, core at 13842–89 feet, *Globotruncana inornata* Zone, Turonian.

8–10 *Gavelinella spissicostata* (Cushman). × 70
Dorsal, peripheral and ventral views. Marabella well no. 34, core at 3512–40 feet, probably *Globotruncana lapparenti tricarinata* Zone, Early Maastrichtian (Marabella facies).

11–13 *Gavelinella* cf. *vesca* (Bykova). × 100
Ventral, peripheral and dorsal views. Marac well no. 1, core at 9578–9621 feet, *Rotalipora appenninica appenninica* Zone, Early Cenomanian.

16–19 *Gyroidina bandyi* (Trujillo). × 80
16–17 Dorsal and peripheral views. Moruga well no. 15, core at 6980–7005 feet, *Globotruncana inornata* Zone, Turonian. **18–19** Peripheral and ventral views. Marac well no. 1, core at 8332–62 feet, *Globotruncana concavata* Zone, Coniacian-Santonian.

20–22 *Gyroidina bollii* (Cushman & Renz). × 40
Dorsal, peripheral and ventral views. Sample Rz389, *Globotruncana stuarti* Zone, Campanian.

23–24 *Gyroidina mauretanica* Carbonnier. × 65

Peripheral and dorsal views. Marac well no. 1, core at 9578–9621 feet, sample at 9602 feet, *Rotalipora appenninica appenninica* Zone, Early Cenomanian.

25–27 *Gyroidina megastoma* (Gzybowski). × 70
25 Dorsal view. Sample K10832, *Globorotalia pseudomenardii* Zone, Late Paleocene. **26** Peripheral view. Sample RM287, *Globorotalia pseudomenardii* Zone, Late Paleocene. **27** Ventral view. Guayaguayare well no. 163, core at 5882–5902 feet, *Globotruncana lapparenti tricarinata* Zone, Early Maastrichtian.

28, *Gyroidina* cf. *noda* Belford. × 65
33–34 Dorsal, peripheral and ventral views. Moruga well no. 15, core at 5427–52 feet, *Globotruncana fornicata* Zone, Santonian.

29–32 *Gyroidina praeglobosa* Brotzen
29–31 Ventral, peripheral and dorsal views. Marac well no. 1, core at 8180–8237 feet, *Globotruncana concavata* Zone, Coniacian-Santonian. **32** Dorsal view. Moruga well no. 15, core at 6519–44 feet, *Globotruncana renzi* Zone, Late Turonian-Coniacian.

35–38 *Linaresia semicribrata* (Beckmann). × 70
35–36 Ventral and peripheral views. Guayaguayare well no. 159, core at 3986–4000 feet, *Globorotalia pseudomenardii* Zone, Late Paleocene. **37** Dorsal view. Sample Hk1831, *Globorotalia subbotinae* Zone, Early Eocene. **38** Pseudoplanispiral variety. Same locality.

39–40 *Rotalia* (?) *hermi* Hillebrandt. × 40
Dorsal and ventral views. A.B. well no. 1, core at 4400–4405 feet, *Globotruncana lapparenti tricarinata* or *G. gansseri* Zone, Maastrichtian.

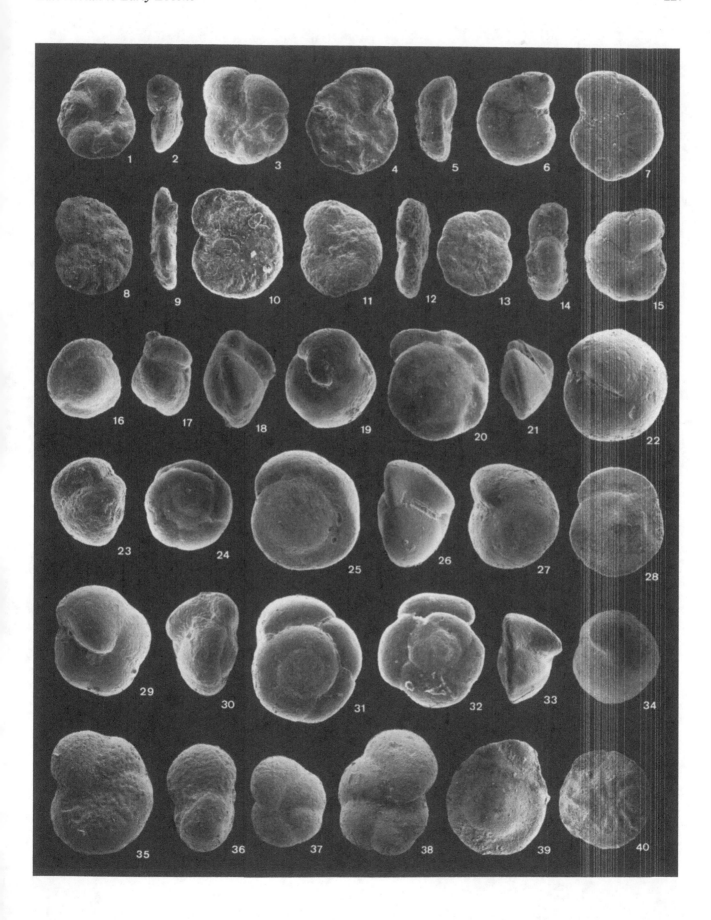

Late Early Eocene to Middle Miocene

Introduction

The selection for this section of taxa and their illustrations for the interval uppermost Early Eocene to Middle Miocene is largely from the three publications by Cushman & Renz (1947b, 1948) and Cushman & Stainforth (1945). In 1943, 1944, and 1945 H. H. Renz and R. M. Stainforth, both at the time paleontologists with Trinidad Leaseholds Ltd in Pointe-à-Pierre, Trinidad, West Indies, issued detailed company reports on the occurrence and distribution of Trinidad foraminifera from Early Cretaceous to Recent. Based on the foraminifera the formations were either subdivided by zonal schemes as in the San Fernando and Cipero formations, or into a combination of members/lentils and zones, as in the Ste. Croix and Brasso formations. The Eocene Navet Formation was subdivided by Cushman & Renz (1948) into seven marl members.

These reports contain detailed maps and sections of the type localities for each of the zones, members and lentils, data on lithology, age, thickness and stratigraphic relations. For each of the units the foraminiferal content was listed and documented in assemblage slides arranged by genera and species, applying a number system to the latter. These detailed reports by H. H. Renz and R. M. Stainforth formed the base for their joint publications with J. A. Cushman.

The foraminiferal fauna of the Cipero Formation published by Cushman & Stainforth (1945) was at the time regarded as wholly belonging to the Oligocene. It was subdivided from bottom to top into the three zones *Globigerina concinna* (now *Globigerina ciperoensis ciperoensis*) Zone, *Globigerinatella insueta* Zone and *Globorotalia fohsi* Zone. The distribution in these zones of a limited number of taxa was shown by these authors on a range chart which extends downwards into the Upper Eocene Hospital Hill Formation (placed by Stainforth (1948) into the *Hantkenina ala-*

bamensis Zone) and upwards into the Lengua calcareous clay (placed by Stainforth, 1948, into the *Globorotalia menardii* Zone). In addition, lists of taxa diagnostic for the three zones are given by Cushman & Stainforth. Cushman & Renz (1947) added to the zonal scheme of Cushman & Stainforth (1945) a *Globigerina dissimilis* Zone, placed between the *Globigerina concinna* Zone below and the *Globigerinatella insueta* Zone above.

Cushman & Renz (1947*b*) published the foraminiferal fauna of the Ste. Croix Formation which in South Trinidad is a shallower water equivalent of the interval *Globigerinatella insueta–Praeorbulina glomerosa* Zone of the Cipero Formation. Compared with the Cipero the Ste. Croix Formation is faunistically characterized by a strong predominance of benthic, mainly calcareous species. Sufficient planktic species including marker species are, however, also present in the formation to allow for its correlation with the planktic foraminiferal zonal scheme.

A close facies equivalent to the Ste. Croix Formation of South Trinidad is the Brasso Formation developed in Central Trinidad. This formation and its foraminiferal fauna has been divided in company reports into nine lithologic and zonal members based on benthic species. In terms of the planktic foraminiferal zonal scheme the formation can be placed between the *Catapsydrax stainforthi* and the *Globorotalia mayeri* Zone. In contrast to the Cipero and Ste. Croix foraminifera those of the Brasso Formation are not published. However, Renz (1948) in his Agua Salada Group study of southeastern Falcon, Venezuela, points to the close relationship between the Agua Salada and Brasso benthic foraminifera. In his Agua Salada monograph he compares the joint occurrence and distribution of 173 taxa in the Agua Salada of Venezuela and the Ste. Croix and Brasso formations of Trinidad.

Cushman & Renz (1948) published the foraminifera of the uppermost Early to Late Eocene Navet and Hospital Hill formations in a similar way as in the previous papers on the Cipero and Ste. Croix formations. The Hospital Hill Formation is today regarded as the uppermost part of the Navet Formation. Like the Cipero Formation the sequence is extremely rich in planktic foraminifera with the benthics at best making up a few per cent only of the total fauna. The marls of the Navet Formation are divided by Cushman & Renz, into seven members. The Ramdat Marl (*Morozovella aragonensis* Zone) is now placed into the Early Eocene upper Lizard Springs Formation. The presence of the individual taxa in the different

members are given in the species description, but no range chart was provided by Cushman & Renz.

The application of a few planktic foraminiferal species as zonal markers in the Cushman *et al.* publications pointed to the possibilitiy of making still better use of them for a closer stratigraphic resolution of the Tertiary, in a way that had already been established for the Late Cretaceous. Intensified studies in Trinidad of this group in the Tertiary (Bolli, 1957*b*, *c*, *d*) eventually led to a much closer zonal subdivision of the formations compared with that applied by Cushman and co-authors. Based on such a closer zonal subdivision of the Tertiary formations rich in planktic foraminifera, it became of interest to know how the distribution of the benthic species compared with the planktic zonal scheme.

Following the unpublished investigations by Renz and Stainforth in the mid-1940s, J. P. Beckmann and J. B. Saunders, both also paleontologists in Pointe-à-Pierre in the 1950s, carried out – again as company reports – an analysis of the benthic species in each of the newly established planktic zones, J. P. Beckmann for the Paleocene and Eocene, J. B. Saunders for the Oligocene to Middle Miocene.

For each zone the reports give data concerning the type localities and lists of benthic foraminifera they contain. The taxa number 272 for the Paleocene-Early Eocene, 386 for the Middle and Late Eocene, and 440 for the Oligocene to Middle Miocene. Each report is accompanied by a chart showing the distribution in stratigraphic order of the benthic species against the established planktic foraminiferal zonal scheme.

Finally, also as a company report, H. M. Bolli and J. B. Saunders in 1957 prepared a stratigraphic atlas where all benthic forms recorded in Trinidad are plotted by taxa against the planktic zonal scheme. The data from these unpublished company reports and the published papers provided the base for the compilation of the late Early Eocene to Middle Miocene benthic foraminifera of Trinidad.

Except for a brief break during the uppermost Eocene, the late Early Eocene to Middle Miocene is developed in South Trinidad by a continuous sequence of deep water marly sediments whose foraminiferal fauna is strongly dominated by planktic forms. Benthic foraminifera are present throughout the sequence but usually number at best a few per cent only of the total fauna. Radiolaria may occur in considerable numbers in the Middle Eocene, to a lesser degree and more sporadically also in parts of the Miocene.

An exception to the strong predominance of

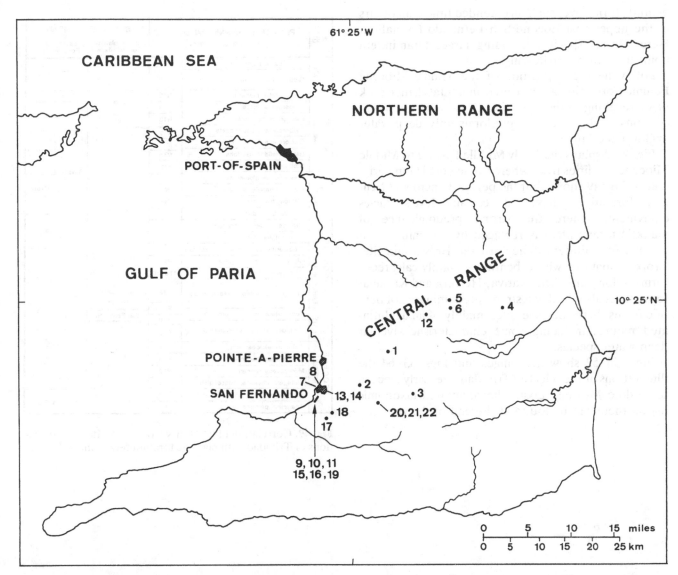

Fig. 48. Map of Trinidad showing the approximate locations of the 22 late Early Eocene to Middle Miocene type localities of the planktic foraminiferal zones dealt with in this volume. Descriptions of them are found for the Eocene localities 1–8 in Bolli (1957*d*). With the exception of the type locality for the *Acarinina pentacamerata* Zone (originally named *Globorotalia palmerae* Zone) which is in a well section (Esmeralda 1), the Eocene locations are outcrops. In addition to those given above most of the Eocene zones also occur as outcrops in tributaries of the Navet River in the Central Range. Their locations are shown in Bolli (1957*d*). The locations of most of the Oligocene and Miocene zonal type localities 9–22 described in Bolli (1957*c*) occur along the Cipero coast. Several of the type localities of the zones listed below are no longer accessible, others may be extremely difficult to reach at the present time. 1 *Acarinina pentacamerata* Zone, 2 *Hantkenina nuttalli* Zone, 3 *Globigerinatheka subconglobata subconglobata* Zone, 4 *Morozovella lehneri* Zone, 5 *Orbulinoides beckmanni* Zone, 6 *Truncorotaloides rohri* Zone, 7 *Globigerinatheka semiinvoluta* Zone, 8 *Turborotalia cerroazulensis* Zone, 9 *Globigerina ampliapertura* Zone, 10 *Globorotalia opima opima* Zone, 11 *Globigerina ciperoensis ciperoensis* Zone, 12 *Globorotalia kugleri* Zone, 13 *Globigerinoides primordius* Zone, 14 *Catapsydrax dissimilis* Zone, 15 *Catapsydrax stainforthi* Zone, 16 *Globigerinatella insueta* Zone, 17 *Globorotalia fohsi peripheroronda* Zone, 18 *Globorotalia fohsi fohsi* Zone, 19 *Globorotalia fohsi lobata* Zone, 20 *Globorotalia fohsi robusta* Zone, 21 *Globorotalia mayeri* Zone, 22 *Globorotalia menardii* Zone.

planktic forms over such an extended time span occurs in the uppermost Eocene San Fernando Formation. Here benthic species, including larger foraminifera point to a strong shallowing.

Following the deposition of the Middle Miocene Lengua Formation a rapid environmental change took place resulting in the overlying Cruse Formation that contains a characteristic predominantly deep-water agglutinated fauna.

Fig. 48 displays the Early Middle Eocene to Middle Miocene localities from where the selected fauna originates. Fig. 49 shows that the pelagic Cipero and Lengua formations have at certain levels facies equivalents where the strong predominance of planktic foraminifera is replaced by a dominance of benthic foraminifera. They are the Brasso and Ste. Croix formations where benthic – mainly calcareous forms – dominate. The Nariva, Herrera and Karamat formations also display a strong predominance of benthic forms; here they are often mainly of an agglutinated nature, including some characteristic alveolar deep water species.

Figs. 50–52 show the ranges and Figs. 53–64 the illustrations of the selected Trinidad late Early Eocene to Middle Miocene taxa. The annotated taxonomic list on them will be found at the end of the volume.

AGE		PLANKTIC FORAMINIFERAL ZONES (from Bolli et al 1985)		S- AND SE- TRINIDAD		
				rich in planktic Foraminifera	predom. benthic Foraminifera	
MIOCENE	LATE	Globorotalia humerosa	N17		Cruse	
		Globorotalia acostaensis	N16			
	MIDDLE	Globorotalia menardii	N15	Lengua	Karamat	
		Globorotalia mayeri	N14			
		Globigerinoides ruber	N13	Hiatus		
		Globorotalia fohsi robusta	N12			
		Globorotalia fohsi lobata	N11		Brasso	Herrera
		Globorotalia fohsi fohsi	N10			
	EARLY	Globorotalia fohsi peripheroronda	N9			Ste. Croix
		Praeorbulina glomerosa	N8	Cipero		
		Globigerinatella insueta	N7			
		Catapsydrax stainforthi	N6			
		Catapsydrax dissimilis	N5		Nariva	
		Globigerinoides primordius	N4			
OLIGOCENE	LATE	Globorotalia kugleri	P22			
		Globigerina ciperoensis ciperoensis	P21			
	EARLY MIDDLE	Globorotalia opima opima				
		Globigerina ampliapertura	P19/20			
		Cassig. chipolensis / Pseud. micra	P18	Hiatus		
EOCENE	LATE	Turborotalia cerroazulensis s.l.	P17			
			P16	San Fernando		
		Globigerinatheka semiinvoluta	P15			
	MIDDLE	Truncorotaloides rohri	P14	Navet		
		Orbulinoides beckmanni	P13			
		Morozovella lehneri	P12			
		Globigerinatheka s. subconglobata	P11			
		Hantkenina nuttalli	P10		Pointe-à-Pierre	
		Acarinina pentacamerata	P9			

Fig. 49. Correlation of late Early Eocene to Miocene formations of Trinidad with planktic foraminiferal zonal schemes.

Fig. 50. Distribution in Trinidad of late Early Eocene to Middle Miocene species of the genera *Bolivina, Bulimina, Uvigerina, Rectuvigerina, Siphogenerina* (illustrated on Figs. 53 and 54).

Bolivina acerosa	17
Bolivina alata	12
Bolivina alazanensis	13
Bolivina alazanensis venezuelana	5
Bolivina anaeriensis carapitana	10
Bolivina byramensis	7
Bolivina danvillensis	4
Bolivina gracilis	3
Bolivina imporcata	15
Bolivina isidroensis	18
Bolivina marginata multicostata	11
Bolivina mexicana aliformis	6
Bolivina pisciformis	1
Bolivina pisciformis optima	8
Bolivina pozonensis	19
Bolivina simplex	16
Bolivina cf. tectiformis	9
Bolivina thalmanni	14
Bolivina tongi	2
Bulimina alazanensis	24
Bulimina bleeckeri	32
Bulimina cf. aspera	21
Bulimina guayabalensis ampla	25
Bulimina illingi	29
Bulimina jacksonensis	26
Bulimina jarvisi	27
Bulimina macilenta	23
Bulimina prolixa	20
Bulimina pupoides	22
Bulimina sculptilis	28
Bulimina striata mexicana	30
Bulimina tuxpamensis	31
Rectuvigerina basispinata	56
Rectuvigerina multicostata	54
Rectuvigerina multicostata optima	55
Siphogenerina kugleri	57
Siphogenerina lamellata	58
Siphogenerina senni	59
Siphogenerina transversa	60
Uvigerina auberiana attenuata	46
Uvigerina capayana	41
Uvigerina carapitana	51
Uvigerina bealli	38
Uvigerina hootsi	44
Uvigerina ciperana	37
Uvigerina curta	50
Uvigerina gallowayi	39
Uvigerina gallowayi basicordata	49
Uvigerina hispidocostata	47
Uvigerina howei	48
Uvigerina isidroensis	45
Uvigerina israelskyi	40
Uvigerina laviculata	43
Uvigerina mantaensis	33
Uvigerina mexicana	42
Uvigerina rustica	52
Uvigerina seriata	34
Uvigerina sparsicostata	53
Uvigerina spinicostata	35
Uvigerina yazooensis	36

1	Bolivina pisciformis
2	Bolivina tongi
3	Bolivina gracilis
4	Bolivina danvillensis
5	Bolivina alazanensis venezuelana
6	Bolivina mexicana aliformis
7	Bolivina byramensis
8	Bolivina pisciformis optima
9	Bolivina cf. tectiformis
10	Bolivina anaeriensis carapitana
11	Bolivina marginata multicostata
12	Bolivina alata
13	Bolivina alazanensis
14	Bolivina thalmanni
15	Bolivina imporcata
16	Bolivina simplex
17	Bolivina acerosa
18	Bolivina isidroensis
19	Bolivina pozonensis
20	Bulimina prolixa
21	Bulimina cf. aspera
22	Bulimina pupoides
23	Bulimina macilenta
24	Bulimina alazanensis
25	Bulimina guayabalensis ampla
26	Bulimina jacksonensis
27	Bulimina jarvisi
28	Bulimina sculptilis
29	Bulimina illingi
30	Bulimina striata mexicana
31	Bulimina tuxpamensis
32	Bulimina bleeckeri
33	Uvigerina mantaensis
34	Uvigerina seriata
35	Uvigerina spinicostata
36	Uvigerina yazooensis
37	Uvigerina ciperana
38	Uvigerina bealli
39	Uvigerina gallowayi
40	Uvigerina israelskyi
41	Uvigerina capayana
42	Uvigerina mexicana
43	Uvigerina laviculata
44	Uvigerina hootsi
45	Uvigerina isidroensis
46	Uvigerina auberiana attenuata
47	Uvigerina hispidocostata
48	Uvigerina howei
49	Uvigerina gallowayi basicordata
50	Uvigerina curta
51	Uvigerina carapitana
52	Uvigerina rustica
53	Uvigerina sparsicostata
54	Rectuvigerina multicostata
55	Rectuvigerina multicostata optima
56	Rectuvigerina basispinata
57	Siphogenerina kugleri
58	Siphogenerina lamellata
59	Siphogenerina senni
60	Siphogenerina transversa

Fig. 51. Distribution in Trinidad of late Early Eocene to Middle Miocene calcareous trochospiral species (illustrated on Figs. 55–61).

Distribution in Trinidad of late Early Eocene to Middle Miocene calcareous trochospiral species illustrated on Figures 55 - 61

Abbreviation of Formations:
SF: San Fernando
Le: Lengua
SC: Ste. Croix

Chart column headers — Series: EOCENE (E M L), OLIGOCENE (E M L), MIOCENE (E M; SC), AGE, FM. Formations: Navet, SF, Cipero, Le. PLANKTIC FORAMINIFERAL ZONES.

Planktic foraminiferal zone species (vertical headers):
Acarinina pentacamerata · Hantkenina nuttalli · Globigerinatheka s. subconglobata · Morozovella lehneri · Orbulinoides beckmanni · Truncorotaloides rohri · Globigerinatheka semiinvoluta · Turborotalia cerroazulensis s.l. · Cassig. chipolensis / Pseud. micra · Globigerina ampliapertura · Globorotalia opima opima · Globigerina ciperoensis ciperoensis · Globorotalia kugleri · Catapsydrax dissimilis · Catapsydrax stainforthi · Globigerinoides primordius · Globigerinatella insueta · Praeorbulina glomerosa · Globorotalia fohsi peripheroronda · Globorotalia fohsi fohsi · Globorotalia fohsi lobata · Globorotalia fohsi robusta · Globigerinoides ruber · Globorotalia mayeri · Globorotalia menardii

Species (alphabetical)	No.
Alabamina atlantisae dissonata	2
Anomalinoides abuillotensis	19
Anomalinoides affinis	17
Anomalinoides alazanensis	52
Anomalinoides alazanensis spissiformis	8
Anomalinoides bilateralis	3
Anomalinoides cicatricosus	16
Anomalinoides dorri aragonensis	4
Anomalinoides mectapecensis	43
Anomalinoides pompilioides	10
Anomalinoides subbadeniensis	56
Anomalinoides trinitatensis	48
Baggina cojimarensis	29
Cancris cubensis	22
Cancris mauryae	26
Cibicidina antiqua	28
Cibicidina cushmani	18
Cibicidoides cf. atratiensis	33
Cibicidoides cf. pseudoungerianus	6
Cibicidoides cookei	20
Cibicidoides grimsdalei	14
Cibicidoides parianus	54
Cibicidoides subspiratus	39
Cibicidoides ungerianus	36
"Discorbis" ciperensis	42
Eponides byramensis	31
Eponides cf. vicksburgensis	13
Eponides crebbsi	51
Eponides elevatus	1
Eponides parantillanum	38
Gyroidina jarvisi	21
Gyroidinoides altispirus	44
Gyroidinoides complanatus	40
Gyroidinoides dissimilis	47
Gyroidinoides girardanus	12
Gyroidinoides girardanus peramplus	35
Gyroidinoides planulatus	46
Gyroidinoides soldanii octocameratus	5
Hanzawaia carstensi	50
Hanzawaia illingi	15
Hanzawaia mantaensis	53
Hanzawaia mississippiensis	27
Heterolepa mexicana	37
Hoeglundia elegans	34
Laticarinina bullbrooki	45
Nuttallides crassaformis	23
Nuttallides truempii	7
Oridorsalis umbonatus	11
Osangularia mexicana	9
Planulina marialana	25
Planulina renzi	41
Planulina woellerstorfi	30
Sisphonina pulchra	32
Valvulineria georgiana	55
Valvulineria palmerae	49
Valvulineria venezuelana	24

No.	Species (stratigraphic order)
1	Eponides elevatus
2	Alabamina atlantisae dissonata
3	Anomalinoides bilateralis
4	Anomalinoides dorri aragonensis
5	Gyroidinoides soldanii octocameratus
6	Cibicidoides cf. pseudoungerianus
7	Nuttallides truempii
8	Anomalinoides alazanensis spissiformis
9	Osangularia mexicana
10	Anomalinoides pompilioides
11	Oridorsalis umbonatus
12	Gyroidinoides girardanus
13	Eponides cf. vicksburgensis
14	Cibicidoides grimsdalei
15	Hanzawaia illingi
16	Anomalinoides cicatricosus
17	Anomalinoides affinis
18	Cibicidina cushmani
19	Anomalinoides abuillotensis
20	Cibicidoides cookei
21	Gyroidina jarvisi
22	Cancris cubensis
23	Nuttallides crassaformis
24	Valvulineria venezuelana
25	Planulina marialana
26	Cancris mauryae
27	Hanzawaia mississippiensis
28	Cibicidina antiqua
29	Baggina cojimarensis
30	Planulina woellerstorfi
31	Eponides byramensis
32	Sisphonina pulchra
33	Cibicidoides cf. atratiensis
34	Hoeglundia elegans
35	Gyroidinoides girardanus peramplus
36	Cibicidoides ungerianus
37	Heterolepa mexicana
38	Eponides parantillanum
39	Cibicidoides subspiratus
40	Gyroidinoides complanatus
41	Planulina renzi
42	"Discorbis" ciperensis
43	Anomalinoides mectapecensis
44	Gyroidinoides altispirus
45	Laticarinina bullbrooki
46	Gyroidinoides planulatus
47	Gyroidinoides dissimilis
48	Anomalinoides trinitatensis
49	Valvulineria palmerae
50	Hanzawaia carstensi
51	Eponides crebbsi
52	Anomalinoides alazanensis
53	Hanzawaia mantaensis
54	Cibicidoides parianus
55	Valvulineria georgiana
56	Anomalinoides subbadeniensis

Fig. 52. Distribution in Trinidad of additional late Early Eocene to Middle Miocene species (illustrated on Figs. 62–63).

Distribution in Trinidad of selected late Early Eocene to Middle Miocene species illustrated on Figures 62 - 63

Abbreviation of Formations:
SF: San Fernando
Le: Lengua
SC: Ste. Croix

Chart column headers (ages): EOCENE (E | M | L) · OLIGOCENE (E | M | L) · MIOCENE (E | SC | M) · AGE · FM

Formations: Navet | SF | Cipero | Le

Planktic foraminiferal species (chart top, left to right):
Acarinina pentacamerata · Hankenina nuttalli · Globigerinatheka s. subconglobata · Morozovella lehneri · Orbulinoides beckmanni · Truncorotaloides rohri · Globigerinatheka semiinvoluta · Cassig. chipolensis / Pseud. micra · Turborotalia cerroazulensis s.l. · Globigerina ampliapertura · Globigerina ciperoensis ciperoensis · Globorotalia opima opima · Globorotalia kugleri · Globigerinoides primordus · Catapsydrax dissimilis · Catapsydrax stainforthi · Globigerinatella insueta · Praeorbulina glomerosa · Globorotalia fohsi peripheroronda · Globorotalia fohsi fohsi · Globorotalia fohsi lobata · Globorotalia fohsi robusta · Globigerinoides ruber · Globorotalia mayeri · Globorotalia menardii

Species (alphabetical)	No.
Ammobaculites cubensis	14
Ammovertella retrorsa	2
Cassidulina spinifer	50
Cassidulinoides bradyi	32
Chrysalogonium breviloculum	53
Chrysalogonium ciperense	34
Chrysalogonium elongatum	6
Chrysalogonium lanceolum	7
Chrysalogonium longicostatum	8
Chrysalogonium tenuicostatum	43
Dentalina semilaevis	3
Dorothia brevis	37
Ellipsoglandulina multicostata	36
Ellipsolagena barri	22
Entosolenia flintiana indomita	45
Entosolenia flintiana plicatura	44
Entosolenia formosa	41
Entosolenia kugleri	46
Entosolenia orbignyana clathrata	40
Entosolenia spinulolaminata	52
Gaudryina pseudocollinsi	17
Karreriella alticamera	38
Karreriella subcylindrica	25
Lagena asperoides	10
Lagena ciperensis	18
Lagena crenata capistrata	48
Lagena pulcherrima	51
Lagena pulcherrima enitens	42
Lagena striata basisenta	9
Lituotuba navetensis	24
Martinottiella petrosa	26
Nodosarella mappa	59
Nodosaria longiscata	11
Orthomorphina rohri	4
Plectina trinitatensis	13
Plectofrondicularia alazanensis	39
Plectofrondicularia cookei	31
Plectofrondicularia lirata	29
Plectofrondicularia mexicana	33
Plectofrondicularia morreyae	49
Plectofrondicularia nuttalli	54
Plectofrondicularia spinifera	55
Plectofrondicularia trinitatensis	30
Plectofrondicularia vaughani	28
Pseudoclavulina trinitatensis	1
Pseudoglandulina gallowayi	58
Pyramidulina lamellata	57
Pyramidulina modesta	27
Pyramidulina stainforthi	47
Rectoguembelina inopinata	16
Rectoguembelina trinitatensis	12
Siphonodosaria nuttalli	21
Siphonodosaria nuttalli aculeata	20
Siphonodosaria nuttalli gracillima	5
Siphonodosaria recta	23
Siphonodosaria verneuili paucistriata	35
Spiroplectammina trinitatensis	15
Stilostomella subspinosa	19
Textularia leuzingeri	56

No.	Species (by number)
1	Pseudoclavulina trinitatensis
2	Ammovertella retrorsa
3	Dentalina semilaevis
4	Orthomorphina rohri
5	Siphonodosaria nuttalli gracillima
6	Chrysalogonium elongatum
7	Chrysalogonium lanceolum
8	Chrysalogonium longicostatum
9	Lagena striata basisenta
10	Lagena asperoides
11	Nodosaria longiscata
12	Rectoguembelina trinitatensis
13	Plectina trinitatensis
14	Ammobaculites cubensis
15	Spiroplectammina trinitatensis
16	Rectoguembelina inopinata
17	Gaudryina pseudocollinsi
18	Lagena ciperensis
19	Stilostomella subspinosa
20	Siphonodosaria nuttalli aculeata
21	Siphonodosaria nuttalli
22	Ellipsolagena barri
23	Siphonodosaria recta
24	Lituotuba navetensis
25	Karreriella subcylindrica
26	Martinottiella petrosa
27	Pyramidulina modesta
28	Plectofrondicularia vaughani
29	Plectofrondicularia lirata
30	Plectofrondicularia trinitatensis
31	Plectofrondicularia cookei
32	Cassidulinoides bradyi
33	Plectofrondicularia mexicana
34	Chrysalogonium ciperense
35	Siphonodosaria verneuili paucistriata
36	Ellipsoglandulina multicostata
37	Dorothia brevis
38	Karreriella alticamera
39	Plectofrondicularia alazanensis
40	Entosolenia orbignyana clathrata
41	Entosolenia formosa
42	Lagena pulcherrima enitens
43	Chrysalogonium tenuicostatum
44	Entosolenia flintiana plicatura
45	Entosolenia flintiana indomita
46	Entosolenia kugleri
47	Pyramidulina stainforthi
48	Lagena crenata capistrata
49	Plectofrondicularia morreyae
50	Cassidulina spinifer
51	Lagena pulcherrima
52	Entosolenia spinulolaminata
53	Chrysalogonium breviloculum
54	Plectofrondicularia nuttalli
55	Plectofrondicularia spinifera
56	Textularia leuzingeri
57	Pyramidulina lamellata
58	Pseudoglandulina gallowayi
59	Nodosarella mappa

Fig. 53.

1 *Bolivina pisciformis* Galloway & Morrey
Holotype, × 45.

2–3 *Bolivina alazanensis* Cushman
2 Holotype, × 50; 3 from Cushman & Renz, 1947*b*,
pl. 6, fig. 6 (as *Bolivina pisciformis*), × 56.

4–5 *Bolivina mexicana aliformis* Cushman
4 Holotype, × 50; 5 from Cushman & Renz, 1947*b*,
pl. 6, fig. 11 (as *Bolivina* cf. *alata* Seguenza), ×
56.

6–7 *Bolivina alazanensis venezuelana* Hedberg
6 Holotype, × 90; 7 from Cushman & Renz, 1948,
pl.5, fig. 19 (as *Bolivina alazanensis*), × 60.

8–9 *Bolivina marginata multicostata* Cushman
8 Holotype, × 40; 9 from Cushman & Renz, 1947*b*,
pl. 6, fig. 10, × 56.

10–11 *Bolivina acerosa* Cushman
10 Holotype, × 83; 11 from Cushman & Renz,
1947, pl. 6, fig. 7, × 56.

12 *Bolivina pisciformis optima* Cushman
Holotype, × 80.

13–14 *Bolivina* cf. *tectiformis* Cushman
13 from Cushman & Stainforth, 1945, pl. 7, fig.
12, × 55; 14 Holotype, × 75.

15 *Bolivina aenariensis carapitana* Hedberg
Holotype, × 120.

16 *Bolivina imporcata* Cushman & Renz
from Cushman & Renz, 1947*b*, pl. 6, fig. 9, × 56.

17 *Bolivina gracilis* Cushman & Applin
Holotype, × 52.

18 *Bolivina danvillensis* Howe & Wallace
Holotype, × 40.

19 *Bulimina* cf. *aspera* Cushman & Parker
from Cushman & Renz, 1946, pl. 6, fig. 10, × 40.

20 *Bulimina illingi* Cushman & Stainforth
Holotype, × 25.

21 *Bulimina pupoides* d'Orbigny
from Cushman & Renz, 1948, pl. 5, fig. 16, × 30.

22 *Bulimina tuxpamensis* Cole
from Cushman & Stainforth, 1945, pl. 6, fig. 6, ×
25.

23 *Bulimina striata mexicana* Cushman
from Cushman & Parker, 1947, pl. 28, fig. 4, ×
35.

24 *Bulimina macilenta* Cushman & Parker
from Cushman & Stainforth, 1945, pl. 6, fig. 4, ×
40.

25–26 *Bulimina bleeckeri* Hedberg
25 Holotype, × 40; 26 from Cushman & Stainforth,
1945, pl. 6, fig. 4, × 40.

27 *Bulimina alazanensis* Cushman
from Cushman & Stainforth, 1945, pl. 6, fig. 2, ×
55.

28 *Bulimina prolixa* Cushman & Parker
Holotype, × 110.

29 *Bulimina guayabalensis ampla* Cushman & Parker
Holotype, × 55.

30a–b *Bulimina jarvisi* Cushman & Parker
Holotype, × 50.

31–33 *Bulimina jacksonensis* Cushman
31 from Cushman & Renz, 1948, pl. 5, fig. 13,
× 30; 32 Holotype, re-figured from Cushman &
Parker, 1947, pl. 22, fig. 16, × 65.; 33 from Cushman & Parker, 1947, pl. 22, fig. 14, × 45.

34–35 *Bulimina sculptilis* Cushman
34 Holotype, × 65; 35 from Cushman & Parker,
1947, pl. 24, fig. 12a, × 50.

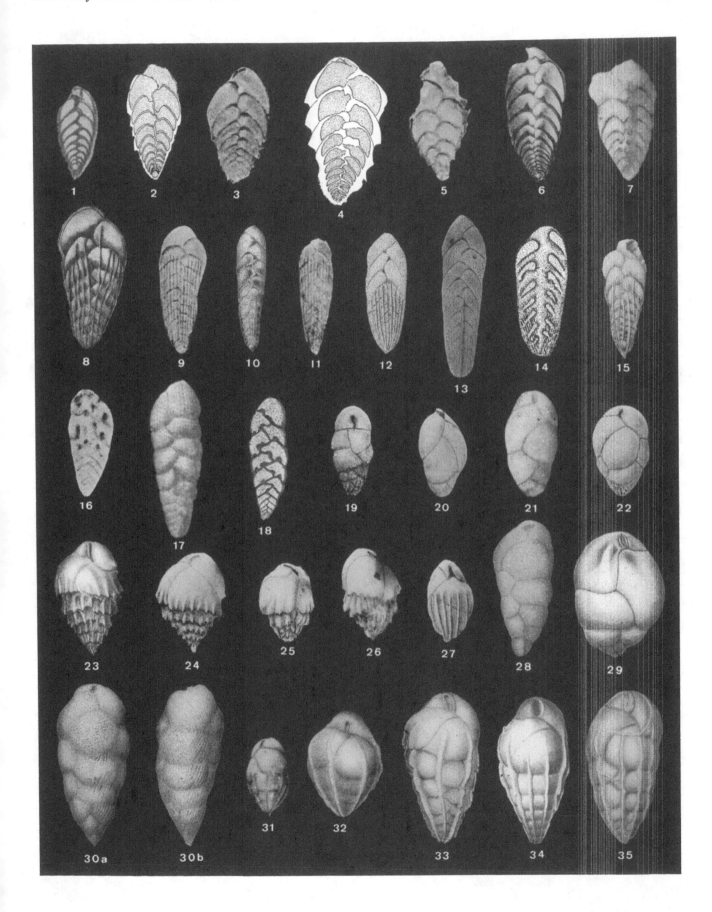

Fig. 54.

1–2 *Uvigerina hootsi* Rankin
1 Holotype, × 50; **2** Paratype, × 50.

3–4 *Uvigerina carapitana* Hedberg
3 Holotype, × 90; **4** from Cushman & Renz, 1947*b*, pl. 6, fig. 15, × 56.

5 *Uvigerina laviculata* Coryell & Rivero
Holotype, × 30.

6 *Uvigerina rustica* Cushman & Edwards
from Cushman & Stainforth, 1945, pl. 7, fig. 13, × 25.

7 *Uvigerina auberiana attenuata* Cushman & Renz
from Cushman & Stainforth, 1945, pl 7, fig. 18, × 55.

8 *Uvigerina mantaensis* Cushman & Edwards
from Cushman & Stainforth, 1945, pl. 7, fig. 17, × 55.

9–10 *Uvigerina capayana* Hedberg
9 Holotype, × 120; **10** from Cushman & Stainforth, 1945, pl. 7, fig. 15, × 55.

11 *Uvigerina isidroensis* Cushman & Renz
from Cushman & Renz, 1947*b*, pl. 6, fig. 18, × 56.

12 *Uvigerina howei* Garrett
from Cushman & Renz, 1947*b*, pl. 6, fig. 19, × 56.

13 *Uvigerina gallowayi basicordata* Cushman & Renz
from Cushman & Stainforth, 1945, pl. 7, fig. 14, × 40 (as *Uvigerina gallowayi* Cushman).

14–15 *Uvigerina hispidocostata* Cushman & Todd
14 from Cushman & Renz, 1947*b*, pl. 6, fig. 17, × 35; **15** Holotype, × 55.

16 *Uvigerina bealli* Bermudez
Holotype, × 30.

17–18 *Uvigerina gallowayi* Cushman
17 Holotype, microspheric, × 35; **18** Paratype, macrospheric, × 35.

19–20 *Uvigerina curta* Cushman & Jarvis

two of three original types, no holotype designated, × 55.

21 *Uvigerina mexicana* Nuttall
specimen illustrated by Nuttall, 1928 as *U. pygmaea* d'Orbigny, × 40.

22 *Uvigerina yazooensis* Cushman
Holotype, × 35.

23–24 *Uvigerina ciperana* Cushman & Stainforth
23 Holotype, × 25; **24** Paratype, × 55.

25–26 *Uvigerina spinicostata* Cushman & Jarvis
25 Holotype, × 55; **26** from Cushman & Stainforth, 1945, pl. 7, fig. 16, × 55.

27–28 *Uvigerina seriata* Cushman & Jarvis
27 Holotype, × 55; **28** Paratype, × 55.

29 *Uvigerina israelskyi* Garrett
Holotype, × 54.

30 *Rectuvigerina multicostata* (Cushman & Jarvis)
Holotype, × 55.

31–32 *Rectuvigerina multicostata optima* (Cushman)
from Cushman & Stainforth, 1945, pl. 8, figs. 1–2 (fig. 1 as *Siphogenerina multicostata*, fig. 2 as *Siphogenerina seriata*), × 55.

33–36 *Rectuvigerina basispinata* (Cushman & Jarvis)
33 from Cushman & Stainforth, 1945, pl. 8, fig. 3, × 40; **34** from Cushman & Renz, 1947*b*, pl. 7, fig. 1, × 28; **35** Holotype, × 55; **36** Paratype, × 55.

37 *Siphogenerina senni* Cushman & Renz
from Cushman & Renz, 1947*b*, pl. 7, fig. 5, × 35.

38 *Siphogenerina kugleri* Cushman & Renz
from Cushman & Renz, 1947*b*, pl. 7, fig. 4, × 35.

39–40 *Siphogenerina transversa* Cushman
39 microspheric form; **40** macrospheric form; both from Cushman & Renz, 1947*b*, pl. 7, figs. 2, 3, × 28.

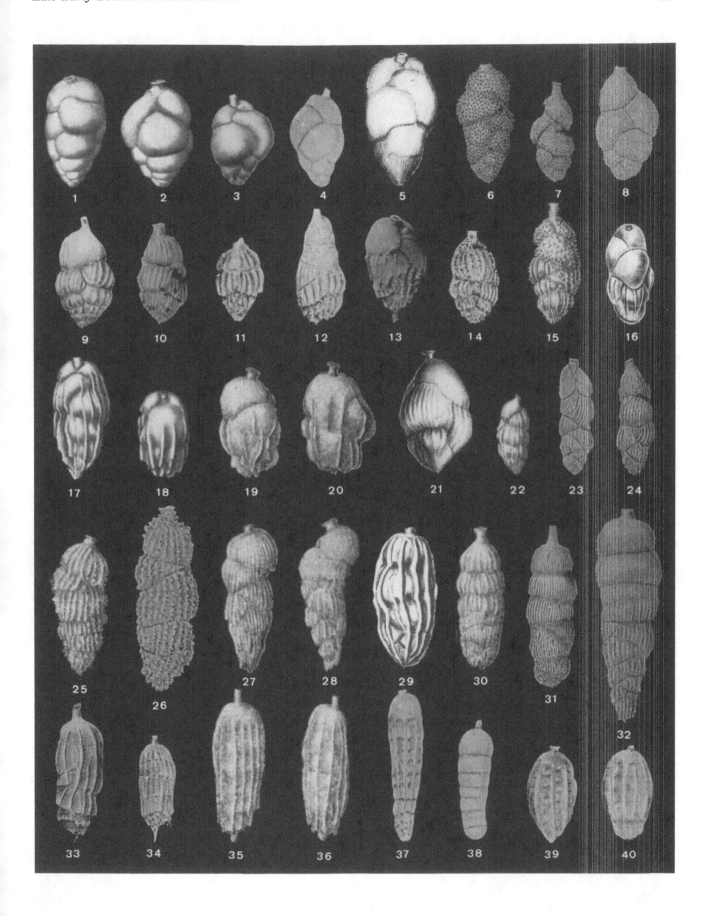

Fig. 55.

1–2 *Baggina cojimarensis* Palmer
1a, b Holotype, × 42; **2** from Cushman & Renz, 1947*b*, pl. 7, fig. 22, × 35.

3a–c *Cancris mauryae* Cushman & Renz
Holotype, × 46.

4–5 *Cancris cubensis* Cushman & Bermudez
4 from Cushman & Renz, 1947*b*, pl. 7, fig. 21, × 35; **5a–c** Holotype, × 22.

6–9 *Hoeglundina elegans* (d'Orbigny)
6 from Cushman & Renz, 1947*b*, pl.7, fig. 20, × 28. **7, 8** from Nuttall, 1928, pl. 7, figs. 9, 10, × 25. **9a–c** after Fornasini, 1906. **9d–e** from Parker, Jones & Brady, 1871, after Soldani.

10–11 *Valvulineria venezuelana* Hedberg
10 from Cushman & Renz, 1947*b*, pl. 7, fig. 16, × 35; **11a–c** Holotype, × 50.

12–13 *Valvulineria georgiana* Cushman
12a–c Holotype, × 33; **13** from Cushman & Renz, 1947*b*, pl. 7, fig. 15, × 35.

14–15 *Valvulineria palmerae* Cushman & Todd
14 from Cushman & Renz, 1947*b*, pl. 7, fig, 13, × 56; **15a–c** Holotype, × 45.

16–17 *Eponides crebbsi* Hedberg
16 from Cushman & Renz, 1947*b*, pl. 7, fig, 19, × 56; **17a–c** Holotype, × 120.

18a–c *Eponides parantillanum* Galloway & Heminway
Holotype, × 37.

19a–b *Eponides byramensis* (Cushman)
from Hedberg, 1937*a*, pl. 92, figs. 2a–b, × 50.

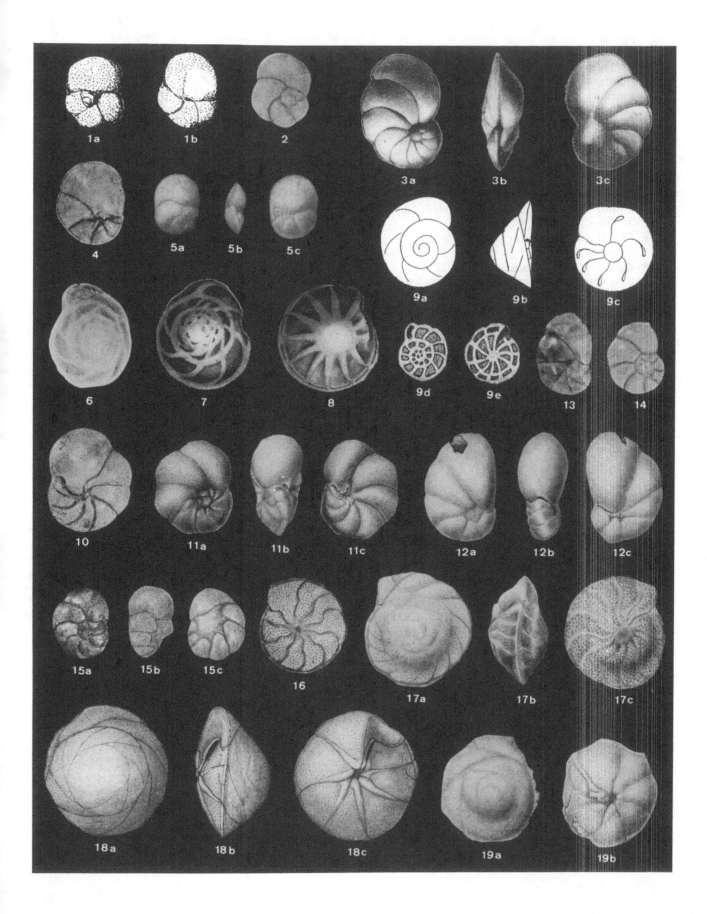

Fig. 56.

1a–c *Eponides vicksburgensis* Cushman & Ellisor
Holotype, × 60.

2a–c *Eponides elevatus* (Plummer)
Holotype, × 50.

3a–c *'Discorbis' ciperensis* Cushman & Stainforth
Holotype, × 40.

4a–b *Siphonina pulchra* Cushman
from Cushman & Stainforth, 1945, pl. 11, figs. 6a–b, × 40.

5–6 *Cibicidoides parianus* (Hedberg)
5 from Cushman & Renz, 1947*b*, pl. 8, fig. 9, × 56. **6a–c** Holotype, × 90.

7a–c *Cibicidoides subspiratus* (Nuttall)
Holotype, × 40.

8 *Cibicidoides* aff. *subspiratus* (Nuttall)
from Cushman & Renz, 1948, pl. 8, fig. 24, × 45.

9–11 *Cibicidoides ungerianus* (d'Orbigny)
9a–c Holotype, × 65; **10**, **11a–b** from Nuttall, 1932, pl. 9, figs. 4–6, × 40.

12a–b *Cibicidoides* cf. *atratiensis* (Tolmachoff)
from Cushman & Stainforth, 1945, pl. 15, figs. 3a–b, × 25.

13a–c *Cibicidoides* cf. *atratiensis* (Tolmachoff)
Holotype, × 50.

14–15 *Cibicidoides cookei* (Cushman & Garrett)
14a–c Holotype, × 38; **15a–b** from Cushman & Stainforth, 1945, pl. 15, figs. 4a–b, × 40.

16a–c *Cibicidoides pippeni* (Cushman & Garrett)
Holotype, × 38.

17–18 *Cibicidoides grimsdalei* (Nuttall)
17a–c from Cushman & Renz, 1948, pl. 8, figs. 17–19, × 30; **18a–c** Holotype, × 40.

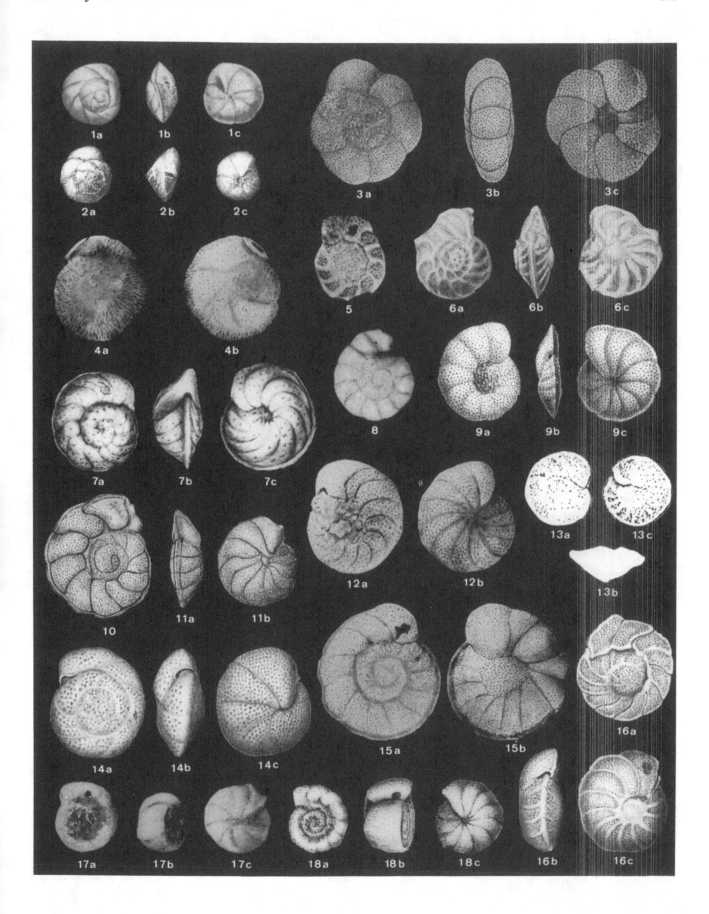

Fig. 57.

1 *Cibicidoides pseudoungerianus* (Cushman)
Holotype, × 66.

2a–b *Laticarinina bullbrooki* Cushman & Todd
from Cushman & Stainforth, 1945, pl. 15, figs. 2a–
b, × 26.

3a–c *Planulina renzi* Cushman & Stainforth
Holotype, × 25.

4–5, *Planulina wuellerstorfi* (Schwager)
7–8 **4a–c** Holotype, × 25; **5a–c** Paratype, × 25. **7** from
Nuttall, 1932, pl. 4, fig. 14, × 25. **8** from Nuttall,
1928, pl. 7, fig. 12, × 10.

6a–b *Planulina* cf. *wuellerstorfi* (Schwager)
from Cushman & Stainforth, 1945, pl. 14, figs. 6a–
b, × 26.

9 *Planulina subtenuissima* (Nuttall)
Holotype, × 20.

10–11 *Planulina marialana* Hadley
10a–c Holotype, × 55; **11a–b** from Cushman &
Stainforth, 1945, pl. 14, figs. 3a–b, × 40.

12–14 *Cibicidina antiqua* (Cushman & Applin)
12 from Cushman & Renz, 1948, pl. 8, fig. 26, ×
45. **13, 14** Syntypes, × 35.

15–16 *Cibicidina cushmani* (Nuttall)
15a–c Holotype, × 40; **16** from Cushman & Renz,
1948, pl. 8, fig. 22, × 45.

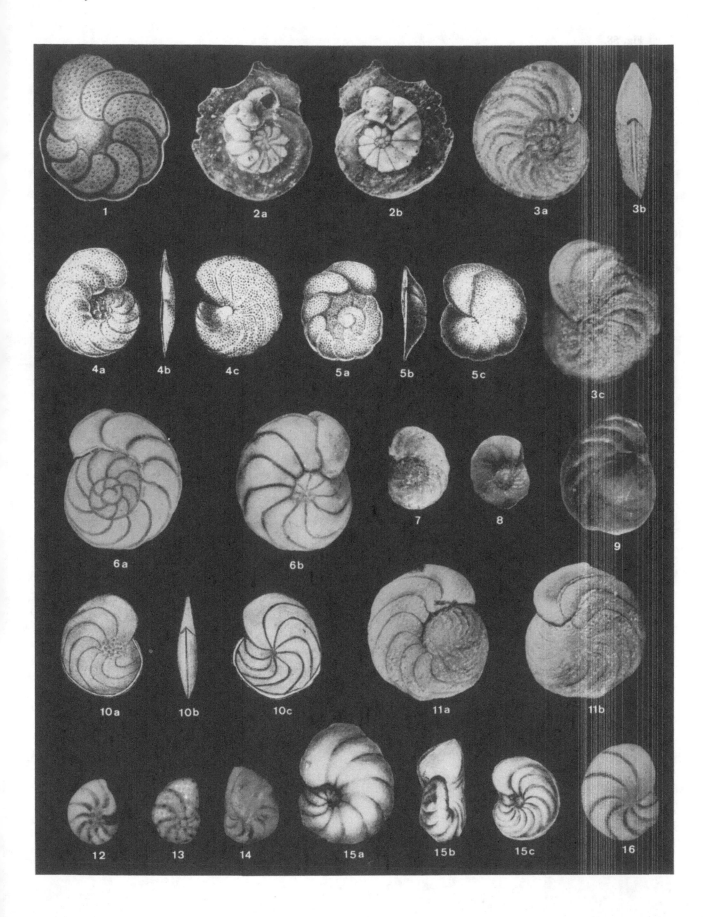

Fig. 58.

1a–c *Nuttallides crassaformis* (Cushman & Siegfus)
Holotype, × 40.

 2–3 *Nuttallides truempyi* (Nuttall)
2a–c Holotype, × 40; **3a–b** from Cushman &
Renz, 1948, pl. 7, figs. 7, 8, × 30.

4a-b *Alabamina atlantisae dissonata* (Cushman & Renz)
Holotype, × 45.

 5–9 *Osangularia mexicana* (Cole)
5 Holotype, × 30. **6** Paratype, × 30. **7a–b** from
Cushman & Stainforth, 1945, pl. 11, figs. 5a–b, ×
40. **8, 9** from Cushman & Renz, 1948, pl. 7, figs. 9,
10, × 45.

10–13 *Oridorsalis umbonatus* (Reuss)
10a–c Holotype, × 60. **11a–b** from Nuttall, 1932,

pl. 6, figs. 4, 5, × 60. **12** from Cushman & Renz,
1948, pl. 7, fig. 6, × 60. **13a–b** from Cushman &
Stainforth, 1945, pl. 11, figs. 4a–b, × 40.

14–15 *Anomalinoides pompilioides* (Galloway & Heminway)
14a–b from Cushman & Stainforth, 1945, pl. 14,
figs. 1a–b, × 25. **15a–c** Holotype, × 76.

16–19 *Anomalinoides subbadenensis* (Pijpers)
16a–b from Cushman & Stainforth, 1945, pl. 14,
figs. 2a–b, × 55. **17** Holotype, × 100. **18, 19** Paratypes, × 100.

20–22 *Anomalinoides abouillotensis* (Bermudez)
20 Holotype, × 25. **21, 22** Paratypes, × 25.

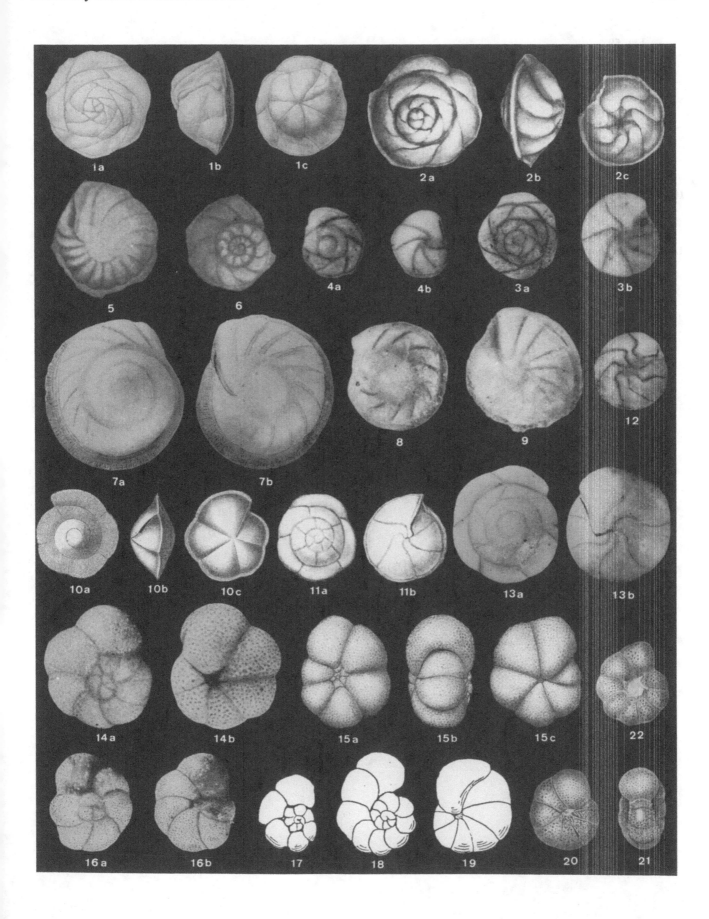

Fig. 59.

1–2 *Anomalinoides dorri aragonensis* (Nuttall)
1 Holotype, × 40; **2** Paratype, × 40.

3–4 *Anomalinoides cicatricosus* (Schwager)
3a–c Holotype, × 33. **4a–b** from Cushman &
Stainforth, 1945, pl. 15, figs. 6a–b, × 40.

5–9 *Anomalinoides alazanensis* (Nuttall)
5a–c from Galloway & Heminway, 1941, pl. 22,
figs. 1a–c, × 40. **6** from Cushman & Renz, 1947*b*,
pl. 8, fig. 5, × 35. **7–9** Syntypes, × 60.

10–12 *Anomalinoides alazanensis spissiformis* (Cushman &
Stainforth)
10, 11 from Cushman & Renz, 1948, pl. 8, figs. 15–
16, × 45. **12a–c** Holotype, × 40.

13–15 *Anomalinoides affinis* (Hantken)

13, 14 from Cushman & Renz, 1948, pl. 8, figs. 9–
10, × 45. **15a–c** Holotype, × 40.

16–18 *Anomalinoides bilateralis* (Cushman)
16, 17 from Cushman & Renz, 1948, pl. 8, figs. 11–
12, × 45. **18a–b** Holotype, × 40.

19–21 *Anomalinoides mecatepecensis* (Nuttall)
19 Holotype, × 25. **20** Paratype, × 25. **21a–b** from
Hedberg, 1937*a*, pl. 92, figs.10a–b, × 50.

22–27 *Anomalinoides trinitatensis* (Nuttall)
22–24 from Cushman, 1946*a*, pl. 8, figs. 9–11, ×
40. **25** from Cushman & Renz, 1947*b*, pl. 8, fig. 6,
× 35. **26** Holotype, × 20. **27** from Nuttall, 1932, pl.
7, fig. 9, × 25.

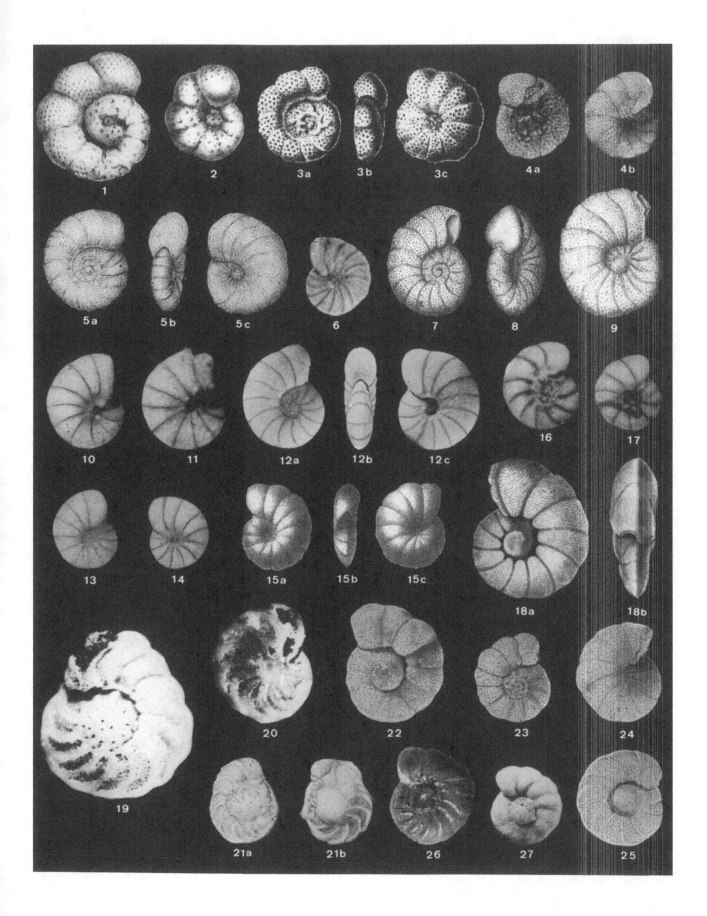

Fig. 60.

1–5 *Heterolepa mexicana* (Nuttall)
1a–c Syntype, × 50. **2a–b** from Cushman & Stainforth, 1945, pl. 15, figs. 5a–b, × 25. **3, 4** from Cushman & Renz, 1948, pl. 8, figs. 20, 21, × 30. **5a–c** from Galloway & Heminway, 1941, pl. 22, figs. 5a–c, × 27.

6a–c *Gyroidinoides complanatus* (Cushman & Stainforth)
Holotype, × 40.

7–8 *Gyroidinoides planulatus* (Cushman & Renz)
7a–c Holotype, × 44. **8** from Cushman & Renz, 1947b, pl. 7, fig. 17, × 56.

9–10 *Gyroidinoides soldanii octocameratus* (Cushman & Hanna)
9a–b from Cushman & Renz, 1948, pl. 7, figs. 2, 3, × 60. **10a–c** Holotype, × 50.

11–12 *Gyroidinoides girardanus peramplus* (Cushman & Stainforth)
11a–b Holotype, × 25. **12** from Cushman & Renz, 1948, pl. 7, fig. 1, × 30.

13–14 *Gyroidinoides girardanus* (Reuss)
13a–c Holotype, × 40. **14a–b** from Cushman & Stainforth, 1945, pl. 10, figs. 18a–b, × 40.

15a–b *Gyroidinoides dissimilis* (Cushman & Renz)
Holotype, × 56.

16a–b *Gyroidinoides altispirus* (Cushman & Stainforth)
Holotype, × 40.

17–18 *Gyroidina jarvisi* Cushman & Stainforth
17a–b from Cushman & Renz, 1948, pl. 7, figs. 4, 5, × 60. **18a–c** Holotype, × 40.

Fig. 61.

1–3 *Hanzawaia mantaensis* (Galloway & Morrey)
  1a–c Holotype, × 54. **2** from Cushman & Renz,
  1947*b*, pl. 8, fig. 7, × 56. **3a–c** from Hedberg,
  1937*a*, pl. 92, figs. 12a–c, × 120.

4–6 *Hanzawaia illingi* (Nuttall)
  4a–b from Cushman & Stainforth, 1945, pl. 14, figs.
  4a–b, × 25. **5** Holotype, × 40; **6** Co-type, × 40.

7–8 *Hanzawaia mississippiensis* (Cushman)
  7a–b Holotype, × 80. **8** from Cushman & Renz,
  1947*b*, pl. 8, fig. 8, × 35.

9–11 *Hanzawaia carstensi* (Cushman & Ellisor)
  9a–c Holotype, × 40. **10**, **11** from Cushman &
  Renz, 1947*b*, pl. 8, figs. 10, 11, both × 56.

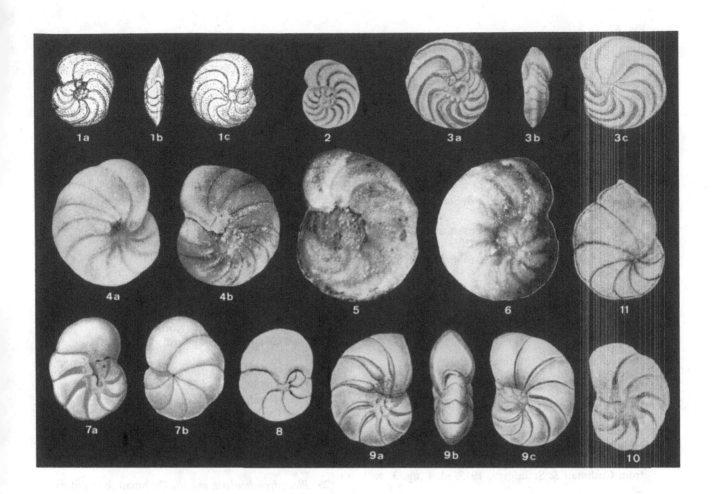

Fig. 62.

1 *Lituotuba navetensis* Cushman & Renz
   Holotype, × 30.

2 *Ammovertella retrorsa* Cushman & Stainforth
   from Cushman & Renz, 1948, pl. 2, fig. 1, × 30.

3 *Ammobaculites cubensis* Cushman & Bermudez
   from Cushman & Renz, 1948, pl. 2, fig. 7, × 30.

4 *Gaudryina pseudocollinsi* Cushman & Stainforth
   Holotype, × 30.

5, 5a *Dorothia brevis* Cushman & Stainforth
   Holotype, × 25.

6, 6a *Karreriella alticamera* Cushman & Stainforth
   Holotype, × 25.

7 *Spiroplectammina trinitatensis* Cushman & Renz
   Paratype, microspheric form, × 30.

8 *Spiroplectammina trinitatensis* Cushman & Renz
   Holotype, macrospheric form, × 30.

9 *Karreriella subcylindrica* (Nuttall)
   from Cushman & Renz, 1948, pl. 3, fig. 17, × 30.

10 *Martinottiella petrosa* (Cushman & Bermudez)
   from Cushman & Stainforth, 1945, pl. 2, fig. 13, ×
   40.

11 *Plectina trinitatensis* Cushman & Renz
   Holotype, × 30.

12 *Textularia leuzingeri* Cushman & Renz
   from Cushman & Stainforth, 1945, pl. 1, fig. 19, ×
   40.

13 *Pseudoclavulina trinitatensis* Cushman & Renz
   Holotype, × 30.

14 *Pseudoglandulina gallowayi* Cushman
   from Cushman & Stainforth, 1945, pl. 4, fig. 3, ×
   25.

15 *Bifarina trinitatensis* (Cushman & Renz)
   Holotype, × 60.

16 *Bifarina inopinata* (Cushman & Stainforth)
   Holotype, × 55.

17 *Orthomorphina rohri* (Cushman & Stainforth)
   from Cushman & Renz, 1948, pl. 7, fig. 11, × 30.

18 *Dentalina semilaevis* Hantken
   from Cushman & Stainforth, 1945, pl. 3, fig. 17, ×
   25.

19 *Ellipsoglandulina multicostata* (Galloway & Morrey)
   from Cushman & Stainforth, 1945, pl. 10, fig. 6, ×
   25.

20 *Plectofrondicularia vaughani* Cushman
   from Cushman & Stainforth, 1945, pl. 5, fig. 13, ×
   40.

21 *Plectofrondicularia spinifera* Cushman & Jarvis
   from Cushman & Jarvis, 1934, pl. 10, fig. 14a, × 35.

22 *Plectofrondicularia lirata* Bermudez
   from Cushman & Renz, 1948, pl. 5, fig. 8, × 30.

23 *Plectofrondicularia cookei* Cushman
   from Cushman & Stainforth, 1945, pl. 5, fig. 12, ×
   55.

24, 24a *Cassidulina spinifera* Cushman & Jarvis
   from Cushman & Stainforth, 1945, pl. 11, fig. 10, ×
   40.

25, 25a *Cassidulinoides bradyi* (Norman)
   from Cushman & Stainforth, 1945, pl. 12, fig. 6, ×
   40.

26 *Plectofrondicularia trinitatensis* Cushman & Jarvis
   Holotype, × 55.

27 *Plectofrondicularia mexicana* (Cushman)
   from Cushman & Stainforth, 1945, pl. 5, fig. 19, ×
   55.

28 *Plectofrondicularia morreyae* Cushman
   from Cushman & Stainforth. 1945, pl. 5, fig. 17, ×
   55.

29 *Plectofrondicularia nuttalli* Cushman & Stainforth
   Holotype, × 40.

30 *Plectofrondicularia alazanensis* Cushman
   from Cushman & Stainforth, 1945, pl. 5, fig. 20, ×
   40.

Fig. 63.

1 *Ellipsolagena barri* Cushman & Stainforth
Holotype, × 25.

2 *Entosolenia spinulolaminata* Cushman & Stainforth
Holotype, × 40.

3 *Entosolenia flintiana indomita* Cushman & Stainforth
Holotype, × 55.

4 *Entosolenia kugleri* Cushman & Stainforth
Holotype, × 25.

5 *Entosolenia pannosa* Cushman & Stainforth
Holotype, × 55.

6 *Entosolenia orbignyana clathrata* (H. B. Brady)
from Cushman & Stainforth, 1945, pl. 6, fig. 20, × 55.

7 *Entosolenia flintiana plicatura* Cushman & Stainforth
Holotype, × 55.

8 *Lagena pulcherrima enitens* Cushman & Stainforth
Holotype, × 25.

9 *Lagena crenata capistrata* Cushman & Stainforth
Holotype, × 55.

10 *Lagena striata basisenta* Cushman & Stainforth
Holotype, × 55.

11 *Lagena ciperensis* Cushman & Stainforth
Holotype, × 55.

12 *Lagena asperoides* Galloway & Morrey
from Cushman & Stainforth, 1945, pl. 4, fig. 9, × 25.

13 *Lagena pulcherrima* Cushman & Jarvis
from Cushman & Stainforth, 1945, pl. 4, fig. 13, × 25.

14 *Pyramidulina modesta* (Bermudez)
from Cushman & Stainforth, 1945, pl. 10, fig. 2, × 25.

15, 15a *Nodosarella mappa* (Cushman & Jarvis)
from Cushman & Stainforth, 1945, pl. 9, figs. 8a, b, × 30, Holotype re-illustrated.

16 *Chrysalogonium elongatum* Cushman & Jarvis
from Cushman & Stainforth, 1945, pl. 16, fig. 4a, × 20.

17 *Nodosaria longiscata* d'Orbigny
from Cushman & Stainforth, 1945, pl. 3, fig. 20, × 25.

18 *Pyramidulina stainforthi* (Cushman & Renz)
from Cushman & Stainforth, 1945, pl. 3, fig. 25, × 25.

19 *Pyramidulina lamellata* (Cushman & Stainforth)
from Cushman & Renz, 1947*b*, pl. 4, fig. 19, × 25.

20 *Siphonodosaria nuttalli gracillima* Cushman & Jarvis
from Cushman & Stainforth, 1945, pl. 9, fig. 14, × 30.

21 *Siphonodosaria recta* (Palmer & Bermudez)
from Cushman & Stainforth, 1945, pl. 10, fig. 5, × 25.

22 *Siphonodosaria nuttalli aculeata* (Cushman & Renz)
Holotype, × 30.

23, 23a *Chrysalogonium longicostatum* Cushman & Jarvis
from Cushman & Stainforth, 1945, pl. 16, figs. 2a–b, × 40, Holotype re-illustrated.

24, 24a *Chrysalogonium breviloculum* Cushman & Jarvis
from Cushman & Stainforth, 1945, pl. 16, figs 1a–b, × 25, Holotype re-illustrated.

25 *Chrysalogonium tenuicostatum* Cushman & Bermudez
from Cushman & Stainforth, 1945, pl. 3, fig. 28, × 25.

26 *Chrysalogonium lanceolum* Cushman & Jarvis
from Cushman & Stainforth, 1945, pl. 3, fig. 29, × 25.

27 *Chrysalogonium ciperense* Cushman & Stainforth
Holotype, × 25.

28 *Siphonodosaria nuttalli* (Cushman & Jarvis)
from Cushman & Stainforth, 1945, pl. 9, fig. 13, × 25.

29 *Siphonodosaria verneuili paucistriata* (Galloway & Morrey)
from Cushman & Stainforth, 1945, pl. 9, fig. 12, × 25.

30, 30a *Stilostomella subspinosa* Cushman
from Cushman & Stainforth, 1945, pl. 9, figs. 10a–b, × 20, Holotype re-illustrated.

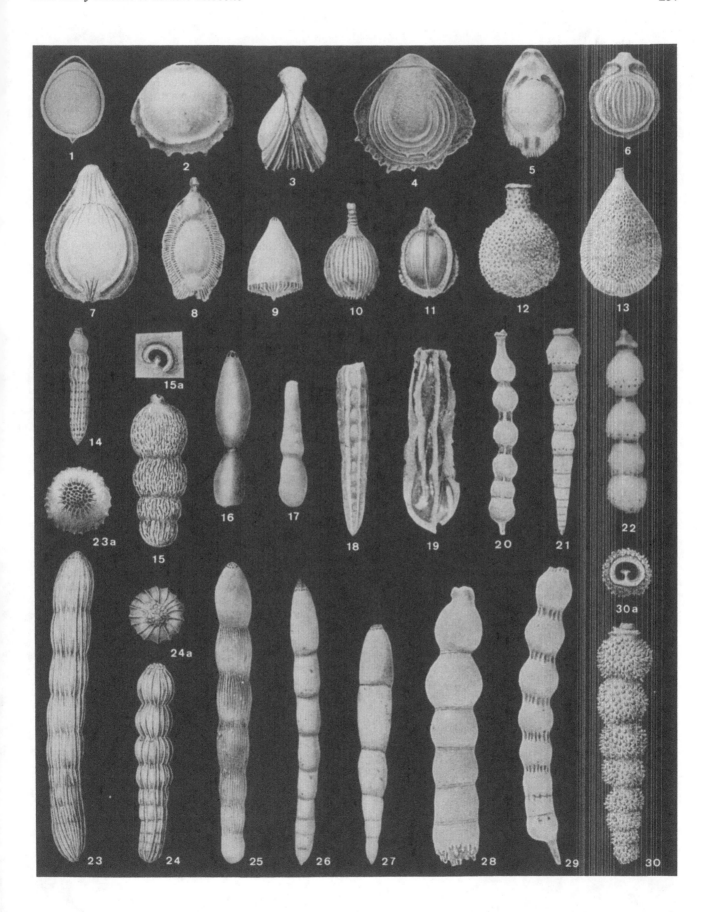

Fig. 64.

1–3 *Discamminoides tobleri* Brönnimann
1a equatorial view, **1b** edge view, from Brönnimann, 1951, textfig. 9a, 9b, × 23.
2 near axial section, end chamber with two straight alveoles, from Brönniman, 1951, textfig. 10a, × 35.
3 broken specimen showing one row of alveoles in the peripheral filling, from Brönnimann, 1951, textfig. 11a, × 35.
Specimens from the Cruse Formation, South Trinidad.

4–7 *Gravellina narivaensis* Brönnimann
from Brönniman, 1953, textfigs. 1e, u, v, s, all × 80.
Specimens from the Nariva Formation, South Trinidad.

8a–b, 9 *Jarvisella karamatensis* Brönnimann
from Brönnimann, 1953, textfigs. 2a, b, × 94, 2e × 35.
Specimens from the Nariva Formation, South Trinidad.

11–14 *Alveovalvulina suteri* Brönnimann
10a–b from Brönniman, 1951, textfigs. 5b, c, × 35,
11, 12 from Brönnimann, 1951, textfigs. 6a, c, × 35,
13 axial section of microspheric specimen, from Brönnimann, 1951, textfig. 8, × 55.
14 from Brönnimann, 1951, textfig. 5f, × 35.
Specimens from the Miocene, South Trinidad.

15–21 *Alveovalvulinella pozonensis* (Cushman & Renz)
15–17 from Brönnimann, 1953, textfigs. 4a, d, f, × 25.
18 uniserial end stage with terminal aperture, from Brönnimann, 1953, textfig. 4h, × 25.
19 uniserial end stage, from Brönnimann, 1953, textfig. 5b, × 70.
20, 21 axial sections, from Brönnimann, 1953, textfig. 7, × 40.
Specimens 15–19 from the Nariva Formation, South Trinidad, 20–21 from the Agua Salada Formation, southeastern Falcon, Venezuela.

22–27 *Guppyella miocenica* (Cushman)
22, 23 exterior views of microspheric specimens, from Brönnimann, 1951, textfigs. 1a, b, × 27. 24–26
24–26 transverse sections through uniserial chambers of microspheric specimens, from Brönnimann, 1951, textfigs. 3a, c, b, × 18.
27 end chamber of microspheric specimen with eroded outer wall showing alveoles, from Brönnimann, 1951, textfig. 4c, × 18.
Specimens from the Miocene of South Trinidad.

28a-b, 29a–b *Eggerella karamatensis* Brönnimann
from Brönnimann, 1953, textfigs. f-i, × 55.
Specimens from the Karamat Formation, South Trinidad.

30, 31, 32a–b *Valvulina flexilis* Cushman & Renz
from Brönnimann, 1953, textfigs. 15h, o, l, n, × 45.
Specimens from the Cruse Formation, South Trinidad.

The agglutinated benthic species illustrated in Fig. 64 occur in small numbers as component constituents of the foraminiferal faunas of rocks that are believed to have been deposited under normal marine conditions. They are also found as a much more dominant part of faunas that suggest turbid water conditions within sediments grouped as flysch or as turbidites. These environments can originate in basins receiving copious land-derived clastics, usually supplied by major rivers. In the areas discussed in the present work, formations that are believed to reflect such conditions include in Trinidad: the Chaudière, Pointe-à-Pierre, Nariva, Herrera, Karamat and Lengua/Cruse; in Venezuela: the Vidoño, Caratas and La Pica.

The stratigraphic distribution for individual species is given in the relevant taxonomic note. The case of *Discamminoides tobleri* illustrates such disjunct distribution as the species occurs sparsely in the open marine, bathyal sediments of the Cipero Formation in Trinidad and in the equivalent-aged, deep water clastics of the Karamat and Nariva formations. In the upper part of the Middle Miocene, it again is found sparsely in the open marine Lengua Formation, becoming commoner as deltaic clastic sediments of the overlying Cruse Formation begin to become dominant. The species illustrated on Fig. 64 are all better adapted to life on a substrate composed of fine-grained sediment, decreasing in numbers as the proportion of sand increases.

The agglutinated foraminiferal faunas that are found in Trinidad during the Middle to Late Miocene are discussed in Batjes (1968). Older flysch-type agglutinated foraminiferal faunas are discussed in the section on Late Albian to Early Eocene faunas and in Kaminski *et al.* (1988).

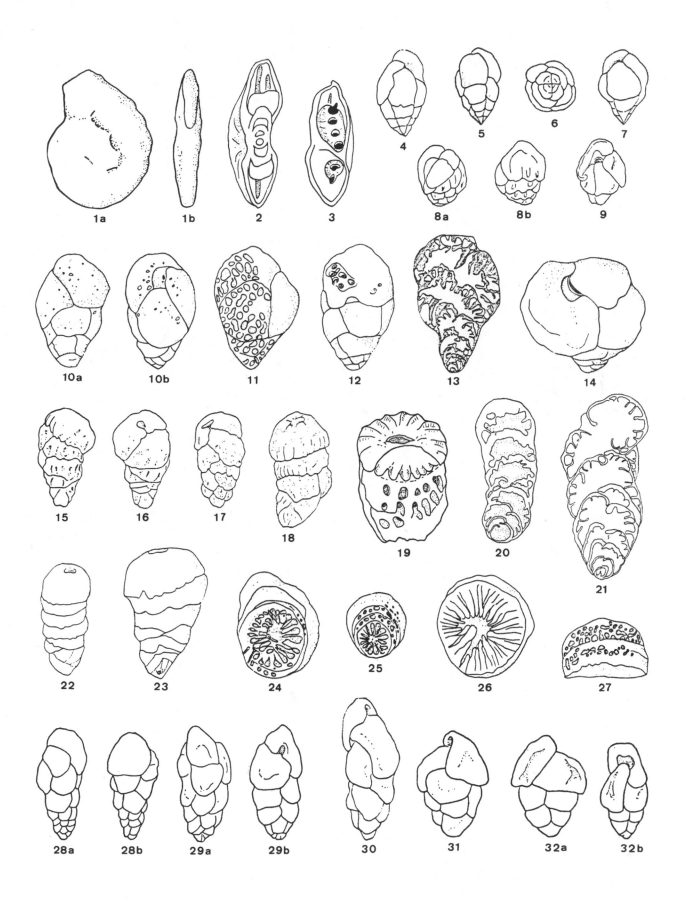

Venezuela

Introduction

Benthic foraminifera from three distinct regions of Venezuela are dealt with in this volume (Fig. 65). They are from the Serrania del Interior area, eastern Venezuela; the Falcon Basin and the Maracaibo Basin, both western Venezuela. Unlike in Trinidad where there exists from Barremian to Middle Miocene a nearly continuous record of described and illustrated planktic and benthic foraminifera and their stratigraphic distribution related to planktic zonal schemes, comparable data from Venezuela are to date less complete.

The Early Cretaceous stratigraphy in eastern Venezuela is well documented, mainly by the investigations of Rod & Maync (1954) and Guillaume, Bolli & Beckmann (1972). Based on numerous surface sections from the Serrania del Interior and the islands off Puerto la Cruz, the latter authors charted the distribution of 101 benthic foraminiferal species from Barremian to Albian, tied into a locally developed predominantly planktic foraminiferal zonal scheme. The benthic taxa were identified partly by direct comparison with types from neighbouring Trinidad or were based on published illustrations. In contrast to the Trinidad Early Cretaceous faunas those from eastern Venezuela were not described and illustrated.

On the Late Cretaceous to Eocene of eastern Venezuela three publications deal with benthic foraminifera. One is by Cushman (1947) describing and illustrating a stratigraphically mixed, predominantly benthic fauna from the Upper Cretaceous to Paleocene Vidoño Formation of the Santa Anita Group. The samples available to Cushman for his study originated from widely separated localities: The Rio Querecual in the southern foothills of the Serrania del Interior, the vicinity of Puerto la Cruz and Borracha Island near Puerto la Cruz. Individual species were compared with forms already known from the Lizard Springs Formation of Trinidad and the Velasco Shale of Mexico, but also from other areas. Cushman considered the fauna to be entirely of Cretaceous age,

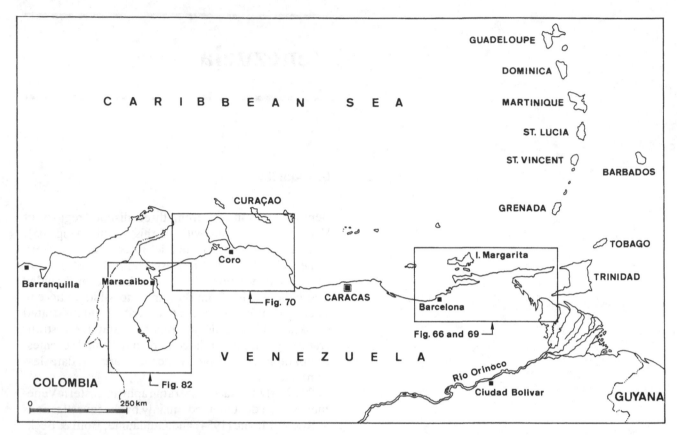

Fig. 65. Map of the northern part of Venezuela showing the three areas dealt with. Detailed maps are found under the Figure numbers shown below the boxes.

while part of it is today placed in the Early Tertiary. No attempt was made at the time by Cushman to fit the faunas into a zonal scheme.

Stanley (1960) investigated the Tertiary Vidoño benthic and planktic foraminifera from the Alcabala section between Barcelona and Puerto La Cruz and from the Rio Querecual type section near Bergantin. He recorded 72 benthic and 14 planktic species which he discusses in some detail. He found the fauna closely comparable to that known from the Lizard Springs Formation of Trinidad. The distribution of the benthic and planktic species in the Alcabala section is shown on range charts, but no illustrations of the recorded species are given. Planktic foraminifera were used by Stanley to subdivide the studied Vidoño sections into a Late Paleocene *Planorotalites pseudomenardii–Morozovella velascoensis* Zone and an Early Eocene *Globorotalia rex* (now *Morozovella subbotinae*) Zone.

Galea-Alvarez (1985) deals in her study with the biostratigraphy and depositional environment of the Late Cretaceous to Eocene Santa Anita Group in the general Barcelona area, western end of the Ser-

rania del Interior. She illustrates smaller benthic and planktic foraminifera from the San Antonio (Santonian to Campanian), Vidoño (Late Maastrichtian to Early Eocene) and Caratas (Early to Middle Eocene) formations. Plotted on a chart are 68 Paleocene Vidoño benthic taxa, their distribution correlated with the standard planktic foraminiferal zones and a local benthic zonal scheme. This chart remains of only limited value, for 27 of the 68 listed taxa have a generic but not a specific name. Furthermore, most of the taxa appearing on the chart are not illustrated.

The Late Cretaceous to Eocene planktic and benthic fauna of the Santa Anita Group of eastern Venezuela still await full taxonomic and stratigraphic documentation which will need to be based on the relatively continuous surface sections that are known to exist.

Studies of the foraminiferal faunas of the Oligocene and Miocene also remain incomplete. This concerns in particular the description, illustration and distribution pattern of both the planktic and benthic species from suitable surface and subsurface sections. The only publications on the Miocene Carapita Formation illustrating the foraminiferal fauna are those by Hedberg (1937*a*) and Franklin (1944). No attempts were at the time made by these authors to tie the distribution of their Carapita taxa into a zonal scheme but both compared them with forms known from other areas such as Trinidad, Falcon, Mexico and Cuba. Efforts have now more recently been made by some authors to apply the planktic foraminiferal zonal scheme to the Oligocene and Miocene of eastern Venezuela. Reference will be made to them in a later section (p. 273).

In western Venezuela it is the Oligocene and Miocene of the Falcon Basin that received early attention, both in the description and illustration of species and also in their application to biostratigraphy. The first Miocene benthic foraminifera described and illustrated from Falcon are those by Cushman (1929*b*) who compared them with closely related faunas from Manta in Ecuador and from Trinidad. This was followed by Cushman & Renz (1941) in a paper on Oligocene and Miocene foraminifera from the Agua Salada Formation of southeastern Falcon, including the description of 54 new taxa, and with the proposal to zone the formation by means of benthic species. Senn (1927) proposed and later (1935 and 1940) developed a zoning of the Agua Salada Formation based on a letter/number system, in its resolution largely comparable to the one later introduced by Cushman & Renz (1941).

In 1948 Renz published his monographic paper 'Stratigraphy and fauna of the Agua Salada Group, State of Falcon, Venezuela'. He describes and illustrates 239 predominantly benthic foraminiferal species and for 205 shows their distribution in relation to a benthic zonal scheme, modified from that of Cushman & Renz (1941).

Blow (1959), in re-studying the Agua Salada foraminifera of southeastern Falcon in a different section, emphasised the planktic forms. As a result he was able to introduce in addition to the benthic scheme the planktic foraminiferal zonal scheme previously developed in Trinidad.

With the two publications by Renz (1948) and Blow (1959) one has at hand a virtually complete coverage of the planktic and benthic foraminifera of the Agua Salada Formation and their distribution related to benthic and planktic zonal schemes. In a more recent study Diaz de Gamero (1985*a*, *b*) published comparable results of the occurrence and distribution of the Agua Salada fauna in two sections in northeastern Falcon.

The planktic and benthic foraminifera of the Oligocene and lower part of the Miocene of central Falcon were studied and partly described and illustrated by Diaz de Gamero (1977*a*) and their distribution related to established planktic zonal schemes in southeastern Falcon. However, in central Falcon suitable continuous and undisturbed sections are less common and preservation of faunas is not as good as in southeastern Falcon.

Publications on planktic and benthic foraminifera from the Cretaceous and Tertiary of the Maracaibo Basin are scarce, though much information on them and their stratigraphic significance rests in unpublished oil company reports.

Cushman (1929*a*) described and illustrated three new *Siphogenerinoides* species from the Upper Cretaceous Colon Formation. In Cushman & Hedberg (1930) on foraminifera from Venezuela and Colombia, three new species are included from western Venezuela. Nuttall (1935) identified 54 foraminifera species, seven of them as new, and most of them benthics, from the upper part of the Middle Eocene Pauji Formation. The taxa are described and many of them illustrated but the locality for the fauna is given only as the region southeast of Lake Maracaibo. A valuable contribution to our knowledge of the foraminiferal fauna in the Maracaibo Basin is the one by Sellier de Civrieux (1952) on the benthic and planktic taxa occurring in the Socuy Member of the Colon Formation. Barbeito *et al.* (1985) and Pittelli Viapiana (1989) are using benthic foraminiferal associations of the Maracaibo Basin primarily for the distinction of changing paleoenvironmental and paleobathymetric conditions through the Eocene of the basin. Fuenmayor (1989) in his 'Manual on the foraminifera of the Maracaibo Basin' published on the planktic and benthic faunas of this basin and their stratigraphic significance. For the subdivision of the Late Cretaceous to Early Miocene formations it offers local benthic zonal schemes which are correlated where possible with the planktic foraminiferal standard zones.

The Early Cretaceous of eastern Venezuela

Introduction

The purpose of this section is to document the stratigraphic distributions of smaller benthic and planktic foraminifera in the Early Cretaceous of eastern Venezuela and their correlation with a planktic foraminiferal zonal scheme. The complexity of the Early Cretaceous in this area as regards lithologies, stratigraphic subdivisions and lateral facies changes has led to differing interpretations by the numerous investigators particularly from the point of view of correlation.

Most of the studies were carried out as oil company projects, some using internal company formational names. Some of these results were later published. One of them is Rod & Maync's (1954) paper 'Revision of Lower Cretaceous stratigraphy of Venezuela'. In the part on eastern Venezuela the authors proposed a number of new lithologic and stratigraphic units and gave lists of foraminifera and ammonites occurring in the different formations. They also proposed a zonal scheme based on foraminifera which will be discussed in more detail below.

Of the published larger scale investigations only in one (Guillaume, Bolli & Beckmann, 1972) was the occurrence and distribution of benthic and planktic foraminifera followed in detail through all lithologic units investigated (Fig. 66) and tied into a predominantly planktic foraminiferal zonal scheme (Fig. 68). This publication is a shortened and revised version of Guillaume's (1961) oil company report. Before considering the foraminiferal results of this publication and report, a brief overview on the Early Cretaceous formations of eastern Venezuela and their faunal content is given below.

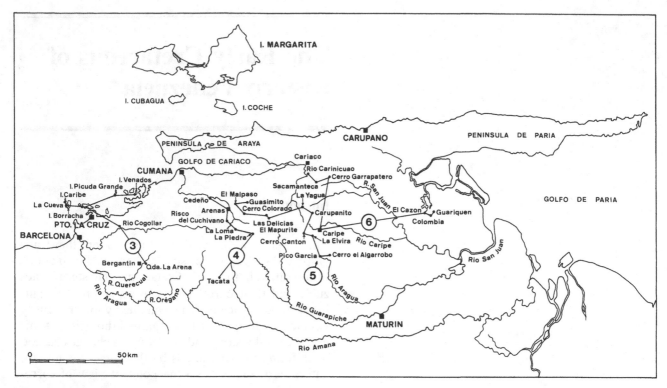

Fig. 66. Transects 3–6 through the Early Cretaceous of the Serrania del Interior, eastern Venezuela (from Guillaume, Bolli & Beckmann 1972). 3 Isla La Borracha to Quebrada La Arena (western part of the Serrania del Interior), 4 Cedeño to Tacata (central part of the Serrania del Interior, west of the San Francisco fault), 5 Guasimito to Cerro del Algarrobo (central part of the Serrania del Interior, east of the San Francisco fault), 6 Rio Carinicuao to Guariquen (eastern part of the Serrania del Interior).

The Early Cretaceous formations

The formations mainly comprise sandstones, limestones and shales (Fig. 67). The latter two as a rule contain rich and diversified faunas characterizing the different environments in which the sediments had formed. Many of the taxa are predominantly facies indicators rather than of stratigraphic significance, particularly those occurring in limestones. Others, like ammonites and certain foraminifera, provide valuable stratigraphic index forms.

The Early Cretaceous lithostratigraphic units comprise the Sucre Group, a term proposed by Hedberg (1950). Its lowermost unit is the Barremian to Lower Aptian Barranquin Formation (Liddle, 1928). Von der Osten (1954) distinguished within the Barranquin Formation from bottom to top the four members: Venados, Morro Blanco, Picuda and Taguarumo. The Barranquin Formation is followed by the Upper Aptian El Cantil Formation (Liddle, 1928), the Albian Chimana Formation (Hedberg & Pyre 1944)

which in turn is overlain by the Querecual Formation of Cenomanian and younger age (Hedberg 1937*a*).

Facies equivalents of the El Cantil Formation are for its lower part the Garcia Member of Rod & Maync (1954), proposed for formation rank by Falcon (1989*b*), and for its higher part the Borracha Formation of Rod & Maync (1954). The Valle Grande Formation of Rod & Maync (1954) is a shaly time equivalent of the El Cantil and Chimana formations.

The Barranquin Formation consists almost exclusively of sandstones with very subordinate limestone and shale layers. Its lower part is regarded as non-marine. The marine higher parts contain long-ranging bivalves, gastropods and corals. Foraminifera are represented by such species as *Pseudocyclammina hedbergi* and *Choffatella decipiens*.

The El Cantil Formation is characterized by alternations of limestones, sandstones and shales. The partial time equivalent of the El Cantil Formation, the Borracha Formation, with its type section on the island of that same name, forms there a thick biostromal limestone containing a shale break in its lower part.

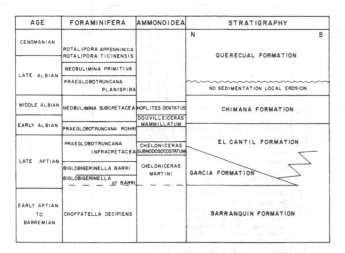

AGE	FORAMINIFERA	AMMONOIDEA	STRATIGRAPHY
CENOMANIAN	ROTALIPORA APPENNINICA ROTALIPORA TICINENSIS		QUERECUAL FORMATION
LATE ALBIAN	NEOBULIMINA PRIMITIVA		
	PRAEGLOBOTRUNCANA PLANISPIRA		NO SEDIMENTATION LOCAL EROSION
MIDDLE ALBIAN	NEOBULIMINA SUBCRETACEA	HOPLITES DENTATUS	CHIMANA FORMATION
EARLY ALBIAN	PRAEGLOBOTRUNCANA ROHRI	DOUVILLEICERAS MAMMILLATUM	
LATE APTIAN	PRAEGLOBOTRUNCANA INFRACRETACEA	CHELONICERAS SUBNODOSOCOSTATUM	EL CANTIL FORMATION
	BIGLOBIGERINELLA BARRI	CHELONICERAS MARTINI	GARCIA FORMATION
	BIGLOBIGERINELLA — — — _cf_ BARRI		
EARLY APTIAN TO BARREMIAN	CHOFFATELLA DECIPIENS		BARRANQUIN FORMATION

Fig. 67. Correlation of Barremian to Cenomanian zonal schemes based on planktic and benthic foraminifera and ammonites with formations in the Serrania del Interior. Based on Guillaume, Bolli & Beckmann (1972), taken from O. Renz (1982).

The limestones of the El Cantil and Borracha formations contain rich faunas of rudists, oysters and Trigonias. *Orbitolina concava texana* is usually present in the middle part of the section. Also reported are *Pseudocyclammina hedbergi* and *Dictyoconus walnutensis* (Maync 1955a). The age of the El Cantil/ Borracha formations comprises the interval Late Aptian to Early Albian, an age inferred from the underlying Upper Aptian Garcia Formation (*Cheloniceras martini* Zone) and the overlying Chimana Formation containing Early and Middle Albian ammonites.

The Chimana Formation overlying the El Cantil Formation consists of shales with subordinate fossiliferous limestones, sandstones and glauconitic beds. The formation yields rich ammonite faunas falling into the Early Albian *Douvilleiceras mammillatum* and the basal Middle Albian *Hoplites dentatus* zones. Benthic and planktic foraminifera are also present. Maync (1955b) reported *Coskinolina sunnilandensis* from the Guacharo Member of the Albian Chimana Formation. The *Neobulimina subcretacea* Zone of Guillaume, Bolli & Beckmann (1972) correlates in part with the *Hoplites dentatus* Zone.

The Garcia Formation consists of a rich fossiliferous shale near the base of the El Cantil Formation. It is correlated by Rod & Maync (1954) with the middle part of the Borracha Formation, while Salvador & Rosales (1960) relate it to the lower part of the El Cantil Formation. The Garcia Formation contains *Trigonia, Exogyra*, ammonites of well- established Late Aptian age (lower part of the *Cheloniceras martini* Zone). The foraminiferal fauna consists of *Choffa-*

tella, Pseudocyclammina and smaller benthic and planktic foraminifera indicative for the *Planomalina maridalensis* Zone.

The Valle Grande Formation, a lateral time equivalent of the El Cantil and Chimana formations, consists predominatly of shales and marls with interbeds of glauconitic limestones and glauconitic sandstones towards the top. The formation contains in its lower part abundant ammonites placed in the upper part of the *Cheloniceras martini* Zone. The foraminifera point to a *Planomalina maridalensis* Zone age of the ammonite bearing layers. They are followed by beds attributed to the *Hedbergella infracretacea* Zone. The upper part of the Valle Grande Formation is placed in the *Hedbergella rohri* Zone and the *Neobulimina subcretacea* Zone of Guillaume, Bolli & Beckmann (1972). The age of the whole Valle Grande Formation would thus embrace the interval Late Aptian to Middle Albian. The Cenomanian Querecual Formation consists of dark calcareous shales comparable to the La Luna Formation of western Venezuela and the Gautier Formation of Trinidad.

Remarks on the occurrence and distribution of benthic and planktic foraminifera in the Early Cretaceous of eastern Venezuela and Trinidad, and the zones based on them

A gradual increase of planktic foraminifera compared to the benthic forms takes place in the Early Cretaceous of eastern Venezuela from the Barremian to Lower Aptian Barranquin Formation to the Cenomanian Querecual Formation. As a rule they are absent or extremely scarce near the base of the Barranquin Formation and present but often not diagnostic higher in the Aptian. In the Early to Middle Albian, planktic foraminifera may in some sections become predominant and provide good markers. They are almost exclusively present in the Upper Albian to Cenomanian Querecual Formation in the sections studied and are apparently equivalent to the rich fauna found in the Gautier Formation of Trinidad. Amongst the benthic foraminifera a few species belonging to the genus *Neobulimina* make a short but distinct appearance in the Middle to Late Albian part of a number of sections of Guillaume, Bolli & Beckmann (1972). According to newer investigations, representatives of this genus also continue through the Cenomanian part of the Querecual Formation, and possibly still higher (personal communication Dr. M. Furrer).

BARRE-MIAN	APTIAN		APTIAN-ALBIAN	ALBIAN			STAGE	ZONE / ZONULE
	EARLY	LATE		EARLY	MIDDLE	LATE		
Choffatella decipiens (Zonule)	Biglobigerinella cf. barri (Zonule)	Biglobigerinella barri	Praeglobotruncana infracretacea	Praeglobotruncana rohri	Neobulimina subcretacea	Praeglobotruncana planispira	Neobulimina primitiva	

Distribution in Eastern Venezuela of Early Cretaceous Barremian to Albian benthic foraminifera after Guillaume, Bolli & Beckmann 1972

1	Coscinolina sp.
2	Cuneolina sp.
3	Conorbina cf. hofkeri (Bartenstein & Brand)
4	Palaeogaudryina cf. magharaensis Said & Barakat
5	Globorotalites sp.
6	Lenticulina cf. macrodisca (Reuss)
7	Turrispirillina cf. subconica Tappan
8	Pseudocyclammina sp.
9	Lenticulina tricarinella (Reuss)
10	Choffatella descipiens Schlumberger
11	Lenticulina nodosa (Reuss)
12	Dentalina linearis (Roemer)
13	Epistomina hechti Bartenstein, Bettenstaedt & Bolli
14	Haplophragmoides cf. rota Nauss
15	Haplophragmoides cf. laevigatus (Lozo)
16	Gaudryina tailleuri (Tappan)
17	Orbitolina concava texana (Roemer)
18	Ammobaculites cf. gratus Cushman & Applin
19	Dentalina guttifera d'Orbigny
20	Discorbis minima Vieaux
21	Epistomina ornata (Roemer)
22	Trochammina depressa Lozo
23	Haplophragmoides concavus (Chapman)
24	Miliammina cf. manitobensis Wickenden
25	Miliammina awunensis Tappan
26	Nodosaria sceptrum Reuss
27	Patellina subcretacea Cushman & Alexander
28	Rectoglandulina mutabilis (Reuss)
29	Saracenaria grandstandensis Tappan
30	Tristix cf. acutangula (Reuss)
31	Vaginulinopsis cf. incurvata (Reuss)
32	Lenticulina cf. turgidula (Reuss)
33	Lingulina nodosaria Reuss
34	Globulina prisca Reuss
35	Gaudryina canadensis Cushman
36	Lenticulina praegaultina Bartenstein, Bettenstaedt & Bolli
37	Nodosaria paupercula Reuss
38	Ammodiscus cf. gaultinus Berthelin
39	Marssonella sp. aff. oxycona (Reuss)
40	Anomalina cf. torcerensis Albritton
41	Dentalina communis d'Orbigny
42	Pyrulina exserta (Berthelin)
43	Lenticulina muensteri (Roemer)
44	Vaginulina recta Reuss
45	Ammobaculites cf. goodlandensis Cushman & Alexander
46	Gaudryina faujasi (Reuss)
47	Glomospira cf. watersi Loeblich
48	Vaginulina cf. arguta Reuss
49	Nodosaria doliiformis Eichenberg
50	Praebulimina nannina Tappan
51	Triplasia emslandensis acuta Bartenstein & Brand
52	Marginulina linearis Reuss
53	Dentalina guttifera d'Orbigny
54	Ramulina laevis Jones
55	Gavelinella strictata Tappan
56	Dentalina nana Reuss
57	Reophax pilulifera H. B. Brady
58	Globorotalites bartensteini s.l. Bettenstaedt
59	Valvulineria gracillima Ten Dam
60	Dentalina distincta Reuss
61	Marginulinopsis collinsi Mellow & Wall
62	Reophax cf. scorpiurus Montfort
63	Valvulineria loetterlei (Tappan)
64	Lenticulina saxocretacea Bartenstein
65	Proteonina ampullacea (H. B. Brady)
66	Dorothia cf. conula (Reuss)
67	Verneulina cretosa Cushman
68	Vaginulinopsis grata (Reuss)
69	Marginulina cf. sulcifera (Reuss)
70	Vaginulina cf. matutina (d'Orbigny)
71	Marssonella subtrochus Bartenstein
72	Vaginulinopsis calliopsis (Reuss)
73	Epistomina cretosa Ten Dam
74	Lenticulina cf. jonesi Sandidge
75	Dentalinopsis cf. nodosaria Tappan
76	Rectoglandulina cf. manifesta (Reuss)
77	Gaudryina cf. foeda (Reuss)
78	Reophax cf. pepperensis Loeblich
79	Gaudryina nanukushensis Tappan
80	Haplophragmoides globosus Lozo
81	Saracenaria cf. frankei Ten Dam
82	Marginulinopsis cf. cephalotes (Reuss)
83	Pleurostomella cf. subnodosa Reuss
84	Dentalina cf. lorneiana d'Orbigny
85	Neobulimina subcretacea (Cushman)
86	Neobulimina wyomingensis (Fox)
87	Gavelinella cf. intermedia (Berthelin)
88	Neobulimina primitiva (Cushman)
89	Gaudryinella cf. sherlocki Bettenstaedt
90	Pyrulina cylindroides (Roemer)
91	Vaginulinopsis complanata (Reuss)
92	Ammobaculites goodlandensis Cushman & Alexander

(a)

Ammobaculites cf. goodlandensis Cushman & Alexander	45
Ammobaculites cf. gratus Cushman & Applin	18
Ammobaculites goodlandensis Cushman & Alexander	92
Ammodiscus cf. gaultinus Berthelin	38
Anomalina cf. torcerensis Albritton	40
Choffatella descipiens Schlumberger	10
Conorbina cf. hofkeri (Bartenstein & Brand)	3
Coscinolina sp.	1
Cuneolina sp.	2
Dentalina cf. lorneiana d'Orbigny	84
Dentalina communis d'Orbigny	41
Dentalina distincta Reuss	60
Dentalina guttifera d'Orbigny	19
Dentalina guttifera d'Orbigny	53
Dentalina linearis (Roemer)	12
Dentalina nana Reuss	56
Dentalinopsis cf. nodosaria Tappan	75
Discorbis minima Vieaux	20
Dorothia cf. conula (Reuss)	66
Epistomina cretosa Ten Dam	73
Epistomina hechti Bartenstein, Bettenstaedt & Bolli	13
Epistomina ornata (Roemer)	21
Gaudryina canadensis Cushman	35
Gaudryina cf. foeda (Reuss)	77
Gaudryina cf. nanukushensis Tappan	79
Gaudryina faujasi (Reuss)	46
Gaudryina tailleuri (Tappan)	16
Gaudryinella cf. sherlocki Bettenstaedt	89
Gavelinella cf. intermedia (Berthelin)	87
Gavelinella strictata Tappan	55
Globorotalites bartensteini s.l. Bettenstaedt	58
Globorotalites sp.	5
Globulina prisca Reuss	34
Glomospira cf. watersi Loeblich	47
Haplophragmoides cf. laevigatus (Lozo)	15
Haplophragmoides cf. rota Nauss	14
Haplophragmoides concavus (Chapman)	23
Haplophragmoides globosus Lozo	80
Lenticulina cf. jonesi Sandidge	74
Lenticulina cf. macrodisca (Reuss)	6
Lenticulina cf. turgidula (Reuss)	32
Lenticulina muensteri (Roemer)	43
Lenticulina nodosa (Reuss)	11
Lenticulina praegaultina Bartenstein, Bettenstaedt & Bolli	36
Lenticulina saxocretacea Bartenstein	64
Lenticulina tricarinella (Reuss)	9

Lingulina nodosaria Reuss	33
Marginulina cf. sulcifera (Reuss)	69
Marginulina linearis Reuss	52
Marginulinopsis cf. cephalotes (Reuss)	82
Marginulinopsis collinsi Mellow & Wall	61
Marssonella sp. aff. oxycona (Reuss)	39
Marssonella subtrochus Bartenstein	71
Miliammina awunensis Tappan	25
Miliammina cf. manitobensis Wickenden	24
Neobulimina primitiva (Cushman)	88
Neobulimina subcretacea (Cushman)	85
Neobulimina wyomingensis (Fox)	86
Nodosaria doliiformis Eichenberg	49
Nodosaria paupercula Reuss	37
Nodosaria sceptrum Reuss	26
Orbitolina concava texana (Roemer)	17
Palaeogaudryina cf. magharaensis Said & Barakat	4
Patellina subcretacea Cushman & Alexander	27
Pleurostomella cf. subnodosa Reuss	83
Praebulimina nannina Tappan	50
Proteonina ampullacea (H. B. Brady)	65
Pseudocyclammina sp.	8
Pyrulina cylindroides (Roemer)	90
Pyrulina exserta (Berthelin)	42
Ramulina laevis Jones	54
Rectoglandulina cf. manifesta (Reuss)	76
Rectoglandulina mutabilis (Reuss)	28
Reophax cf. pepperensis Loeblich	78
Reophax cf. scorpiurus Montfort	62
Reophax pilulifera H. B. Brady	57
Saracenaria cf. frankei Ten Dam	81
Saracenaria grandstandensis Tappan	29
Triplasia emslandensis acuta Bartenstein & Brand	51
Tristix cf. acutangula (Reuss)	30
Trochammina depressa Lozo	22
Turrispirillina cf. subconica Tappan	7
Vaginulina cf. arguta Reuss	48
Vaginulina cf. matutina (d'Orbigny)	70
Vaginulina recta Reuss	44
Vaginulinopsis calliopsis (Reuss)	72
Vaginulinopsis cf. incurvata (Reuss)	31
Vaginulinopsis complanata (Reuss)	91
Vaginulinopsis grata (Reuss)	68
Valvulineria gracillima Ten Dam	59
Valvulineria loetterlei (Tappan)	63
Verneuilina cretosa Cushman	67

(b)

Fig. 68a,b. Distribution in eastern Venezuela of Early Cretaceous Barremian to Albian benthic foraminifera.

Stratigraphically significant Early Cretaceous planktic foraminifera were previously described from Trinidad (Bolli, 1959). Continuous Early Cretaceous sections with diagnostic faunas are absent there. The stratigraphic position in Trinidad of the rare occurrences of rich and diagnostic faunas had therefore to be established based on often widely separated individual outcrops that occur within the normally poorly fossiliferous shales of the thick Toco and Cuche formations.

Based on benthic and planktic foraminifera the

Barremian to Early Albian of Trinidad was zoned by Bolli (1959) from top to bottom into the following five zones:

Praeglobotruncana rohri (changed to *Hedbergella rohri*)

Biglobigerinella barri (changed to *Planomalina maridalensis*)

Leupoldina protuberans

Lenticulina ouachensis ouachensis

Lenticulina barri

The situation within the Early Cretaceous sections in eastern Venezuela is, in contrast to Trinidad, often well exposed over considerable stratigraphic intervals, allowing their zonation based on both foraminifera and ammonites which occur together in some of the sections.

Based on the distribution of planktic and benthic species in these sections Guillaume, Bolli & Beckmann (1972) introduced seven foraminiferal zones and two zonules for the interval Barremian to Early Cenomanian:

Rotalipora ticinensis/appenninica

Neobulimina primitiva

Praeglobotruncana planispira (changed to *Hedbergella planispira*)

Neobulimina subcretacea

Praeglobotruncana rohri (changed to *Hedbergella rohri*)

Praeglobotruncana infracretacea (changed to *Hedbergella infracretacea.*)

Biglobigerinella barri (changed to *Planomalina maridalensis*)

Biglobigerinella cf. barri (Zonule)

Choffatella decipiens (Zonule)

Two of the zones, the *Planomalina maridalensis* Zone and the *Hedbergella rohri* Zone, were adopted from the Trinidad Early Cretaceous zonal scheme. The others, based on local occurrences and distributions, were introduced as new. Whenever possible, zones were based on planktic species. Three, however, had to be established on benthic forms, one of them with *Choffatella decipiens* as a zonule and two as zones with representatives of the genus *Neobulimina* which is characteristically represented by several short ranging species within the Albian.

The zones and zonules are discussed in Guillaume, Bolli & Beckmann (1972) where definitions, type localities and type samples are given for the newly proposed zonal units. Type localities are shown on maps and sections. The stratigraphic sequence of the foraminiferal zonal units shows particularly well in the Isla Chimana Grande section, with a zonal sequence from the *Neobulimina subcretacea* to the *Rotalipora ticinensis/appenninica* Zone, and in the Rio Carinicuao section, situated in the eastern part of the Serrania del Interior. Here, a zonal sequence from the *Planomalina maridalensis* Zone to the *Neobulimina primitiva* Zone is exposed.

Guillaume, Bolli & Beckmann (1972) list the distribution of 23 planktic species within their zonal frame-work, as well as 101 benthic taxa. The ranges of 92 of these benthic species, tied into the established zonal scheme, are here reproduced on Fig. 68. The planktic foraminifera were determined largely on taxa previously identified in Trinidad. Beckmann named the benthic forms, partly based on collections available from Trinidad, partly on illustrations from publications. For several taxa only generic names could be given at the time.

Many of the typical Barremian foraminiferal species that occur in the Toco and Cuche formations of Trinidad, such as *Lenticulina barri, L. ouachensis* and its subspecies could not be found in the sections studied in the Serrania del Interior of eastern Venezeula. Their absence is explained by differing environmental conditions that prevailed during this time in the two areas.

However, further to the west in eastern Venezuela, between Barcelona and Altagracia de Orituco, the conditions in the Barremian sea must have become deeper marine again and therefore comparable to those prevailing at the time of the Cuche Formation in Trinidad. This is indicated by a fauna quoted in Guillaume, Bolli & Beckmann (1972) from a sample collected near El Horno, Quebrada Aguadita, Rio Patatal, Estado Miranda, near the boundary with Estado Guarico. The fauna is of typical Barremian aspect such as known from the Cuche Formation. Besides numerous *Epistomina* specimens it contains the characteristic *Lenticulina ouachensis ouachensis* and *L. ouachensis multicella*, thus falling into the *Lenticulina ouachensis ouachensis* Zone of Trinidad.

Further to the above Barremian outcrops in the western part of the Maturin Basin, Cretaceous sediments also occur as blocks in the Paleocene Guarico Flysch. Their ages range from Late Jurassic to Hauterivian. Furrer (1972) published on the fossil tintinnids found in these reworked blocks. In reworked blocks near Altagracia del Orituco occur also representatives of the genus *Protopeneroplis* (personal communication Dr. M. Furrer), and in Valanginian blocks other foraminifera on which Beck & Furrer (1977) reported.

Guillaume, Bolli & Beckmann's (1972) paper was a first attempt to use isolated smaller benthic and planktic foraminifera collected from shale and marl intervals for age identification, zonation and correlations in the Early Cretaceous of eastern Venezuela. This paper represents a condensed version of the oil company report by H. A. Guillaume (1961) entitled 'Lower Cretaceous stratigraphy of the Serrania del Interior'. His report contains detailed descriptions of the formations including their fossil contents and illustrations of 59 sections with detailed foraminiferal distribution charts for 15 of them.

The investigations of the isolated foraminifera for Guillaume's report, based on some 500 shale and marl samples from 24 sections collected between the island La Borracha to the West and Caño Guariquen to the East were carried out by H. M. Bolli. Foraminiferal range charts from 15 of these sections are included in Guillaume's report. Combined results were presented on two range charts, one showing the distribution of the planktic, the other of the benthic species in relation to the zonal scheme proposed. These two charts and the zonal definitions were published in Guillaume, Bolli & Beckmann (1972).

As Guillaume's report was the first and so far only major attempt to carry out a detailed study of smaller benthic and planktic foraminifera and the correlation of their ranges with a zonal scheme in the Early Cretaceous of eastern Venezuela, it would be regrettable if these data should remain hidden in an oil company report, even though it was written some 30 years ago.

For these reasons we have placed a copy of H. A. Guillaume's unpublished report in the Natural History Museum in Basel. Deposited there is also a representative collection of the planktic and benthic foraminifera plotted in Guillaume, Bolli & Beckmann (1972).

More recently Odehnal & Falcon (1989) published on a rich and well-preserved planktic and benthic foraminiferal fauna that occurs together with ammonites of the genera *Gargasiceras*, *Dufrenoiya* and *Cheloniceras*. The fauna occurs in the lower part of the Garcia Member which was proposed by Falcon (1989a) to be given formation rank. It is exposed in a section along the Caripe–Santa Maria road in the Serrania del Interior.

Based on the planktic foraminiferal association, these authors place the fauna into the Early Aptian *Schackoina cabri* Zone that is age equivalent to the *Leupoldina protuberans* Zone in the upper part of the Cuche Formation of Trinidad. Higher in the exposed section the *Schackoina* species become rapidly replaced by species of the genus *Globigerinelloides*, a change thought by Odehnal & Falcon to have been caused by rapidly changing environmental conditions.

The abundant benthic assemblage in the sample with *Schackoina* is dominated by the genera *Lenticulina*, *Epistomina*, *Buliminella* and *Dentalina* of which some species are listed on a distribution chart (their fig. 3) together with the planktic species.

Odehnal & Falcon mainly investigated the planktic foraminifera and only more tentatively the benthic forms. From the results presented it can be concluded that the section would be suitable for close investigation of the benthic species and their comparison with faunas from other age equivalent sections in the Serrania del Interior, as well as with the benthic fauna of the *Leupoldina protuberans* Zone of Trinidad, which was published in detail by Bartenstein & Bolli (1977). The need for more work is highlighted by Odehnal & Falcon who point out that more detailed sampling of the section will be required to delineate more precisely the boundaries of the *Schackoina cabri* Zone and prove the possible presence of other, younger biozones.

Oligocene and Miocene of eastern Venezuela

Review of foraminiferal investigations

There are two distinct areas in eastern Venezuela where Oligocene, Miocene and younger sediments with benthic and planktic foraminifera occur:

(a) The coastal area from Cabo Blanco in the west to the peninsula of Araya, the islands of Cubagua and Margarita in the east. No Oligocene and almost no Early Miocene occurs in this area. As virtually all sediments are of Late Miocene, Pliocene and younger age they are not considered in this compilation.

(b) The southern foothills of the Serrania del Interior and, mainly in the subsurface, extending to the south into the Llanos (Fig. 69). In this area the formations carrying benthic and planktic foraminifera are of Oligocene and Miocene age.

Planktic and benthic foraminifera have been studied from both areas and the planktic foraminiferal zonal scheme applied widely. Bermudez & Stainforth (1975) provided a detailed chronological overview on the introduction and application of the Tertiary planktic foraminiferal zonal scheme in Venezuela which was largely adopted from that previously developed in neighbouring Trinidad. Bermudez & Farias (1977) in a complementary paper, presented in more detail the individual Cenozoic to Recent planktic foraminiferal zones, with illustrations of the zonal markers.

In contrast to Falcon and Trinidad little published information is available in these two areas on the stratigraphic correlation of benthic foraminifera with the planktic foraminiferal zonal scheme, the prime objective of this compilation. It is therefore only possible to summarize the available data for the above quoted two areas in eastern Venezuela.

As far as the coastal area is concerned, reference is made to the three publications on the Middle Miocene to Recent benthic and planktic foraminifera by Bermudez (1966) Bermudez & Fuenmayor (1968) and Bermudez & Bolli (1969). Key sections where both benthic and planktic foraminifera have been

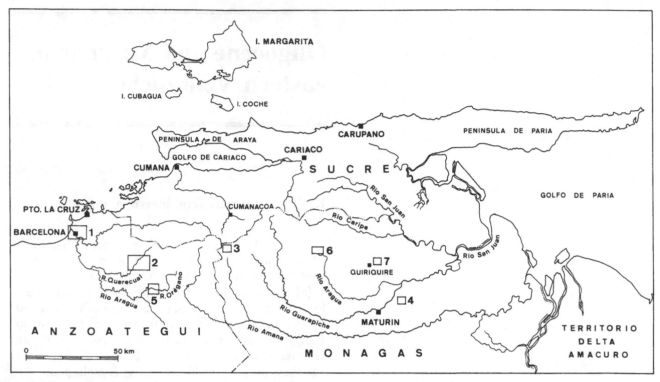

Fig. 69. Map showing type localities of Tertiary formations and areas studied in the southern foothills of the Serrania del Interior, eastern Venezuela. 1 Naricual, Tinajitas, Stanley (1960), Galea-Alvarez (1985); 2 Carapita, Chapapotal, Los Jabillos, Caratas, Vidoño, Hedberg (1937*b*); 3 Areo; 4 La Pica (subsurface); 5 Franklin (1944); 6 Lamb (1964*a*); 7 Quiriquire oil field. See Fig. 3 in this volume and Lexico Estratigrafico de Venezuela (1970) for age and descriptions of the formations.

studied and their distribution charted are the closely cored wells Cubagua-1 and Cubagua-2 on the island of Cubagua. Stratigraphically their Late Miocene *Globorotalia acostaensis* to the Middle Pliocene *Globorotalia miocenica* Zone are thus falling outside the scope of this compilation. For the record and possible future investigations on these younger sediments it may be said that for Cubagua-1 the stratigraphic distribution of 166 benthic and 30 planktic taxa, for Cubagua-2 152 benthic and 26 planktic taxa are included on the published range charts.

According to Bermudez (1966) the only Miocene formation in the coastal area remains the Carenero Formation, outcropping at Carenero on the road between Higuerote and Chirimena, Estado Miranda. Bermudez places it into the late Middle Miocene *Globorotalia menardii* Zone. Based on *Globorotalia acostaensis* and *Globorotalia juanai*, index forms for the Late Miocene *Globorotalia acostaensis* Zone, both appearing on Bermudez' faunal list, the Carenero Formation at its type locality has rather to be placed into the basal Late Miocene *Globorotalia acostaensis* Zone. The presence of *Globigerinoides ruber* and *Glo-*

bigerinoides ruber altiapertura on this faunal list is also disturbing as they are not expected to reappear in numbers prior to the upper part of the *Globorotalia dutertrei* Zone.

Of the Oligocene-Miocene sediments outcropping in the southern foothills of the Serrania del Interior it is in particular the Carapita Formation that carries a rich benthic and planktic foraminiferal fauna, closely comparable with that of the Agua Salada Formation in eastern Falcon and the Cipero/Ste. Croix/Brasso formations in Trinidad. The Carapita Formation has a wide geographic extension in the subsurface, underlying the northern parts of the states of Anzoategui and Monagas and to the east reaching to the oil field of Pedernales at the northwestern end of the Orinoco delta, facing the Gulf of Paria between eastern Venezuela and Trinidad, and extending into the Gulf of Paria, there linking up with the equivalent Trinidad formations mentioned above. Whereas both the benthic and planktic foraminifera of Falcon and Trinidad are well known from surface sections and have been exensively published, no such detailed investigations are to date available for the Carapita Formation.

Comparison with data available from surface sections in Falcon and Trinidad is hampered in eastern Venezuela by the fragmentary and scattered outcrops of the Carapita Formation, such as at its type locality in the Quebrada Carapita (Hedberg 1937*a*), the Rio Oregano (Franklin 1944) or the Rio Aragua section (Lamb 1964*a*). Despite these drawbacks Hedberg and Franklin described and illustrated the characteristic Carapita benthic species. Hedberg's paper is based on the fauna from the type locality in the Quebrada Carapita (Fig. 69), that by Franklin on surface samples from the vicinity of the Rio Oregano (Fig. 69). As both publications appeared prior to the development of a planktic foraminiferal zonal scheme, these authors made no particular attempt to use these forms.

However, following the introduction of the planktic foraminiferal zonal scheme in Trinidad, several authors, Sulek (1961), Lamb (1964*a*) and Lamb & Sulek (1968), reported a zonal subdivision for parts of the Carapita Formation, mainly based on subsurface sections. These authors recognized two zones between the Early Oligocene *Globigerina ampliapertura* and the Middle Miocene *Globorotalia menardii* Zone. From the *Globigerina ampliapertura* Zone in shales of the diapiric Pedernales anticline, Bermudez (1962) listed over a hundred foraminiferal species. An apparently promising section in eastern Venezuela for the study of the distribution of Oligocene to Early Miocene planktic and benthic foraminifera is that published by Lamb (1964*a*) from the Rio Aragua (Fig. 69) some 50 kilometres northwest of Maturin. Here the Carapita Formation is reported as consisting of only a thin sliver of *Catapsydrax dissimilis* Zone age. Based on this section and in combination with subsurface information from the Quiriquire oil field (Fig. 69) where he studied the distribution of planktic foraminifera and benthic index forms, Lamb (1964*a*) concluded that the Naricual Formation which underlies the Carapita Formation could fall into the *Globorotalia kugleri* Zone, though the zonal marker could not be identified in the Rio Aragua section.

The upper part of the Areo Formation, itself underlying the Naricual Formation, following Lamb, would be of *Globigerina ciperoensis ciperoensis* Zone age. According to Dr. M. Furrer (personal communication) the Areo Formation is more likely to be of *Globorotalia opima opima* Zone age because the oldest part of the overlying Carapita Formation is of *Globigerina ciperoensis ciperoensis* Zone age. The lower part of the Areo Formation and the Los Jabillos Formation underlying the Areo Formation are regarded

by Lamb as being of *Globorotalia opima opima* Zone age, while the *Globigerina ampliapertura* Zone falls entirely within the Los Jabillos sandstone.

In the Quiriquire area the *Globigerina ampliapertura* Zone extends according to Lamb down into the Caratas shales below the Los Jabillos Sandstone. Contrary to this view, Dr. M. Furrer (personal communication) is of the opinion that in the Quiriquire area the Caratas Formation does not go higher than the Middle Eocene *Orbulinoides beckmanni* Zone and thus does not extend into the *Globigerina ampliapertura* Zone. Furthermore, also from a personal communication by Dr. M. Furrer, the Tinajitas Formation, situated between the Los Jabillos Formation above and the Caratas Formation below, correlates with the uppermost Middle Eocene *Truncorotaloides rohri* Zone.

Lamb & Sulek (1965) introduced within the upper part of the Carapita Formation a Cachipo Member of turbiditic nature, analogous to the Herrera sands of Trinidad and of *Globorotalia fohsi fohsi* Zone age. Later, Lamb & Sulek (1968) changed the name Cachipo to Chapapotal. Carapita shales occurring below and above the Chapapotal Member are of *Globorotalia fohsi peripheroronda* and *Globorotalia fohsi lobata/robusta* Zone age respectively. The full time interval represented in the thickest development of the Carapita Formation measures between 15 000 and 20 000 feet and corresponds according to these authors to the interval *Globigerina ciperoensis ciperoensis* to *Globorotalia menardii* Zone, with the *Globigerinatella insueta* Zone represented most widely. According to Dr. M. Furrer (personal communication) *Globigerinatella insueta* is comparatively scarce in the Carapita Formation.

A rare occurrence of the Early Oligocene *Cassigerinella chipolensis/Pseudohastigerina micra* Zone has been reported by Galea-Alvarez (1985) from the Roblecito Formation in the Pedrera Quarry, Cerro de Pedrera section, about 60 kilometres east-southeast of Barcelona, Estado Anzoategui (Sample FAGA 473). Besides the zonal markers the fauna consists of a varied assemblage of planktic species, including *Globigerina ampliapertura*, *Catapsydrax dissimilis* and characteristic benthic species with several diagnostic *Uvigerina* species like *Uvigerina jacksonensis* and *U. mexicana*.

With only sparse information available to date concerning the distribution of the benthic foraminifera throughout the Carapita and the other formations quoted above, and concerning the correlation of the formations according to the planktic foraminiferal zonal scheme, it is not yet possible to treat the Oligocene-Miocene eastern Venezuelan faunas in the same way as was possible for those from Falcon and Trinidad.

Many of the benthic species described and illustrated by Hedberg (1937*a*) from the Carapita Formation also occur in Falcon and/or Trinidad. His descriptions and illustrations are therefore partly incorporated in the systematic part, and on the species illustrations in the Trinidad chapter.

Franklin (1944) illustrated amongst the few planktic species a *Globigerina concinna* Reuss (his pl. 48, fig. 5) which clearly corresponds to *Globigerina ciperoensis ciperoensis* Bolli. Another figured planktic specimen is identified as *Globorotalia spinulosa* Cushman, obviously reworked from the Eocene. If not also reworked, the *Globigerina ciperoensis ciperoensis* illustrated by Franklin points to a Middle to Late Oligocene age for at least part of his figured benthic fauna from the Rio Oregano. This would be close to the Early Oligocene age assigned by that author and correlated by him with the Lower Oligocene Alazan Formation of Mexico and the Finca Adelina Formation of Cuba. Some of the species illustrated by Franklin are also reproduced in the species illustrations of the Trinidad late Early Eocene to Middle Miocene chapter.

The Falcon Basin

Introduction

The Falcon Basin (Fig. 70) is one of the classical regions where both Oligocene and Miocene benthic and planktic foraminifera and their correlation with established planktic zonal schemes have been studied. The rich and well-preserved Miocene foraminiferal faunas, in particular those of the eastern part of the basin, have since the 1920s attracted the attention of micropaleontologists and stratigraphers. Their work was initiated and supported mainly by interests of the oil industry.

Following earlier investigations, mostly unpublished, it was in particular the monographic studies by Renz (1948) and Blow (1959) that provided the base for a correlation of the distribution of the benthic and planktic foraminifera in the continuous Miocene sections of southeastern Falcon.

The most recent investigations on foraminifera in Falcon are those published by Diaz de Gamero (1985a, b) from the northeastern part of the basin, an area to the North of that previously studied by Renz (1948) and Blow (1959).

The benthic and planktic foraminifera in the central part of the Falcon Basin, ranging in age from Middle Oligocene to Early Miocene were studied by Diaz de Gamero (1977). Here conditions are not nearly as favourable as in the northeastern part of the basin. Despite these drawbacks Diaz de Gamero was able to obtain results comparable to those of eastern Falcon, though on a reduced scale.

The above quoted investigations by Renz (1948), Blow, (1959) and Diaz de Gamero (1977a) complemented by unpublished observations by one of the present authors (HMB) form the base for this chapter.

For an overview of the development, facies distribution and stratigraphy of the Falcon Basin reference is made to Wheeler (1960, 1963). He describes the different facies developments during the marine transgression and gives an account of the associated formations, including lists of selected benthic foraminifera. While deeper water sediments prevail in the central

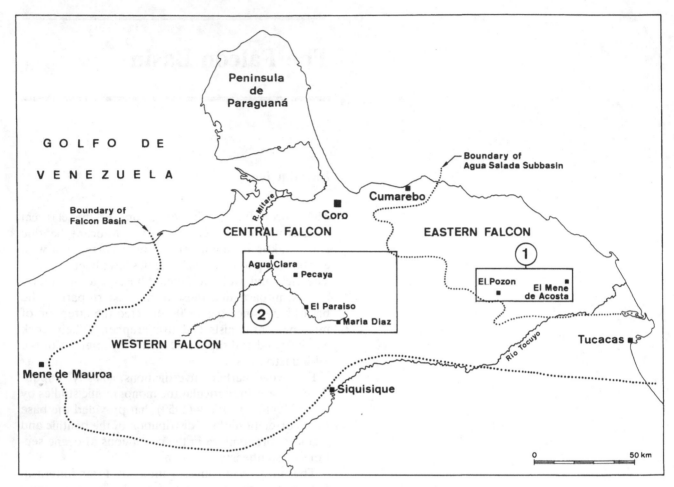

Fig. 70. Map of Falcon showing boundary of Falcon Basin and Agua Salada Sub-basin, and study areas: (1) in south- eastern Falcon of Renz (1948) and Blow (1959), (2) in central Falcon of Diaz de Gamero (1977*a*).

part of the basin, they grade into shallow water facies towards its southern margin. Wheeler provides a number of cross sections and maps displaying the north–south and east–west correlation between the individual formations, with reconstructed areal limits of the formations as well as facies and isopach maps for some of the stratigraphic units. Wheeler's 1960 paper is in Spanish, the 1963 one in English; the contents of both, including illustrations, are similar.

Changes to Wheeler's interpretations for some intervals were proposed by Diaz de Gamero (1977*b*) for central Falcon, and by Diaz de Gamero (1989) for the Early and Middle Miocene of northern Falcon.

Eastern Falcon

Diaz de Gamero (1985*a*) re-defined the term 'Agua Salada Group' to include the Guacharaca Formation.

She also introduces the term 'Agua Salada Formation' for northeastern Falcon, as an equivalent to the above re-defined Agua Salada Group without, however, the distinction of the various shale units, but maintaining apart the term 'El Salto Member', representing the clastic unit within the formation.

The Miocene San Lorenzo and Pozon formations in southeastern Falcon, the Agua Salada Group, have long been known for their rich, well-preserved and diversified foraminiferal fauna of predominantly smaller benthic but also planktic forms. Because of the exploration for oil in the area and the comparative lithologic monotony of the stratigraphic sequence that consists mainly of clays and marls, the foraminifera became an important tool for a biostratigraphic sub-division of the Agua Salada Group measuring some 1500 metres in thickness.

First unpublished reports on the geology and paleontology of the Agua Salada Group were issued as early as 1919 when W. F. Penny, a geologist from

Trinidad, began to show the usefulness of foraminifera for dating and correlating deposits of widely separated areas such as Trinidad and Falcon in Venezuela. In 1926 P. W. Jarvis, also of Trinidad, studied further Agua Salada faunas, correlating them with beds in the Brasso Formation of Trinidad. More systematic studies on smaller foraminifera throughout the Agua Salada region were initiated in 1927 by A. Senn whose work resulted in a first biostratigraphic subdivision and correlation of the lithologically uniform sediments. His subdivision into six zones with age assignments was published in Senn (1935, 1940) but without definitions of the zones (Fig. 71).

Cushman (1929*b*) was the first to describe and illustrate a foraminiferal fauna from the Agua Salada Group from a sample outcropping on a sea cliff near Aguide. It was followed by a publication by Gravell (1933) on Venezuelan Tertiary larger foraminifera that contained forms from Falcon. Hedberg (1937*a*) described the foraminiferal fauna of the Carapita Formation of eastern Venezuela and compared it with that of Senn in his Agua Salada Zone A4.

The first major publication on the Agua Salada Group was that by Cushman & Renz (1941) in which they described 42 new species and 12 new varieties, all benthic forms. These authors stated that in addition to the new taxa they described, the Agua Salada contains approximately another 135 species and varieties already known; however, these were not listed . They considered the sequence as being of Late Oligocene to Middle and questionably partly Late Miocene age. Based on the distribution of the foraminifera they subdivided the Agua Salada sequence into seven foraminiferal zones (see Fig. 71). These zones were, like those of Senn (1935), not defined whereas the ranges of the foraminifera described from within them were listed in the text.

Renz (1942) in his paper 'Stratigraphy of northern South America, Trinidad, and Barbados' listed numerous foraminifera from the Agua Salada Group which at the time he considered to be of Late Oligocene to Miocene age.

In 1948 Renz published his comprehensive study 'Stratigraphy and fauna of the Agua Salada Group, State of Falcon, Venezuela'. It contains a discussion of the geologic and stratigraphic setting and a historical review of the paleontologic and biostratigraphic studies previously published. Renz discusses and describes in considerable detail the lithostratigraphy of the Agua Salada Group with its formations San Lorenzo and Pozon, and their members. In the chapter on biostratigraphy he defines the newly introduced biostratigraphic terms. The proposed zonal sequence

which consists of seven zones and nine zonules supplements that published in Cushman & Renz (1941) where no zonal descriptions or definitions were given. With the exception of one, the *Globorotalia fohsi* Zone, all zones are based on benthic forms.

Each zone is dealt with in considerable detail, including descriptions of type locality, thickness, lithology, characters of the foraminiferal content such as key species, occurrences of the Agua Salada forms outside the area, ecology, and age of the zonal intervals. Pie-diagrams for each zonal unit illustrate the percentages of the foraminiferal families making up the total of the zonal assemblage. In addition to the zonal scheme, Renz also introduces the three stage names Acostian, Araguatian and Lucian as further subdivisions of the Agua Salada Group. However, later authors dealing with the Agua Salada Group, such as Blow (1959), did not make use of these stage names.

The ranges of 205 taxa are plotted by Renz (1948) on a range chart, his table 4, also showing the frequency of occurrence. They are based on some 270 surface samples collected through the type sections of the San Lorenzo and Pozon formations that form the Agua Salada Group. On his table 5, pp. 31–37, Renz presents in addition to the detailed chart on his table 4 a more condensed range chart of virtually the same taxa, here showing their presence and frequency in the individual zones. Further, in the species descriptions on pp. 113–179, Renz gives under 'Stratigraphic occurrence' presence and frequency again, here still more generalized for the stages Acostian, Araguatian and Lucian, and in addition selected occurrences of the individual species outside Falcon.

A comparison of the ranges on Renz' table 4 (detailed ranges by samples) and table 5 (distribution by zones) shows that in many instances the ranges shown on table 5 by zones are more extended than those on table 4. On the range charts of the 152 taxa dealt with by us in the present work (Fig. 73, 74) the ranges shown in full lines are those taken from Renz' detailed table 4, while dotted lines show the extended ranges as given in Renz' table 5.

Renz (pp. 76–111) also deals with the correlation of the Agua Salada faunas with age equivalent ones in the Caribbean region, tropical America and California. This is based on a selected number of taxa showing their joint distribution in the areas under discussion. A comparison of the Agua Salada fauna with the members of Trinidad's Brasso and Ste. Croix formations is dealt with in more detail, showing the occurrence and distribution of some 170 taxa, almost all benthics. They are shown for the Agua Salada

Venezuela

Fig. 71. Agua Salada Group, southeastern Falcon: proposed ages, stratigraphic units, zonal schemes based on benthic and planktic foraminifera and calcareous nannofossils. * Guacharaca Formation, Oligocene.

AGES

Senn 1940	Cushman & Renz 1941	Renz 1948	Blow 1959, this paper
Miocene: Upper / Middle / Lower	Miocene: Middle-Upper / Middle / Lower to Middle / lowermost	Miocene: Middle / Lower / lowermost / topmost	Miocene: Late / Middle / Early
Oligocene: Upper	Oligocene: uppermost / Upper	Oligocene: Upper / Mid.-Upp. / Middle	*

STRATIGRAPHIC UNITS (Renz 1948)

GROUP	STAGE	FORMATION	MEMBER
Agua Salada	Lucian	Pozon	Huso Clay
			Husito marly Clay
	Araguatian		
	Acostian		Menecito Clay
		San Lorenzo *	Poli-carpio
			El Salto

Senn 1935, 1940: A1, A1a, A2, A3, A 3-4, A4

BENTHIC FORAMINIFERA — ZONES

Cushman & Renz 1941:
- Rotalia becarii / Elphid. poeyanum
- Textularia panamensis
- Trochammina sp.
- Marginulina basispinosus / Robulus senni
- Cibicides kugleri
- Siphogenerina multicostata / Gaudryina thalmanni
- Marginulina wallacei

Renz 1948 (ZONES):
- Robulus senni
- Marginulinopsis basispinosus
- Valvulineria herricki
- Globorotalia fohsi
- Siphogenerina transversa
- Robulus wallacei
- "Uvigerinella" sparsicostata

Renz 1948 (ZONULE):
- Elph. poeyanum - Reuss. spinulosa
- Textularia panamensis
- V. sup. - T. cf. pac.
- Robulus nuttalli - Cibicides mantaensis
- Valvulineria venezuelana
- Plan. venezuelana - Saracenaria senni
- Uvigerina gallowayi basicordata
- Ammobaculites cf. strathearensis
- Bolivina alazanensis

PLANKTIC FORAMINIFERA — ZONES

Bolli 1957, emended Blow 1959	Bolli 1957, emended Bolli & Bermudez 1965, Blow & Banner 1966, Bolli 1966	Banner & Blow 1965 (N)
Globigerina bulloides	Globorotalia acostaensis	16
Sphaeroidinellopsis seminulina		15
Globorotalia menardii / Globigerina nepenthes	Globorotalia menardii	14
Gr. mayeri/Gg. nepenthes		13
Gr. mayeri/ Gr. lenguaensis	Globorotalia mayeri	12
Globorotalia fohsi robusta	Globorotalia fohsi robusta	11
Globorotalia fohsi lobata	Globorotalia fohsi lobata	10
Globorotalia fohsi fohsi	Globorotalia fohsi fohsi	9
Globorotalia fohsi barisanensis	Globorotalia fohsi peripheroronda	8
Globigerinatella insueta / Globigerinoides bisphericus	Praeorbulina glomerosa	7
Globigerinatella insueta / Globigerinoides trilobus	Globigerinatella insueta	6
Catapsydrax stainforthi	Catapsydrax stainforthi	5
Catapsydrax dissimilis	Catapsydrax dissimilis	

CALCAREOUS NANNOFOSSILS (Martini 1971)

NN	Nannofossil zone
11	Discoaster quinqueramus
10	Discoaster calcaris
9	Discoaster hamatus
8	Discoaster coalitus
7	Discoaster kugleri
6	Discoaster exilis
5	Sphenolithus heteromorphus
4	Helicosphaera ampliaperta
3	Sphenolithus belemnos
2	Discoaster druggii

zones and the age comparable Brasso and Ste. Croix stratigraphic units. These observations form a good base for eventual more detailed investigations on the occurrence and stratigraphic distribution of benthic taxa in the Brasso and Ste. Croix formations of Trinidad and other areas discussed by Renz.

Renz described and re-illustrated not only the taxa originally dealt with as new in Cushman & Renz (1941) but added another four new species and 180 taxa occurring in the Agua Salada but described earlier from other areas such as from Mexico, the Gulf Coast, California, Trinidad and eastern Venezuela. Of a total of 239 species described from the Agua Salada Group, 59 are new. Most of these species are illustrated in Renz (1948), all excellent pen drawings by Bramine Caudri. Of the 239 species only 8 were planktics.

Though the Agua Salada foraminiferal faunas are strongly dominated by benthic forms, the ratio of 231 benthic to only 8 planktic species in Renz (1948) does not reflect the true percentage and number of planktic species as recognized today. According to the pie-diagrams provided by Renz for each of his zonal units, planktics amount to between 1.5 and 15.4 percent with a general trend towards an increase from the base to the top of the sequence.

An unpublished preliminary investigation by one of the present authors (HMB) of the planktic foraminifera in the same samples as used by Renz revealed that almost all taxa recorded in Trinidad from the age equivalent stratigraphic intervals, including zonal markers, are also present in the Agua Salada Group. The detailed investigations in particular on the benthic forms by Renz on the Agua Salada foraminifera therefore provided an ideal base for a re-study of the associated planktic forms.

Such a study was eventually undertaken by Blow (1959) primarily based on a different set of samples but also making use of those utilized by Renz (1948). Based on some 700 samples collected by R. Muehlemann along the Pozon–El Mene road section (Blow's map 4) between Caiman and Buena Vista (Blow's map 3), ranging from the *Robulus wallacei* Zone in the North to the *Elphidium poeyanum-Reussella spinulosa* Zone in the South, Blow was able to establish a close correlation of the benthic zonal scheme of Renz with the planktic zonal scheme proposed by Bolli (1957c) for the Cipero and Lengua formations of Trinidad. For the correlation of the two faunal groups Blow not only identified and described 72 planktic species and subspecies but also investigated 174 benthic taxa, most of them corresponding to species in

Renz (1948) and plotted them against the benthic and planktic zonal schemes (chart 2 of Blow 1959).

While for ecologic reasons in Trinidad the highest zone based on planktic foraminifera is the *Globorotalia menardii* Zone, planktic foraminifera continue in the Agua Salada slightly higher stratigraphically. Blow (1959) therefore proposed above the *Globorotalia menardii* Zone the two Late Miocene zones *Sphaeroidinella seminulina* and *Globigerina bulloides*. They were combined into Zone N16 by Banner & Blow (1965) and regarded by Bolli & Bermudez (1965) as equivalent to their *Globorotalia acostaensis* Zone. Blow (1959) also proposed a subdivision of the *Globorotalia mayeri* Zone of Bolli (1957c) into a lower *Globorotalia mayeri/Globorotalia lenguaensis* and an upper *Globorotalia mayeri/Globigerina nepenthes* Zone. Furthermore, he subdivided the *Globigerinatella insueta* Zone of Bolli (1957c) into a lower *Globigerinatella insueta/Globigerinoides trilobus* and an upper *Globigerinatella insueta/Globigerinoides bisphaericus* Zone. The latter is equivalent to the *Praeorbulina glomerosa* Zone of Bolli (1966). Furthermore, the *Globorotalia fohsi barisanensis* Zone of Bolli (1957c) also applied by Blow (1959) was changed to *Globorotalia fohsi peripheroronda* Zone by Blow & Banner (1966).

From the 239 taxa dealt with by Renz (1948) 152 have been chosen for the present volume. They were selected on their stratigraphic, morphologic and ecologic significance and within some genera also for comparative reasons, to distinguish them from other, similar forms. Of several genera all species reported have been selected. They are the calcareous genera *Bolivina, Bulimina, Uvigerina, Rectuvigerina, Siphogenerina, Nodosaria, Saracenaria, Lenticulina* and all calcareous rotalid species in Renz (1948). Of the arenaceous groups all *Textularia* species were chosen and in addition a limited number of characteristic other forms. On Figs. 76–80 the genera and species are arranged in a way to facilitate easy comparison between morphologically similar forms. Figs. 73 and 74 show the stratigraphic ranges of these taxa

The dating of the sediments of the Agua Salada Group, including the placing of the Oligocene-Miocene boundary, based on smaller benthic and planktic foraminifera, has been differently interpreted by the various authors. As shown on Fig. 71, Senn (1940) placed the Oligocene-Miocene boundary between his zones A3–4 and A3, without offering an explanation. Cushman & Renz (1941) lowered it slightly, between their *Marginulina wallacei* Zone below and the *Siphogenerina multicostata/Gaudryina thalmanni* Zone above, or between Senn's zones A4 and A3–4. Following the criteria applied to the placing of the Oligocene-Miocene boundary in the Cipero and Brasso formations of Trinidad, Renz (1948) not only placed the whole of the San Lorenzo Formation into the Oligocene but also a considerable part of the Pozon Formation. His *Globorotalia fohsi* Zone became topmost Oligocene, his *Valvulineria herricki* Zone lowermost Miocene.

With the introduction in Trinidad of a zonal scheme based on planktic foraminifera by Bolli (1957c), the Oligocene-Miocene boundary within the Agua Salada Group was lowered again drastically, that is to the first appearance of the genus *Globigerinoides*, or in zonal terms between the *Globorotalia kugleri* Zone below and the *Catapsydrax dissimilis* Zone above. As a consequence Blow (1959) placed the whole Agua Salada Group, ranging from the *Catapsydrax dissimilis* Zone to the *Globigerina bulloides* Zone into the Miocene. Only the Guacharaca Formation underlying the San Lorenzo Formation is today regarded as Oligocene.

In eastern Falcon the San Lorenzo Formation of the Agua Salada Group is conformably underlain by the Guacharaca Formation which measures some 235 metres. In turn the Guacharaca Formation overlies the Late Eocene Cerro Mision Formation which has an assumed thickness of about 400 metres. According to Renz (1948) the boundary between the San Lorenzo Formation and the Guacharaca Formation does not coincide with the base of the lowest biostratigraphic unit, the *Uvigerinella sparsicostata* Zone. This zone thus contains, according to Renz, about 80 metres of sediments already attributable to the Guacharaca Formation. The *Catapsydrax dissimilis* Zone, part of the Guacharaca Formation, is Early Miocene.

The underlying parts of the Guacharaca Formation, between the base of the *Catapsydrax dissimilis* Zone and the Eocene Cerro Mision Formation may correspond to the Oligocene planktic foraminiferal zones. However, according to Renz no attempt at the time was made to subdivide these parts of the Guacharaca Formation. It therefore is not known whether they really represent the whole or only parts of the Oligocene planktic foraminiferal sequence. An evaluation of the distribution of the Oligocene smaller benthic foraminifera in these parts of the Guacharaca Formation is therefore not yet possible.

After Renz (1948), the following views on the Guacharaca Formation were expressed by Wheeler (1960): inclusion of the orbitoidal beds of the formation into the underlying Cerro Mision Formation and the rest into the Late Oligocene which in turn he correlated with the Pecaya Formation and the lower

orbitoidal member with the El Paraiso Formation, both of central Falcon. Later, Wheeler (1963) considered the Guacharaca Formation as representing the whole Oligocene and correlated it as a whole with the El Paraiso Formation of central Falcon.

A more recent study on Oligocene-Miocene foraminifera in northeastern Falcon was published by Diaz de Gamero (1985a, b). It follows those by Renz (1948) and Blow (1959) and is an investigation of the Oligocene to Pliocene in an area to the North of that dealt with by Renz and Blow. Her 1985b paper deals with the stratigraphy and sedimentology, the 1985a one with the micropaleontology of the area, in particular the foraminifera and their distribution. The two publications are in part based on a number of unpublished investigations by students of the Universidad Central de Venezuela.

Diaz de Gamero maintains both terms: an Agua graphic units as defined in the southeastern part of the sub-basin (Renz, Blow) cannot be recognized in the northeastern part. She therefore proposes a redefinition of the Agua Salada Group to include also the underlying Guacharaca Formation whose marine clays belong to the same sedimentological cycle. As in southeastern Falcon the Agua Salada Formation remains the main Oligocene-Miocene lithostratigraphic unit also in northeastern Falcon. The formation also includes the El Salto Member of arenaceous character, but not the partly age-equivalent Capadare Formation consisting of discontinuous limestone masses (Fig. 72).

Diaz de Gamero maintains both terms, an Agua Salada Group with the Guacharaca Formation in its lower part and an Agua Salada Formation which in northeastern Falcon is regarded as equivalent to the undifferentiated Agua Salada Group. The sedimentological and micropaleontological investigations point to a predominantly deep marine, bathyal environment of the Agua Salada Formation in northeastern Falcon where it is unconformably overlain by the Pliocene Punta Gavilan Formation.

The distribution of the rich planktic and benthic foraminiferal fauna in northeastern Falcon presented in Diaz de Gamero (1985a) is based on two sections. One is situated East of the Rio Hueque, ranging from the Oligocene *Globorotalia opima opima* Zone to the Middle Miocene *Globorotalia menardii* Zone. The other, on the western side of the river, embraces the interval from the basal Middle Miocene *Globorotalia fohsi peripheroronda* Zone to the Late Miocene *Globorotalia acostaensis* Zone. Because of the poor reproduction in the publication of plates and range charts, results of this paper by Diaz de Gamero could not be incorporated in the Falcon Basin chapter.

This is regrettable as in particular the section east of Rio Hueque appears to represent a continuous sequence comparable in its stratigraphic extent to the combined sections in central and southeastern Falcon. Furthermore, a direct comparison of the presence and ranges of the individual taxa between these areas would contribute towards a still better understanding of the paleoecology of the Falcon Basin during Oligocene and Miocene time.

Central Falcon

Outcrops suitable for the evaluation of the distribution of Oligocene to Early Miocene benthic and planktic foraminifera occur south of Coro in central Falcon. Diaz de Gamero (1977) studied both the smaller benthic and the planktic foraminifera and tied them into the planktic foraminiferal zonal scheme of Bolli (1957c). Compared with the Agua Salada Group in eastern Falcon where faunas are excellently preserved in a virtually undisturbed and continuous sequence, conditions are not nearly as favourable in central Falcon. Because of structural disturbance facies changes and poorly preserved faunas, recognition of the planktic foraminiferal zones often becomes difficult or impossible. As a result, the Early to Middle Oligocene *Cassigerinella chipolensis/Pseudohastigerina micra* and the *Globigerina ampliapertura* zones could not be recorded. The same is true for the latest Oligocene because of the scarcity or absence of *Globorotalia kugleri*.

The Oligocene sedimentation began in the central Falcon Basin with the El Paraiso Formation which presumably discordantly overlies the Middle to Upper Eocene Jarillal/Santa Rita formations. It consists of clays, sands and coal and measures some 1000 metres. The microfauna is scarce and shows little variation, and is typical for sediments laid down in shallow, at times brackish waters. While the lower part of the formation is barren the top part contains distinctive planktic foraminifera together with the benthics, making it possible to place these beds in the lower part of the *Globorotalia opima opima* Zone.

The overlying Pecaya Formation has a wide extension in the central Falcon Basin, measuring at least 2000 metres in thickness. Stratigraphically it extends from the Middle Oligocene *Globorotalia opima opima* Zone to the Early Miocene *Catapsydrax dissimilis* Zone, or to the level of the upper part of the Guacharaca Formation in northeastern Falcon. Compared with the underlying El Paraiso Formation, the Pecaya Formation contains a rich and varied microfauna

indicative according to Diaz de Gamero (1977a), of a bathyal water depth of about 1000 metres . A particular, more local facies development occurs within the upper part of the Pecaya Formation where the Pedregoso Formation consisting of turbidites originating from the Coro platform is distinguished. Its age falls into the Early Miocene *Globigerinoides primordius* and *Catapsydrax dissimilis* zones and measures some 700 metres.

In addition to the foraminifera, Diaz de Gamero also investigated the calcareous nannofossils. Because their preservation has been affected even more than that of the foraminifera, results were poor. Stratigraphically it has been possible, however, to recognize the *Sphenolithus distentus* Zone or younger, equivalent to parts of the *Globorotalia opima opima/Globigerina ciperoensis ciperoensis* zones and the *Discoaster druggi* Zone equivalent to the *Globigerinoides primordius* and *Catapsydrax dissimilis* zonal intervals.

Diaz de Gamero investigated the distribution of the benthic and planktic foraminifera in three sections through the El Paraiso and Pecaya/Pedregoso formations. Section I is the most complete and covers the El Paraiso and most of the Pecaya Formation (*Globorotalia opima opima* to *Globigerinoides primordius* Zone). It runs along the Churuguara–Coro road. Section II situated along the Pedregal–Purureche road and the Quebrada Maica covers the interval *Globorotalia opima opima* to *Catapsydrax dissimilis* Zone, but is strongly disturbed structurally and contains only a poorly preserved fauna. Section II, situated in the southern part of the basin, allows for a faunal comparison with the more centrally situated Section I. Section III, positioned between Section I and II, is shorter and represents the Pecaya Formation including its member San Juan de la Vega which is almost void of foraminifera, and the Pedregoso Formation.

Diaz de Gamero distinguishes 174 foraminiferal taxa of which 26 are planktic. On her tables 1–3 she plots the distribution of the planktic foraminifera and the representatives of the genera *Siphogenerina* and *Uvigerina*. Her tables 7–9 contain the distribution in order of last occurrence of the benthic species found in Sections I to III. A selected number of taxa are included in the stratigraphic distribution table (Fig. 75) and illustrated on Fig. 81.

Environmental changes within the formations strongly affect the presence and distribution of both the planktic and benthic fauna. Thus for instance in Section I the lower part of the El Paraiso Formation is – except for a few arenaceous species indicative of brackish water conditions – void of foraminifera. Higher in the same section a considerable interval

AGE		PLANKTIC FORAMINFERAL ZONES	SE FALCON Renz 1948, Blow 1959			NE FALCON Diaz de Gamero 1985a, b	CENTRAL FALCON Diaz de Gamero 1977	N-CENTR. FALCON Diaz de Gamero 1989
			Gr	Fm	Member			
Miocene	Late	Gr. acostaensis	Agua Salada	Pozon	Huso Clay	Agua Salada / Capadare	La Vela	Urumaco
		Gr. menardii			Husito Clay		Caujarao	?
		Gr. mayeri						Socorro
	Middle	Gr. fohsi robusta					Socorro	
		Gr. fohsi lobata						
		Gr. fohsi fohsi						
		Gr. fohsi peripheroronda					Cerro Pelado	
		Po. glomerosa		San Lorenzo	Policarpio	El Salto	Agua Clara	Querales
	Early	Ga. insueta			Menecito Clay	Agua Salada		Cerro Pelado
		Ca. stainforthi						Agua Clara
		Ca. dissimilis			El Salto	?	Pedregoso	Pedregoso
		Gg. primordius			Guacharaca			
Oligocene	Late	Gr. kugleri / Gg. cip. ciperoensis				Agua Salada	Pecaya	Pecaya
	Middle	Gr. opima opima						
		Gg. ampliapertura			?	?	El Paraiso	
	Early	Ca. chipolensis / Ps. micra					?	
Eocene	Late	Tr. cerroazulensis s.l.			Cerro Mision		Jarillal Santa Rita	

between the *Globorotalia opima opima* and the *Globigerinoides primordius* Zone apparently representing the *Globigerina ciperoensis ciperoensis* zonal interval is altogether void of planktic foraminifera. Benthic taxa are scarce, both in number of species and specimens.

For ecologic reasons many of the benthic species listed by Diaz de Gamero occur only sporadically and in low numbers and are therefore not suitable for stratigraphic purposes. For this reason Diaz de Gamero quotes beside the planktic foraminifera only species of the genera *Siphogenerina* and *Uvigerina* as being of distinct stratigraphic significance throughout the three sections.

However, it is worth mentioning that apart from the planktic foraminifera and these two genera, a number of benthic taxa make their first appearance in the three investigated sections at or close to the Oligocene-Miocene boundary, or with the first appearance of *Globigerinoides* (Gamero's table 7). This is in contrast to the Middle/Late Oligocene boundary, between the *Globorotalia opima opima* Zone and the *Globigerina ciperoensis ciperoensis* Zone, where in Sections I to III only a few benthic species became extinct or made their first appearance. About the only exception to this is the first occurrence at this level of *Martinottiella communis* and *Dorothia cylindrica*, though the latter begins already in the *Globorotalia opima opima* Zone.

Additional forms present in central Falcon whose stratigraphic significance has been proven in the Miocene of southeastern Falcon and in the Oligocene/Miocene of Trinidad are *Gravellina narivaensis* (scarce in the *Globigerina ciperoensis ciperoensis* Zone) and *Alveovalvulina pozonensis* in the uppermost part of the *Globigerina ciperoensis ciperoensis* Zone to the *Globigerinoides primordius* Zone. *Ammospirata mexicana* is restricted in central Falcon to the *Globorotalia opima opima* Zone.

In addition to the species listed in table 7 of Diaz de Gamero a considerable number of benthic species, many ranging up from the Oligocene *Globorotalia opima opima* Zone, become extinct within the *Globigerinoides primordius* Zone and do not continue into the *Catapsydrax dissimilis* Zone. They are (in alphabetical order):

Ammodiscus muehlemanni Blow

Bolivina vaceki Schubert

Cassidulina subglobosa Brady

Fig. 72. Correlations of formations in southeastern, northeastern, central and northcentral Falcon with the planktic foraminiferal zonal scheme.

Chilostomella czizeki Reuss
Fursenkoina pontoni (Cushman)
Guttulina jarvisi Cushman & Ozawa
Haplophragmoides carinatus Cushman & Renz
Karreriella subcylindrica (Nuttall)
Lenticulina americana grandis (Cushman)
Lenticulina subpapillosa (Nuttall)
Liebusella crassa Cushman & Renz
Nodosaria raphanistrum caribbeana Hedberg
Oridorsalis ecuadorensis (Galloway & Morrey)
Plectofrondicularia vaughani Cushman
Saracenaria italica carapitana Franklin
Stilostomella verneuili (d'Orbigny)
Textularia nipeensis Keijzer
Trochammina cf. *pacifica* Cushman
Valvulina flexilis Cushman & Renz

Considering the fact that many of the benthic taxa encountered in Sections I to III range from the Oligocene *Globorotalia opima opima* Zone through to the *Catapsydrax dissimilis* Zone it appears that at least for some of the above-listed species their extinction within the *Globigerinoides primordius* Zone in the three sections is of stratigraphic significance.

The 17 benthic species originating and 19 disappearing in the central Falcon sections within the *Globigerinoides primordius* Zone point to a remarkable change in the benthic association within this interval, at least in this part of the Falcon Basin.

Summarizing, it can be said that despite the many adverse conditions encountered, like structurally disturbed sections, facies changes and poor preservation in part of the faunas, it was possible to zone the Oligocene to Early Miocene in Central Falcon on planktic foraminifera and calcareous nannofossils. Furthermore, the stratigraphic value of several species of the genera *Uvigerina* and *Siphogenerina* could be confirmed. The marked change in the benthic fauna within the basal Miocene *Globigerinoides primordius* Zone is another notable result of Diaz de Gamero's investigations. Its stratigraphic significance is certainly valid at least for central Falcon. Comparative studies in other areas will have to prove its applicability over a less restricted area.

This problem can be solved only if the ecologic significance of the benthic faunas is also more thoroughly investigated. It would not be surprising if at least some of the above-discussed appearances and disappearances in central Falcon would eventually prove to be ecologically controlled. This applies particularly to the many 'contemporaneous' appearances at the Oligocene/Miocene boundary.

Distribution in SE- Falcon of Miocene species of the genera Bolivina, Bulimina, Uvigerina, Rectuvigerina, Siphogenerina and calcareous trochospiral rotalid species

Ranges combined from Table 4 and 5 of Renz, 1948

— Ranges based on individual samples as given by Renz on his Table 4

- - additional ranges by zones as given by Renz on his Table 5

Species genus groups	Renz 1948 ZONE / ZONULE	Bolli 1957, emend. Blow 1959 ZONE	No.	Species
"Uvigerinella" sparsicostata — Bol. alazanensis		Ca. dissimilis	1	Bolivina alazanensis
Robulus wallacei — Amm. cf. strathearensis / Uvig. gallowayi basicordata		Ca. stainforthi	2	Bolivina caudriae
— Pl. venezuelana - Sar. senni		Ga. insueta	3	Bolivina tongi
Siphogenerina transversa — Valv. venezuelana		Po. glomerosa	4	Bolivina imporcata
Globorotalia Tohsi — Rob. nuttalli - Cib. mantaensis		Gr. fohsi peripheroronda	5	Bolivina simplex
— Gr. fohsi fohsi		Gr. fohsi fohsi	6	Bolivina pisciformis
Valvulineria herricki — Gr. fohsi lobata		Gr. fohsi lobata	7	Bolivina byramensis
Marginulinopsis basispinosus — Gr. fohsi robusta		Gr. fohsi robusta	8	Bolivina suteri
— Gr.mayeri/Gg.nepenthes			9	Bolivina marginata multicostata
Robulus senni — Gr. menardii / Gg. nepenthes			10	Bolivina cf. cochei
V. sup. - T. cf. pac. — Ss. semmulina			11	Bolivina inconspicua
Text. panamensis			12	Bolivina alata
Elph. poeyanum - Reuss. spinulosa — Gg. bulloides		Gg. bulloides	13	Bolivina rudderi

Species list:

1 Bolivina alazanensis
2 Bolivina caudriae
3 Bolivina tongi
4 Bolivina imporcata
5 Bolivina simplex
6 Bolivina pisciformis
7 Bolivina byramensis
8 Bolivina suteri
9 Bolivina marginata multicostata
10 Bolivina cf. cochei
11 Bolivina inconspicua
12 Bolivina alata
13 Bolivina rudderi
14 Bolivina isidroensis
15 Bolivina thalmanni
16 Bolivina pozonensis
17 Bolivina advena
18 Bulimina inflata alligata
19 Bulimina pupoides
20 Bulimina perversa
21 Bulimina alazanensis
22 Bulimina cf. inflata
23 Bulimina falconensis
24 Uvigerina sparsicostata
25 Uvigerina smithi
26 Uvigerina gallowayi basicordata
27 Uvigerina capayana
28 Uvigerina carapitana
29 Uvigerina isidroensis
30 Uvigerina cf. beccarii
31 Uvigerina auberiana attenuta
32 Uvigerina rustica
33 Uvigerina cf. hannai
34 Rectuvigerina multicostata
35 Siphogenerina transversa
36 Siphogenerina lamellata
37 Siphogenerina senni
38 Siphogenerina kugleri
39 Heterolepa perlucida
40 Osangularia culter
41 Hanzawaia mantaensis
42 Cibicidoides matanzaensis
43 Gyroidinoides altiformis
44 Oridorsalis umbonatus ecuadorensis
45 Gyroidinoides parvus
46 Hanzawaia carstensi
47 Heterolepa compressa
48 Gyroidinoides cf. soldanii
49 Valvulineria venezuelana
50 Eponides crebbsi
51 Valvulineria inaequalis lobata
52 Cibicidoides falconensis
53 Gyroidinoides byramensis campester
54 Anomalinoides trinitatensis
55 Gyroidinoides venezuelanus
56 Siphonina pozonensis
57 Hanzawaia americana
58 Planulina subtenuissima
59 Gyroidinoides planulatus
60 Anomalinoides alazanensis
61 Planulina marialana
62 Osangularia jarvisi
63 Cibicorbis concentricus
64 Planulina dohertyi
65 Cibicorbis herricki
66 Cancris panamensis
67 Cancris sagra
68 Cibicorbis isidroensis
69 Planulina cf. mexicana
70 Ammonia beccarii

(a)

Ammonia beccarii	70	Gyroidinoides byramensis campester	53
Anomalinoides alazanensis	60	Gyroidinoides cf. soldanii	48
Anomalinoides trinitatensis	54	Gyroidinoides parvus	45
Bolivina advena	17	Gyroidinoides planulatus	59
Bolivina alata	12	Gyroidinoides venezuelanus	55
Bolivina alazanensis	1	Hanzawaia americana	57
Bolivina byramensis	7	Hanzawaia carstensi	46
Bolivina caudriae	2	Hanzawaia mantaensis	41
Bolivina cf. cochei	10	Heterolepa compressa	47
Bolivina imporcata	4	Heterolepa perlucida	39
Bolivina inconspicua	11	Oridorsalis umbonatus ecuadorensis	44
Bolivina isidroensis	14	Osangularia culter	40
Bolivina marginata multicostata	9	Osangularia jarvisi	62
Bolivina pisciformis	6	Planulina cf. mexicana	69
Bolivina pozonensis	16	Planulina dohertyi	64
Bolivina rudderi	13	Planulina marialana	61
Bolivina simplex	5	Planulina subtenuissima	58
Bolivina suteri	8	Rectuvigerina multicostata	34
Bolivina thalmanni	15	Siphogenerina kugleri	38
Bolivina tongi	3	Siphogenerina lamellata	36
Bulimina alazanensis	21	Siphogenerina senni	37
Bulimina cf. inflata	22	Siphogenerina transversa	35
Bulimina falconensis	23	Siphonina pozonensis	56
Bulimina inflata alligata	18	Uvigerina auberiana attenuta	31
Bulimina perversa	20	Uvigerina capayana	27
Bulimina pupoides	19	Uvigerina carapitana	28
Cancris panamensis	66	Uvigerina cf. beccarii	30
Cancris sagra	67	Uvigerina cf. hannai	33
Cibicidoides falconensis	52	Uvigerina gallowayi basicordata	26
Cibicidoides matanzaensis	42	Uvigerina isidroensis	29
Cibicorbis concentricus	63	Uvigerina rustica	32
Cibicorbis herricki	65	Uvigerina smithi	25
Cibicorbis isidroensis	68	Uvigerina sparsicostata	24
Eponides crebbsi	50	Valvulineria inaequalis lobata	51
Gyroidinoides altiformis	43	Valvulineria venezuelana	49

(b)

Fig. 73a, b. Distribution in southeastern Falcon of Miocene species of the genera *Bolivina, Bulimina, Uvigerina, Rectuvigerina, Siphogenerina* and calcareous trochospiral species (illustrated on Figs. 78–80).

Distribution in SE-Falcon of Miocene species of the genera Textularia, Lenticulina and selected arenaceous and calcareous species

Ranges combined from Table 4 and 5 of Renz, 1948

— ranges based on individual samples as given by Renz on his Table 4
- - - additional ranges by zones as given by Renz on his Table 5

Zone / genus columns (left to right):

Genus / assemblage	Renz 1948 ZONE	Renz 1948 ZONULE	Bolli 1957, emend., Blow 1959 ZONE
"Uvigerinella" sparsicostata		Bol. alazanensis	Ca. dissimilis
Robulus wallacei		Amm. cf. strathearensis / Uvg. gallowayi basilordata	Ca. stainforthi
Siphogenerina transversa		Pl. venezuelana - Sar. senni	Ga. insueta
Siphogenerina transversa		Valv. venezuelana	Po. glomerosa
Globorotalia fohsi		Rob. nuttalli - Cib. mantaensis	Gr. fohsi peripheroronda
Globorotalia fohsi			Gr. fohsi fohsi
Globorotalia fohsi			Gr. fohsi lobata
Valvulineria herricki			Gr. fohsi robusta
Marginulinopsis basispinosus			Gr.mayeri / lenguaensis; Gr.mayeri/Og.nepenthes
Marginulinopsis basispinosus			Gr. menardii / Gg. nepenthes
Robulus senni		V. sup. - T. cf. pac.	Ss. seminulina
Text. panamensis			
Elph. poeyanum - Reuss. spinulosa			Gg. bulloides

Species (numbered):

1	Textularia isidroensis
2	Textularia mississippiensis alazanensis
3	Textularia cf. mexicana
4	Textularia leuzingeri
5	Textularia excavata
6	Textularia lalikeri
7	Textularia kugleri
8	Textularia falconensis
9	Textularia abbreviata
10	Textularia crassisepta
11	Textularia panamensis
12	Textularia pozonensis
13	Valvulina flexilis
14	Ammobaculites cf. strathearensis
15	Haplophragmoides coronatus
16	Cyclammina cancellata
17	Haplophragmoides carinatus
18	Ammodiscus incertus
19	Schenkiella pallida
20	Gaudryina jacksonensis abnormis
21	Textuariella miocenica
22	Textuariella miocenica brevis
23	Liebusella pozonensis
24	Haplophragmoides emaciatus
25	Clavulina carinata
26	Gaudryina bullbrooki
27	Gaudryina leuzingeri
28	Liebusella pozonensis crassa
29	Trochamina pacifica
30	Lenticulina wallacei
31	Lenticulina subaculeata glabrata
32	Lenticulina clerici
33	Lenticulina formosa
34	Lenticulina melvilli
35	Lenticulina americana
36	Lenticulina occidentalis torrida
37	Lenticulina hedbergi
38	Lenticulina arcuata-striata caroliniana
39	Lenticulina nuttalli
40	Lenticulina suteri
41	Lenticulina iota
42	Lenticulina americana grandis
43	Lenticulina americana spinosa
44	Lenticulina calcar
45	Lenticulina protuberans
46	Lenticulina vortex
47	Lenticulina nuttalli obliquiloculata
48	Lenticulina senni
49	Pyramidulina nuttalli
50	Nodosaria longiscata
51	Pyramidulina raphinastrum caribbeana
52	Nodosaria schlichti
53	Pyramidulina vertebralis
54	Pyramidulina stainforthi
55	Plectofrondicularia floridiana
56	Cassidulina carapitana
57	Amphistegina lessonii
58	Lingulina grimsdalei
59	Planularia venezuelana
60	Pseudoglandulina comatula
61	Guttulina irregularis
62	Darbyella subkubinyii
63	Planularia arbenzi
64	Plectofrondicularia cf. californica
65	Lingulina prolata
66	Ehrenbergina caribbea
67	Guttulina jarvisi
68	Gypsina aff. vesicularis
69	Trifarina bradyi
70	Cassidulina delicata
71	Planularia clara
72	Pseudoglandulina gallowayi paucicostata
73	Plectofrondicularia cf. longistriata
74	Cassidulinoides erectus
75	Plectofrondicularia mansfieldi
76	Angulogerina illingi
77	Elphidium poeyanum
78	Elphidium sagrum
79	Reussella spinulosa

(a)

Ammobaculites cf. strathearensis	14
Ammodiscus incertus	18
Amphistegina lessonii	57
Angulogerina illingi	76
Cassidulina carapitana	56
Cassidulina delicata	70
Cassidulinoides erectus	74
Clavulina carinata	25
Cyclammina cancellata	16
Darbyella subkubinyii	62
Ehrenbergina caribbea	66
Elphidium poeyanum	77
Elphidium sagrum	78
Gaudryina bullbrooki	26
Gaudryina jacksonensis abnormis	20
Gaudryina leuzingeri	27
Guttulina irregularis	61
Guttulina jarvisi	67
Gypsina aff. vesicularis	68
Haplophragmoides carinatus	17
Haplophragmoides coronatus	15
Haplophragmoides emaciatus	24
Lenticulina americana	35
Lenticulina americana grandis	42
Lenticulina americana spinosa	43
Lenticulina arcuata-striata caroliniana	38
Lenticulina calcar	44
Lenticulina clericii	32
Lenticulina formosa	33
Lenticulina hedbergi	37
Lenticulina iota	41
Lenticulina melvilli	34
Lenticulina nuttalli	39
Lenticulina nuttalli obliquiloculata	47
Lenticulina occidentalis torrida	36
Lenticulina protuberans	45
Lenticulina senni	48
Lenticulina subaculeata glabrata	31
Lenticulina suteri	40
Lenticulina vortex	46

(b)

Lenticulina wallacei	30
Liebusella pozonensis	23
Liebusella pozonensis crassa	28
Lingulina grimsdalei	58
Lingulina prolata	65
Nodosaria longiscata	50
Nodosaria schlichti	52
Planularia arbenzi	63
Planularia clara	71
Planularia venezuelana	59
Plectofrondicularia cf. californica	64
Plectofrondicularia cf. longistriata	73
Plectofrondicularia floridiana	55
Plectofrondicularia mansfieldi	75
Pseudoglandulina comatula	60
Pseudoglandulina gallowayi paucicostata	72
Pyramidulina nuttalli	49
Pyramidulina raphinastrum caribbeana	51
Pyramidulina stainforthi	54
Pyramidulina vertebralis	53
Reussella spinulosa	79
Schenkiella pallida	19
Textuariella miocenica	21
Textuariella miocenica brevis	22
Textularia abbreviata	9
Textularia cf. mexicana	3
Textularia crassisepta	10
Textularia excavata	5
Textularia falconenesis	8
Textularia isidroensis	1
Textularia kugleri	7
Textularia lalikeri	6
Textularia leuzingeri	4
Textularia mississippiensis alazanensis	2
Textularia panamensis	11
Textularia pozonensis	12
Trifarina bradyi	69
Trochamina pacifica	29
Valvulina flexilis	13

Fig. 74a, b. Distribution in southeastern Falcon of Miocene species of the genera *Textularia, Lenticulina, Nodosaria* and selected agglutinated and calcareous species (illustrated on Fig. 76–77, 80).

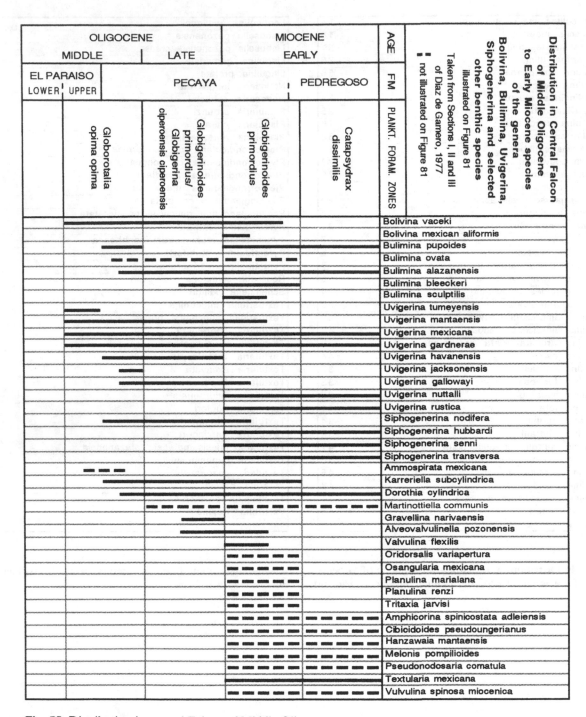

Fig. 75. Distribution in central Falcon of Middle Oligocene to Early Miocene species of the genera *Bolivina*, *Bulimina*, *Uvigerina*, *Siphogenerina* and selected other benthic species (illustrated on Fig. 81).

Figs. 76–80

Fig. 76.

  1 *Textularia mississippiensis alazanensis* Nuttall
     from Renz, 1948, pl. 1, fig. 1, × 33.

  2 *Textularia* cf. *mexicana* Cushman
     from Renz, 1948, pl. 12, fig. 2, × 33.

3a–b *Textularia leuzingeri* Cushman & Renz
     Holotype, × 33.

  4 *Textularia excavata* Cushman
     from Renz, 1948, pl. 1, fig. 15, × 22.5.

  5 *Textularia crassisepta* Cushman
     from Renz, 1948, pl. 1, fig. 12, × 33.

6a–b *Textularia isidroensis* Cushman & Renz
     Holotype, × 22.5.

 7–8 *Textularia lalickeri* Cushman & Renz
     7a–b Holotype, × 22.5, **8** Paratype from Renz,
     1948, pl. 1, fig. 20, × 33.

 9–10 *Textularia abbreviata* Cushman
     from Renz, 1948, pl. 1, figs. 10, 11, × 12.5.

11–12 *Textularia panamensis* Cushman
     from Renz, 1948, pl. 1, figs. 21, 22, × 33.

13a–b *Textularia falconensis* Cushman & Renz
     Holotype, × 33.

14–15 *Textularia kugleri* Cushman & Renz
     14 Paratype, × 22.5, **15** Holotype, × 33.5.

16a–b *Textularia pozonensis* Cushman & Renz
     Holotype, × 22.5.

17–18 *Valvulina flexilis* Cushman & Renz
     17 Paratype, from Renz, 1948, pl. 2, fig. 11, ×
     22.5, **18a–c** Holotype, × 22.5.

19–20 *Martinottiella pallida* (Cushman)
     from Renz, 1948, pl. 2, figs. 17, 18, × 22.5.

21a–b *Clavulina carinata* Cushman & Renz
     Holotype, × 22.5.

22–23 *Gaudryina jacksonensis abnormis* Cushman &
     Renz

22a–b Holotype, × 22.5, **23** Paratype, from Renz,
     1948, pl. 2, fig. 4, × 22.5.

  24 *Gaudryina bullbrooki* Cushman
     from Renz, 1948, pl. 2, fig. 6, × 22.5.

  25 *Gaudryina leuzingeri* Cushman & Renz
     Holotype, × 22.5.

26a–b *Textulariella miocenica brevis* Cushman & Renz
     Holotype, × 22.5.

  27 *Textulariella miocenica* Cushman
     from Renz, 1948, pl. 2, fig. 14, × 22.5.

28–29 *Alveovalvulinella pozonensis* (Cushman & Renz)
     28 Paratype, from Renz, 1948, pl. 2, fig. 20, ×
     22.5, **29a–b** Holotype, × 22.5.

30–31 *Alveovalvulinella pozonensis crassa* (Cushman &
     Renz)
     30 Paratype, from Renz, 1948, pl. 2, fig. 22, ×
     22.5, **31** Holotype, × 22.5.

  32 *Haplophragmoides coronatus* (Brady)
     from Renz, 1948, pl. 1, fig. 5, × 22.5.

33a–b *Haplophragmoides carinatus* Cushman & Renz
     Holotype, × 33.

34a–b *Haplophragmoides emaciatus* (Brady)
     from Renz, 1948, pl. 1, figs. 6a–b, × 22.5.

35a–b *Trochammina pacifica* Cushman
     from Renz, 1948, pl. 3, figs. 4a–b, × 49.

36–37 *Ammobaculites* cf. *strathearnensis* Cushman &
     LeRoy
     from Renz, 1948, pl. 1, figs. 7, 8, × 22.5.

38–39 *Ammodiscus incertus* d'Orbigny
     from Renz, 1948, pl. 1, figs. 1, 2, × 22.5.

  40 *Cyclammina cancellata* Brady
     from Renz, 1948, pl. 1, fig. 9, × 33.

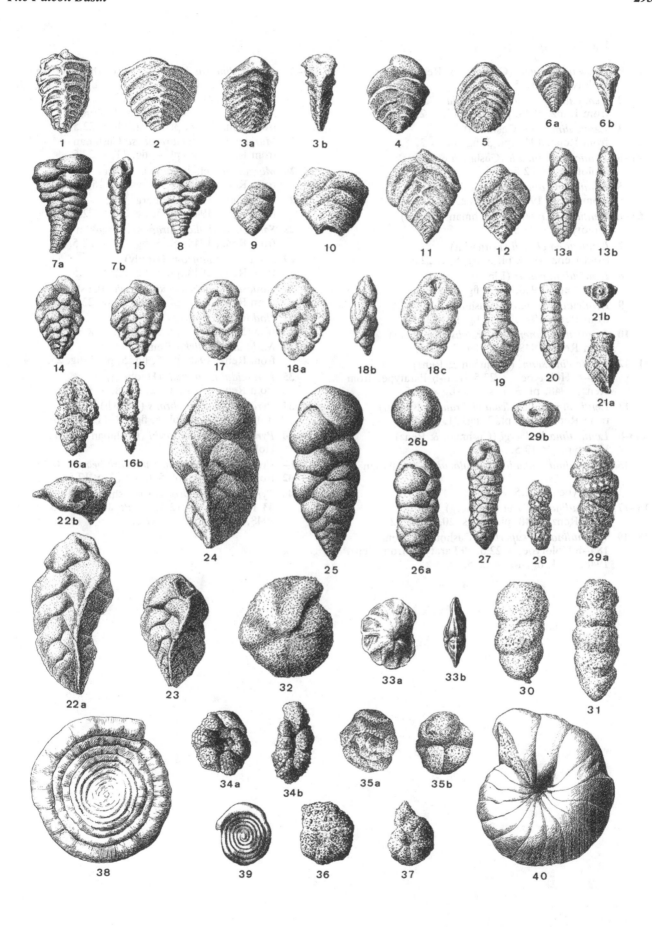

Fig. 77.

1a–b *Lenticulina suteri* (Cushman & Renz)
Holotype, × 12.5.

2 *Lenticulina occidentalis torrida* (Cushman)
from Renz, 1948, pl. 3, fig. 17, × 22.5.

3 *Lenticulina iota* (Cushman)
from Renz, 1948, pl. 3, fig. 14, × 22.5.

4a–b *Lenticulina nuttalli* (Cushman & Renz)
Holotype, × 22.5.

5 *Lenticulina americana* (Cushman)
from Renz, 1948, pl. 12, fig. 3, × 22.5.

6a–b *Lenticulina melvilli* (Cushman & Renz)
Holotype, × 22.5.

7 *Lenticulina clericii* (Fornasini)
from Renz, 1948, pl. 3, fig. 8, × 22.5.

8 *Lenticulina calcar* (Linné)
from Renz, 1948, pl. 3, fig. 6, × 22.5.

9 *Lenticulina formosa* (Cushman)
from Renz, 1948, pl. 3, fig. 9, × 22.5.

10 *Lenticulina americana grandis* (Cushman)
from Renz, 1948, pl. 3, fig. 7, × 22.5.

11–12 *Lenticulina senni* (Cushman & Renz)
11a–b Holotype, × 22.5, **12a–b** Paratype, from
Renz, 1948, pl. 3, figs. 16a–b, × 22.5.

13 *Lenticulina subaculeata glabrata* (Cushman)
from Renz, 1948, pl. 3, fig. 21, × 12.5.

14a–b *Lenticulina hedbergi* (Cushman & Renz)
Holotype, × 12.5.

15 *Lenticulina nuttalli obliquiloculata* (Cushman &
Renz)
Holotype, × 22.5.

16–17 *Lenticulina wallacei* (Hedberg)
from Renz, 1948, pl. 4, figs. 20, 19, × 22.5.

18–19 *Vaginulinopsis superbus* (Cushman & Renz)
18a–b Holotype, × 22.5, **19** Paratype, from Renz,
1948, pl. 4, fig. 18, × 22.5.

20–22 *Marginulinopsis basispinosus* (Cushman & Renz)
20, 21 Paratypes, from Renz, 1948, pl. 4, figs. 9,
10, × 22.5, **22** Holotype, × 22.5.

23 *Astacolus ovatus* Galloway & Heminway
from Renz, 1948, pl. 4, fig. 11, × 22.5.

24 *Marginulina* cf. *glabra obesa* Cushman
from Renz, 1948, pl. 4, fig. 11, × 22.5.

25–26 *Marginulina* cf. *striatula* Cushman
from Renz, 1948, pl. 4, figs. 15, 16, × 33.

27 *Saracenaria senni* Hedberg
from Renz, 1948, pl. 5, fig. 21, × 22.5.

28 *Saracenaria italica carapitana* Franklin
from Renz, 1948, pl. 5, fig. 18, × 22.5.

29 *Saracenaria latifrons* (Brady)
from Renz, 1948, pl. 5, fig. 22, × 22.5.

30 *Saracenaria italica acutocarinata* (Cushman)
from Renz, 1948, pl. 5, fig. 19, × 22.5.

31–32 *Nodosaria longiscata* d'Orbigny
from Renz, 1948, pl. 5, figs. 1, 2, × 12.5.

33–34 *Nodosaria schlichti* Reuss
from Renz, 1948, pl. 5, fig. 5; pl. 4, fig. 25, × 33.

35 *Pyramidulina nuttalli* (Hedberg)
from Renz, 1948, pl. 12, fig. 8, × 12.5.

36–37 *Pyramidulina vertebralis* (Batsch)
from Renz, 1948, pl. 5, figs. 8–11, × 12.5.

39–40 *Pyramidulina stainforthi* (Cushman & Renz)
Holotype, × 22.5.

38, 41– *Pyramidulina raphanistrum caribbeana* (Hedberg)
42 from Renz, 1948, pl. 5, figs. 6, 7, × 12.5.

43–44 *Pyramidulina isidroensis* (Cushman & Renz)
43 Holotype, × 12.5, **44** Paratype, from Renz,
1948, pl. 4, fig. 29, × 12.5.

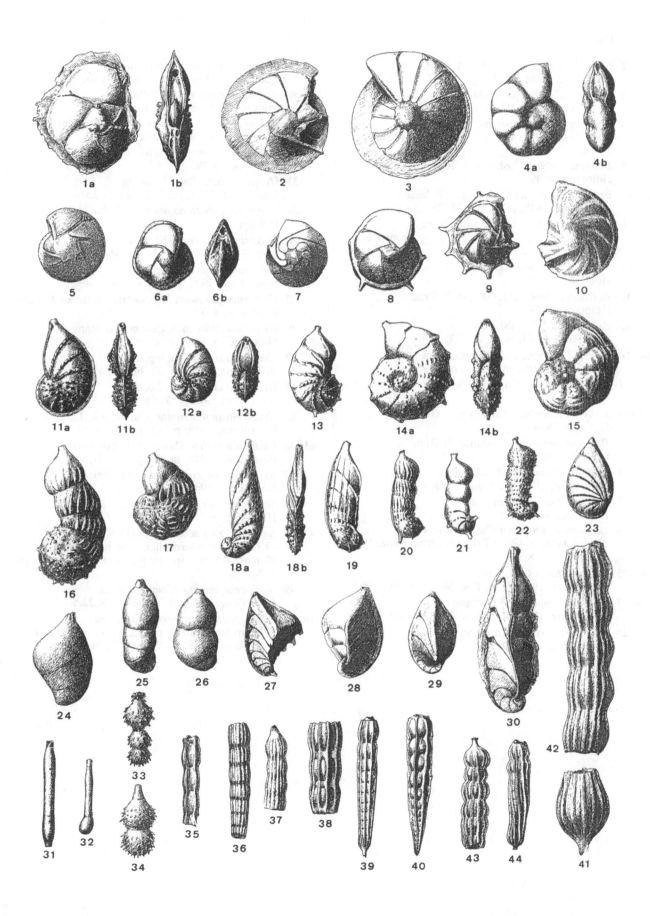

Fig. 78.

1–2 *Bolivina pisciformis* Galloway & Morrey
from Renz, 1948, pl. 7, figs. 12, 11, × 49.

3 *Bolivina alazanensis* Cushman
from Renz, 1948, pl. 12, fig. 7, × 49.

4–5 *Bolivina alata* (Seguenza)
from Renz, 1948, **4** pl. 6, fig. 26, **5** pl. 12, fig. 12.
× 49.

6 *Bolivina rudderi* Cushman & Renz
Holotype, × 49.

7–8 *Bolivina* cf. *cochei* Cushman & Adams
from Renz, 1948, pl. 6, fig. 27, 28, × 49.

9–10 *Bolivina caudriae* Cushman & Renz
9a–b Holotype, × 49, **10** Paratype, from Renz,
1948, pl. 7, fig. 2, × 49.

11 *Bolivina pozonensis* Cushman & Renz
Holotype, × 49.

12 *Bolivina isidroensis* Cushman & Renz
Holotype, × 49.

13 *Bolivina advena* Cushman
from Renz, 1948, pl. 6, fig. 23, × 49.

14–16 *Bolivina marginata multicostata* Cushman
from Renz, 1948, pl. 7, figs. 6–8, × 49.

17 *Bolivina simplex* Cushman & Renz
Holotype, × 49.

18a–b *Bolivina thalmanni* Cushman & Renz
Holotype, × 49.

19 *Bolivina imporcata* Cushman & Renz
Holotype, × 49.

20 *Bolivina byramensis* Cushman
from Renz, 1948, pl. 6, fig. 22, × 49.

21 *Bolivina suteri* Cushman & Renz
Holotype, × 49.

22–23 *Bolivina inconspicua* Cushman & Renz
22 Holotype, × 49, **23** Paratype, from Renz, 1948,
pl. 7, fig. 10, × 49.

24–25 *Bolivina tongi* Cushman
from Renz, 1948, pl. 6, figs. 24, 25, × 49.

26–27 *Bulimina pupoides* d'Orbigny
from Renz, 1948, pl. 6, figs. 11, 12, × 49.

28 *Bulimina perversa* Cushman
from Renz, 1948, pl. 6, fig. 16, × 33.

29 *Bulimina* cf. *inflata* Seguenza
from Renz, 1948, pl. 12, figs. 14a–b, × 49.

30 *Bulimina alazanensis* Cushman
from Renz, 1948, pl. 6, fig. 14, × 49.

31 *Bulimina inflata alligata* Cushman & Laiming
from Renz, 1948, pl. 6, fig. 13, × 49.

32 *Bulimina falconensis* Renz
Holotype, × 49.

33 *Uvigerina carapitana* Hedberg
from Renz, 1948, pl. 7, fig. 21, × 33.

34 *Uvigerina auberiana attenuta* Cushman & Renz
Holotype, × 49.

35–36 *Uvigerina* cf. *hannai* Kleinpell
from Renz, 1948, pl. 12, figs. 16, 17, × 49.

37 *Uvigerina capayana* Hedberg
from Renz, 1948, pl. 12, fig. 15, × 49.

38 *Uvigerina gallowayi basicordata* Cushman & Renz
Holotype, × 33.

39 *Uvigerina isidroensis* Cushman & Renz
Holotype, × 49.

40 *Uvigerina smithi* (Kleinpell)
from Renz, 1948, pl. 12, fig. 18, × 33.

41 *Uvigerina* cf. *beccarii* Fornasini
from Renz 1948 pl. 7, fig. 22, × 33

42 *Uvigerina sparsicostata* (Cushman & Laiming)
from Renz, 1948, pl. 12, fig. 20, × 49.

43–44 *Uvigerina rustica* Cushman & Edwards
from Renz, 1948, pl. 7, figs. 23, 24, × 33.

45 *Rectuvigerina multicostata* (Cushman & Jarvis)
from Renz, 1948, pl. 7, fig. 26, × 22.5.

46 *Siphogenerina kugleri* Cushman & Renz
Holotype, × 22.5.

47–48 *Siphogenerina senni* Cushman & Renz
47 Holotype, microspheric specimen, × 22.5,
48 macrospheric specimen, from Renz, 1948, pl.
7, fig. 30, × 22.5.

49 *Siphogenerina lamellata* Cushman
from Renz, 1948, pl. 7, fig. 25, × 22.5.

50–52 *Siphogenerina transversa* Cushman
50 from Renz, 1948, pl. 12, fig. 6, × 22.5, **51**, **52**
from Renz, 1948, pl. 7, figs. 27, 28, × 22.5.

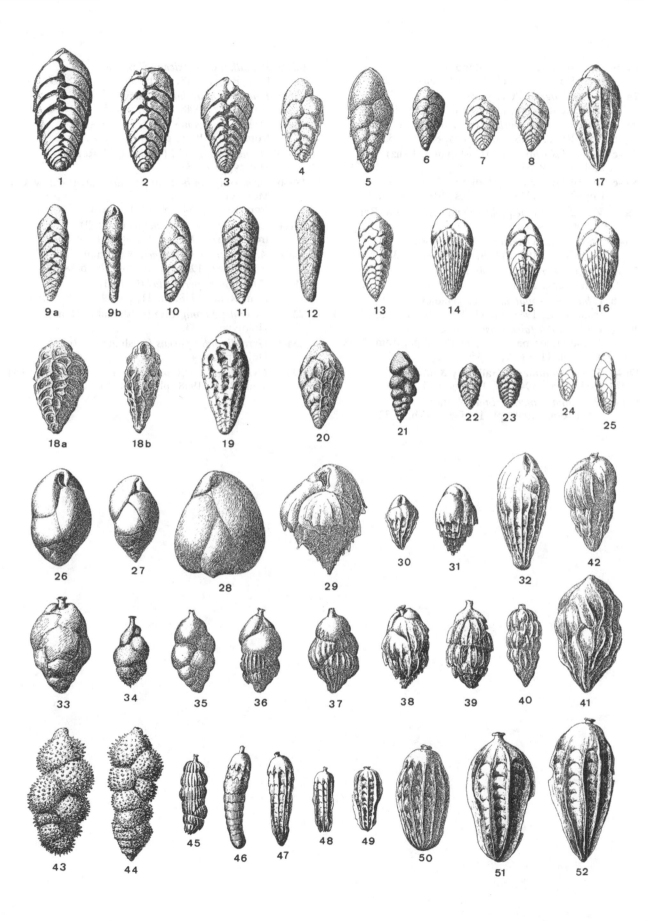

Fig. 79.

1a–b *Cancris panamensis* Natland
from Renz, 1948, pl. 12, figs. 22a–b, × 33.

2a–c *Cancris sagra* (d'Orbigny)
from Renz, 1948, pl. 9, figs. 3a–c, × 33.

3a–c *Cibicorbis concentricus* (Cushman)
from Renz, 1948, pl. 10, figs. 8a–c, × 33.

4a–c *Cibicorbis isidroensis* (Cushman & Renz)
Holotype, × 33.

5a–c *Cibicorbis herricki* (Hadley)
from Renz, 1948, pl. 8, figs. 10a–c, × 33.

6a–c *Valvulineria inaequalis lobata* Cushman & Renz
Holotype, × 33.

7a–c *Valvulineria venezuelana* Hedberg
from Renz, 1948, pl. 8, figs. 9a–c, × 33.

8a–b *Eponides crebbsi* (Hedberg)
from Renz, 1948, pl. 12, figs. 26a–b, × 33.

9a–b *Cibicidoides matanzaensis* (Hadley)
from Renz, 1948, pl. 11, figs. 12a–b, × 33.

10–11 *Cibicidoides falconensis* (Renz)
10a–c Holotype, × 33, **11** Paratype, from Renz,
1948, pl. 11, fig. 7, × 33.

12a–b *Planulina dohertyi* Galloway & Morrey
from Renz, 1948, pl. 10, figs. 6a-b, × 33.

13a–b *Planulina subtenuissima* (Nuttall)
from Renz, 1948, pl. 11, figs. 4a–b, × 33.

14a–b *Planulina* cf. *mexicana* Cushman
from Renz, 1948, pl. 11, figs. 5a–b, × 33.

15a–b *Planulina marialana* Hadley
from Renz, 1948, pl. 10, figs. 5a–b, × 33.

16a–c *Osangularia culter* (Parker & Jones)
from Renz, 1948, pl. 9, figs. 6a–c, × 33.

17a–c *Osangularia jarvisi* (Cushman & Renz)
Holotype, × 33.

18a–b *Oridorsalis umbonatus ecuadorensis* (Galloway &
Morrey)
from Renz, 1948, pl. 12, figs. 25a–b, × 33.

19a–c *Anomalinoides trinitatensis* (Nuttall)
from Renz, 1948, pl. 10, figs. 11a–c, × 33.

20a–b *Anomalinoides alazanensis* (Nuttall)
from Renz, 1948, pl. 10, figs. 7a–b, × 33.

21a–b *Cibicidoides perlucidus* (Nuttall)
from Renz, 1948, pl. 11, figs. 9a–b, × 33.

22a–c *Heterolepa compressa* (Cushman & Renz)
Holotype, × 33.

23a–c *Gyroidinoides parvus* (Cushman & Renz)
Holotype, × 49.

24a–c *Gyroidinoides altiformis* (R. E. & K. C. Stewart)
from Renz, 1948, pl. 8, figs. 13a–c, × 33.

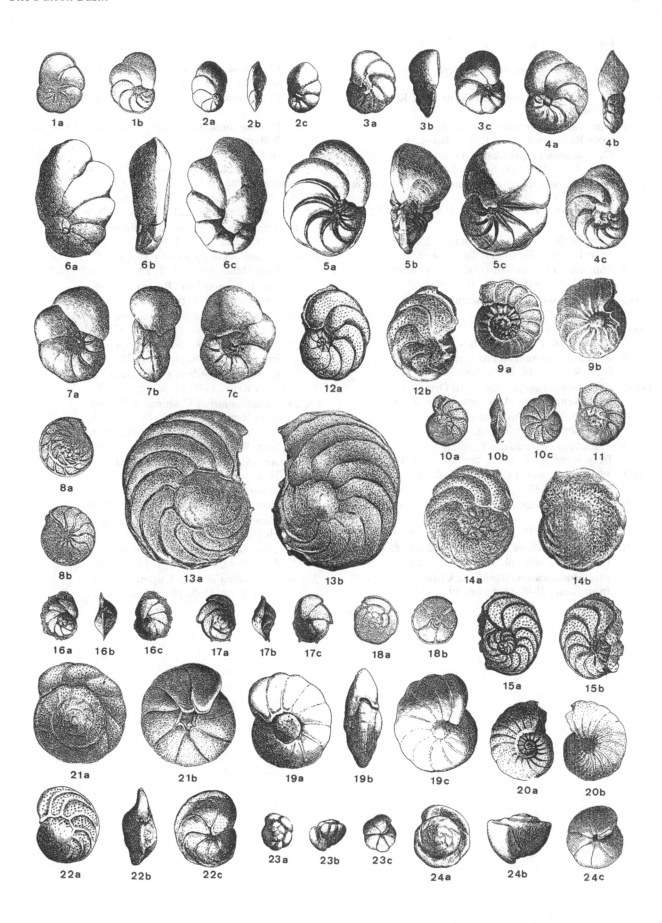

Fig. 80.

1a–c *Gyroidinoides venezuelanus* Renz
Holotype, × 49.

2a–c *Gyroidinoides planulatus* (Cushman & Renz)
Holotype, × 33.

3a–c *Hanzawaia americana* (Cushman)
from Renz, 1948, pl. 11, figs. 10a–c, × 33.

4a–c *Gyroidinoides* cf. *soldanii* (d'Orbigny)
from Renz, 1948, pl. 8, figs. 14a–c, × 33.

5a–c *Gyroidinoides byramensis campester* (Palmer &
Bermudez)
from Renz, 1948, pl. 8, figs. 15a–b; pl. 9, fig. 1,
× 22.5.

6a–c *Hanzawaia carstensi* (Cushman & Ellisor)
from Renz, 1948, pl. 11, figs. 11a–c, × 33.

7a–b *Hanzawaia mantaensis* (Galloway & Morrey)
from Renz, 1948, pl. 11, figs. 8a–b, × 33.

8a–b *Siphonina pozonensis* Cushman & Renz
Holotype, × 33.

9a–c *Ammonia beccarii* (Linné)
from Renz, 1948, pl. 9, figs. 2a–c, × 33.

10a–b *Elphidium sagrum* (d'Orbigny)
from Renz, 1948, pl. 6, figs. 7a–b, × 49.

11a–b *Cribroelphidium poeyanum* (d'Orbigny)
from Renz, 1948, pl. 6, figs. 6a–b, × 49.

12 *Planularia clara* Cushman & Jarvis
from Renz, 1948, pl. 4, fig. 4, × 33.

13a–b *Planularia venezuelana* Hedberg
from Renz, 1948, pl. 4, figs. 5a–b, × 33.

14a–b *Planularia arbenzi* Cushman & Renz
Holotype, × 33.

15 *Plectofrondicularia* cf. *californica* Cushman &
Stewart
from Renz, 1948, pl. 12, fig. 10, × 33.

16 *Plectofrondicularia mansfieldi* Cushman & Ponton
from Renz, 1948, pl. 12, fig. 11, × 33.

17 *Plectofrondicularia floridiana* Cushman
from Renz, 1948, pl. 6, fig. 19, × 33.

18 *Plectofrondicularia* cf. *longistriata* LeRoy
from Renz, 1948, pl. 6, fig. 21, × 33.

19a–b *Tollmannia grimsdalei* (Cushman & Renz)
Holotype, × 12.5.

20 *Pseudoglandulina gallowayi paucicostata* Cushman
& Renz
Holotype, × 22.5.

21 *Pseudoglandulina comatula* (Cushman)
from Renz, 1948, pl. 5, fig. 12, × 22.5.

22a–b *Darbyella subkubinyii* (Nuttall)
from Renz, 1948, pl. 4, figs. 3a–b, × 33.

23 *Amphistegina lessonii* d'Orbigny
from Renz, 1948, pl. 9, fig. 4, × 33.

24a–b *Gonatosphaera prolata* (Guppy)
from Renz, 1948, pl. 5, figs. 24a–b, × 12.5.

25 *Reussella spinulosa* (Reuss)
from Renz, 1948, pl. 7, fig. 16, × 33.

26a–b *Ehrenbergina caribbea* Galloway & Heminway
from Renz, 1948, pl. 9, figs. 17a–b, × 49.

27 *Pseudoglandulina incisa* (Neugeboren)
from Renz, 1948, pl. 5, fig. 16, × 22.5.

28 *Guttulina irregularis* (d'Orbigny)
from Renz, 1948, pl. 6, fig. 1, × 33.

29 *Guttulina jarvisi* Cushman & Ozawa
from Renz, 1948, pl. 6, fig. 2, × 33.

30a–b *Gypsina* aff. *vesicularis* (Parker & Jones)
from Renz, 1948, pl. 10, figs. 12a–b, × 33.

31a–b *Angulogerina illingi* Cushman & Renz
Holotype, × 49.

32 *Trifarina bradyi* Cushman
from Renz, 1948, pl. 7, fig. 33, × 49.

33a–b *Cassidulinoides erecta* Cushman & Renz
Holotype, × 49.

34a–b *Cassidulina carapitana* Hedberg
from Renz, 1948, pl. 9, figs. 8a–b, × 49.

35a–b *Cassidulina delicata* Cushman
from Renz, 1948, pl. 9, figs. 10a–b, × 49.

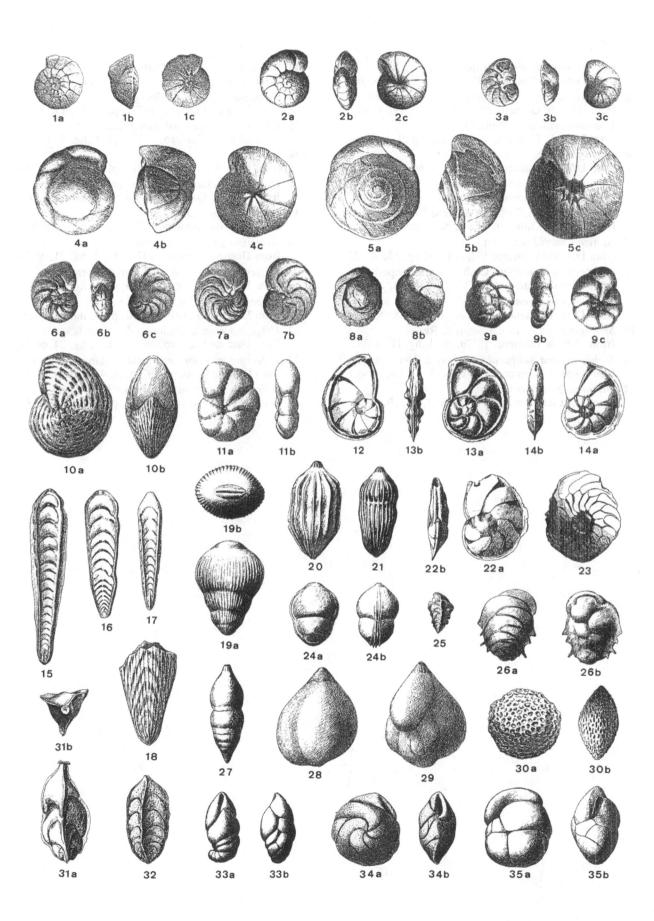

Fig. 81.

1 *Gravellina narivaensis* Brönnimann
   from Diaz de Gamero, 1977*a*, pl. 3, fig. 12, × 36.

2 *Alveovalvulinella pozonensis* (Cushman & Renz)
   from Diaz de Gamero,1977*a*, pl. 3, fig. 14, × 36.

3 *Dorothia cylindrica* (Nuttall)
   from Diaz de Gamero, 1977*a*, pl.3, fig. 9, × 18.

4 *Karreriella subcylindrica* (Nuttall)
   from Diaz de Gamero, 1977*a*, pl.3, fig. 10, × 36.

5 *Textularia mexicana* Cushman
   from Diaz de Gamero, 1977*a*, pl. 3, fig. 6, × 36.

6 *Valvulina flexilis* Cushman & Renz
   from Diaz de Gamero, 1977*a*, pl.3, fig. 13, × 36.

7 *Bolivina vaceki* Schubert
   from Diaz de Gamero, 1977*a*, pl. 3, fig. 15, × 72.

8 *Siphogenerina nodifera* Cushman & Kleinpell
   from Diaz de Gamero, 1977*a*, pl. 4, fig. 8, × 38.7.

9 *Siphogenerina transversa* Cushman
   from Diaz de Gamero, 1977*a*, pl. 4, fig. 8, × 19.

10 *Siphogenerina senni* Cushman & Renz
    from Diaz de Gamero, 1977*a*, pl. 4, fig. 11, × 38.7.

11 *Siphogenerina hubbardi* Galloway & Heminway
    from Diaz de Gamero, 1977*a*, pl. 4, fig. 10, × 38.7.

12 *Bulimina sculptilis* Cushman
    from Diaz de Gamero, 1977*a*, pl. 3, fig. 19, × 36.

13 *Bulimina alazanensis* Cushman
    from Diaz de Gamero, 1977*a*, pl. 3, fig. 16, × 72.

14 *Bulimina bleeckeri* Hedberg
    from Diaz de Gamero, 1977*a*, pl. 3, fig. 17, × 36.

15 *Uvigerina rustica* Cushman & Edwards
    from Diaz de Gamero, 1977*a*, pl. 4, fig. 5, × 38.7.

16 *Uvigerina gallowayi* Cushman
    from Diaz de Gamero, 1977*a*, pl. 3, fig. 20, × 36.

17 *Uvigerina jacksonensis* Cushman
    from Diaz de Gamero, 1977*a*, pl. 3, fig. 18, × 6.

18 *Uvigerina tumeyensis* Lamb
    from Diaz de Gamero, 1977*a*, pl. 4, fig. 6, × 38.7.

19 *Uvigerina gardnerae* Cushman
    from Diaz de Gamero, 1977*a*, pl. 3, fig. 21, × 36.

20 *Uvigerina havanensis* Cushman & Bermudez
    from Diaz de Gamero, 1977*a*, pl. 3, fig. 22, × 36.

21 *Uvigerina mexicana* Nuttall
    from Diaz de Gamero, 1977*a*, pl. 3, fig. 23, × 36.

22 *Uvigerina nuttalli* Cushman & Edwards
    from Diaz de Gamero, 1977*a*, pl. 3, fig. 24, × 36.

23 *Uvigerina auberiana attenuata* Cushman & Renz
    from Diaz de Gamero, 1977*a*, pl. 4, fig. 1, × 77.5,

24 *Uvigerina mantaensis* Cushman & Edwards
    from Diaz de Gamero, 1977*a*, pl. 4, fig. 3, × 38.7.

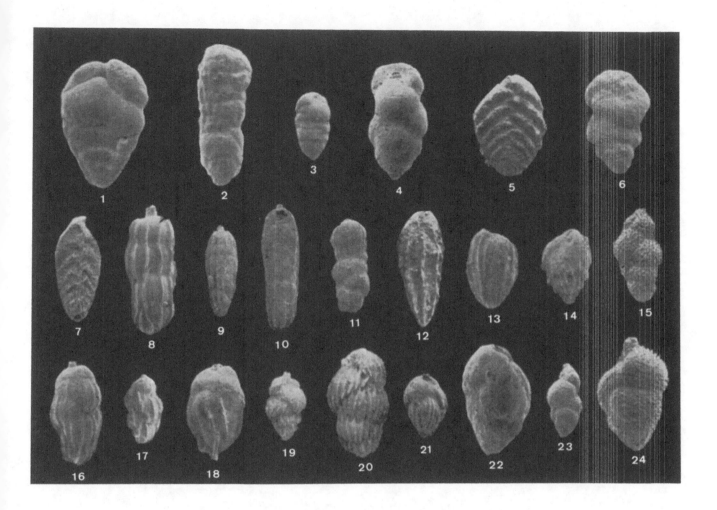

The Maracaibo Basin

Introduction

The Maracaibo Basin (Fig. 82) is the major oil producing area of Venezuela. Extensive micropaleontological studies by the oil industry have, based primarily on foraminifera and palynomorphs, greatly contributed towards the understanding and interpretation of the stratigraphy, correlations and changing environmental conditions of the Cretaceous and Tertiary sediments of the area. Much of this information has remained unpublished. Exceptions to this are the palynological investigations by Kuyl *et al.* (1955) and Fuenmayor's (1989) 'Manual on the foraminifera of the Maracaibo Basin'. In addition there exist publications by Barbeito *et al.* (1985) and Pittelli Viapiana (1989) on the paleoenvironmental significance of Eocene benthic foraminifera in the Maracaibo Basin.

Barbeito *et al.* (1985) distinguish in the Mara-Maracaibo area (Fig. 82) 14 associations mainly based on benthic foraminifera, each characterizing a certain paleoenvironment, primarily water depth but also on variations in salinity. The range is from shoreline (alluvial) to bathyal. One of the 14 associations is characterized by a strong presence of *Globigerina* spp. amongst a dwarfed calcareous benthic fauna. On their table 2, intervals of the 14 associations are plotted against the 10 investigated well sections within the area Mara Maracaibo.

Pittelli Viapiana (1989) distinguishes five paleo-bathymetric foraminiferal assemblages through the Early Eocene to late Middle Eocene of the eastern and northwestern parts of the Maracaibo Basin based on well and surface evidence. They range from lower alluvial, deltaic, nearshore, inner mid-neritic to outer neritic/upper bathyal.

Fuenmayor's (1989) manual, together with the 'Lexico Estratigrafico de Venezuela' (1970 edition) forms the base for the following outlines. It draws extensively from Maraven's micropaleontological data. This company, formerly the Compania Shell de Venezuela and now a subsidiary of Petroleos de Venezuela has kindly released this information for publication.

Fig. 82. Maracaibo Basin locality map.

The manual contains a biostratigraphic/chronostratigraphic chart for the general Lake Maracaibo area, showing the correlation of ages with the locally established benthic foraminiferal zones and the standard planktic zonal scheme. The benthic zonal markers are listed by their generic names, but with the company numbering system in place of species names. An emended version of this chart, containing in addition also the scientific species names and the names of formations, is reproduced here as Fig. 83.

Fuenmayor illustrates by scanning electron micrographs planktic and benthic Cretaceous species. Shown on a chart is the distribution in the Upper Cretaceous La Luna, Colon and Mito Juan formations of planktic and benthic taxa in relation to the established benthic zonal schemes. Similarly are illustrated planktic and benthic species from the Lower to Middle Eocene Misoa (uppermost benthic zone only) and the Middle Eocene Pauji Formation. The Cretaceous is divided into five, the Eocene into four benthic zones. The Guasare Formation representing the much reduced Paleocene in the Maracaibo Basin is divided into three benthic zones.

Within the Maracaibo Basin foraminiferal associations are affected by changing environmental conditions which locally necessitate the use of different benthic zonal markers. This is shown here on Fig. 83 for the Late Cretaceous zones applicable throughout most of the Lake area including the oil fields from Cabimas to Mene Grande along the eastern shore of the Lake, further west of Maracaibo (Mara, La Paz, La Concepcion, Boscan) and the coastal area south of Mene Grande on the southeastern side of the Lake (Motatan, Ceuta, Barua, Tomoporo). The markers for corresponding Late Cretaceous benthic foraminiferal zones had to be replaced as shown on Fig. 84 in the Colon District, southwest of the Lake (Rosario, Los Manueles, Tarra). Fuenmayor also demonstrates for four sections from different Basin areas the irregular presence and changing facies of Tertiary formations and their zonal subdivision by different benthic markers.

The Late Cretaceous to Miocene formations

Following is a brief review on the lithology, zones and principal faunal contents of those Late Cretaceous to Miocene formations and members in the Maracaibo Basin whose benthic and planktic foraminifera are the subject of Fuenmayor's manual. Their age sequence and benthic and planktic foraminiferal zonal subdivision as shown on Fig. 83 and 84, is also discussed below. For more information on the individual formations reference can be made to the Lexico Estratigrafico de Venezuela (1970).

The oldest formation here treated is the Cenomanian to Santonian La Luna Formation. It consists of dark grey to black limestones and shales laid down under euxinic conditions. It is regarded as being the most prolific oil source rock in Venezuela. The formation is developed throughout the Maracaibo Basin, extending from there into the State of Lara to the east and into the Guajira peninsula to the north. Its thickness varies from 100 to 300 metres. In the Lake area the formation overlies the Maraca Formation, the uppermost unit of the Aptian-Albian Cogollo Group.

The La Luna Formation is characterized by a rich micro- and macro-fauna. Due to the euxinic conditions under which the formation was deposited, planktic-living forms strongly dominate. Planktic foraminifera are present in both the shales and the limestones. Among the megafossils it is in particular the ammonite fauna which dominates. It was ably described and illustrated by O. Renz (1982) who also gives a valuable account of the characteristics and geographic extension of the La Luna Formation in the Maracaibo Basin and neighbouring areas.

The La Luna Formation is biostratigraphically not subdivided in Fuenmayor's manual where the whole formation corresponds to the 'Molds of planktic foraminifera' on Fig. 83 and the *Cibicides* 28/*Anomalina redmondi* Zone on Fig. 84. In terms of the planktic foraminiferal zonal scheme these zones represent the Cenomanian to Santonian interval *Rotalipora brotzeni* Zone to *Dicarinella asymetrica* Zone.

The Colon Formation overlying the La Luna Formation measuring some 900 metres in thickness occurs throughout the Maracaibo Basin. Lithologically it is characterized by uniform dark shales. Locally, in the southwestern part of the Maracaibo Basin, the base of the Colon Formation is marked by a conspicuous glauconitic interval of 2–5 metres, named the Tres Esquinas Member. A Socuy Member of the Colon Formation consisting of marly limestone, also situated at its base and measuring some 40 metres, occurs throughout most of the Maracaibo Basin.

The formation is rich in microfossils with benthic and planktic foraminifera strongly dominating. Several authors, Cushman & Hedberg (1941), Sutton (1946), Sellier de Civrieux (1952) and Key (1960), published on them. Of these mention is made in particular of Sellier de Civrieux's detailed study on the benthic and planktic foraminifera of the Socuy Member situated at the base of the Colon Formation. The numerous species are illustrated on 10 plates. In

Age / Stages		Fm. Mb.	Zones	
			Benthic Foraminifera (Maraven and scientific terminology)	Planktic Foraminifera
TERTIARY	Early Miocene	La Rosa	Textularia 19 Textularia falconensis	Globigerinatella insueta Globigerinoides primordius
	Middle Eocene	Pauji	Trochammina 1 Trochammina teasi	Truncorotaloides rohri
			Cibicides 5 Cibicides sp.	
			Textularia 5 Textularia saggitula	
		Misoa	Haplophragmoides 1 var. Haplophragmoides sp.	Globigerinatheka kugleri
			Trochammina 8 / superior Trochammina sp.	
			Bolivina 27 Brizalina sp.	
			Quinqueloculina Quinqueloculina cf. moremanni	
	Early Eocene		Trochammina 8 / inferior Trochammina sp.	
	Paleocene	Guasare	Discorbis 4 Discorbis sp.	
			Rotalia 6 / superior Eponides haidingeri	
			Rotalia 6 / inferior Eponides haidingeri	
CRETACEOUS	Maastrichtian	Colon	Guembelina 2, Heterohelix sp. / Marginulina 4, Marginulina silicula	Abathomphalus mayaroensis
			Bolivina 10 Brizalina incrassata	Gansserina gansseri
				G. aegyptiaca / G. havanensis
	Campanian	Socuy	Eponides 17 Eponides simplex	Globotruncanita calcarata
				Globotruncanita elevata
	Santonian | Cenomanian	La Luna	Molds of planktic foraminifera	Dicarinella asymetrica Rotalipora brotzeni

Fig. 83. Correlation of age, stages and formations/members of benthic and planktic foraminiferal zones in the Lake Maracaibo area. Benthic zonal markers are in Maraven genus/number terminology; scientific species names after Fuenmayor. Emended from Fuenmayor (1989).

Age / Stages		Fm. Mb.	Zones	
			Benthic Foraminifera (Maraven and scientific terminology)	Planktic Foraminifera
LATE CRETACEOUS	Maastrichtian	Mito Juan	Ammobaculites 6, Ammomarginulina colombiana / Guembelitria 1, Guembelitria cretacea	Abathomphalus mayaroensis
		Colon	Siphogenerinoides 2 Siphogenerinoides parva	Gansserina gansseri | Globotruncana aegyptiaca
			Cibicides 16 Planulina spissicostata	Globotruncanella havanensis | Globotruncanita calcarata
	Campanian	Tres Esquinas	Gyroidina 12 Gyroidina globosa	Globotruncana ventricosa | Globotruncanita elevata
	Santonian	La Luna	Cibicides 28 Anomalina redmondi	Dicarinella asymetrica | Dicarinella concavata

Fig. 84. Correlation of age, stages, formations/members of Late Cretaceous benthic and planktic foraminiferal zones in the Colon district, southwest of Lake Maracaibo (Rosario, Los Manueles, Campo Tarra). Benthic zonal markers in Maraven genus/number terminology; scientific species names after Fuenmayor. Emended from Fuenmayor (1989).

the systematic part one finds, in addition to the synonymy lists, annotations on each taxon dealt with. Furthermore, on several charts the occurrence of the described Socuy fauna is compared not only with the overlying Colon Formation but also with formations from other regions such as Colombia, eastern Venezuela, Trinidad and the Gulf Coast area. Among megafossils are to be mentioned ammonites belonging to a number of different genera, quoted in O. Renz (1959).

The Colon Formation, including its members Tres Esquinas and Socuy, corresponds to the Campanian and Maastrichtian. The interval is divided into three benthic zones which correspond to the planktic interval *Globotruncanita elevata* Zone to *Abathomphalus mayaroensis* Zone. Fig. 84 shows the benthic zonal subdivision of the formation in the Colon District. Including the Tres Esquinas Member it is also divided into three zones but based on different zonal markers. They correspond to the planktic zonal interval *Globotruncanita elevata* Zone to *Gansserina gansseri* Zone.

The Mito Juan Formation (Fig. 84) is distinguished in the western and southern part of the Maracaibo Basin. Lithologically it is similar to the Colon Formation from which it differs in an interstratification of thin arenaceous, sometimes glauconitic layers and limestones. Because of its similarity with the underlying Colon Formation its base may be difficult to establish. For separation from the overlying Paleocene Guasare Formation it is possible to use the distinctive basal arenaceous layer of that formation.

The foraminifera include benthic and planktic forms which are reduced in number compared to the underlying Colon fauna. The entire Mito Juan Formation is placed into one benthic zone which correlates with the Late Maastrichtian *Abathomphalus mayaroensis* Zone.

The Guasare Formation with a thickness varying between some 120 and 450 metres extends throughout most of the Maracaibo Basin. Lithologically it consists of fossiliferous limestones and calcareous sands, interbedded with mainly fine-grained, delicately stratified sands and locally glauconitic or calcareous shales. The contact with the underlying Mito Juan Formation is normally transitional.

The fossil content of the Guasare Formation displays a great variety ranging from molluscs to foraminifera, and to palynomorphs in its brackish to non-marine parts. The formation is divided into three zones in Fuenmayor's benthic zonal scheme (Fig. 83). A correlation of them with the planktic foraminiferal zonal scheme is not given because of the paucity or total absence of such forms. No foraminiferal distribution chart is given by Fuenmayor for the Guasare Formation.

The Misoa Formation, of variable thickness, measures at its type locality some 5000 metres, to become reduced to 1000 metres or less in the oil producing areas of the Maracaibo Basin where it contains the most important oil resources of the Basin. Lithologically the formation consists of sands and siltstones and contains in its lower part intercalated shales with some limestone layers. The contact of the Misoa Formation with the underlying Guasare Formation is discordant in the Lake Maracaibo subsurface, but it is concordant with the overlying Pauji Formation.

Some molluscs and rare foraminifera have been reported from the shale layers of the Misoa Formation. Of biostratigraphic interest are larger foraminiferal species published by Raadshoven (1951) from limestone layers.

Fuenmayor defines in the Early to Middle Eocene Misoa Formation five benthic zones. The topmost one, the *Haplophragmoides* 1 var. Zone, is correlated with the Middle Eocene planktic foraminiferal zonal scheme, while for the other four no such correlation is offered, apparently due to the paucity or total absence of diagnostic species.

The Pauji Formation measures some 1200 metres at its type section in the foothills east of Lake Maracaibo from where it extends over large parts of the Basin.

Lithologically the formation consists of a sequence of grey shales clearly distinct from the sands of the underlying Misoa Formation.

The shales carry a rich foraminiferal fauna, published by Nuttall (1935). Based on the associated planktic species *Globigerina senni*, '*Hastigerina*' *bolivariana*, *Morozovella spinulosa* and *Truncorotaloides rohri*, the Pauji Formation falls into the Middle Eocene.

In Fuenmayor (1989) the formation is divided into three benthic foraminiferal zones, correlated with the Middle Eocene planktic zonal interval *Truncorotaloides rohri* to *Globigerinatheka kugleri* Zone without, however, offering a closer correlation between the three benthic and the individual planktic zones of this interval (Fig. 83).

The Miocene La Rosa Formation is of variable thickness, measuring up to 250 metres and in most places discordantly overlying the Middle Eocene Pauji Formation. It occurs in the subsurface of Lake Maracaibo, in the Boscan field southwest of Maracaibo and in the eastern coastal area. It consists primarily of greenish, laminated shales and is the only open marine Miocene unit in the Maracaibo Basin. The formation is richly fossiliferous, containing abundant molluscs. Hoffmeister (1938) mentions the presence of the genus *Sorites*. Other larger foraminifera that occur in the La Rosa Formation such as *Amphistegina*, *Miogypsina*, *Peneroplis* are quoted in Haas & Hubman (1937).

The age of the formation, earlier regarded as Oligocene but later referred to the Middle Miocene, is now considered to fall into the Early Miocene on planktic foraminiferal evidence. Fuenmayor (1989) places the whole formation into the *Textularia* 19 (*Textularia falconensis*) Zone which is correlated as shown on Fig. 83 with the planktic foraminiferal interval *Globigerinoides primordius* to *Globigerinatella insueta* Zone.

Barbados

Introduction

Studies on the geology of Barbados go back to the last century with the first account of the Oceanic Formation sediments being given by Jukes-Browne & Harrison (1892). They recognized that what they called the basal *Globigerina* marls had been deposited at a water depth of 900 m to 1800 m while the radiolarian earths were evidence of a depth from 3600 m to 5400 m. These were unusual findings as it was thought at that time to be highly unlikely that such deep sea sediments would be found on land.

The first modern comprehensive study of the sediments grouped as the Oceanic Formation of Barbados was carried out by Alfred Senn at the time that he was also producing a map of the pre-Pleistocene rocks of the island. His publications (Senn 1940, 1947) gave field information on the outcrops of the Oceanic Formation but no detail of the foraminiferal faunas, though these were used by him for dating. Beckmann (1953) used Senn's samples for a detailed analysis of the foraminiferal fauna and this has been the basis on which the ranges of the benthics have been plotted here (Fig. 86).

The earlier studies were hampered by a lack of understanding of the nature of accretionary prisms and their relationship to subduction zones at plate boundaries. The new generation of work undertaken by R. C. Speed and his co-workers (Speed 1981, 1988; Torrini 1988) has provided a sound framework in which to place rock sequences, the dating of which by the help of microfossils has, in return, been essential for the understanding of the structural complexities.

Senn's continued use of the term Oceanic Formation for the partly indurated calcareous and siliceous oozes was correct in that they form mappable units. What was not realized is that they comprise deep sea sediments that have been scraped off the surface of the Atlantic Plate and from fore-arc basins and piled up in a series of imbricated nappes to form part of an accretionary prism. The existence of the prism is the result of the subduction of the Atlantic plate below

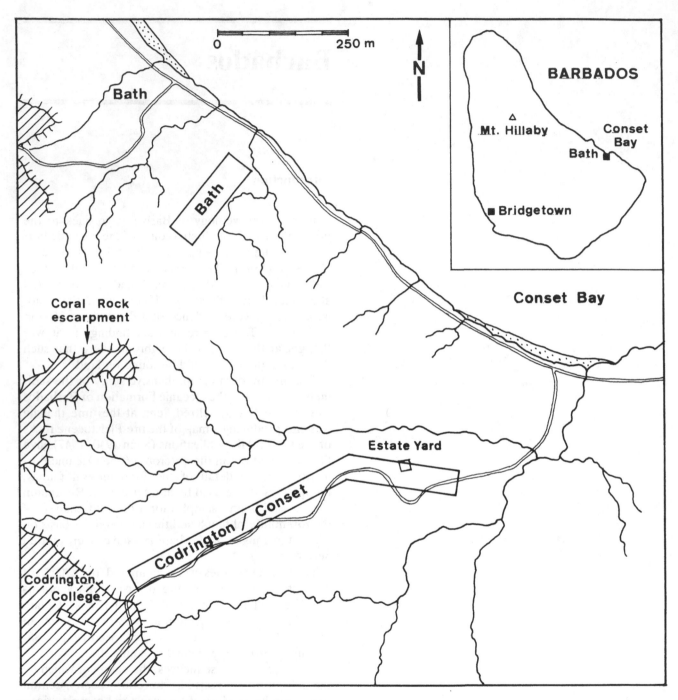

Fig. 85. Map of Barbados showing location of stratigraphic sections.

the Caribbean plate as the two continue to collide along the line of the West Indian Arc. Such accretionary prisms are widely known as the result of deep ocean drilling. The present day toe of the Barbados prism has been penetrated by the drill in two legs of the Ocean Drilling Project (Leg 78A: Biju Duval,

Moore *et al*. 1984. Leg 110: Moore, Mascle *et al*. 1990).

The imbricated packages of Oceanic Formation sediments that now form discrete nappes within the prism have come from different locations on the ocean floor and thus reflect differing original environments.

Particularly important is the paleo-water depth at various times during the life of any particular package as the proximity to the carbonate compensation depth (CCD) will strongly affect the richness of the calcareous microfauna. In this respect, Saunders *et al.* (1984) estimated for the section at Bath Cliff a loss of 90% or more of the calcareous fauna in the Eocene decreasing to a loss of nearer 50% in the Oligocene. The loss of tests of both planktic and benthic foraminifera by solution affects the ranges of the species and will be different for the various sections where Oceanic sediments can be studied. Using the results from Mount Hillaby combined with those from Bath and Codrington helps to overcome this difficulty.

Saunders *et al.* (1984) estimated a water depth at the time of deposition of the Oceanic sediments in the Bath Cliff section to be almost certainly greater than 2000 m and probably more than 2800 m.

The position of the CCD in the Late Eocene in the area of Barbados is considered by Saunders *et al.* (1984) to have been no deeper than 4800 m by back-tracking the position of DSDP Site 543, drilled on the abyssal plain in front of the toe of the accretionary prism on Leg 78A (see Biju-Duval *et al.* 1984). The Late Eocene levels in this hole are entirely siliceous, all calcareous components having been dissolved.

The difficulty of being certain of the depth of the CCD means that a possible range of water depth for the Late Eocene in the Bath Cliff section is from 2800 m to 4800 m.

A probable paleo-depth of 2500 m to 3000 m for the Oceanic Formation was given by Wood *et al.* (1985) but this is not well constrained as samples from all sections are combined whereas it is now realised that sediments from more than one nappe represent different original paleoenvironments, and particularly different paleo-water depths.

Middle Eocene to base Miocene

To create the range chart of benthic foraminifera (Fig. 86), those Senn samples held at the Basel Natural History Museum which were used by Beckmann in his work (1953) have been re-studied to verify the age determinations. In addition, we have used the work of Saunders & Cordey (1968) and Saunders *et al.* (1984) in the Bath and Codrington outcrop areas.

The location of the sections used is shown on Figure 85. The Mount Hillaby section is indicated on the inset map only as a point while the Bath and Codrington College/Conset section is shown on a larger scale. More detail of the lithologies and sample positions is to be found in Senn (1940) and Beckmann (1953) with additional information in Saunders & Cordey (1968) and Saunders *et al.* (1984). A detailed geological map of the overall Conset Bay area, with a modern interpretation of the complicated structure of the various nappes forming this part of the accretionary prism, is given in Torrini *et al.* (1985). On fig. 2 of their paper it can be seen that both the Bath profile and the major part of the Conset profile originally sampled by Senn and used by Beckmann fall within the same nappe, though with a zone of melange (earlier called mudflow) within it. The coherence of the section is important in indicating a common origin for the sediment within it.

New dating of the samples used to create the range chart of Figure 86 according to the planktic zonation used in Bolli *et al.* (1985) has been attempted. Any uncertainty is caused by the lack of foraminifera at some levels due to the nearness of the CCD at the time of deposition. It is the lower part of the section, within the Late and Middle Eocene, that suffers most in this respect.

The oldest stratigraphic level included in the present study is thought to be the *Morozovella lehneri* Zone though the evidence for this on foraminifera in the Bath material used by Beckmann is sparse due to solution of the fauna. The situation in the Mount Hillaby section is worse and so, for this reason, a query has been placed against this zone on Figure 86.

The topmost samples, which are found close to Codrington College (Fig. 85), fall within the *Globorotalia kugleri* Zone very close to the boundary between the Oligocene and the Miocene. The top of the *Cassigerinella chipolensis/Pseudohastigerina micra* Zone, which forms the lowest part of the Oligocene, is not well constrained due to a paucity of the delicate tests of the marker species, but sediments representing the zone are certainly present.

Fig. 86. Distribution of selected Eocene to Oligocene benthic foraminifera from the Oceanic Formation of Barbados.

Alabamina dissonata	42
Anomalinoides alazanensis spissiformis	49
Anomalinoides badenensis	60
Anomalinoides bilateralis	47
Anomalinoides dorri aragonensis	43
Bifarina inopinata	37
Bolivina cf. tectiformis	35
Bulimina jarvisi	17
Bulimina macilenta	20
Bulimina ovata	23
Chrysalogonium lanceolum	14
Cibicidoides cookei	62
Cibicidoides grimsdalei	50
Cibicidoides havanensis	51
Cibicidoides martinezensis	44
Cibicidoides perlucidus	58
Cibicidoides pseudoungerianus	61
Cibididoides cf. pseudoungerianus	46
Cibicidoides robertsonianus haitiensis	68
Cibicidoides subapiratus limbatus	41
Cibicidoides trincherasensis	63
Dorothia brevis	12
Dorothia colei	2
Dorothia nipeensis	8
Ellipsodimorphina subcompacta	13
Gaudryina jacksonensis	1
Gaudryina pseudocollinei	6
Guttulina lehneri	26
Gyroidinoides altiformis	52
Gyroidinoides girardanus	53
Gyroidinoides girardanus peramplus	54
Gyroidinoides planulatus	55
Gyroidinoides soldanii octocameratus	59
Hanzawaia illingi	64
Heterolepa mexicana	67
Karreriella baccata	10
Karreriella hantkeniana	7
Karreriella subcylindrica	5
Karreriella washingtonensis	9
Lagena asperoides	18
Lagena ciperensis	15
Lagena flintiana indomita	19
Lagena pulcherrima enitens	32
Laticarinina bullbrooki	66
Linaresia pompilioides semicribrata	57
Martinottiella petrosa	11
Nodosarella mappa	27
Nuttallides truempyi	45
Oridorsalis umbonatus ecuadorensis	56
Orthomorphina havanensis	38
Osangularia mexicana	48
Planulina renzi	65
Plectofrondicularia lirata	22
Plectofrondicularia trinitatensis	21
Plectofrondicularia vaughani	33
Pyrulinoides antilleanus	16
Sigmomorphina trinitatensis	24
Siphogenerina aff. transversa	31
Siphonodosaria annulifera	28
Siphonodosaria consobrina	29
Siphonodosara curvatura	25
Siphonodosaria curvatura spinea	34
Spiroplectammina trinitatensis	3
Stilostomella matanzana	30
Stilostomella modesta	39
Uvigerina nuttalli	40
Uvigerina spinicostata	36
Vulvulina spinosa	4

Fig. 87.

1, 1a, 2 *Spiroplectammina trinitatensis* (Cushman & Renz)
from Beckmann, 1953, pl. 17, figs. 1, 2, × 18. **1,1a**
macroscopic form; **2** microspheric form.

3, 3a *Vulvulina spinosa* Cushman
from Beckmann, 1953, pl. 17, fig. 6, × 18.

4, 4a *Gaudryina jacksonensis* Cushman
from Beckmann, 1953, pl. 17, fig. 9, × 18.

5, 5a *Gaudryina pseudocollinsi* Cushman & Stainforth
from Beckmann, 1953, pl. 17, fig. 8, × 18.

6, 6a *Dorothia brevis* Cushman & Stainforth
from Beckmann, 1953, pl. 17, fig. 14, × 18.

7, 7a *Dorothia nipeensis* Keijzer
from Beckmann, 1953, pl. 17, fig. 19, × 18.

8, 8a *Dorothia colei* (Nuttall)
from Beckmann, 1953, pl. 17, fig. 18, × 18.

9 *Martinottiella petrosa* (Cushman & Bermudez)
from Beckmann, 1953, pl. 17, fig. 28, × 18.

10, 10a *Karreriella subcylindrica* (Nuttall)
from Beckmann, 1953, pl. 17, fig. 25, × 18

11, 11a *Karreriella baccata* (Schwager)
from Beckmann, 1953, pl. 17, fig. 21, × 18.

12, 12a *Karreriella hantkeniana* Cushman
from Beckmann, 1953, pl. 17, fig. 24, × 18.

13 *Karreriella washingtonensis* Rau
from Beckmann, 1953, pl. 17, fig. 27, × 18.

14 *Bifarina inopinata* (Cushman & Stainforth)
from Beckmann, 1953, pl. 21, fig. 3, × 50.

15 *Plectofrondicularia lirata* Bermudez
from Beckmann, 1953, pl. 21, fig. 4, × 18.

16, 16a *Plectofrondicularia trinitatensis* Cushman & Jarvis
from Beckmann, 1953, pl. 21, fig. 5, × 10.

17 *Bolivina* cf. *tectiformis* Cushman
from Beckmann, 1953, pl. 21, fig. 17, × 35.

18 *Bulimina ovata* d'Orbigny
from Beckmann, 1953, pl. 21, fig. 12, × 18.

19 *Bulimina jarvisi* Cushman & Parker
from Beckmann, 1953, pl. 21, fig. 10, × 35.

20 *Bulimina macilenta* Cushman & Parker
from Beckmann, 1953, pl. 21, fig. 11, × 35.

21 *Uvigerina nuttalli* Cushman & Edwards
from Beckmann, 1953, pl. 21, fig. 18, × 35.

22 *Uvigerina spinicostata* Cushman & Jarvis
from Beckmann, 1953, pl. 21, fig. 19, × 35.

23, 23a *Siphogenerina* aff. *transversa* Cushman
from Beckmann, 1953, pl. 21, fig. 21, × 18.

24 *Plectofrondicularia vaughani* Cushman
from Beckmann, 1953, pl. 21, fig. 6, × 35.

25, 25a *Lagena pulcherrima enitens* Cushman & Stainforth
from Beckmann, 1953, pl. 20, fig. 7, × 18.

26, 26a *Lagena flintiana indomita* (Cushman & Stainforth)
from Beckmann, 1953, pl. 19, fig. 36, × 25.

27 *Lagena asperoides* Galloway & Morrey
from Beckmann, 1953, pl. 19, fig. 28, × 18.

28, 28a *Lagena ciperensis* Cushman & Stainforth
from Beckmann, 1953, pl. 19, fig. 32, × 25.

29, 29a *Pyrulinoides antilleanus* Beckmann
from Beckmann, 1953, pl. 20, fig. 19, × 18.

30, 30a *Sigmomorphina trinitatensis* Cushman & Ozawa
from Beckmann, 1953, pl. 20, fig. 27, × 18.

31, 31a *Guttulina lehneri* Cushman & Ozawa
from Beckmann, 1953, pl. 20, fig. 14, × 18.

32–34 *Ellipsodimorphina subcompacta* Liebus
from Beckmann, 1953, pl. 23 figs. 4, 3, 2, × 25.
32 uniserial, **33** with biserial early stage, **34** with
triserial early stage.

35 *Chrysalogonium lanceolum* Cushman & Jarvis
from Beckmann, 1953, pl. 19, fig. 9, × 18.

36 *Orthomorphina havanensis* (Cushman & Bermudez)
from Beckmann, 1953, pl. 21, fig. 7, × 35

37, 37a *Nodosarella mappa* (Cushman & Jarvis)
from Beckmann, 1953, pl. 22, fig. 22, × 18.

38, 38a *Stilostomella modesta* (Bermudez)
from Beckmann, 1953, pl. 21, fig. 32. **38** × 35, **38a**
× 50.

39 *Stilostomella matanzana* (Palmer & Bermudez)
from Beckmann, 1953, pl. 21, fig. 31, × 35.

40, 40a *Siphonodosaria annulifera* (Cushman & Bermudez)
from Beckmann, 1953, pl. 21, fig. 23. **40** × 10, **40a**
× 25.

41 *Siphonodosaria consobrina* (d'Orbigny)
from Beckmann, 1953, pl. 21, fig. 25, × 35.

42 *Siphonodosaria curvatura* (Cushman)
from Beckmann, 1953, pl. 21, fig. 26, × 10.

43 *Siphonodosaria curvatura spinea* (Cushman)
from Beckmann, 1953, pl. 21, fig. 28, × 18.

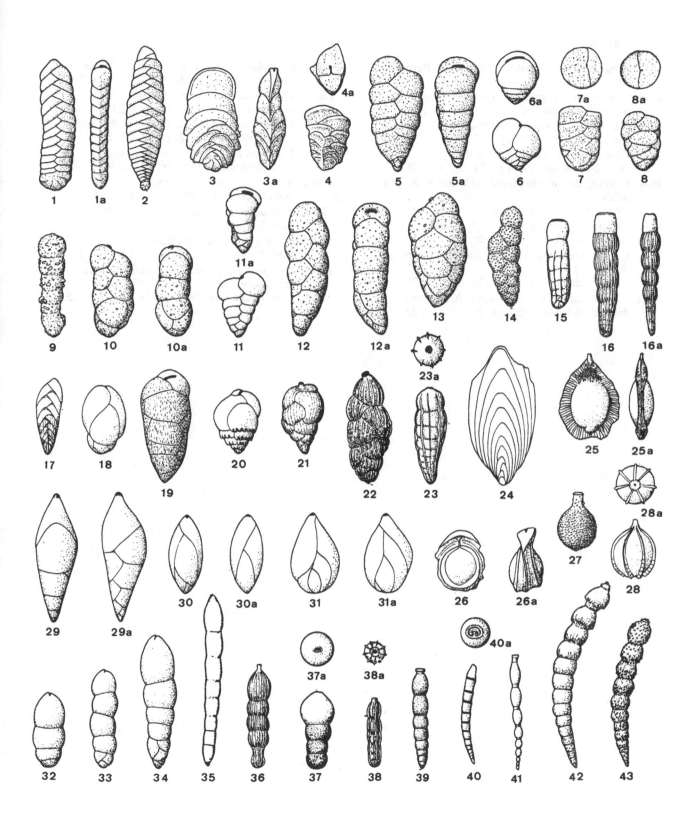

Fig. 88.

1a–c *Cibicidoides havanensis* (Cushman & Bermudez)
from Beckmann, 1953, pl. 27, fig. 8, × 25.

2a–c *Cibicidoides* cf. *pseudoungerianus* (Cushman)
from Beckmann, 1953, pl. 28, fig. 10, × 25.

3a–c *Cibicidoides perlucidus* (Nuttall)
from Beckmann, 1953, pl. 28, fig. 2, × 25.

4a–c *Cibicidoides grimsdalei* (Nuttall)
from Beckmann, 1953, pl. 27, fig. 7, × 25.

5a–c *Cibicidoides martinezensis* (Cushman & Barksdale)
from Beckmann, 1953, pl. 27, fig. 9, × 35.

6a–c *Cibicidoides trincherasensis* (Bermudez)
from Beckmann, 1953, pl. 28, fig. 8, × 25.

7a–c *Cibicidoides robertsonianus haitensis* (Coryell
& Rivero)
from Beckmann, 1953, pl. 28, fig. 5, × 25.

8a–c *Cibicidoides pseudoungerianus* (Cushman)
from Beckmann, 1953, pl. 28, fig. 4, × 25.

9a–c *Cibicidoides cookei* (Cushman & Garrett)
from Beckmann, 1953, pl. 27, fig. 6, × 25.

10a–c *Cibicidoides subspiratus limbatus* (Cita)
from Beckmann, 1953, pl. 28, fig.7, × 25.

11a–c *Laticarinina bullbrooki* Cushman & Todd
from Beckmann, 1953, pl. 27, fig. 5, × 25.

12a–c *Planulina renzi* Cushman & Stainforth
from Beckmann, 1953, pl. 27, fig. 4, × 18.

13a–c *Nuttallides truempyi* (Nuttall)
from Beckmann, 1953, pl. 24, fig. 2, × 25.

14a–c *Alabamina dissonata* (Cushman & Renz)
from Beckmann, 1953, pl. 24, fig. 4, × 35.

15a–c *Osangularia mexicana* (Cole)
from Beckmann, 1953, pl. 24, fig. 1, × 25.

16a–c *Oridorsalis umbonatus ecuadorensis* (Galloway
& Morrey)
from Beckmann, 1953, pl. 23, fig. 27, × 25.

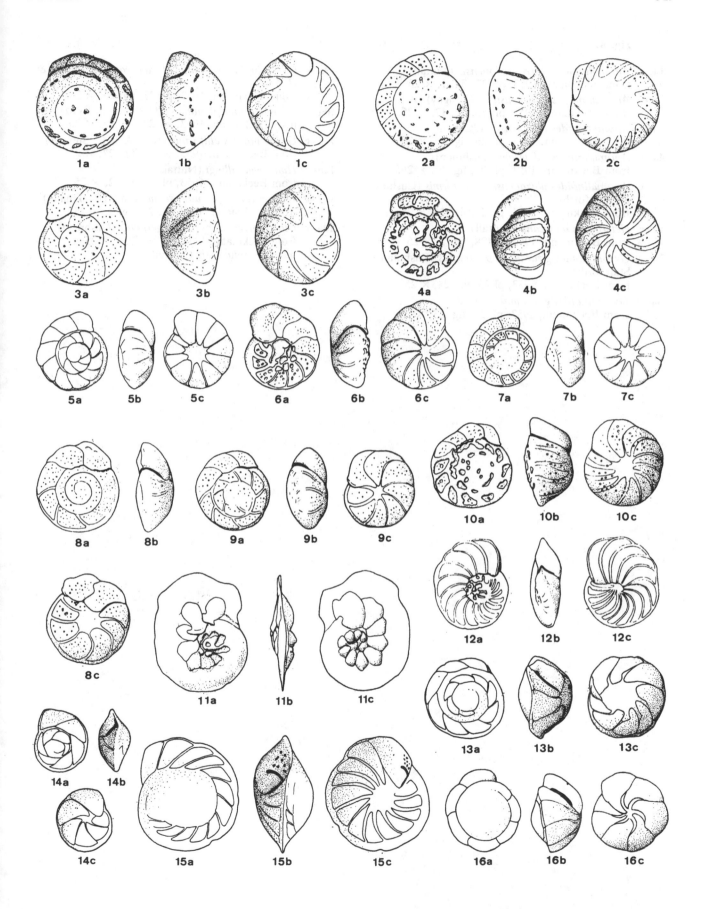

1a 1b 1c 2a 2b 2c

3a 3b 3c 4a 4b 4c

5a 5b 5c 6a 6b 6c 7a 7b 7c

8a 8b 9a 9b 9c 10a 10b 10c

8c 11a 11b 11c 12a 12b 12c

13a 13b 13c

14a 14b 14c 15a 15b 15c 16a 16b 16c

Fig. 89.

1a–c *Anomalinoides dorri aragonensis* (Nuttall)
from Beckmann, 1953, pl. 27, fig. 1, × 25.

2a–c *Anomalinoides badenensis* (d'Orbigny)
from Beckmann, 1953, pl. 26, fig. 17, × 25.

3a–c *Anomalinoides badenensis* (d'Orbigny)
from Beckmann, 1953, pl. 26, fig. 16, × 25

4a–c *Anomalinoides bilateralis* (Cushman)
from Beckmann, 1953, pl. 26, fig. 15, × 26.

5a–c *Anomalinoides alazanensis spissiformis* (Cushman
& Stainforth)
from Beckmann, 1953, pl. 26, fig. 14, × 25.

6a–c *Heterolepa mexicana* (Nuttall)
from Beckmann, 1953, pl. 28, fig. 1, × 25.

7a–c *Gyroidinoides girardanus peramplus* (Cushman &
Stainforth)
from Beckmann, 1953, pl.23, fig. 24, × 25.

8a–c *Gyroidinoides girardanus* (Reuss)
from Beckmann, 1953, pl. 23, fig. 23, × 25.

9a–c *Gyroidinoides soldanii octocameratus* (Cushman &
Hanna)
from Beckmann, 1953, pl. 23, fig. 26, × 25.

10a–c *Gyroidinoides planulatus* (Cushman & Renz)
from Beckmann, 1953, pl. 23, fig. 25, × 25.

11a–c *Gyroidinoides altiformis* (R.E. & K.C. Stewart)
from Beckmann, 1953, pl. 23, fig. 22, × 25.

12a–c *Hanzawaia illingi* (Nuttall)
from Beckmann, 1953, pl. 27, fig. 2, × 25.

13a–c *Linaresia pompilioides semicribrata* (Beckmann)
from Beckmann, 1953, pl. 27, fig. 3, × 25.

14a–c *Cibicidoides* cf. *pseudoungerianus* (Cushman)
from Beckmann, 1953, pl. 28, fig. 9, × 25, as *Cibicides* cf. *trinitatensis* (Nuttall).

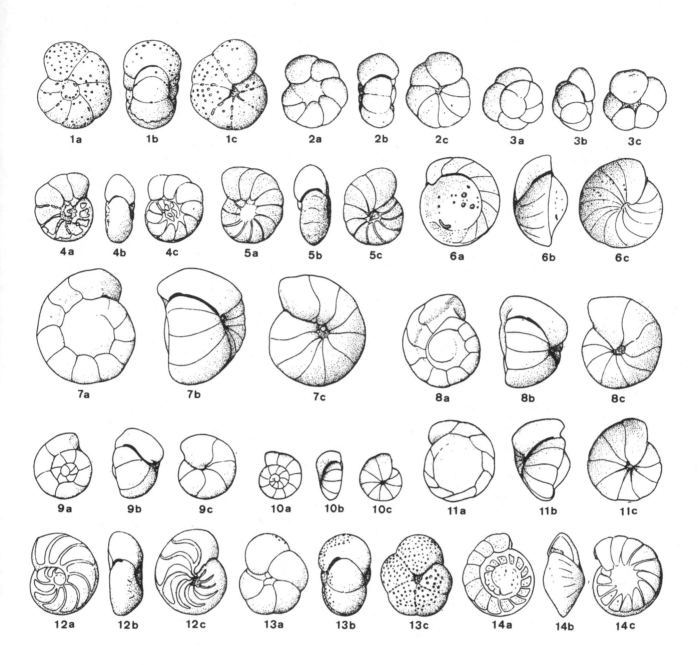

1a 1b 1c 2a 2b 2c 3a 3b 3c

4a 4b 4c 5a 5b 5c 6a 6b 6c

7a 7b 7c 8a 8b 8c

9a 9b 9c 10a 10b 10c 11a 11b 11c

12a 12b 12c 13a 13b 13c 14a 14b 14c

Annotated taxonomic list of selected late Early Eocene to Middle Miocene benthic foraminifera of Trinidad, Venezuela (Falcon) and Barbados

Introductory Remarks

To avoid repetitions in the treatment of the numerous taxa which are identical in the three areas covering Trinidad, Venezuela (Falcon) and Barbados, the systematics of all chosen late Early Eocene to Middle Miocene taxa are combined in a single chapter. Their ranges and illustrations are however attached to the individual parts: Trinidad (Figs. 50–52, 53–64), Falcon (Figs. 73–75, 76–81) and Barbados (Figs. 86, 87–89).

The selected late Early Eocene to Middle Miocene benthic foraminifera included here were described and illustrated from Trinidad by Cushman & Renz (1947b, 1948) and Cushman & Stainforth (1945), from Falcon by Cushman & Renz (1941), Renz (1948), Blow (1959), Diaz de Gamero (1977a) and from Barbados by Beckmann (1953), Saunders et al. (1984) and Wood et al. (1985). The incorporation in the present book of all the taxa published by these authors from the three areas would have grown beyond the limits set. A selection similar to that for the Barremian to Early Albian of Trinidad had therefore to be made.

Instead of randomly choosing taxa it was decided rather to concentrate on certain characteristic genera represented in each area, many of them of stratigraphic significance, and only in second place to include taxa from additional genera. The following are the genera chosen whose species are followed more closely in each of the investigated areas:

Bolivina
Bulimina
Uvigerina
Rectuvigerina
Siphogenerina
calcareous trochospiral genera

Introductory comments on these genera where necessary are found in the systematic part at the head of the respective species descriptions. Annotations on some species from this selected group of genera discuss for instance the distinctions of morphologically closely related forms or consist of comparisons of the illustrated Caribbean taxa with the corresponding type species which in such cases are also illustrated.

While the taxa of the above listed genera are included for all three areas – Trinidad, Falcon, Barbados – the additionally chosen species were selected more for their local significance. For Trinidad these are illustrated on Figs. 62–64 and their ranges – except for those on Fig. 64 – are shown on Fig. 52. As can be deduced from Fig. 62 and 63 some preference is given to representatives of certain genera such as *Plectofrondicularia* (Fig. 62.20–30), *Ellipsolagena–Entosolenia–Lagena* (Fig. 63.1–13), and characteristic nodosariids (Fig. 63.14–30).

Illustrated on Fig. 64.1–32 are agglutinated species that play an important role in the Trinidad Oligo-Miocene as stratigraphic and environmental markers.

In the Agua Salada Formation of southeastern Falcon numerous *Textularia* species as well as some other agglutinated taxa are typical for the benthic part of that fauna. Some are of stratigraphic, others more of environmental significance. They are illustrated on Fig. 76 and their ranges plotted on Fig. 74.

In contrast to their occurrence in the corresponding Trinidad formations, lenticulinids are often frequent and represented in the Agua Salada Formation by some characteristic, though mostly long ranging species. They are illustrated on Fig. 77.1–30, and their ranges shown on Fig. 74. In addition to the lenticulinids, many species of nodosariids also characterize the Agua Salada fauna. Some of the more typical ones are illustrated on Fig. 77.31–42 and their ranges plotted on Fig. 74.

In addition to the species of the genera *Bolivina*, *Bulimina*, *Uvigerina* and *Siphogenerina*, only a few characteristic agglutinated species are illustrated from central Falcon on Fig. 81.1–6, with their ranges shown on Fig. 75.

The selection of the benthic taxa from Barbados is focussed in first place on the 29 calcareous trochospi-

ral species which are illustrated on Fig. 88 and 89 and with their ranges plotted on Fig. 86. Fig. 87 contains a selection of other benthic species characteristic for the investigated Barbados sediments, some with quite restricted ranges as can be seen on Fig. 86. On Fig. 87.17–23 are also illustrated the few representatives in the Oceanic Formation of the genera *Bolivina*, *Bulimina*, *Uvigerina* and *Siphogenerina*.

Agglutinated taxa

Family Ammodiscidae

AMMODISCUS INCERTUS d'Orbigny

Figure **76.38, 39**; 74

Ammodiscus incertus d'Orbigny 1839, p. 49, pl. 6, figs. 16, 17, – Renz 1948, p. 113, pl. 1, figs. 1,2.

AMMOVERTELLA RETRORSA Cushman & Stainforth
Figure **62.2**; 52

Ammovertella retrorsa Cushman & Stainforth 1945, p. 14, pl. 1, fig. 5. – Cushman & Renz 1948, p. 7, pl. 2, fig. 1.

Family Haplophragmoididae

HAPLOPHRAGMOIDES CARINATUS Cushman & Renz
Figure **76.33a–b**; 74

Haplophragmoides carinatum Cushman & Renz 1941, p. 2, pl. 1, fig. 1. – Renz 1948, p. 141, pl. 1, figs. 4a–b.

HAPLOPHRAGMOIDES CORONATUS (Brady)
Figure **76.32**; 74

Trochammina coronata Brady 1879, Quart. Jour. Micr. Sci. London, 19, p. 58, pl. 5, fig. 15.
Haplophragmoides coronatum (Brady), Renz 1948, p. 141, pl. 1, fig. 5.

HAPLOPHRAGMOIDES EMACIATUS
(Brady)
Figure **76.34a–b; 74**

Haplophragmium emaciatum Brady 1884,
  p. 305, pl. 33, figs. 26–28.
Haplophragmoides emaciatum (Brady), Renz
  1948, p. 142, pl. 1, figs. 6a–b.

Family Lituolidae

LITUOTUBA NAVETENSIS Cushman &
Renz
Figure **62.1; 52**

Lituotuba eocenica Cushman & Renz 1948,
  p. 6, pl. 1, fig. 20 (holotype), fig. 21
  (paratype, figured here).
Lituotuba navetensis Cushman & Renz, new
  name, 1950, p. 45.

AMMOBACULITES CUBENSIS Cushman
& Bermudez
Figure **62.3; 52**

Ammobaculites cubensis Cushman &
  Bermudez 1937, Contrib. Cushman Lab.
  foramin. Res., 13, p. 106, pl. 16, fig. 4. –
  Cushman & Renz 1948, p. 10, pl. 2, fig. 2.

AMMOBACULITES cf.
STRATHEARNENSIS Cushman & LeRoy
Figure **76.36,37; 74**

Ammobaculites strathearnensis Cushman &
  LeRoy 1938, J. Paleontol., 12, p. 122,
  pl. 22, figs. 1a–b. – Renz 1948, pl. 113, p. 1,
  figs. 7–8.

DISCAMMINOIDES TOBLERI
Broennimann
Figure **64.1a–b, 2, 3**

Discamminoides tobleri Broennimann 1951,
  p. 103, pl. 11, fig. 7, textfigs. 9–12.

The species is planispiral with a finely agglutinated test that may become irregular in the final
chambers, with a tendency to become plani-

spiral. The surface is smooth, the interior of the
walls is alveolar.

Common at the top of the Lengua Formation
where this passes gradually upwards into the
non-calcareous clay facies of the Lower Cruse
Formation. As a species confined to silty clay or
pure clay, it may occur again in younger beds
when the lithology is suitable. Found also in the
Early Miocene Karamat and Nariva clays, and
in the Cipero Formation where it sometimes
shows a coarser test wall.

Family Cyclamminidae

CYCLAMMINA CANCELLATA Brady
Figure **76.40; 74**

Cyclammina cancellata Brady 1884, p. 351,
  pl. 37, figs. 8–16. – Renz 1948, p.129, pl.1,
  fig. 9.

Family Spiroplectamminidae

SPIROPLECTAMMINA
TRINITATENSIS Cushman & Renz
Figure **62.7–8; 52. Figure 87.1, 1a, 2; 86**

Spiroplectammina trinitatensis Cushman &
  Renz 1948, p. 11, pl. 2, figs. 13, 14. – Saunders
  et al. 1984, p. 404, pl. 2, figs. 1, 2.
Bolivinopsis trinitatensis (Cushman & Renz),
  Beckmann 1953, p. 339, pl. 17, figs. 1, 2.

VULVULINA SPINOSA Cushman
Figure **87.3, 3a; 86**

Vulvulina spinosa Cushman 1927, Contrib.
  Cushman Lab. foramin. Res., 3, p. 111, pl. 23,
  fig. 1. – Beckmann 1953, p. 340, pl. 17, figs.
  6, 7. – Saunders *et al.* 1984, p. 404, pl. 2, figs.
  1, 2. – Wood *et al.* 1985, pl. 1, figs. 1, 2, 5.

Family Trochamminidae

TROCHAMMINA PACIFICA Cushman
Figure **76.35a–b**, 74

Trochammina pacifica Cushman 1925, Contrib.
    Cushman Lab. foramin. Res., 1, p. 39, pl. 6,
    fig. 3. – Renz 1948, p. 172, pl. 3, figs. 4a–b,
    5a–b.

Family Prolixoplectidae

PLECTINA TRINITATENSIS Cushman &
Renz
Figure **62.11;** 52

Plectina trinitatensis Cushman & Renz, 1948,
    p. 15, pl. 3, fig. 12.

Family Verneuilinidae

GAUDRYINA BULLBROOKI Cushman
Figure **76.24;** 74

Gaudryina (Pseudogaudryina) bullbrooki
    Cushman 1936, Cushman Lab. foramin. Res.,
    Spec. Publ., 6, p. 16, pl. 2, fig. 16. – Renz
    1948, p. 135, pl. 2, fig. 6.

GAUDRYINA JACKSONENSIS Cushman
Figure **87.4, 4a;** 86

Gaudryina jacksonensis Cushman 1926,
    Contrib. Cushman Lab. foramin. Res., 2,
    p. 33, pl. 5, fig. 1. – Beckmann 1953, p. 340,
    pl. 17, fig. 9.

**GAUDRYINA JACKSONENSIS
ABNORMIS** Cushman & Renz
Figure **76.22a–b, 23;** 74

Gaudryina (Pseudogaudryina) jacksonensis
    var. *irregularis* Cushman & Renz 1941, p. 6,
    pl. 1, figs. 11–12.
Gaudryina (Pseudogaudryina) jacksonensis
    var. *abnormis* Cushman & Renz, new
    name, 1944, Contrib. Cushman Lab.

foramin. Res., 20, p. 78. – Renz 1948,
    p. 136, pl. 2, figs. 4, 5a–b.

GAUDRYINA LEUZINGERI Cushman &
Renz
Figure **76.25;** 74

Gaudryina leuzingeri Cushman & Renz 1941,
    p. 6, pl. 1, fig. 13. – Renz 1948, p. 135, pl. 2,
    figs. 7a–b.

GAUDRYINA PSEUDOCOLLINSI
Cushman & Stainforth
Figure **62.4;** 52. Figure **87.5, 5a;** 86

Gaudryina pseudocollinsi Cushman &
    Stainforth 1945, p. 17, pl. 2, figs. 1–3 (1
    holotype, 2, 3 paratypes) – Beckmann 1953,
    p. 341, pl. 17, fig. 8.

Family Globotextulariidae

GRAVELLINA NARIVAENSIS
Broennimann
Figure **64.4–7.** Figure **81.1;** 75

Gravellina narivaensis Broennimann 1953,
    p. 87, pl. 15, fig. 9, textfig. 1. – Diaz de
    Gamero 1977, p. 40, pl. 3, fig. 12.

This is a valvulinid form characterized by a
quadriserial arrangement of chambers throughout the test; apparent even in crushed forms.
The wall is finelly agglutinated, the surface
smooth.
    The species is characteristic for the Late
Oligocene to Early Miocene Nariva clays.

JARVISELLA KARAMATENSIS
Broennimann
Figure **64.8a–b**

Jarvisella karamatensis Broennimann 1953,
    p. 88, pl. 15, fig. 7, textfigs. 2, 3.

The species possesses a finely agglutinated test
with smooth surface. The characteristic infolding
of the chamber walls gives an undulating appearance even in crushed specimens.
    Confined to and characteristic of the Karamat

clays of Middle Miocene age in Trinidad. In Venezuela probably in comparable formations.

Family Textulariellidae

ALVEOVALVULINA SUTERI
Broennimann
Figure **64.10a–b, 11–14**

Alveovalvulina suteri Broennimann 1951, p. 102, pl. 11, fig. 5, textfigs. 5–8.

The test is rapidly increasing in size with usually three chambers in the last whorl occupying a large proportion. The wall is finely agglutinated with alveolar cavities in the adult chambers, characteristically visible on the surface as mottling or even as holes in slightly eroded shells.

The taxon is recorded from the Late Miocene of Trinidad, continuing into the Pliocene. Present also in age equivalent beds in eastern Venezuela.

GUPPYELLA MIOCENICA (Cushman)
Figure **64.22–27**

Goesella miocenica Cushman 1936, Cushman Lab. foramin. Res., Spec. Publ., 6, p. 33, pl. 5, figs. 9a–b.
Guppyella miocenica (Cushman), Broennimann, 1951, p. 99, pl. 11, fig. 6, textfigs. 1–4.

The test begins with four or more chambers to a whorl but rapidly reducing to become uniserial in the adult. The wall is finely agglutinated. The bases of the chambers are folded due to alveoles which also may show as holes in eroded specimens.

Common in clays of the Late Miocene and Pliocene Cruse and Forest formations in Trinidad; Also in the Late Miocene Karamat and questionably in the Early Miocene Nariva beds.

TEXTULARIELLA MIOCENICA Cushman
Figure **76.27; 74**

Textulariella miocenica Cushman 1936, Cushman Lab. foramin. Res., Spec. Publ., 6, p. 45, pl. 6, figs. 17, 19. – Renz 1948, p. 171, pl. 2, fig. 14.

ALVEOVALVULINELLA POZONENSIS
(Cushman & Renz)
Figure **64.15–21**. Figure **76.28, 29a–b**. Figure **81.2; 75**

Liebusella pozonensis Cushman & Renz 1941, p. 9, pl. 2, figs. 1, 2. – Renz 1948, p. 144, pl. 2, figs. 19a–b, 20.
Alveovalvulinella pozonensis (Cushman & Renz), Broennimann 1953, p. 91, pl. 15, fig. 3, textfigs. 3–7. – Diaz de Gamero 1977, p. 40, pl. 3, fig. 14.

The test is similar to that of *Guppyella* but with the uniserial portion occupying only the last one or two chambers. A separation between *Alveovalvulinella* and *Guppyella* may not be warranted. *A. pozonensis* has been recorded from the Miocene Agua Salada Formation of Venezuela and the Early Miocene of Trinidad.

ALVEOVALVULINELLA POZONENSIS
CRASSA (Cushman & Renz)
Figure **76.30, 31**

Liebusella pozonensis crassa Cushman & Renz 1941, p. 10, pl. 2, figs. 3–4. – Renz 1948, p. 144, pl. 2, figs. 21, 22.
Alveovalvulinella pozonensis crassa (Cushman & Renz), Broennimann 1953, p. 91 (regarded by Broennimann as synonym of *A. pozonensis*)

Family Eggerellidae

DOROTHIA BREVIS Cushman & Stainforth
Figure **62.5, 5a; 52**. Figure **87.6, 6a; 86**

Dorothia brevis Cushman & Stainforth 1945, p.18, pl. 2, figs. 5a–b. – Beckmann 1953, p. 342, pl. 17, fig. 14.

DOROTHIA COLEI (Nuttall)
Figure **87.8, 8a; 86**

Gaudryina colei Nuttall 1932, p. 7, pl. 2, fig. 6.
Dorothia colei (Nuttall), Beckmann 1953, p. 343, pl. 17, fig. 18.

DOROTHIA CYLINDRICA (Nuttall)
Figure **81.3**; 75.

Gaudryina cylindrica Nuttall 1932, p. 7, pl. 2,
  fig. 7.
Dorothia cylindrica (Nuttall), Diaz de Gamero
  1977a, p. 40, pl. 3, fig. 9.

DOROTHIA NIPEENSIS Keijzer
Figure **87.7, 7a**; 86.

Dorothia nipeensis Keijzer 1945, Geogr. Geol.
  Meded. Physiogr.- Geol. Reeks, Ser. 2,
  No. 6, p. 208, pl. 1, fig. 14. – Beckmann
  1953, p. 343, pl. 17, fig. 19.

EGGERELLA KARAMATENSIS
Broennimann
Figure **64.28a–b, 29a–b.**

Eggerella karamatensis Broennimann 1953,
  p. 92, pl. 15, fig. 2, textfigs. 5f-i.

The species is a characteristically elongate form
with up to five chambers forming the initial
whorls and three in the adult ones. The agglutinated wall does not show any convolutions or
alveoles; it is not strong and the tests are therefore often flattened.
   In Trinidad the species appears to be confined
to the Middle Miocene Karamat Formation.

KARRERIELLA ALTICAMERA Cushman
& Stainforth
Figure **62.6, 6a**; 52.

Karreriella alticamera Cushman & Stainforth
  1945, p. 19, pl. 2, figs. 10a–b.

KARRERIELLA BACCATA (Schwager)
Figure **87.11, 11a**; 86.

Gaudryina baccata Schwager 1866, Novara
  Exped., Geol. Theil, 2, p. 200, pl. 4, fig. 12.
Karreriella baccata (Schwager), Beckmann
  1953, p. 344, pl. 17, fig. 21.
Dorothia chapapotensis (Cole) part, Saunders
  et al. 1984, p. 404, pl. 2, figs. 3, 4.

KARRERIELLA HANTKENIANA
Cushman
Figure **87.12, 12a**; 86.

Karreriella hantkeniana Cushman 1936,
  Cushman Lab. foramin. Res., Spec. Publ.,
  6, p. 36, pl.5, fig. 19. – Beckmann 1953,
  p. 344, pl. 17, fig. 24.

KARRERIELLA SUBCYLINDRICA
(Nuttall)
Figure **62.9**; 52. Figure **81.4**; 75. Figure **87.10, 10a**;
86

Gaudryina subcylindrica Nuttall 1928, p. 76,
  pl. 3, figs. 17, 18.
Karreriella subcylindrica (Nuttall), Cushman &
  Renz 1948, p. 18. pl. 3, fig. 17. – Beckmann
  1953, p. 345, pl. 17, figs. 25, 26. – Diaz de
  Gamero 1977a, p. 40, pl. 3, figs. 10, 11. –
  Saunders *et al.* 1984, p. 404, pl. 2, fig. 5.

KARRERIELLA WASHINGTONENSIS
Rau
Figure **87.13**; 86

Karreriella washingtonensis Rau 1948, J.
  Paleontol., 22, p. 158, pl. 27, figs. 5, 6. –
  Beckmann 1953, p. 345, pl. 17, fig. 27.

MARTINOTTIELLA PALLIDA (Cushman)
Figure **76.19, 20**

Clavulina communis d'Orbigny var. *pallida*
  Cushman 1927, Bull. Calif. Univ., Scripps
  Inst. Oceanogr., Tech. Ser., 1, pl. 138, pl. 2,
  fig. 1.
Schenckiella pallida (Cushman), Renz 1948,
  p. 163, pl. 2, fig. 17–18.

MARTINOTTIELLA PETROSA (Cushman
& Bermudez)
Figure **62.10**; 52. Figure **87.9**; 86

Listerella petrosa Cushman & Bermudez 1937,
  Contrib. Cushman Lab. foramin. Res., 13,
  p. 5, pl. 1, figs. 24–26.
Schenckiella petrosa (Cushman & Bermudez),
  Cushman & Stainforth 1945, p. 19, pl. 2,
  fig. 13. – Beckmann 1953, p. 345, pl. 17,
  fig. 28.

Family Textulariidae

TEXTULARIA ABBREVIATA d' Orbigny
Figure **76.9–10;** 74

Textularia abbreviata d'Orbigny 1846, p. 249,
   pl. 15, figs. 7–12. – Renz 1948, p. 168, pl. 1,
   figs. 10–11.

TEXTULARIA CRASSISEPTA Cushman
Figure **76.5;** 74

Textularia crassisepta Cushman 1911, Bull.
   U.S. natl. Mus., 71, p. 24, textfig. 41. – Renz
   1948, p. 169, pl. 1, fig. 12.

TEXTULARIA EXCAVATA Cushman
Figure **76.4;** 74

Textularia excavata Cushman 1913, Proc. U.S.
   natl. Mus., 44, no. 1973, p. 634, pl. 79, fig. 5.
   – Renz 1948, p. 169, pl. 1, fig. 15.

TEXTULARIA FALCONENSIS Cushman &
Renz
Figure **76.13a–b;** 74

Textularia falconensis Cushman & Renz 1941,
   p. 3, pl. 1, fig. 3. – Renz 1948, p. 169, pl. 1,
   figs. 14a–c.

TEXTULARIA ISIDROENSIS Cushman &
Renz
Figure **76.6a–b;** 74

Textularia isidroensis Cushman & Renz 1941,
   p. 4, pl. 1, fig. 7. – Renz 1948, p. 169, pl. 1,
   figs. 13a–c.

TEXTULARIA KUGLERI Cushman & Renz
Figure **76.14, 15;** 74

Textularia kugleri Cushman & Renz 1941, p. 5,
   pl. 1, figs. 8, 9. – Renz 1948, p. 171, pl. 1,
   figs. 23a–c.

TEXTULARIA LALICKERI Cushman &
Renz
Figure **76.7a–b, 8;** 74

Textularia lalickeri Cushman & Renz 1941,
   p. 3, pl. 1, figs. 4, 5. – Renz 1948, p. 169,
   pl. 1, figs. 19a–c, 20.

TEXTULARIA LEUZINGERI Cushman &
Renz
Figure **62.12;** 52. Figure **76.3a–b;** 74

Textularia leuzingeri Cushman & Renz 1941,
   p. 3, pl. 1, fig. 2. – Renz 1948, p. 169, pl. 1,
   fig. 15.

TEXTULARIA cf. MEXICANA Cushman
Figure **76.2;** 74. Figure **81.5;** 75

Textularia mexicana Cushman 1922, Bull. U.S
   natl. Mus., 104, p. 17, pl. 2, fig. 9. –
   Hedberg 1937a, p. 666, pl. 90, fig. 4. – Renz
   1948, p. 170, pl. 12, fig. 2. – Diaz de Gamero
   1977a, p. 39, pl. 3, fig. 6.

TEXTULARIA MISSISSIPPIENSIS
ALAZANENSIS Nuttall
Figure **76.1;** 74

Textularia mississipiensis var. alazanensis
   Nuttall 1932, p. 5, pl. 1, fig. 5. – Renz 1948,
   p. 170, pl. 12, fig. 1.

TEXTULARIA PANAMENSIS Cushman
Figure **76.11, 12;** 74

Textularia panamensis Cushman 1918, Bull.
   U.S. natl. Mus., 103, p. 53, pl. 20, fig. 1. –
   Renz 1948, p. 170, pl. 1, figs. 21–22.

TEXTULARIA POZONESIS Cushman &
Renz
Figure **76.16a–b;** 74

Textularia pozonensis Cushman & Renz 1941,
   p. 5, pl. 1, figs. 8, 9. – Renz 1948, 32, p. 171,
   pl. 1, figs. 23a–c.

Family Pseudogaudryinidae

PSEUDOCLAVULINA TRINITATENSIS
Cushman & Renz
Figure **62.13;** 52

Pseudoclavulina trinitatensis Cushman & Renz
1948, p. 13, pl. 3, fig. 5.

Family Valvulinidae

CLAVULINA CARINATA Cushman & Renz
Figure **76.21a–b;** 74

Clavulina carinata Cushman & Renz 1941, p. 8,
pl. 1, fig. 18. – Renz 1948, p. 129, pl. 1, figs.
25a–b.

VALVULINA FLEXILIS Cushman & Renz
Figure **64.30–31, 32a–b.** Figure **76.17, 18a–c;** 74.
Figure **81.6;** 75

Valvulina flexilis Cushman & Renz 1941, p. 7,
pl. 1, figs. 16–17. – Renz 1948, p. 177, pl. 2,
figs. 11, 12a–c. – Brönnimann 1953, p. 95,
pl. 15, fig. 1, textfigs. 15h–o. – Diaz de
Gamero 1977*a*, p. 40, pl. 3, fig. 13.

Calcareous non-trochospiral taxa

Family Nodosariidae

CHRYSALOGONIUM BREVILOCULUM
Cushman & Jarvis
Figure **63.24, 24a;** 52

Chrysalogonium breviloculum Cushman &
Jarvis 1934, p. 74, pl. 10, fig. 13. – Cushman
& Stainforth 1945, p. 26, pl. 3, fig. 27; pl. 16,
fig. 1.

CHRYSALOGONIUM CIPERENSE
Cushman & Stainforth
Figure **63.27;** 52

Chrysalogonium ciperense Cushman &
Stainforth 1945, p. 26, pl. 3, fig. 31.

CHRYSALOGONIUM ELONGATUM
Cushman & Jarvis
Figure **63.16;** 2

Chrysalogonium elongatum Cushman & Jarvis
1934, p. 73, pl. 10, figs. 10, 11. – Cushman
& Stainforth 1945, p. 26, pl. 3, fig. 30; pl. 16,
figs. 3, 4.

CHRYSALOGONIUM LANCEOLUM
Cushman & Jarvis
Figure **63.26;** 52. Figure **87.35;** 86

Chrysalogonium lanceolum Cushman & Jarvis
1934, p. 75, pl. 10, fig. 16. – Cushman &
Stainforth 1945, p. 25, pl. 3, fig. 29; pl. 16,
fig. 5. – Beckmann 1953, p. 352, pl. 19,
fig. 9.

CHRYSALOGONIUM
LONGICOSTATUM Cushman & Jarvis
Figure **63.23, 23a;** 52

Chrysalogonium longicostatum Cushman &
Jarvis 1934, p. 74, pl. 10, fig. 12. – Cushman
& Stainforth 1945, p. 25, pl. 3, fig. 26; pl. 16,
fig. 2.

CHRYSALOGONIUM
TENUICOSTATUM Cushman & Bermudez
Figure **63.25;** 52

Chrysalogonium tenuicostatum Cushman &
Bermudez 1936, Contrib. Cushman Lab.
foramin. Res., 12, p. 27, pl. 5, figs. 3–5. –
Cushman & Stainforth 1945, p. 25, pl. 3,
fig. 28.

DENTALINA SEMILAEVIS Hantken
Figure **62.18;** 52

Dentalina semilaevis Hantken 1875, Magyar
kir. földt. int. evkön., 4, pl. 32, pl. 4, fig. 6;
pl. 12, fig. 13. – Cushman & Stainforth 1945,
p. 24, pl. 3, figs. 15–17.

NODOSARIA LONGISCATA d'Orbigny
Figure **63.17**; 52. Figure **77.31–32**; 74

Nodosaria longiscata d'Orbigny 1846, p. 32, pl. 1, figs. 10–12. – Cushman & Stainforth 1945, p. 24, pl. 3, figs. 19–21. – Renz 1948, p. 146, pl. 5, figs. 1–4.

NODOSARIA SCHLICHTI Reuss
Figure **77.33–34**; 74

Nodosaria schlichti Reuss 1870, Sitzungsber. K. Akad. Wiss. Wien, math.-naturwiss. Kl. Vienna, 62, p. 472, pl. 6, figs. 29–31. – Renz 1948, p. 147, pl. 4, fig. 25; pl. 5, fig. 5.

PSEUDOGLANDULINA COMATULA
(Cushman)
Figure **80.21**; 74

Nodosaria comatula Cushman 1923, U. S. Nat. Mus. Bull., 104, pt. 3, p. 83, pl. 14, fig. 5.
Pseudoglandulina comatula (Cushman), Hedberg 1937a, p. 673, pl. 91, figs. 9–10. – Renz 1948, p. 153, pl. 5, fig. 12.

PSEUDOGLANDULINA GALLOWAYI
Cushman
Figure **62.14**; 52

Pseudoglandulina gallowayi Cushman 1929, Contrib. Cushman Lab. foramin. Res., 5, p. 87, pl. 13, fig. 13. – Cushman & Stainforth 1945, p. 27, pl. 4, fig. 3.

PSEUDOGLANDULINA GALLOWAYI
PAUCICOSTATA Cushman & Renz
Figure **80.20**; 74

Pseudoglandulina gallowayi Cushman var. *paucicostata* Cushman & Renz 1941, p. 16, pl. 3, fig. 5. – Renz 1948, p. 153, pl. 5, fig. 13.

PSEUDOGLANDULINA INCISA
(Neugeboren)
Figure **80.27**

Glandulina incisa Neugeboren 1850, Verh. & Mitt. Siebenbürg. Ver. Naturwiss., 1, p. 52, pl. 1, figs. 7a–b.

Pseudoglandulina incisa (Neugeboren), Renz 1948, p. 154, pl. 5, fig. 16.

The range of *P. incisa* is not shown on Fig. 74. According to Renz (1948) the species is rare in the *Catapsydrax dissimilis* Zone, present in the *Globigerinatella insueta* Zone, rare in the *Globorotalia fohsi robusta* and *Globorotalia mayeri* zones.

PYRAMIDULINA ISIDROENSIS
(Cushman & Renz)
Figure **77.43–44**

Dentalina isidroensis Cushman & Renz 1941, p. 15, pl. 3, figs. 2–3. – Renz 1948, p. 130, pl. 4, figs. 28–29.

The range of *P. isidroensis* is not shown on Fig. 74. According to Renz (1948) the species is rare in the *Globigerinatella insueta* Zone.

PYRAMIDULINA LAMELLATA (Cushman
& Stainforth)
Figure **63.19**; 52

Nodosaria lamellata Cushman & Stainforth 1945, p. 24, pl. 3, figs. 23, 24.

PYRAMIDULINA MODESTA (Bermudez)
Figure **63.14**; 52

Ellipsonodosaria modesta Bermudez 1937, Mem. Soc. Cubana Hist. Nat., 11, p. 238, pl. 20, fig. 3. – Cushman & Stainforth 1945, p. 57, pl. 10, fig. 2.

PYRAMIDULINA NUTTALLI (Hedberg)
Figure **77.35**; 74

Nodosaria nuttalli Hedberg 1937a, p. 673, pl. 91, fig. 6. – Renz 1948, p. 146, pl. 12, fig. 8.

PYRAMIDULINA RAPHANISTRUM
CARIBBEANA Hedberg
Figure **77.38, 41–42**; 74

Nodosaria raphanistrum (Linné) var. *caribbeana* Hedberg 1937a, p. 671, pl. 91, fig. 1. – Renz 1948, p. 147, pl. 5, figs. 6, 7.

PYRAMIDULINA STAINFORTHI
(Cushman & Renz)
Figure **63.18**; 52. Figure **77.39–40**; 74

Nodosaria stainforthi Cushman & Renz 1941,
   p. 15, pl. 3, fig. 4. – Cushman & Stainforth
   1945, p. 25, pl. 3, fig. 25. – Renz 1948, p. 147,
   pl. 4, figs. 31a–b.

PYRAMIDULINA VERTEBRALIS
(Batsch)
Figure **77.36–37**; 74

Nautilus (Orthoceras) vertebralis Batsch 1791,
   Conchylien des Seesandes, pt. 3, no. 6, pl. 2,
   figs. 6a–b.
Nodosaria vertebralis (Batsch), Renz 1948,
   p. 147, pl. 5, figs. 8–11.

GONATOSPHAERA PROLATA Guppy
Figure **80.24a–b**

Gonatosphaera prolata Guppy 1894, Zool. Soc.
   London Proc., p. 651, pl. 41, figs. 14–19.
Lingulina prolata (Guppy), Renz 1948, p. 144,
   pl. 5, figs. 24a–b.

The species is scarce from the *Globigerinatella
insueta* Zone to the *Globorotalia fohsi peripheroronda* Zone of the the Agua Salada Formation, southeastern Falcon.

TOLLMANNIA GRIMSDALEI (Cushman & Renz)
Figure **80.19a–b**

Lingulina grimsdalei Cushman & Renz 1941,
   p. 14, pl. 3, fig. 1. – Renz 1948, p. 144, pl. 5,
   figs. 23a–b.

The species is scarce in the *Catapsydrax stainforthi* and the *Globigerinatella insueta* zones of the
Agua Salada Formation, southeastern Falcon.

PLECTOFRONDICULARIA ALAZANENSIS Cushman
Figure **62.30**; 52

Plectofrondicularia alazanensis Cushman 1927,
   Contrib. Cushman Lab. foramin. Res., 3,
   p. 113, pl. 22, fig. 12. – Cushman &
   Stainforth 1945, p. 38, pl. 5, fig. 20.

PLECTOFRONDICULARIA cf. CALIFORNICA Cushman & Stewart
Figure **80.15**; 74

Plectofrondicularia californica Cushman &
   Stewart 1926, Contrib. Cushman Lab.
   foramin. Res., 2, p. 39, pl. 6, figs. 9–11. –
   Renz 1948, p. 151, pl. 12, fig. 10.

PLECTOFRONDICULARIA COOKEI
Cushman
Figure **62.23**; 52

Plectofrondicularia cookei Cushman 1933,
   Contrib. Cushman. Lab. foramin. Res., 9,
   p. 11, pl. 1, fig. 26. – Cushman & Stainforth
   1945, p. 36, pl. 5, fig. 12.

PLECTOFRONDICULARIA FLORIDANA Cushman
Figure **80.17**; 74

Plectofrondicularia floridana Cushman 1930,
   Florida Geol. Survey Bull., 4, p. 41, pl. 8,
   fig. 1. – Renz 1948, p. 152, pl. 6, fig. 19.

PLECTOFRONDICULARIA LIRATA
Bermudez
Figure **62.22**; 52. Figure **87.15**; 86

Plectofrondicularia lirata Bermudez 1937,
   Mem. Soc. Cubana Hist. Nat., 11, p. 240,
   pl. 20, fig. 6. – Cushman & Renz 1948, p. 24,
   pl. 5, fig. 8. – Beckmann 1953, p. 364, pl. 21,
   fig. 4.

PLECTOFRONDICULARIA cf. LONGISTRIATA Le Roy
Figure **80.18**; 74

Plectofrondicularia longistriata Le Roy 1939,
   Natuurk. tijdschr. Ned.-Indie, afl. 6, deel 9,
   p. 241, pl. 5, figs. 4–6.
Plectofrondicularia cf. *longistriata* Le Roy,
   Renz 1948, p. 152, pl. 6, fig. 21.

PLECTOFRONDICULARIA
MANSFIELDI Cushman & Ponton
Figure **80.16**; 74

Plectofrondicularia mansfieldi Cushman &
Ponton 1931, Contrib. Cushman Lab.
foramin. Res., 7, p. 60, pl. 8, figs. 1a–b. –
Renz 1948, p. 152, pl. 12, fig. 11

PLECTOFRONDICULARIA MEXICANA
(Cushman)
Figure **62.27**; 52

Frondicularia mexicana Cushman 1926,
Contrib. Cushman Lab. foramin. Res., 1,
p. 88, pl. 13, fig. 5.
Plectofrondicularia mexicana (Cushman),
Cushman & Stainforth 1945, p. 38, pl. 5,
fig. 19.

PLECTOFRONDICULARIA
MORREYAE Cushman
Figure **62.28**; 52

Plectofrondicularia morreyae Cushman 1929,
Contrib. Cushman Lab. foramin. Res., 5,
p. 92, pl. 13, fig. 23. – Cushman & Stainforth
1945, p. 37, pl. 5, figs. 15–17.

PLECTOFRONDICULARIA NUTTALLI
Cushman & Stainforth
Figure **62.29**; 52

Plectofrondicularia nuttalli Cushman &
Stainforth 1945, p. 38, pl. 5, figs. 21–23 (22
holotype).

PLECTOFRONDICULARIA SPINIFERA
Cushman & Jarvis
Figure **62.21**; 52

Plectofrondicularia spinifera Cushman & Jarvis
1929, p. 10, pl. 2, figs. 1–3. – Cushman &
Stainforth 1945, p. 36, pl. 5, fig. 11; pl. 16,
figs. 6, 7.

PLECTOFRONDICULARIA
TRINITATENSIS Cushman & Jarvis
Figure **62.26**; 52. Figure **87.16, 16a**; 86

Plectofrondicularia trinitatensis Cushman &
Jarvis 1929, p. 11, pl. 2, fig. 16. – Beckmann
1953, p. 365, pl. 21, fig. 5.
Plectofrondicularia paucicostata Cushman &
Jarvis (part), Saunders *et al.* 1984, p. 406,
pl. 3, fig. 3.

PLECTOFRONDICULARIA VAUGHANI
Cushman
Figure **62.20**; 52. Figure **87.24**; 86

Plectofrondicularia vaughani Cushman 1927,
Contrib. Cushman Lab. foramin. Res., 3,
p. 112, pl. 23, fig. 3. – Cushman & Stainforth
1945, p. 36, pl. 5, fig. 13. – Beckmann 1953,
p. 365, pl. 21, fig. 6.

Family Vaginulinidae

LENTICULINA AMERICANA (Cushman)
Figure **77.5**; 74

Cristellaria americana Cushman 1918, Bull.
U. S. Geol. Surv., 676, p. 50, pl. 10, figs. 5–
6.
Robulus americanus (Cushman), Renz 1948,
p. 50, pl. 12, fig. 3.

LENTICULINA AMERICANA GRANDIS
(Cushman)
Figure **77.10**; 74

Cristellaria americana var. *grandis* Cushman
1920, Prof. Pap. U. S. geol. Surv., 128-B,
p. 68, pl. 11, fig. 2.
Robulus americanus grandis (Cushman), Renz
1948, p. 157, pl. 3, fig. 7.

LENTICULINA CALCAR (Linné)
Figure **77.8**; 74

Nautilus calcar Linnaeus 1758, Systema naturae
Ed., 10, tomus 1, p. 709.
Robulus calcar (Linné), Renz 1948, p. 158,
pl. 3, fig. 6.

LENTICULINA CLERICII (Fornasini)
Figure 77.7; 74

Cristellaria clericii Fornasini 1895, Mem. R.
Acad. Sci. Bologna, ser. 5, 9, 1901, p. 65,
fig. 17 (in text). – Nuttall 1928, p. 87, pl. 5,
fig. 10.
Robulus clericii (Fornasini), Renz 1948, p. 84,
pl. 12, figs. 17, 17a.

LENTICULINA FORMOSA (Cushman)
Fig. 77.9; 74

Cristellaria formosa Cushman 1923, Bull. U. S.
natl. Mus., 104, pt. 4, p. 110, pl. 29, fig. 1;
pl. 30, fig. 6.
Robulus formosus (Cushman), Renz 1948,
p. 158, pl. 3, fig. 9.

LENTICULINA HEDBERGI (Cushman &
Renz)
Figure 77.14a–b; 74

Robulus hedbergi Cushman & Renz 1941,
p. 10, pl. 2, fig. 9. – Renz 1948, p. 159, pl. 3,
figs. 10a–b.

LENTICULINA IOTA (Cushman)
Figure 77.3; 74

Cristellaria iota Cushman 1923, Bull. U.S. natl.
Mus., 104, p. 111, pl. 29, fig. 2; pl. 30, fig. 1.
Robulus iotus (Cushman), Renz 1948, p. 159,
pl. 3, fig. 14.

LENTICULINA MELVILLI (Cushman &
Renz)
Figure 77.6a–b; 74

Robulus melvilli Cushman & Renz 1941, p. 12,
pl. 2, fig. 12. – Renz 1948, p. 159, pl. 3, figs.
14a–b.

LENTICULINA NUTTALLI (Cushman &
Renz)
Figure 77.4a–b; 74

Robulus nuttalli Cushman & Renz 1941, p. 11,
pl. 2, fig. 10. – Renz 1948, p. 159, pl. 3, figs.
12a–b.

LENTICULINA NUTTALLI
OBLIQUILOCULATA (Cushman & Renz)
Figure 77.15; 74

Robulus nuttalli var. *obliquiloculata* Cushman
& Renz 1941, p. 11, pl. 2, fig. 11. – Renz 1948,
p. 160, pl. 3, fig. 13.

LENTICULINA OCCIDENTALIS
TORRIDA (Cushman)
Figure 77.2; 74

Cristellaria occidentalis var. *torrida* Cushman
1923, Bull. U. S. natl. Mus., 104, p. 105,
pl. 25, fig. 1.
Robulus occidentalis var. *torridus* (Cushman),
Cushman & Jarvis 1930, p. 357, pl. 32, figs.
8a–b. – Renz 1948, p. 160, pl. 3, fig. 17.

LENTICULINA SENNI (Cushman & Renz)
Figure 77.11a–b, 12a–b; 74

Robulus senni Cushman & Renz 1941, p. 12,
pl. 2, figs. 14–15. – Renz 1948, p. 160, pl. 3,
figs. 15a–b, 16a–b.

LENTICULINA SUBACULEATA
GLABRATA (Cushman)
Figure 77.13; 74

Cristellaria subaculeata var. *glabrata* Cushman
1923, Bull. U.S. natl. Mus., 104, p. 124,
pl. 32, fig. 4.
Robulus subaculeatus var. *glabratus*
(Cushman), Renz 1948, p. 160, pl. 3, figs. 20–
21.

LENTICULINA SUTERI (Cushman & Renz)
Figure 77.1a–b; 74

Robulus suteri Cushman & Renz 1941, p. 10,
pl. 2, figs. 5–8. – Renz 1948, p. 161, pl. 3,
figs, 18a–b; pl. 4, figs. 1–2.

LENTICULINA WALLACEI (Hedberg)
Figure 77.16, 17; 74

Marginulina wallacei Hedberg 1937a, p. 670,
pl. 90, figs. 15–17.
Robulus wallacei (Hedberg), Renz 1948,
p. 161, pl. 4, figs. 19–20.

DARBYELLA SUBKUBINYII (Nuttall)
Figure **80.22a–b; 74**

Cristellaria subkubinyii Nuttall 1932, p. 11,
pl. 1, fig. 16.
Darbyella subkubinyii (Nuttall), Renz 1948,
p. 130, pl. 4, figs. 3a–b.

MARGINULINOPSIS BASISPINOSUS
(Cushman & Renz)
Figure **77.20–22**

Marginulina basispinosa Cushman & Renz
1941, p. 13, pl. 2, figs. 16–18.
Marginulinopsis basispinosus (Cushman &
Renz), Renz 1948, p. 145, pl. 4, figs. 8a–b,
9–10.

In the Agua Salada Formation of southeastern
Falcon the characteristic species has been
recorded from the *Globorotalia menardii/Glo-
bigerina nepenthes* Zone.

SARACENARIA ITALICA
ACUTOCARINATA (Cushman)
Figure **77.30**

Cristellaria italica (Defrance) var. *acutocarinata*
Cushman 1917, Proc. U.S. natl. Mus., 51,
no. 2172, p. 661.
Saracenaria italica Defrance var. *acutocarinata*
(Cushman), Renz 1948, p. 162, pl. 5, fig. 19.

The subspecies occurs in the *Globigerinatella
insueta* and *Globorotalia fohsi peripheroronda*
zones of the Agua Salada Formation, southeastern Falcon.

SARACENARIA ITALICA
CARAPITANA Franklin
Figure **77.28**

Saracenaria italica carapitana Franklin 1944,
p. 312, pl. 45, fig. 14.
Saracenaria cf. *italica* Defrance var. *carapitana*
Franklin, Renz 1948, p. 162, pl. 5, fig. 17.

The subspecies is scarce from the *Globigerinatella insueta* to the *Globorotalia fohsi robusta*
zones of the Agua Salada Formation, southeastern Falcon.

SARACENARIA LATIFRONS (Brady)
Figure **77.29**

Cristellaria latifrons Brady 1884, Rept.
Challenger, Zoology, 9, p. 544, pl. 113, figs.
11a–b.
Saracenaria latifrons (Brady), Renz 1948,
p. 132, pl. 5, fig. 22.

The species is scarce in the *Globorotalia fohsi
fohsi* Zone of the Agua Salada Formation of
southeastern Falcon.

SARACENARIA SENNI Hedberg
Figure **77.27**

Saracenaria senni Hedberg 1937*a*, p. 674,
pl. 90, figs. 18a–b. – Renz 1948, p. 163, pl. 5,
fig. 21.

The characteristic species occurs in the Agua
Salada Formation of southeastern Falcon from
the *Catapsydrax dissimilis* Zone to the *Globigerinatella insueta* Zone.

ASTACOLUS OVATUS Galloway &
Heminway
Figure **77.23**

Astacolus ovatus Galloway & Heminway 1941,
New York Acad. Sci., 3, p. 334, pl. 8, figs.
10a–b. – Renz 1948, p. 115, pl. 4, fig. 11.

The species is scarce from the *Catapsydrax stainforthi* Zone to the *Globorotalia fohsi peripheroronda* Zone of the Agua Salada Formation,
southeastern Falcon.

MARGINULINA cf. GLABRA OBESA
Cushman
Figure **77.24**

Marginulina glabra d'Orbigny var. *obesa*
Cushman 1923, Bull. U. S. natl Mus., 104,
p. 128, pl. 37, fig. 1.
Marginulina cf. *glabra* d'Orbigny var. *obesa*
Cushman, Renz, 1948, p. 145, pl. 4, fig. 21.

The subspecies occurs in the Agua Salada Formation of southeastern Falcon from the *Catapsydrax stainforthi* Zone to the *Globigerinatella
insueta* Zone.

MARGINULINA cf. **STRIATULA**
Cushman
Figure **77.25, 26**

Marginulina striatula Cushman 1913, Bull. U.S.
    natl. Mus., 71, pt. 3, p. 79, pl. 23, fig. 4.
Marginulina cf. *striatula* Cushman, Renz 1948,
    p. 145, pl. 4, figs. 15, 16.

The species occurs in the Agua Salada Formation of southeastern Falcon from the *Catapsydrax dissimilis* Zone to the *Globorotalia fohsi
peripheroronda* Zone.

VAGINULINOPSIS SUPERBUS (Cushman
& Renz)
Figure **77.18a–b, 19**

Marginulina superba Cushman & Renz 1941,
    p. 14, pl. 2, figs. 19–20.
Vaginulinopsis superbus (Cushman & Renz),
    Renz 1948, p. 177, pl. 4, figs. 17a–b, 18.

The species occurs in the Agua Salada Formation of southeastern Falcon in the *Globorotalia
menardii/Globigerina nepenthes* and *Sphaeroidinellopsis seminulina* zones.

PLANULARIA ARBENZI Cushman & Renz
Figure **80.14a–b;** 74

Planularia arbenzi Cushman & Renz 1941,
    p. 13, pl. 2, fig. 13. – Renz 1948, p. 150,
    pl. 4, figs. 6a–b.

PLANULARIA CLARA Cushman & Jarvis
Figure **80.12;** 74

Planularia clara Cushman & Jarvis 1929, p. 7,
    pl. 2, figs. 14–15. – Renz 1948, p. 150, pl. 4,
    fig. 4.

PLANULARIA VENEZUELANA Hedberg
Figure **80.13a–b;** 74

Planularia venezuelana Hedberg 1937*a*, p. 670,
    pl. 90, figs. 5a–b.

Family Lagenidae

LAGENA ASPEROIDES Galloway & Morrey
Figure **63.12;** 52. Figure **87.27;** 86

Lagena asperoides Galloway & Morrey 1929,
    Bull. Am. Paleontol., 15, p. 19, pl. 2, fig. 6.
    – Cushman & Stainforth 1945, p. 28, pl. 4,
    fig. 9. – Beckmann 1953, p. 315, pl. 19, fig. 28.
Lagena asperoides group (part), Saunders *et al.*
    1984, p. 405, pl. 2, fig. 16.

LAGENA CIPERENSIS Cushman &
Stainforth
Figure **63.11;** 52. Figure **87.28, 28a;** 86

Lagena ciperensis Cushman & Stainforth 1945,
    p. 30, pl. 4, fig. 18. – Beckmann 1953, p. 356,
    pl. 19, fig. 32. – Saunders *et al.* 1984, p. 405,
    pl. 2, fig. 12.

LAGENA CRENATA CAPISTRATA
Cushman & Stainforth
Figure **63.9;** 52

Lagena crenata capistrata Cushman &
    Stainforth 1945, p. 30, pl. 4, fig. 16.

LAGENA FLINTIANA INDOMITA
(Cushman & Stainforth)
Figure **87.26, 26a;** 86

Entosolenia flintiana (Cushman) var. *indomita*
    Cushman & Stainforth 1945, p. 43, pl. 6,
    fig. 16.
Lagena flintiana Cushman, var. *indomita*
    (Cushman & Stainforth), Beckmann 1953,
    p. 357, pl. 19, fig. 36.

LAGENA PULCHERRIMA Cushman &
Jarvis
Figure **63.13;** 52

Lagena pulcherrima Cushman & Jarvis 1929,
    p. 8, pl. 2, fig. 10. – Cushman & Stainforth
    1945, p. 29, pl. 4, fig. 12.

LAGENA PULCHERRIMA ENITENS
Cushman & Stainforth
Figure **63.8**, 52. Figure **87.25, 25a**; 86

Lagena pulcherrima var. *enitens* Cushman &
Stainforth 1945, p. 29, pl. 4, fig. 13. –
Beckmann 1953, p. 359, pl. 20, fig. 7.

LAGENA STRIATA BASISENTA Cushman
& Stainforth
Figure **63.10**; 52

Lagena striata var. *basisenta* Cushman &
Stainforth 1945, p. 29, pl. 4, fig. 15.

Family Polymorphinidae

GUTTULINA IRREGULARIS (d'Orbigny)
Figure **80.28**; 74

Globulina irregularis d'Orbigny 1846, p. 226,
pl. 13, figs. 9–10.
Guttulina irregularis (d'Orbigny), Renz 1948,
p. 138, pl. 6, fig. 1.

GUTTULINA JARVISI Cushman & Ozawa
Figure **80.29**; 74

Guttulina jarvisi Cushman & Ozawa 1930,
p. 39, pl. 7, figs. 4–5. – Renz 1948, p. 138,
pl. 6, fig. 2.

GUTTULINA LEHNERI Cushman & Ozawa
Figure **87.31, 31a**; 86

Guttulina lehneri Cushman & Ozawa 1930,
p. 39, pl. 8, figs. 1, 2. – Beckmann 1953,
p. 360, pl. 20, fig. 14.

PYRULINOIDES ANTILLEANUS
Beckmann
Figure **87.29, 29a**; 86

Pyrulinoides antillaneus Beckmann 1953,
p. 361, pl. 20, fig. 19.

SIGMOMORPHINA TRINITATENSIS
Cushman & Ozawa
Figure **87.30, 30a**; 86

Sigmomorphina trinitatensis Cushman &
Ozawa 1930, p. 134, pl. 36, figs. 1, 2. –
Beckmann 1953, p. 363, pl. 20, fig. 27. –
Saunders *et al.* 1984, p. 406, pl. 2, fig. 11.

Family Ellipsolagenidae

ENTOSOLENIA FLINTIANA
INDOMITA Cushman & Stainforth
Figure **63.3**; 52

Entosolenia flintiana var. *indomita* Cushman &
Stainforth 1945, p. 43, pl. 6, fig. 16.

ENTOSOLENIA FLINTIANA
PLICATURA Cushman & Stainforth
Figure **63.7**; 52

Entosolenia flintiana var. *plicatura* Cushman &
Stainforth 1945, p. 42, pl. 6, fig. 15.

ENTOSOLENIA KUGLERI Cushman &
Stainforth
Figure **63.4**; 52

Entosolenia kugleri Cushman & Stainforth
1945, p. 45, pl. 7, fig. 5.

ENTOSOLENIA ORBIGNYANA
CLATHRATA (H. B. Brady)
Figure **63.6**; 52

Lagena clathrata H. B. Brady 1884, p. 485,
pl. 60, fig. 4.
Entosolenia orbignyana (Seguenza) var.
clathrata (H. B. Brady), Cushman &
Stainforth 1945, p. 44, pl. 6, fig. 20.

ENTOSOLENIA PANNOSA Cushman &
Stainforth
Figure **63.5**

Entosolenia pannosa Cushman & Stainforth
1945, p. 43, pl. 6, fig. 18.

The species has been recorded in the Cipero Formation of Trinidad from the *Globorotalia opima*

opima Zone to the *Globorotalia fohsi fohsi* Zone.

ENTOSOLENIA SPINULOLAMINATA Cushman & Stainforth
Figure **63.2**; 52

Entosolenia spinulolaminata Cushman & Stainforth 1945, p. 43, pl. 6, fig. 17.

ELLIPSOLAGENA BARRI Cushman & Stainforth
Figure **63.1**; 52

Ellipsolagena barri Cushman & Stainforth 1945, p. 59, pl. 10, fig. 14.

Family Heterohelicidae

BIFARINA INOPINATA (Cushman & Stainforth)
Figure **62.16**. Figure **87.14**; 86

Rectoguembelina inopinata Cushman & Stainforth 1945, p. 35, pl. 5, fig. 27a, b. – Beckmann 1953, p. 364, pl. 21, fig. 3.

The species has been recorded in Trinidad from the Eocene *Hantkenina nuttalli* Zone to the Oligocene *Globorotalia opima opima* Zone.

BIFARINA TRINITATENSIS (Cushman & Renz)
Figure **62.15**

Rectoguembelina trinitatensis Cushman & Renz 1948, p. 23, pl. 5, fig. 7.

The species occurs in Trinidad throughout the Eocene Navet Formation.

Family Bolivinidae

Genus *Bolivina* d'Orbigny 1839

Note: The difference between the genera *Bolivina* d'Orbigny, 1839 and *Brizalina* O. G. Costa, 1856, consists mainly in the presence or absence of a more

or less distinct surface ornamentation which when present obscures the intercameral septa. Intermediate forms, difficult to assign to one or the other genus may make a clear distinction between the two genera difficult. Because it is common practice among the authors – with one exception – who published in the presently discussed areas of the Caribbean to adhere to the genus *Bolivina*, this practice is also followed here; this despite the fact that, strictly speaking, only a few of the strongly ornamented species like *B. byramensis*, *B. danvillensis*, *B. imporcata* and *B. thalmanni* would be assigned the genus *Bolivina*.

The genus *Bolivina* in the sense used here is represented by numerous species in the Agua Salada Formation of southeastern Falcon, Venezuela, and in the Navet – San Fernando – Cipero – Ste. Croix – Brasso – Lengua formations of Trinidad. Few species are reported from central Falcon and only a single species was described from Barbados.

The 19 *Bolivina* species included from Trinidad are shown on range chart Fig. 50 and illustrated partly on Fig. 53.1–18, partly on Fig. 78 (southeastern Falcon). Only nine species were illustrated previously from Trinidad by Cushman & Stainforth (1945) and Cushman & Renz (1947*b*, 1948). They are again figured here partly together with the corresponding holotypes. Of the remaining species only the holotypes but no corresponding Trinidad specimens are here illustrated. These taxa were compared either directly with the holotype as far as deposited in the Cushman collection in Washington, or with the holotype illustrations.

Of the 17 *Bolivina* species treated here from the Agua Salada Formation in southeastern Falcon (range chart Fig. 73, illustrations Fig. 78.1–25), nine are holotypes described from this formation by Cushman & Renz (1941). Their original illustrations are here re-figured. Illustrations of the remaining eight species are taken from Renz (1948).

Of the four *Bolivina* species listed from central Falcon, Diaz de Gamero (1977*a*) only illustrated one, as *Bolivina vaceki*. It is here reproduced on Fig. 81.7. For illustrations of the other three central Falcon taxa reference is made to the corresponding illustrations from Trinidad (Fig. 53) and southeastern Falcon (Fig. 78). The ranges of two of the four listed central Falcon species are shown on Fig. 75.

Only one *Bolivina* species from Barbados, *B.* cf. *tectiformis,* appears in Beckmann (1953). It is illustrated here on Fig. 87.17.

Comparing the ranges of the *Bolivina* species occurring jointly in the Oligocene and Miocene of southeastern Falcon, Venezuela (Fig. 73) and Trinidad

(Fig. 50) shows that in almost all cases they do not fully coincide. Compared with Trinidad, several species continue in the Agua Salada Formation of Falcon into stratigraphically younger zones. This is ascribed to more favourable environmental conditions for Bolivinas in Falcon, in particular from about the *Globorotalia fohsi peripheroronda* Zone onwards.

BOLIVINA ADVENA Cushman
Figure **78.13**; 73

Bolivina advena Cushman 1925, Contrib.
  Cushman Found. foramin. Res., 1, p. 29,
  pl. 5, fig. 1. – Renz 1948, p. 116, pl. 6, fig. 23.

The specimen figured from southeastern Falcon (Fig. 78.13) differs from the holotype illustration in the intercameral sutures of the later growth stage being more strongly curved and with slight depressions showing at their culmination points. Furthermore, compared with the holotype the Falcon specimen has a slightly narrower, more pointed early stage.

In SE-Falcon the species is present in the late Middle Miocene, possibly extending into the Late Miocene.

BOLIVINA ACEROSA Cushman
Figure **53.10–11**; 50

Bolivina acerosa Cushman 1936, Cushman Lab.
  foramin. Res., Spec. Publ., 6, p. 54, pl. 8,
  fig. 1. – Cushman & Renz 1947*b*, p. 26, pl. 6,
  fig. 7.

While the illustrated Trinidad specimen (Fig. 53.11) possesses about 12 fine longitudinal costae, developing from the early stage but ending below the penultimate and ultimate chambers, the holotype (Fig. 53.10) displays such costae only in its earlier part with the surface of the later half remaining smooth. The chambers of the Trinidad specimen also broaden more rapidly as added, thus giving the test compared with the holotype a less slender aspect. However, specimens that in all aspects are very close to identical with the holotype are also known from Trinidad. See also taxonomic notes under *Bolivina caudriae*.

The species ranges in Trinidad from the late Early to the early Middle Miocene.

BOLIVINA AENARIENSIS CARAPITANA Hedberg
Figure **53.15**; 50

Bolivina aenariensis (Costa) var. *carapitana*
  Hedberg 1937*a*, p. 676, pl. 91, fig. 16.

This is a characteristic taxon with 5–6 distinct longitudinal costae that as a rule do not continue on the last formed 3–4 chambers. Compared with this subspecies the original type described from the Pliocene near Siena, Italy, possesses a less sharply constricted initial part and possesses fewer costae. Compared with *Bolivina marginata multicostata* the subspecies *carapitana* possesses fewer costae (5–6 against 9–13). See also annotations under *Bolivina marginata multicostata*.

In Trinidad the subspecies has been recorded only from the Early Miocene *Catapsydrax stainforthi* Zone.

BOLIVINA ALATA (Seguenza)
Figure **78.4–5**; 73

Vulvulina alata Seguenza 1862, Atti Accad.
  Gioenia Sci. Nat. ser. 2., 18, p. 115, figs. 5a,
  b.
Bolivina alata (Seguenza), Cushman 1937,
  Cushman Lab. foramin. Res., Spec. Publ.,
  9, p. 106, pl. 13, figs. 3–11. – Renz 1948,
  p. 116, pl. 6, fig. 26; pl. 12, fig. l2.

The species was originally described from the Pleistocene of Sicily. It differs from *Bolivina pisciformis* and *Bolivina alazanensis* in possessing fewer chambers which are higher and appear slightly inflated. Further, the sharply keeled periphery shows in the basal portion of each chamber a spinose projection turned towards the initial part of the test. The morphology of the SE-Falcon specimens placed by Renz into *B. alata*, though occurring there as in Trinidad in the Early Miocene, apparently conform closely with Seguenza's species.

BOLIVINA ALAZANENSIS Cushman
Figure **53.2–3**; 50. Figure **78.3**; 73

Bolivina alazanensis Cushman 1926, Contrib.
  Cushman Lab. foramin. Res., 1, p. 82,
  pl. 12, figs. 1a, b. – Renz 1948, p. 117, pl. 12,
  fig. 20.

For annotation see *Bolivina pisciformis*. *Bolivina mexicana* Cushman appears to fall within the variability of *B. alazanensis*, like the specimen figured under that name by Bermudez (1949, pl. 12, fig. 20).

BOLIVINA ALAZANENSIS VENEZUELANA Hedberg
Figure **53.6–7**; 50

Bolivina alazanensis Cushman var. *venezuelana* Hedberg 1937a, p. 676, pl. 91, fig. 14. – Cushman & Renz 1947b, p. 26, pl. 6, fig. 8.
Bolivina alazanensis Cushman 1926, Cushman & Renz 1948, p. 27, pl. 5, fig. 19.

The subspecies differs from the species in possessing a thickening along the axis, giving it a more rhomboidal appearance. Further, the individual chambers are not as high and the sutures more angled downwards. The subspecies is described as one of the most common and distinctive taxa in the Carapita Formation of eastern Venezuela.

In Trinidad the subspecies ranges from Late Eocene into the Middle Oligocene.

BOLIVINA BYRAMENSIS Cushman
Figure **78.20**; 73

Bolivina caelata Cushman var. *byramensis* Cushman 1923, Prof. Paper, U.S. geol. Surv., 133, p. 19, pl. 1, fig. 9.
Bolivina byramensis Cushman, Cushman & Renz 1947b, p. 26. – Renz 1948, p. 117, p. 6, fig. 22. – Diaz de Gamero 1977a, p. 41.

The characteristic reticulate surface ornamentation present in *Bolivina caelata* is only poorly and incompletely developed in *Bolivina byramensis*, originally regarded as subspecies of that species. Despite its weak surface ornamentation the species is characteristic, but in general reported as rare.

In Trinidad the species has been recorded from the Middle Oligocene to Middle Miocene, in southeastern Falcon from the Early and Middle Miocene.

BOLIVINA CAUDRIAE Cushman & Renz
Figure **78.9a–b,10**; 73

Bolivina caudriae Cushman & Renz 1941, p. 19, pl. 3, fig. 13. – Renz 1948, p. 117, pl.7, figs. 1, 2.
Brizalina caudriae (Cushman & Renz), Diaz de Gamero 1977a, p. 41.

A slender test shape with rounded periphery and with chambers only slowly increasing in size as added characterizes *Bolivina caudriae* and *B. isidroensis*. The latter differs from the former in the very fine longitudinal costae present in the earlier part of the test. *Bolivina pozonensis* differs from both species in a slightly broader test shape and a more acute periphery.

All three species were originally described from southeastern Falcon, where *Bolivina caudriae* is restricted to the Early Miocene and lower part of the Middle Miocene, *Bolivina isidroensis* to the Middle Miocene and *Bolivina pozonensis* to the upper part of the Middle Miocene.

The finely costate *Bolivina isidroensis* compares very closely with the holotype of *Bolivina acerosa* (Fig. 53.10) and judging from the holotype illustration appears to be synonymous. However, Cushman & Renz (1941) in the description of *Bolivina isidroensis* state that it differs from *Bolivina acerosa* in its larger size, less slender shape and more coarsely perforated wall.

BOLIVINA cf. COCHEI Cushman & Adams
Figure **78.7–8**; 73

Bolivina cochei Cushman & Adams 1935, Contrib. Cushman Lab. foramin. Res., 11, p. 19, pl. 3, fig. 6. – Renz 1948, p.117, pl. 6, figs. 27, 28.

The species appears to be very close to *Bolivina pisciformis* of which it might represent the juvenile growth stage.

In southeastern Falcon both species appear first in the Early Miocene. While *Bolivina pisciformis* disappears within the Middle Miocene, *Bolivina* cf. *cochei* continues at least to the top of the Middle Miocene.

BOLIVINA DANVILLENSIS Howe &
Wallace
Figure **53.18**; 50

Bolivina danvillensis Howe & Wallace 1932,
  Geol. Bull. Louisiana Dept. Conserv., 2,
  p. 56, pl. 11, fig. 8.

The elongate, slightly compressed test with a
rounded periphery is readily identifiable by its
characteristic intercameral sutures which in the
adult two-thirds of the test are depressed and
display a peculiar lobate pattern.

The species, apparently a short-ranged index
form, was originally described from the Late
Eocene of Louisiana. In Trinidad the species is
restricted to the same age.

BOLIVINA GRACILIS Cushman & Applin
Figure **53.17**; 50

Bolivina gracilis Cushman & Applin 1926,
  Bull., Am. Assoc. Petr. Geol., 10, p. 167,
  pl. 7, fig. 1.

A comparison of Trinidad specimens of *Bolivina
gracilis* (Fig. 53.17) with *Bolivina caudriae* from
southeastern Falcon (Fig. 78.9, 10) point to a
close similarity of the two taxa. In comparing
holotype illustrations the individual chambers of
Bolivina gracilis appear more inflated and the
intercameral sutures not as strongly oblique as
in the holotype of *Bolivina caudriae*.

The species was originally described from
Late Eocene beds of Texas. In Trinidad it is
reported to occur from late Middle Eocene to
Middle Oligocene, while *Bolivina caudriae* is
known in southeastern Falcon from Early
Miocene to the basal part of Middle Miocene.

BOLIVINA IMPORCATA Cushman & Renz
Figure **53.16**; 50. Figure **78.19**; 73

Bolivina floridana Cushman var. *regularis*
  Cushman & Renz 1941, p. 17, pl. 3, fig. 7.
Bolivina floridana Cushman var. *importata*,
  new name, Cushman & Renz 1944, Contrib.
  Cushman Lab. foramin. Res., 20, p. 78.
Bolivina importata Cushman & Renz 1947*b*,
  p. 26, pl. 6, fig. 9. – Renz 1948, p. 118, pl. 7,
  figs. 3a, b.

The deeply pitted surface is the characteristic
feature of the species.

In Trinidad it ranges from the upper part of
the Early Miocene through the Middle Miocene,
in southeastern Falcon apparently throughout
the Miocene. In eastern Venezuela the species
continues in the Cubagua Formation on the
Arraya Peninsula into the Pliocene (*Globorotalia margaritae margaritae* Zone).

BOLIVINA INCONSPICUA Cushman &
Renz
Figure **78.22–23**; 73

Bolivina inconspicua Cushman & Renz 1941,
  p. 1, pl. 3, figs. 10, 11. – Renz 1948, p. 118,
  pl. 7, figs. 9a, b, 10.

This species with an acute, occasionally slightly
serrate periphery differs from other morphologically similar forms like *Bolivina pisciformis* and
Bolivina cf. *cochei* in its small size and slightly
inflated chambers.

In southeastern Falcon the species has been
recorded only from the uppermost Early
Miocene and lower part of the Middle Miocene.

BOLIVINA ISIDROENSIS Cushman & Renz
Figure **78.12**; 73

Bolivina isidroensis Cushman & Renz 1941,
  p. 17, pl. 3, fig. 8. – Renz 1948, p. 118, pl. 7,
  figs. 5a, b.

For annotations see *Bolivina caudriae*.

BOLIVINA MARGINATA
MULTICOSTATA Cushman
Figure **53.8–9**; 50. Figure **78.14–16**; 73

Bolivina aenariensis (Costa) Brady var.
  multicostata Cushman 1919, Bull. U.S. geol.
  Surv., no. 676, p. 48, pl. 10, fig. 2. –
  Cushman & Renz 1947*b*, p. 27, pl. 6, fig. 10.
  – Renz 1948, p. 118, pl. 7, figs. 6–8.

The taxon *multicostata*, originally regarded by
Cushman as a variety of *Bolivina aenariensis*, was
later changed by him to a variety of *B. marginata*
Cushman (Cushman 1930, Bull. Florida State
Geol. Surv, 4, p. 46, pl. 8, figs. 13, 14).

There exists considerable variation in specimens assigned to the subspecies. Compared with
the holotype (Fig. 53.8), the initial portion of
the Trinidad specimen (Fig. 53.9) tends to have
a more strongly tapering early stage. Compared

with the holotype, tests of the specimens from the Agua Salada Formation of southeastern Falcon (Fig. 78.14–16) are not as slender as are the Trinidad forms, but more like the holotype. The Trinidad specimens tend to have finer and more numerous costae, about 12–13 against 9–10 in the holotype.

A comparison of ranges of the specimens assigned to *B. marginata multicostata* in Trinidad and Falcon shows that all appear first in the *Catapsydrax stainforthi* Zone. In Trinidad the taxon is only known from this zone and the next younger *Globigerinatella insueta* Zone, while in Falcon it continues into the *Globorotalia acostaensis* Zone. A closer investigation of the Trinidad and Falcon specimens might show that the forms from the two areas might eventually be distinguished taxonomically.

BOLIVINA MEXICANA ALIFORMIS
Cushman
Figure **53.4–5**; 50

Bolivina mexicana var. *aliformis* Cushman
      1926, Contrib. Cushman Lab. foramin. Res.,
      1, p. 82, pl. 12, figs. 4a, b.
Bolivina cf. *alata* Seguenza, Cushman & Renz
      1947*b*, p. 27, pl. 6, fig. 11.
Brizalina mexicana aliformis (Cushman), Diaz
      de Gamero 1977*a*, p. 41.

The subspecies differs from *Bolivina alata* in its distinctly wider test with a broad, acute and transparent keel without spinose projections as typical for *Bolivina alata*. The subspecies was originally described from the Alazan shales of Mexico.

In Trinidad the characteristic subspecies is restricted to the Middle and Late Oligocene.

BOLIVINA PISCIFORMIS Galloway &
Morrey
Figure **53.1**; 50. Figure **78.1–2**; 73

Bolivina pisciformis Galloway & Morrey 1929,
      Bull. Am. Paleontol.,15, no. 55, p. 36, pl. 5,
      figs. 10a, b. – Cushman & Renz 1947*b*, p. 25,
      pl. 6, fig. 6. – Renz 1948, p. 119, pl. 7, figs.
      11, 12.

The species was originally described from the probable Late Eocene of Ecuador. Closely comparable specimens occur in Trinidad through the late Early Eocene to the top Eocene. The holo-

types of *Bolivina pisciformis* (Fig. 53.1) and *Bolivina alazanensis* (Fig. 53.2) appear to be close morphologically. Comparing the holotype illustrations, *B. pisciformis* is slightly smaller in size and the intercameral sutures are fairly straight compared with *B. alazanensis* where sutures appear more curved towards the centre. Further, chambers of *B. alazanensis* increase more rapidly throughout the growth of the test while in *B. pisciformis* the increase of the last few pairs of chambers is comparatively slow.

The specimen illustrated by Cushman & Renz (1947*b*, pl. 6, fig, 6; here Fig. 53.3) as *B. pisciformis* from the Lower to Middle Miocene Ste. Croix Formation of Trinidad, compares much closer to the holotype of *B. alazanensis*. Compared with the comparatively small *B. pisciformis,* apparently restricted to the Eocene, *B. alazanensis* is known in Trinidad from the Early Miocene *Catapsydrax stainforthi* to the *Praeorbulina glomerosa* Zone.

Closely comparable forms, illustrated by Renz (1948) as *B. pisciformis,* range in southeastern Falcon up into the *Globorotalia fohsi lobata* Zone. On the other hand the somewhat smaller specimens identified by Renz as *B. alazanensis* and ranging in Falcon through the *Catapsydrax dissimilis* and *Catapsydrax stainforthi* zones appear intermediate between the typical *B. pisciformis* and *B. alazanensis*. Renz (1948) regards *B. alazanensis* as the ancestor of *B. pisciformis*.

Admittedly, the two species are morphologically close but from their stratigraphic occurrences the smaller and older *B. pisciformis* is rather to be regarded as the ancestral form of the Early to Middle Miocene *B. alazanensis*. The Falcon forms described by Renz (1948) as *B. alazanensis* from the Early Miocene probably represent intermediate forms, linking the Eocene *B. pisciformis* with the Early to Middle Miocene *B. alazanensis*.

BOLIVINA PISCIFORMIS OPTIMA
Cushman
Figure **53.12**; 50

Bolivina pisciformis Galloway & Morrey var.
      optima Cushman 1943, Contrib. Cushman
      Lab. foramin. Res., 19, p. 91, pl. 16, figs.
      2a, b.

The subspecies differs from the species (Fig. 53.1) in having most of the early half of the test covered by numerous, fine longitudinal costae, while the surface of the later part of the test remains smooth.

This characteristic taxon ranges in Trinidad from base Middle Oligocene to top Middle Miocene.

BOLIVINA POZONENSIS Cushman & Renz
Figure **78.11**; 73

Bolivina pozonensis Cushman & Renz 1941, p. 16, pl. 3, fig. 6. – Renz 1948, p. 119, pl. 7, figs. 14a, b.

For annotations see *Bolivina caudriae.*

BOLIVINA RUDDERI Cushman & Renz
Figure **78.6**; 73

Bolivina rudderi Cushman & Renz 1941, p. 19, pl. 3, fig. 12. – Renz 1948, p. 119, pl. 7, figs. 13a, b.

Characteristic for the small species is the strongly tapering test. Except for this feature it is close to the also small *Bolivina* cf. *cochei* (Fig. 78.7, 8) and to the larger *Bolivina pisciformis* (Figure 53.1; 78.1,2).

In contrast to *Bolivina* cf. *cochei* and *Bolivina pisciformis* which in SE-Falcon first appear within the Early Miocene, the occurrence there of *Bolivina rudderi* is restricted to the Middle Miocene.

BOLIVINA SIMPLEX Cushman & Renz
Figure **78.17**; 73

Bolivina interjuncta var. *simplex* Cushman & Renz 1941, p. 20, pl. 3, fig. 15. – Renz 1948, p. 119, pl. 7, figs. 4a, b.

Bolivina simplex differs from other costate *Bolivina* like *B. marginata multicostata* in the fewer and more strongly developed longitudinal costae, usually about four on each side. Specimens with only two costae running along the whole test close to the centre are known from the Agua Salada Formation in Falcon.

The specimen figured as *Bolivina interjuncta* var. *bicostata* Cushman by Bermudez (1949, pl. 12, fig. 24) with two well-developed costae, appears closely comparable to such forms in the Agua Salada. In contrast the holotype of this variety (Cushman 1937c, pl. 22, fig. 23) shows

four distinct longitudinal costae as are also characteristic for *B. simplex.* Often, in particular in the Trinidad specimens of *B. simplex,* the longitudinal costae do not extend across the second last or/and the last pair of chambers.

While the characteristic species is recorded in Trinidad from the *Globigerinatella insueta* Zone to the *Globorotalia fohsi lobata* Zone, it ranges in Falcon from the *Catapsydrax dissimilis* Zone to the *Globorotalia acostaensis* Zone. The species is also present in the Early Pliocene *Globorotalia margaritae* Zone on the Araya peninsula in eastern Venezuela. In these areas the range of *B. simplex* is apparently strongly controlled by environmental conditions.

BOLIVINA SUTERI Cushman & Renz
Figure **78.21**; 73

Bolivina suteri Cushman & Renz 1941, p. 18, pl. 3, fig. 9. – Renz 1948, p. 119, pl. 7, figs. 15a, b.

This small species is easily identifiable by its inflated chambers which in the later growth stage of the test rapidly increase in size, and by the intercameral sutures running at nearly right angles to the longitudinal axis, resulting in a distinctly lobate periphery.

The species occurs in southeastern Falcon throughout most of the Early and Middle Miocene.

BOLIVINA cf.TECTIFORMIS Cushman
Figure **53.13–14**; 50. Figure **87.17**; 86

Bolivina tectiformis Cushman 1926, Contrib. Cushman Lab. foramin. Res., 1, p. 83, pl. 12, figs. 6a, b. – Cushman & Stainforth 1945, p. 47, pl. 7, fig. 12.
Bolivina cf. *tectiformis* Cushman, Beckmann 1953, p. 367, pl. 21, figs. 16, 17.

Cushman & Stainforth (1945) report specimens from the lower and middle part of the Cipero Formation of Trinidad as belonging to *Bolivina tectiformis,* a species originally described from the Alazan Formation of Mexico. The specimen illustrated by these authors (Fig. 53.13) however differs significantly in several respects from the here also reproduced holotype figure (Fig. 53.14). The characteristics for the species are according to its author Cushman the distinct, much thickened intercameral sutures that fuse in the median line to form a thick, rounded ridge. These features are clearly visible on the holotype figure. The longer and more slender Trinidad specimen illustrated by Cushman & Stainforth does in contrast not display such features. All it shows is a delicate longitudinal ridge confined to the lower half of the specimen. The illustrated Trinidad specimen should therefore not be included in *Bolivina tectiformis.*

The specimen illustrated by Beckmann (1953) from the Oceanic Formation of Barbados as *Bolivina* cf. *tectiformis* (Fig. 87.17) compares with the illustration in Cushman & Stainforth (1945) where general test shape and weak striation in the early part of the test are concerned. It should for the same reasons also not be included in *Bolivina tectiformis.*

On the other hand the specimen illustrated by Gamero (1977a) as *Brizalina vaceki* Schubert (Fig. 81.7) from the Oligocene and lowermost Early Miocene of central Falcon, appears to be closely comparable with Cushman's *Bolivina tectiformis.* However, in contrast to the massive sutures and longitudinal ridge characterizing *Bolivina tectiformis* the specimen of *Bolivina vaceki* illustrated by Schubert (1901) lacks such features completely. Instead its test is, compared with *Bolivina tectiformis,* more distinctly tapering, and its whole surface is very finely striate.

BOLIVINA THALMANNI Renz
Figure **78.18a–b;** 73

Bolivina thalmanni Renz 1948, p. 120, pl. 12. figs. 13a–c.

Typical for the species are the characteristic surface features that consist of two sharply edged longitudinal ridges running about parallel to the periphery. These two median and the peripheral ridges are interconnected by sharply raised lamellar ridges, separated by irregular deep depressions, generally numbering 5–10 between initial and apertural ends, giving the test its characteristic reticulate appearance.

In southeastern Falcon the species has a restricted range, confined to the Middle Miocene *Globorotalia fohsi robusta* and the *Globorotalia mayeri/Globorotalia lenguaensis* zones.

BOLIVINA TONGI Cushman
Figure **78.24–25;** 73

Bolivina tongi Cushman 1929b, p. 93, pl. 13, fig. 29. – Renz 1948, p. 120, pl. 6, figs. 24, 25.

This small species is characterized by its truncate sides and few plate-like costae near the borders. It was originally described from a Miocene sample collected from a sea cliff southeast of the cemetery of Aguide, Falcon, close to the Agua Salada type section.

In this section the species is recorded from the Early Miocene *Catapsydrax dissimilis* Zone to the basal part of the Middle Miocene *Globorotalia fohsi fohsi* Zone; in Trinidad from the Middle Eocene *Orbulinoides beckmanni* Zone to the Middle Oligocene *Globigerina ampliapertura* Zone.

BOLIVINA VACEKI Schubert
Figure **81.7;** 75

Bolivina vaceki Schubert 1901, Beitr. Geol. Pal. Oesterr.-Ungar. Orients, Vienna, 14, p. 25, pl. 1, fig. 29.
Brizalina vaceki (Schubert), Diaz de Gamero 1977a, p. 41, pl. 3, fig. 15.

For annotations see *Bolivina tectiformis.*

Family Cassidulinidae

CASSIDULINA CARAPITANA Hedberg
Figure **80.34a–b;** 74

Cassidulina carapitana Hedberg 1937a, p. 680, pl. 96, figs. 6a–b. – Renz 1948, p. 124, pl. 9, figs. 8a–b.

CASSIDULINA DELICATA Cushman
Figure **80.35a–b;** 74

Cassidulina delicata Cushman 1927, Calif. Univ. Scripps Inst. Oceanogr. Bull. Tech. Ser., 1, p. 169, pl. 6, fig. 5. – Renz 1948, p. 125, pl. 9, figs. 10a–b.

CASSIDULINA SPINIFERA Cushman & Jarvis
Figure **62.24, 24a;** 52

Cassidulina spinifera Cushman & Jarvis 1929, p. 17, pl. 3, fig. 1. – Cushman & Stainforth 1945, p. 64, pl. 11, fig. 10.

CASSIDULINOIDES BRADYI (Norman)
Figure **62.25, 25a;** 52

Cassidulina bradyi Norman 1880, Ms. in Wright, Proc. Belfast Nat. Field Club, App., p. 152.
Cassidulinoides bradyi (Norman), Cushman & Stainforth 1945, p. 65, pl. 12, fig. 6.

CASSIDULINOIDES ERECTA Cushman & Renz
Figure **80.33a–b;** 74

Cassidulinoides erecta Cushman & Renz 1941, p. 25, pl. 4, figs. 6–7. – Renz 1948, p. 126, pl. 9, figs. 15a–c.

EHRENBERGINA CARIBBEA Galloway & Heminway
Figure **80.26a–b;** 74

Ehrenbergina caribbea Galloway & Heminway 1941, p. 426, pl. 32, figs. 4a–d. – Renz 1948, p. 131, pl. 9, figs. 17a–b.

Family Siphogenerinoididae

Genus *Rectuvigerina* Mathews 1945

The chamber arrangement of the genus *Rectuvigerina* is characterized by a short initial triserial stage, followed by a distinct biserial one and an uniserial end-stage comprising two to four chambers. The genus is intermediate between *Siphogenerina* which displays a very short triserial initial stage, often difficult to discern, followed by a dominating uniserial stage, and *Uvigerina* which is triserial throughout. Certain taxa may occupy intermediate positions and be difficult to be clearly assigned to one or the other of the three genera.

See also the annotations on the genus *Siphogenerina*.

RECTUVIGERINA BASISPINATA (Cushman & Jarvis)
Figure **54.33–36;** 50

Siphogenerina basispinata Cushman & Jarvis 1929, p. 13, pl. 3, figs. 4, 5. – Cushman & Stainforth 1945, p. 49, pl. 8, fig. 3. – Cushman & Renz 1947*b*, p. 29, pl. 7, fig. 1.

RECTUVIGERINA MULTICOSTATA (Cushman & Jarvis)
Figure **54.30;** 50. Figure **78.45;** 73

Siphogenerina multicostata Cushman & Jarvis 1929, p. 14, pl. 3, fig. 6. – Cushman & Stainforth 1945, p. 49, pl. 8, figs. 1, 2. – Renz 1948, p. 165, pl. 7, fig. 26.

RECTUVIGERINA MULTICOSTATA OPTIMA (Cushman)
Figure **54.31–32;** 50

Siphogenerina multicostata var. optima Cushman 1943, Contrib. Cushman Lab. foramin. Res., 19, p. 91, pl. 16, figs. 9, 10 (9 holotype).
Siphogenerina multicostata Cushman & Jarvis, Cushman & Stainforth 1945, p. 49, pl. 8, fig. 1.
Siphogenerina seriata (Cushman & Jarvis), Cushman & Stainforth 1945, p. 50, pl. 8, fig. 2.
Uvigerina seriata Cushman & Jarvis 1929, p. 13, pl. 3, figs. 11, 12.

Genus *Siphogenerina* Schlumberger 1883

The genera *Siphogenerina* and *Rectuvigerina* are represented in the Oligocene and Miocene of the Falcon Basin, Venezuela and Trinidad by a number of characteristic and stratigraphically significant taxa. Only one species has been reported from the Oceanic Formation of Barbados.

Renz (1948) described and illustrated six *Siphogenerina* species from southeastern Falcon of which one (*S. multicostata*) is here transferred to *Rectuvigerina*. Diaz de Gamero (1977*a*) figured four species of *Sipho-*

generina from central Falcon. Of the seven *Siphogenerina* species described by Cushman & Stainforth (1945) and Cushman & Renz (1947*b*) from Trinidad, three are now placed in *Rectuvigerina (basispinata, multicostata, seriata)*.

The stratigraphically oldest of these forms, *Siphogenerina nodifera,* was reported from central Falcon where it ranges from the *Globorotalia opima opima* Zone to the *Globigerinoides primordius* Zone. The other three central Falcon species *S. hubbardi, S. senni* and *S. transversa* appear there at the base of the *Globigerinoides primordius* Zone to continue into the *Catapsydrax dissimilis* Zone, the youngest zone investigated in central Falcon by Diaz de Gamero. Two of them, *Siphogenerina transversa* and *S. senni,* continue in southeastern Falcon into the basal *Globorotalia fohsi fohsi* and *Globorotalia mayeri* zones respectively.

Of the other four species reported by Renz (1948) from southeastern Falcon as *Siphogenerina,* his *S. smithi* is here placed in *Uvigerina* (Fig. 78.40–41) and regarded as very close to '*Uvigerinella*' *sparsicostata,* which also has the same range (*Catapsydrax dissimilis* to *Globigerinatella insueta* Zone). His *Siphogenerina multicostata* now falls into *Rectuvigerina*. The remaining two species *S. lamellata* and *S. kugleri* range from the *Catapsydrax dissimilis* into the *Globorotalia acostaensis* Zone, and from the *Globigerinatella insueta* to the *Globorotalia fohsi fohsi* Zone respectively.

All *Siphogenerina* and *Rectuvigerina* species reported from southeastern Falcon are also known from Trinidad. Here, the four *Siphogenerina* species and *Rectuvigerina multicostata* are restricted to the *Globigerinatella insueta* and the overlying *Praeorbulina glomerosa* zones. The other two *Rectuvigerina* species, *R. multicostata optima* and *R. basispinata* continue well into the Middle Miocene.

The simultaneous appearance in Trinidad of all *Siphogenerina* and *Rectuvigerina* species in the *Globigerinatella insueta* Zone, that is considerably later compared with the Falcon Basin in Venezuela, is controlled by environment. Only with the onset of shallower water conditions as typical for the Agua Salada Formation in Falcon and the Ste. Croix and Brasso formations in Trinidad did these forms migrate to Trinidad, while they remained absent in the lower part of the deeper water Cipero Formation.

SIPHOGENERINA HUBBARDI Galloway & Heminway
Figure **81.11;** 75

Siphogenerina hubbardi Galloway & Heminway 1941, p. 434, pl. 34, fig. 2. – Diaz de Gamero 1977*a*, p. 43, pl. 4, fig. 10.

SIPHOGENERINA KUGLERI Cushman & Renz
Figure **54.38;** 50. Figure **78.46;** 73

Siphogenerina kugleri Cushman & Renz 1941, p. 22, pl. 3, figs. 21–22. – Cushman & Renz 1947*b*, p. 31, pl. 7, fig. 4. – Renz 1948, p. 164, pl. 7, fig., 28.

SIPHOGENERINA LAMELLATA Cushman
Figure **78.49;** 73

Siphogenerina lamellata Cushman 1918, Bull. U.S. geol. Surv., 676, p. 55, pl. 12, fig. 3. – Renz 1948, p. 165, pl. 7, fig. 25.

SIPHOGENERINA NODIFERA Cushman & Kleinpell
Figure **81.8;** 75

Siphogenerina nodifera Cushman & Kleinpell 1934, Contrib. Cushman Lab. foramin. Res., 10, p. 13, pl. 2, figs. 15, 16. – Diaz de Gamero 1977*a*, p. 43, pl. 4, fig. 8.

SIPHOGENERINA SENNI Cushman & Renz
Figure **54.37;** 50. Figure **78.47–48;** 73. Figure **81.10;** 75

Siphogenerina senni Cushman & Renz 1941, p. 22, pl. 3, figs. 21, 22. – Cushman & Renz 1947*b*, p. 30, pl. 7, fig. 5. – Renz 1948, p. 165, pl. 7, figs. 29, 30. – Diaz de Gamero 1977*a*, p. 43, pl. 4, fig. 11.

SIPHOGENERINA TRANSVERSA
Cushman
Figure **54.39–40**; 50. Figure **78.50–52**; 73. Figure
81.9; 75. Figure **87.23, 23a**; 86

Siphogenerina raphanus (Parker & Jones) var.
transversus Cushman 1918, Bull. U.S. natl.
Mus., 103, p. 64, pl. 22, fig. 8.
Siphogenerina transversa Cushman, Cushman
& Renz 1947*b*, p. 30, pl. 7, figs. 2, 3. – Renz
1948, p. 166, pl. 7, figs. 27, 28. – Diaz de
Gamero 1977*a*, p. 43, pl. 4, fig. 8.
Siphogenerina sp. aff. *transversa* Cushman
1926, Beckmann 1953, p. 368, pl. 21, fig. 21.

Family Buliminidae

Genus *Bulimina* d'Orbigny, 1826

Representatives of the genus *Bulimina* are present in
both the deeper-water Navet, Cipero and Lengua formations of Trinidad and in the shallower Pecaya Formation of central Falcon, the Agua Salada Formation
of southeastern Falcon and the Ste. Croix and Brasso
formations of Trinidad. A few species that occur in
Trinidad and Falcon are also recorded from the deep-
water Oceanic Formation of Barbados.

A total of 18 *Bulimina* species are listed and illustrated of which 13 occur in Trinidad (ranges Fig. 50,
illustrations Fig.53.19–35), six in southeastern Falcon
(ranges Fig. 73, illustrations Fig. 78.26–32), five in
central Falcon (ranges Fig. 75, illustrations Fig.
81.12–14), and three in Barbados (ranges Fig. 86,
illustrations Fig. 87.18–20).

Several of the taxa occur in two or more of the four
areas under review. While certain species such as *B.
alazanensis* seem more facies tolerant others, like *B.
pupoides,* appear to be more facies restricted. In
Trinidad four species range up from the Early Eocene
of which two become extinct in the Middle Eocene
Orbulinoides beckmanni Zone. Another two, *B. guayabalensis ampla* and the characteristic and short-
ranged *B. jacksonensis,* disappear in the Late Eocene
Turborotalia cerroazulensis s. l. Zone. *B. alazanensis,*
the most extended species, ranges e.g. in Trinidad
from the Early Eocene *Morozovella subbotinae* Zone
to the Middle Miocene *Globorotalia menardii* Zone.

Certain anomalies concerning the occurrence and
stratigraphic distribution in the Oligo-Miocene of the
areas under review are noted for some species as can
be seen from the respective range charts. As an
example may serve *Bulimina pupoides* that in Trinidad ranges from the Late Paleocene *Morozovella velascoensis* Zone to the Late Oligocene *Globigerina
ciperoensis ciperoensis* Zone, but continues in the
Agua Salada Formation of southeastern Falcon from
the *Catapsydrax dissimilis* Zone to the *Globorotalia
acostaensis* Zone.

Shorter-ranging species in Trinidad are *Bulimina
sculptilis, B. striata mexicana, B. tuxpamensis* and *B.
bleeckeri.* In southeastern Falcon the recorded *Bulimina* species occur throughout most of the Agua Salada
with the only exception of *B. falconensis* which is
restricted to the interval *Globorotalia fohsi fohsi* to
Globorotalia mayeri Zone.

BULIMINA ALAZANENSIS Cushman
Figure **53.27**; 50. Figure **78.30**; 73. Figure **81.13**; 75

Bulimina alazanensis Cushman 1927, J.
Paleontol., 1, p. 161, pl. 25, fig. 4. –
Cushman & Stainforth 1945, p. 40, pl. 6,
fig. 2. – Cushman & Renz 1948, p. 25, pl. 5,
figs. 14, 15. – Renz 1948, p. 120, pl. 6, fig. 14.
– Diaz de Gamero 1977*a*, p. 41, pl. 3, fig. 16.

The small species with its numerous prominent
longitudinal costae is readily recognizable.

Originally described from the Alazan Clay,
Mexico, the species has a long range, in Trinidad
from Middle Eocene to top Middle Miocene, in
central and southeastern Falcon through most of
the measured sections.

BULIMINA cf. ASPERA Cushman & Parker
Figure **53.19**; 50

Bulimina aspera Cushman & Parker 1940,
Contrib. Cushman Lab. foramin. Res., 16,
p. 44, pl. 8, figs. 18, 19. – Cushman & Renz
1946, p. 37, pl. 6l, fig. 10.

The holotype of *Bulimina aspera* (Cushman &
Parker 1940, pl. 8, fig. 18) is described from the
Late Cretaceous Taylor Formation of Texas and
today is regarded as falling into *Praebulimina.*
Cushman & Renz (1946) illustrated a specimen
from the Paleocene Lower Lizard Springs Formation of Trinidad (here illustrated on Fig.
53.19) identified by them as *B. aspera.* Instead of
having one or two basal spines and a roughened
surface with distinct perforations in the initial
part, the Trinidad specimen shows a few short

vertical ribs and some spinose sutures; it is probably not conspecific with *B. aspera*.

Superficially similar specimens without distinct spines are found in the Early and Middle Eocene of Trinidad and may be close to *B. tuxpamensis* Cole.

BULIMINA BLEECKERI Hedberg
Figure **53.25–26**; 50. Figure **81.14**; 75

Bulimina bleeckeri Hedberg 1937*a*, p. 675, pl. 91, figs. 12, 13. – Cushman & Stainforth 1945, p. 41, pl. 6, fig. 4. – Diaz de Gamero 1977*a*, p. 41, pl. 3, fig. 17.

For annotations see *B. striata mexicana*.

BULIMINA FALCONENSIS Renz
Figure **78.32**; 73

Bulimina falconensis Renz 1948, p. 121, pl. 6, figs. 15a, b.

The species shows a certain resemblence to *B. sculptilis* but differs in a more slender test shape, more and finer longitudinal costae and the test usually being made up of more chambers.

In the Agua Salada Formation of southeastern Falcon the species is restricted to the lower part of the Middle Miocene.

BULIMINA GUAYABALENSIS AMPLA
Cushman & Parker
Figure **53.29**; 50

Bulimina guayabalensis var. *ampla* Cushman & Parker 1936, Contrib. Cushman Lab. foramin. Res., 12, p. 43, pl. 8, figs. 1a–c.

The illustrated holotype of the subspecies differs from the species and the other here illustrated taxa with smooth surfaces (Fig. 53.20, 21, 22; Fig. 78.26–27) in its larger size and in being broader in relation to its length.

The species was originally described from the Eocene of California. In Trinidad it is recorded from late Middle to Late Eocene.

BULIMINA ILLINGI Cushman & Stainforth
Figure **53.20**; 50

Bulimina (Desinobulimina) illingi Cushman & Stainforth 1945, p. 41, pl. 6, fig. 7.

The here illustrated species assigned to *B. illingi*, *B. ovata*, *B. pupoides*, *B. tuxpamensis*, further to *B. guayabalensis ampla* and *B. perversa* are characterized by their smooth surfaces. Their specific differences lie in size and test shape determined by number and growth rates of successive chambers. Extreme forms like *B. perversa* (Fig. 78.28) or *B. guayabalensis ampla* (Fig. 53.29) are readily identifiable. Others, like *B. illingi*, *B. ovata*, *B. pupoides* and *B. tuxpamensis* may be more difficult to differentiate from each other.

The species ranges in Trinidad from Middle Oligocene to lower part of Middle Miocene.

BULIMINA cf. INFLATA Seguenza
Figure **78.29**; 73

Bulimina inflata Seguenza 1862, Atti. Accad. Gioenia Sci. Nat. ser., 2, 18, p. 109, pl. 1, fig. 10. – Renz 1948, p. 121.
Bulimina cf. *inflata* Seguenza, Renz 1948, p. 122, pl. 12, figs. 14a–b.

For annotations see *B. striata mexicana*.

BULIMINA INFLATA ALLIGATA
Cushman & Laiming
Figure **78.31**; 73

Bulimina inflata Seguenza var. *alligata* Cushman & Laiming 1930, J. Paleontol., 4, p. 362, pl. 33, fig. 5. – Renz 1948, p. 122, pl. 6, fig. 13.

For annotations see *B. striata mexicana*.

BULIMINA JACKSONENSIS Cushman
Figure **53.31–33**; 50

Bulimina jacksonensis Cushman 1925, Contrib. Cushman Lab. foramin. Res., 1, p. 6, pl. 1, figs. 6, 7. – Cushman & Parker, 1947, Prof. Pap. U.S. geol. Surv., 210-D, p. 97, pl. 22, figs. 14–16.

The species is characterized by the rapidly increasing chambers, typically seen in Fig. 53.32, and few (6–8) but very prominent longitudinal costae.

The species is an excellent index for the Late Eocene. In Trinidad it occurs in the *Globigerinatheka semiinvoluta* and *Turborotalia cerroazulensis* zones.

BULIMINA JARVISI Cushman & Parker
Figure **53.30a–b; 50**. Figure **87.19; 86**

Bulimina jarvisi Cushman & Parker 1936,
  Contrib. Cushman Lab. foramin. Res., 12,
  p. 39, pl. 7, figs. 1a–c. – Cushman &
  Stainforth 1945, p. 41, pl. 6, fig. 5. –
  Beckmann 1953, p. 366, pl. 21, fig. 10.

The overall test shape of *B. jarvisi* resembles
that of *B. prolixa*. It however differs from that
species in that most chambers, in particular the
earlier ones, are covered by fine somewhat
irregular costae. Furthermore, the test is usually
larger and of somewhat angular appearance
compared with the rounder *B. prolixa*.
  B. jarvisi ranges in Trinidad from Late Eocene
to within the Middle Miocene. *B. prolixa* on the
other hand is restricted to the Early and Middle
Eocene. In Barbados *B. jarvisi* occurs in the
Oceanic Formation from Middle Eocene to Late
Oligocene.

BULIMINA MACILENTA Cushman &
Parker
Figure **53.24; 50**. Figure **87.20; 86**

Bulimina denticulata Cushman & Parker 1936,
  Contrib. Cushman Lab. foramin. Res., 12,
  p. 42, pl. 7, figs. 7a–c, 8a–c.
Bulimina macilenta, new name, Cushman &
  Parker 1939, Contrib. Cushman Lab. foramin.
  Res., 15, p. 93. – Cushman & Stainforth
  1945, p. 40, pl. 6, fig. 4. – Beckmann 1953,
  p. 366, pl. 21, fig. 10. – Wood *et al.* 1985,
  pl. 1, fig. 10.

For annotations see *B. striata mexicana*.
  The species ranges in Trinidad from late Early
Eocene to Early Miocene. In Barbados it is
reported from the Late Eocene and Oligocene
of the Oceanic Formation.

BULIMINA OVATA d'Orbigny
Figure **87.18; 86**

Bulimina ovata d'Orbigny 1846, p. 185, pl. 11,
  figs. 13, 14. – Beckmann 1953, p. 366, pl. 21,
  fig. 12.

The specimen illustrated from the Late Eocene
to Oligocene Oceanic Formation of Barbados
differs from d'Orbigny's type from the Vienna
Basin in having fewer chambers making up the
early part of the test. In the Barbados specimen

it is followed by three large and broad final
chambers, giving the test, compared with the
type specimen, a shorter and wider shape,
against a longer and more delicately built one.
  For annotations see also *B. pupoides*.

BULIMINA PERVERSA Cushman
Figure **78.28; 73**

Bulimina pyrula d'Orbigny var. *perversa*
  Cushman 1921, Bull. U.S. natl. Mus., 100,
  p.163, textfigs. 2a–c.
Bulimina (Globobulimina) perversa Cushman,
  Renz 1948, p.122, pl. 6, fig. 16.

The illustrated Falcon specimen differs from the
holotype in having a nearly flat base compared
to a more rounded one in the type species which
is a Recent form.
  The southeastern Falcon specimens occur virtually throughout the Agua Salada Formation.

BULIMINA PROLIXA Cushman & Parker
Figure **53.28; 50**

Bulimina prolixa Cushman & Parker 1935,
  Contrib. Cushman Lab. foramin. Res., 11,
  p. 98, pl. 15, figs. 5a, b.

Eocene to Miocene specimens identified in
Trinidad as *B. prolixa* were thought to be closely
comparable to this species originally described
from the Cretaceous Selma Chalk, Tennessee.
Re-examination now indicates that those Trinidad specimens identified as *B. prolixa* are morphologically closer to the younger *B. jarvisi* and
possibly represent its ancestral form. The Cretaceous *B. prolixa* is today placed in *Praebulimina* and has apparently no relation to the Eocene
specimens identified in Trinidad under this
name.
  For further annotations see *B. jarvisi*.

BULIMINA PUPOIDES d'Orbigny
Figure **53.21; 50**. Figure **78.26–27; 73**

Bulimina pupoides d'Orbigny 1846, p. 185,
  pl. 11, figs. 11–12. – Renz 1948, p. 122,
  pl. 6, figs. 11, 12. – Cushman & Renz 1948,
  p. 25, pl. 5, fig. 16.

Compared with the type the illustrated Trinidad
specimen (Fig. 53.21) appears more compact
and less tapering at the initial end, while the two

specimens from Falcon (Fig. 78.26, 27) in comparison are distinctly broader in their middle parts. The specimens here placed in *B. pupoides* are probably synonymous with *Praeglobobulimina ovata* of the Early Eocene Lizard Springs Formation of Trinidad.

The species ranges in Trinidad from Early Eocene to Late Oligocene. It virtually ranges throughout the Miocene Agua Salada Formation of southeastern Falcon.

BULIMINA SCULPTILIS Cushman
Figure **53.34–35**; 50. Figure **81.12**; 75

Bulimina sculptilis Cushman 1923, Prof. Pap. U.S. geol. Surv., 133, p. 23, pl. 3, fig. 19. – Diaz de Gamero 1977*a*, p. 42, pl. 3, fig. 19.

The species resembles *B. jacksonensis* but differs in a more delicate test with chambers less rapidly increasing in size (Fig. 53.35), and with the longitudinal costae being slightly more numerous (about 10).

In Trinidad the species occurs from the Late Eocene *Turborotalia cerroazulensis* s. l. Zone to the Middle Oligocene *Globigerina ciperoensis ciperoensis* Zone, thus has a longer range than *B. jacksonensis*.

BULIMINA STRIATA MEXICANA Cushman
Figure **53.23**; 50

Bulimina inflata Seguenza var. *mexicana* Cushman 1922, Bull. U.S. natl. Mus., 104, pt. 3, p. 95, pl. 21, fig. 2.
Bulimina striata d'Orbigny var. *mexicana* Cushman, Cushman & Parker 1947, Prof. Pap., U.S. geol. Surv., 210-D, p. 119, pl. 28, fig. 4.

The five here illustrated species *B. striata mexicana*, *B. macilenta*, *B. bleeckeri*, *B. inflata alligata* and *B.* cf. *inflata* are similar in their overall test shape and characteristic fine spines situated at the base of each chamber. Specimens placed in Trinidad in *B. striata mexicana* are very small, those in *B. bleeckeri* comparatively large, while specimens placed in *B. macilenta* are intermediate in size. The *B. inflata alligata* identified in southeastern Falcon (Fig. 78.31) fall where their sizes are concerned between *B. striata mexicana* and *B. macilenta* of Trinidad. *B.* cf. *inflata* (Fig.

78.29) is again comparable with *B. macilenta* (Fig. 53.24) of Trinidad.

There exists considerable variability within this whole group. Specimens may therefore be difficult to assign to a certain species.

In Trinidad and southeastern Falcon these taxa occur throughout the Miocene, with the exception of the small *B. macilenta* which in Trinidad appears as from the late Early Eocene.

BULIMINA TUXPAMENSIS Cole
Figure **53.22**; 50

Bulimina tuxpamensis Cole 1928, Bull. Am. Paleontol., 14, no. 53, p. 212, pl. 1, fig. 23. – Cushman & Stainforth 1945, p. 41, pl. 6, fig. 6.

Compared with the holotype described from the Eocene of Mexico, the illustrated Trinidad specimen from the Cipero Formation (late Middle to early Late Miocene) has a more rounded initial part and higher chambers.

Closely similar specimens also occur in the Eocene of Trinidad (see *B.* cf. *aspera*).

Family Uvigerinidae

Genus *Uvigerina* d'Orbigny 1826

Numerous *Uvigerina* species occur in Trinidad between the Middle Eocene and the Middle Miocene. Apart from *Uvigerina mantaensis* which ranges throughout this interval, most of the other 21 species here selected from Trinidad have much more restricted ranges as can be seen on Fig. 50. Several of the species extend from the Eocene into the Oligocene to disappear there at about the *Globigerina ampliapertura* Zone level. A few short-ranged species are restricted to various zones in the Oligocene and Early Miocene. Of significance is the approximately joint appearance of seven taxa in the *Catapsydrax stainforthi* Zone, with some of them being restricted to it. Others continue into higher levels. The reason for this simultaneous appearance lies in the more favourable shallower water facies as represented by the Ste. Croix and Brasso formations which in Trinidad set in at about *Catapsydrax stainforthi* Zone time. As is the case with other benthic forms this reflects the strong dependence on environment also of the *Uvigerina* species. Consequently, the *Uvigerina* ranges as they occur in Trinidad have to be viewed in this light.

Of the 21 *Uvigerina* species selected from Trinidad (ranges Fig. 50, illustrations Fig.54.1–29), 11 have been published previously in Cushman & Renz (1947*b*, 1948) and in Cushman & Stainforth (1945). They are here re-illustrated while the additional seven species are figured by their holotypes and in some cases also by their paratypes.

In the Agua Salada Formation of southeastern Falcon, which represents most of the Miocene, Cushman & Renz (1941) and Renz (1948) listed and figured ten *Uvigerina* species (ranges Fig. 73, illustrations Fig. 78.33–44). Of these, six are also known from Trinidad. In the Oligocene to Early Miocene of central Falcon, Diaz de Gamero (1977*a*) illustrated nine *Uvigerina* species (Fig. 81.15–24, ranges Fig. 75).

Beckmann (1953) described and illustrated two *Uvigerina* species from the Oceanic Formation of Barbados (Fig. 87.21–22, ranges Fig. 86).

The 'Handbook of common Tertiary *Uvigerina*' by Boersma (1984) treats 61 species occurring in different parts of the world. In addition to scanning electron micrographs of several specimens for each taxon, description, discussion, similar species, geographic variation, ecology, stratigraphic ranges tied into the planktic foraminiferal P and N zonal scheme, and type references are given for each species. A considerable number of Boersma's species are reported from the Caribbean-Gulf of Mexico area, and 13 of the 29 species here discussed are also described and illustrated by Boersma. They are here listed in the synonymy list for each taxon concerned and commented on in the species annotations.

Based on the different surface ornamentations the 29 *Uvigerina* species here discussed from Trinidad, Falcon (Venezuela) and Barbados are broadly grouped as follows:

1. smooth throughout, occasionally finely costate in early stage *(carapitana, hootsi, hannai)*
2. smooth with early stage distinctly costate *(bealli)*
3. hispid throughout *(auberiana, attenuta, mantaensis, rustica)*
4. early stage costate, late stage hispid *(hispidocostata, howei, gardnerae)*
5. finely costate throughout *(laviculata, nuttalli)*
6. finely to medium costate throughout, test slender-elongate *(ciperana, havanensis, seriata, spinicostata, yazooensis)*
7. medium-sized costae *(capayana, isidroensis, jacksonensis, smithi, sparsicostata, mexicana)*

8. coarse-sized costae *(cf. beccarii, curta, gallowayi, gallowayi basicordata, israelskyi, tumeyensis)*

UVIGERINA AUBERIANA ATTENUATA Cushman & Renz
Figure **54.7**; 50. Figure **78.34**; 73; Figure **81.23**

Uvigerina auberiana d'Orbigny var. *attenuata* Cushman & Renz 1941, p. 21, pl. 3, fig. 17a, b. – Cushman & Stainforth 1945, p. 49, pl. 7, fig. 18. – Renz 1948, p. 173, pl. 7, figs. 20a, b.

For annotations see *U. rustica*.

UVIGERINA BEALLI Bermudez
Figure **54.16**; 50

Uvigerina bealli Bermudez 1949, p. 201, pl. 13, fig. 5.

The species with its distinctly costate early and smooth later part, typical for Group 2, was not previously described from Trinidad where, however, typical specimens occur from Late Eocene to early Middle Oligocene. The species was originally described from the Middle Oligocene of the Dominican Republic.

UVIGERINA cf. BECCARII Fornasini
Figure **78.41**; 73

Uvigerina beccarii Fornasini 1898, Rend. Accad. Sci. Bologna, 2, p. 12, pl. 1, fig. 5.
Uvigerina cf. *beccarii* Fornasini, Renz, 1948, p. 174, pl. 7, fig. 22.

Renz (1948) identified his illustrated specimen (Fig. 78.41) from SE-Falcon as close to Fornasini's which was described from the Pliocene near Bologna, while the SE-Falcon forms occur in the Early and Middle Miocene. The illustrated southeastern Falcon specimen differs from the holotype illustration in the few distinct costae being present also in the final chambers while in the type specimen the costae are restricted to the lower part of the test with the later part remaining smooth.

UVIGERINA CAPAYANA Hedberg
Figure **54.9–10**; 50. Figure **78.37**; 53

Uvigerina pigmea d'Orbigny var. *capayana*
    Hedberg 1937a, p. 677, pl. 91, fig. 19. –
    Cushman & Stainforth 1945, p. 48, pl. 7,
    fig. 15. – Renz 1948, p. 173, pl. 12, fig. 15.

To the *Uvigerina* species with medium-sized costae here dealt with (Group 7) belong *U. capayana* Hedberg (Fig. 54.9–10, Fig. 78.37), *U. isidroensis* Cushman & Renz (Fig. 54.11, Fig. 78.39), *U. jacksonensis* Cushman (Fig. 81.17), *U. mexicana* Nuttall (Fig. 54.21, Fig. 81.21), *U. smithi* (Kleinpell) (Fig. 78.40) and *U. sparsicostata* (Cushman & Laiming) (Fig. 78.42).

Numerous *Uvigerina* species with medium sized costae have been described world-wide. The names used for the more frequent and typical ones in the South Caribbean area are *U. capayana* and *U. isidroensis*. It is mainly *U. isidroensis* that under favourable conditions may appear in floods. As the illustrations show it differs from *U. capayana* in the costae being more acute and pronounced. In Trinidad *U. isidroensis* ranges throughout the Miocene, while *U. capayana*, possibly the forerunner of *U. isidroensis*, occurs in the Oligocene.

Boersma (1984, p. 25, pl. 1) illustrates four specimens from the Late Miocene and Pliocene of Venezuela. Compared with the holotype from eastern Venezuela (Fig. 54.9), the specimens here illustrated from Trinidad (Fig. 54.10) and southeastern Falcon (Fig. 78.37) differ in possessing fewer costae.

Boersma (1984, p. 85, pl. 1, figs. 1, 3–6) illustrates specimens of *U. isidroensis* from the Miocene and Pliocene of Venezuela. They seem to fall within the variability of the holotype from southeastern Falcon (Fig. 78.39) and the specimen illustrated from Trinidad (Fig. 54.11).

Diaz de Gamero (1977a) illustrates as *U. jacksonensis* a small specimen from the Middle Oligocene of central Falcon. Compared with the holotype from the Late Eocene Jackson Formation of Alabama the specimen has to be regarded a juvenile form and its assignment to this species remains doubtful.

Boersma (1984) illustrates on p. 87, pls. 1 and 2, nine specimens of *U. jacksonensis* from the Late Eocene to Early Oligocene of California, the southern United States, Mexico and Cuba. Compared with the illustration of the Late Eocene holotype from Alabama all specimens

illustrated by Boersma seem to fall within the variability of the species.

Specimens assigned to *U. sparsicostata* and *U. smithi* were illustrated by Renz (1948) from the Agua Salada Formation of southeastern Falcon where the two taxa have the same restricted range in the Early Miocene *Catapsydrax dissimilis* and lower part of the *Catapsydrax stainforthi* zones. As the name indicates *U. sparsicostata* has fewer and less acute costae compared for instance with *U. isidroensis* which also possesses a more pointed end stage of the test, against a broader and more blunt one in *U. sparsicostata*. The specimen illustrated by Renz (1948) as *U. smithi* appears closely related to *U. sparsicostata* as also the identical ranges indicate. Character and arrangement of costae also compare well. The test of *U. smithi,* however, appears more slender in that the last chambers increase less rapidly in size.

UVIGERINA CARAPITANA Hedberg
Figure **54.3–4**; 50. Figure **78.33**; 73

Uvigerina carapitana Hedberg 1937a, p. 677,
    pl. 91, fig. 20. – Cushman & Renz 1947b,
    p. 29, pl. 6, fig. 15. – Renz 1948, p. 174, pl. 7,
    fig. 21. – Boersma 1984, p. 28, pl. 1, figs. 1–
    5.

U. carapitana is a typical representative of Group 1: 'smooth surface throughout, occasionally finely costate in early stage'. Both the Trinidad (Fig. 54.3–4) and the southeastern Falcon (Fig. 78.33) specimens possess smooth surfaces throughout with little indication of faint longitudinal striations in the early chambers present in some of the eastern Venezuelan specimens from the Carapita Formation.

Boersma's (1984, p. 28, pl. 1) illustrations of five specimens of *U. carapitana* from Venezuela and the Dominican Republic compare with the here illustrated holotype (Fig. 54.4) and southeastern Falcon specimen (Fig. 78.33).

U. hootsi Rankin (Fig. 54.1–2) has very much the same features, described as having a smooth test with the earliest portion in some specimens slightly costate. A comparison of the holotypes shows that *U. hootsi* (Fig. 54.1) possesses a flaring lip which in *U. carapitana* (Fig. 54.3) is absent.

The specimens illustrated from southeastern Falcon as *U.* cf. *hannai* Kleinpell (Fig. 78.35–

36) show basically the same features as *U. carapitana* and *U. hootsi,* the first with a smooth surface throughout, the other with the early portion slightly costate, which is in agreement with the original description of the taxon.

UVIGERINA CIPERANA Cushman & Stainforth
Figure **54.23–24;** 50

Uvigerina ciperana Cushman & Stainforth 1945, p. 49, pl. 7, figs. 19a, b, 20.

For annotations see *U. seriata.*

UVIGERINA CURTA Cushman & Jarvis
Figure **54.19–20;** 50

Uvigerina curta Cushman & Jarvis 1929, p. 13, pl. 3, figs. 13–15. – Cushman & Renz 1948, p. 27. – Boersma 1984, p. 41, pl. 1, figs. 2–3.

In Group 8 of the *Uvigerina* species here discussed are included those with 'coarse-sized costae'. They are *U. curta* Cushman & Jarvis, *U. gallowayi* Cushman, *U. gallowayi basicordata* Cushman & Renz, *U. israelskyi* Garrett and *U. tumeyensis* Lamb.

U. curta (Fig. 54.19–20) possesses a very short and stout test. The high and thin costae are few, with the surface of the final chamber remaining smooth. The species described from Trinidad was originally said to occur in the Eocene. Its known range today is from the Early Miocene *Catapsydrax stainforthi* Zone to the Middle Miocene *Globorotalia fohsi peripheroronda* Zone.

Boersma's (1984, p. 41, pl. 1) illustrations of two specimens of *U. curta* from Trinidad agree with those shown here from this island (Fig. 54.19–20).

U. gallowayi (Fig. 54.17–18, Fig. 81.16) and *U. gallowayi basicordata* (Fig. 54.13, Fig. 78.38) have similarly sized and arranged pronounced costae as *U. curta,* also leaving the last chamber smooth. Their tests differ from that of *U. curta* in being distinctly more elongate. The subspecies *basicordata* is said to differ in the test being shorter and broader with a greater concentration of costae at the base. These differences are apparent when comparing the respective holotype illustrations (Fig. 54.17 and Fig. 78.38) but in practice the two taxa may be difficult to differ-

entiate. *U. gallowayi* occurs in Trinidad in the Oligocene and Early Miocene, *U. gallowayi basicordata* in the late Early and early Middle Miocene, in southeastern Falcon in the Early Miocene.

Boersma's (1984, p. 62, pl. 1) illustrations of specimens from the Late Miocene of Ecuador compare well with the here illustrated holo- and paratypes (Fig. 54.17, 18), also from Ecuador.

UVIGERINA GALLOWAYI Cushman
Figure **54.17–18;** 50. Figure **81.16;** 75

Uvigerina gallowayi Cushman 1929*b*, Contrib. Cushman Lab. foramin. Res., 5, p. 94, pl. 13, figs. 33, 34. – Diaz de Gamero 1977*a*, p. 42, pl. 3, fig. 20. – Boersma 1984, p. 62, pl. 1, figs. 1–5.

For annotations see *U. curta.*

UVIGERINA GALLOWAYI BASICORDATA Cushman & Renz
Figure **54.13;** 50. **78.38;** 73

Uvigerina gallowayi var. *basicordata* Cushman & Renz 1941, p. 21, pl. 3, figs. 18a, b. – Renz 1948, p. 174, pl. 3, figs. 19a, b.
Uvigerina gallowayi Cushman, Cushman & Stainforth 1945, p. 48, pl. 7, fig. 14.

For annotations see *U. curta.*

UVIGERINA GARDNERAE Cushman
Figure **81.19;** 75

Uvigerina gardnerae Cushman 1926, Bull. Am. Assoc. Petr. Geol., 10, p. 175, pl. 8, figs. 16, 17. – Diaz de Gamero 1977*a*, p. 42, pl. 3, fig. 21.

For annotations see *U. hispidocostata.*

UVIGERINA cf. HANNAI Kleinpell
Figure **78.35–36;** 73

Uvigerina californica Hanna (not Cushman) 1928, Bull. Am. Assoc. Petr. Geol., 12, pl. 9, fig. 3.
Uvigerina hannai Kleinpell 1938, Miocene Stratigraphy of California, p. 294.
Uvigerina cf. *hannai* Kleinpell, Renz 1948, p. 174, pl. 12, figs. 16a, b, 17.

For annotations see *U. carapitana.*

UVIGERINA HAVANENSIS Cushman & Bermudez
Figure **81.20**; 75

Uvigerina havanensis Cushman & Bermudez 1936, Contrib. Cushman Lab. foramin. Res., 12, p. 59, pl. 10, figs. 19, 20 (19 holotype). – Diaz de Gamero 1977*a*, p. 42, pl. 3, fig. 22.

For annotations see *U. seriata*.

UVIGERINA HISPIDOCOSTATA Cushman & Todd
Figure **54.14–15**; 50

Uvigerina hispido-costata Cushman & Todd 1945, p. 51, pl. 7, fig. 31. – Cushman & Renz 1947*b*, p. 28, pl. 6, fig. 17.

Uvigerina hispidocostata Cushman & Todd and *U. howei* Garrett belong to Group 4 'early stage costate, late stage hispid'. The holotype of *U. hispidocostata* (Fig. 54.15) displays the early costate and the late hispid stage well. According to the authors different specimens vary greatly in proportions of costae to serrate costae to spines. In comparison the illustrated Trinidad specimen (Fig. 54.14) is less characteristic but can be placed within the variability of the species.

The specimen figured by Cushman & Renz (1947*b*) as *U. howei* (Fig. 54.12) differs from the holotype illustration in having a distinctly hispid late part whereas the holotype appears costate throughout, though its authors point out that in some specimens final chambers are somewhat hispid, the same variability as also found in *U. hispidocostata*. The occurrence of the two taxa as found in the Cipero Formation of Trinidad is restricted for both to the Early Miocene *Catapsydrax stainforthi* Zone. It would therefore appear that the two taxa are conspecific, with *U. howei*, though somewhat less typical, taking priority between the two.

Diaz de Gamero (1977*a*) identified and illustrated from central Falcon an *U. gardnerae* Cushman (Fig. 81.19). The holotype illustration of this species shows a specimen with a more compact, spindlelike-shaped test, originating from the Late Eocene Jackson Formation, while according to Diaz de Gamero the species ranges in central Falcon from Middle Oligocene to Early Miocene. For above considerations the inclusion of the southeastern Falcon specimen in *U. gardnerae* does not seem correct.

UVIGERINA HOOTSI Rankin
Figure **54.1–2**; 50

Uvigerina hootsi Rankin 1934, Contrib. Cushman Lab. foramin. Res., 10, p. 22, pl. 3, figs. 8, 9.

For annotations see *U. carapitana*.

UVIGERINA HOWEI Garrett
Figure **54.12**; 50

Uvigerina howei Garrett 1939, J. Paleontol., 13, p. 577, pl. 65, figs. 16, 17 (16 holotype). – Cushman & Renz 1947*b*, p. 29, pl. 6, fig. 19.

For annotations see *U. hispidocostata*.

UVIGERINA ISIDROENSIS Cushman & Renz
Figure **54.11**; 50. Figure **78.39**; 73

Uvigerina isidroensis Cushman & Renz 1941, p. 20, pl. 3, figs. 16a, b. – Cushman & Renz 1947*b*, p. 28, pl. 6, fig. 18. – Renz 1948, p. 175, pl. 7, figs. 18a, b. – Boersma 1984, p. 85, pl. 1, figs. 1, 3–6.

For annotations see *U. capayana*.

UVIGERINA ISRAELSKYI Garrett
Figure **54.29**; 50

Uvigerina israelskyi Garrett 1939, J. Paleontol., 13, p. 577, pl. 65, fig. 15.

U. israelskyi, with its well-developed high costae traversing the entire length of the test, is a characteristic species, originally described from the Middle Tertiary *Heterostegina* Zone of Texas.

U. tumeyensis described by Lamb (1964*b*) from the Oligocene *Globorotalia opima opima* Zone of the Pleito Formation, California, appears close to identical with *U. israelskyi* (see comments by W. A. Akers in Lamb 1964*b*, p. 464). This species was however placed by Lamb into synonymy with *Siphogenerina nodifera* Cushman & Kleinpell (1934) originally described from the Miocene Temblor Formation, California. It appears that the three species *Uvigerina israelskyi, tumeyensis and nodifera* are in fact very closely related, possibly representing one species only. Reference is made to the detailed comments on these taxa by Lamb (1964*b*).

Lamb (1964*b*), in his textfig. 2, shows the ranges in Venezuela, California and the Gulf coastal region for *U. tumeyensis* and *S. nodifera*. The former is according to this chart restricted to the lower part of the Oligocene *Globorotalia opima opima* Zone, becoming replaced in the higher part of the same zone by *S. nodifera* which continues into the basal part of the *Globorotalia kugleri* Zone, where it gives way to *Siphogenerina transversa* Cushman. From this distribution pattern it would appear that one is dealing with an evolutionary sequence beginning with the forerunner *U. jacksonensis* in the Late Eocene *Turborotalia cerroazulensis* s.l. Zone to *U. tumeyensis/U. israelskyi* to *S. nodifera* and eventually to *S. transversa* in the Late Oligocene-Early Miocene.

The specimen illustrated by Diaz de Gamero (1977*a*) as *U. tumeyensis* from the Early Miocene *Globigerinoides primordius* Zone in central Falcon (Fig. 81.18) belongs with its costae extending across chambers to the *U. israelskyi/U. tumeyensis* complex.

UVIGERINA JACKSONENSIS Cushman
Figure **81.17**; 75

Uvigerina jacksonensis Cushman 1925, Contrib. Cushman Lab. foramin. Res., 1, p. 67, pl. 10, fig. 13. – Diaz de Gamero 1977*a*, p. 42, pl. 3, fig. 18. – Boersma 1984, p. 87, pl. 1, figs. 1–5, pl. 2, figs. 1–4.

For annotations see *U. capayana*.

UVIGERINA LAVICULATA Coryell & Rivero
Figure **54.5**; 50

Uvigerina laviculata Coryell & Rivero 1940, J. Paleontol., 14, p. 343, pl. 44, fig. 24. – Boersma 1984, p. 91. pl. 1, figs. 1, 3.

Group 5 contains the characteristic, very finely striate to costate *U. laviculata* Coryell & Rivero (Fig. 54.5) and *U. nuttalli* Cushman & Edwards (Fig. 81.22, Fig. 87.21), two species judged from their holotypes as closely related. The age for *U. laviculata* was given as late Middle Miocene (Haiti), that for *U. nuttalli* as Early Oligocene (Alazan shales, Mexico). *U. laviculata* is confined in Trinidad to the Middle Oligocene *Globorotalia opima opima* Zone, *U. nuttalli* in Central Falcon to the Early Miocene.

Boersma's (1984, p. 91, figs. 1, 3) specimens from northeastern Venezuela and Haiti, both from the Late Miocene, compare well with the holotype from Haiti.

UVIGERINA MANTAENSIS Cushman & Edwards
Figure **54.8**; 50. Figure **81.24**; 75

Uvigerina mantaensis Cushman & Edwards 1938, Contrib. Cushman Lab. foramin. Res., 14, p. 84, pl. 14, figs. 8a, b. – Cushman & Stainforth 1945, p. 47, pl. 7, fig. 17. – Cushman & Renz 1948, p. 27. – Diaz de Gamero 1977*a*, p. 42, pl. 4, figs. 1–3. – Boersma 1984, p. 101, pl. 1, figs. 1– 4, 6.

For annotations see *U. rustica*.

UVIGERINA MEXICANA Nuttall
Figure **54.21**; 50. Figure **81.21**; 75

Uvigerina pygmaea d'Orbigny var. Nuttall 1928, p. 94, pl. 6, fig. 13.
Uvigerina mexicana Nuttall 1932, p. 22, pl. 5, fig. 12. – Cushman & Renz 1947*b*, p. 28, pl. 6, fig. 16. – Diaz de Gamero 1977*a*, p. 43, pl. 3, fig. 23. – Boersma 1984, p. 110, pl. 1, figs. 1–2.

With its medium-sized costae the species is here placed in Group 7 (see annotations under *U. capayana*). *U. mexicana* is a characteristic species clearly distinguishable from other taxa placed in this group. Its test is short and stout, usually widest in its middle part. The surface of the chambers appears polished and each shows several low profile costae, which remain absent in the final chamber. Nuttall (1928) first named and illustrated this form from Trinidad as *U. pygmaea* d'Orbigny var., shown here on Fig. 54.21. Later, in 1932, he described and illustrated the same form as a new species, *U. mexicana,* from the Early Oligocene Alazan shale, Tampico Embayment, Mexico.

In Trinidad the species occurs from the Middle Oligocene *Globigerina ampliapertura* to the Early Miocene *Globigerinatella insueta* Zone for which interval it is considered a reliable index. Its range in Central Falcon is given as *Globorotalia opima opima* to *Catapsydrax dissimilis* Zone.

Boersma (1984) illustrates specimens identified as *U. mexicana*. Those from the Early

Miocene of the Gulf of Mexico, Eureka Core 151 (figs. 1 and 2) closely conform with the holotype. More slender test shapes and more pronounced costae of the specimens shown on her figs. 4 and 5 from the Late Oligocene of the Blake Plateau (JOIDES 5) are not regarded as typical for the species.

For further annotations see *U. capayana*.

UVIGERINA NUTTALLI Cushman & Edwards
Figure **81.22**; 75. Figure **87.21**; 86

Uvigerina canariensis d'Orbigny, var. Nuttall 1932, p. 22, pl. 5, fig. 9.
Uvigerina nuttalli Cushman & Edwards 1938, Contrib. Cushman Lab. foramin. Res., 14, p. 82, pl. 14, figs. 3–5. – Diaz de Gamero 1977*a*, p. 43, pl. 3, fig. 24. – Beckmann 1953, p. 368, pl. 21, fig. 19.

The *U. canariensis* d'Orbigny var. by Nuttall (1932, p. 22, pl. 5, fig. 9) was as *U. nuttalli* subsequently given species rank by Cushman & Edwards (1938). The holotype from the Early Alazan shales, Mexico, is characterized by very low and fine longitudinal costae. In the two paratypes illustrated by Cushman & Edwards they appear more distinct.

The specimens illustrated by Diaz de Gamero (1977*a*) from the Early Miocene of central Falcon (Fig. 81.22) and by Beckmann (1953) from the Oligocene of the Oceanic Formation, Barbados (Fig. 87.21) seem to conform well with the species concept.

Boersma (1984) illustrates on p.119, pl. 1, specimens as *U. nuttalli* from the Gulf Coast area. The specimen shown on her fig. 1 originates from the Early Oligocene Red Bluff clay, Mississippi, that on fig. 2 from the Blake Plateau Oligocene and the one on fig. 5 from the Middle Eocene Cook Mauntaim Formation, Mississippi. Compared with the holotype illustration all three, in particular the one illustrated on fig.1, display distinctly fewer and more strongly developed costae. They are therefore regarded as not falling within the species concept of *U. nuttalli*.

For further annotations see *U. laviculata*.

UVIGERINA RUSTICA Cushman & Edwards
Figure **54.6**; 50. Figure **78.43–44**; 73. Figure **81.15**;75

Uvigerina rustica Cushman & Edwards 1938, Contrib. Cushman Lab. foramin. Res., 14, p. 83, pl. 14, fig. 6. – Cushman & Stainforth 1945, p. 47, pl. 7, fig. 13. – Renz 1948, p. 175, pl. 7, figs. 23, 24. – Diaz de Gamero 1977*a*, p. 43, pl. 4, figs. 4, 5.

U. rustica Cushman & Edwards belongs together with *U. auberiana attenuta* Cushman & Renz and *U. mantaensis* Cushman & Edwards to Group 3: 'surface throughout hispid to finely spinose'. As interpreted in Trinidad and Falcon, *U. rustica* (Fig. 54.6, Fig. 78.43–44, Fig. 81.15) is, compared with the two other taxa *U. auberiana attenuta* (Fig. 54.7, Fig. 78.34, Fig. 81.23) and *U. mantaensis* (Fig. 54.8, Fig. 81.24), distinctly larger, more robust and with thicker spines. Concerning the above criteria *U. mantaensis* may be regarded as intermediate between the large and robust *U. rustica* and the small delicate and very finely hispid *U. auberiana attenuta*.

Boersma (1984, p. 103, pl. 1, figs. 1–3, 4, 6) illustrates five specimens of *U. mantaensis* from the Late Miocene of Ecuador. They compare well with the specimens here shown from Trinidad (Fig. 54.8) and central Falcon (Fig. 81.24).

UVIGERINA SERIATA Cushman & Jarvis
Figure **54.27–28**, 50

Uvigerina seriata Cushman & Jarvis 1929, p. 13, pl. 3, figs. 11, 12.

Species belonging to Group 6 are characterized by distinctly slender/elongate, finely to medium-costate tests. In the area here under discussion taxa belonging to this group were recorded in Trinidad and Barbados from within the Middle Eocene to Middle Oligocene.

Characteristic and morphologically closely related species are *U. seriata* Cushman & Jarvis (Fig. 54.27–28) and *U. spinicostata* Cushman & Jarvis (Fig. 54.25–26), both originally described from Trinidad. The distinctly elongate test shape and number of costae are about the same in both. *U. spinicostata* differs from *U. seriata* in particular in the early portion of the test, but sometimes also in the later chambers in that in *U. spinicostata*, as the name indicates, the costae are broken up into fine spinose projections. In comparison, costae in *U. seriata* are somewhat less distinctly developed and spines are either absent or only present in the basalmost portion. Another indication that the two species are closely related is their identical range in Trinidad from Middle Eocene to lower part of Middle Oligocene.

Boersma (1984, p. 156, pl. 1, figs 1–2) illustrates two specimens of *U. spinicostata* from the Cipero Formation of Trinidad and one (fig. 3) from the Late Eocene, Blake Plateau, both falling within the variability range of the holotype (Fig. 54.25). The other specimens illustrated by Boersma as *U. spinicostata* are from the Late Eocene Blue Clay of Biarritz (fig. 4), the Early Oligocene of Denmark (fig. 5), DSDP Site 529, Angola Basin (fig. 6) and DSDP Site 526a, Walvis Ridge (fig. 7). They all fall within the concept of this species which points to a wide global distribution of the taxon without significant local variations.

Another characteristic and short-ranged species with a slender and elongate test shape is *U. ciperana* Cushman & Stainforth (Fig. 54.23–24), also originally described from Trinidad. Compared with *U. seriata* and *U. spinicostata* this species is smaller, with finer costae and without any spinose projections replacing parts of the costae. In Trinidad the species has a restricted range from the Late Eocene *Turborotalia cerroazulensis* s.l. Zone to the Middle Oligocene *Globigerina ampliapertura* Zone. Other species close to *U. ciperana* are the Late Eocene *U. gardnerae texana* Cushman & Applin and the Eocene *U. havanensis* Cushman & Bermudez.

The specimen figured by Diaz de Gamero (1977*a*) as *U. havanensis* from the Oligocene of central Falcon (Fig. 81.20) differs from the holotype in being considerably less slender and elongate. Boersma (1984, p. 70, pl. 1, fig. 4) illustrates a specimen identified as *U. havanensis* from the Early Oligocene of Mexico. It shows little similarity to the holotype in that like Diaz de Gamero's specimen illustrated from central Falcon (Fig. 81.20) it is less slender/elongate. Furthermore, the costae in Boersma's specimen are fewer, blade-like and higher.

U. yazooensis Cushman (Fig. 54.22) originally described from the Late Eocene Jackson Formation of Alabama is another species to be placed in this group of slender taxa with frequent costae. In Trinidad its range is given as Middle and Late Eocene. Boersma (1984, p.189, pl. 1,

figs. 1–3) illustrates three specimens of *U. yazooensis* from the Late Eocene Yazoo Clay, Mississippi. Her figs. 1 and 2 represent elongate, fusiform specimens that compare well with the holotype (Fig. 54.22). The third specimen, her fig. 3, is on the other hand distinctly shorter and stout, with fewer costae. Its placing in *U. yazooensis* is therefore to be regarded with reservation.

UVIGERINA SMITHI (Kleinpell)
Figure **78.40–41**; 73

Siphogenerina smithi Kleinpell 1938, Miocene Stratigraphy of California, p. 304, pl. 6, figs. 1, 2. – Renz 1948, p. 165, p. 12, figs. 18, 19.

For annotations see *U. capayana*.

UVIGERINA SPARSICOSTATA (Cushman & Laiming)
Figure **78.42**; 73, 50

Uvigerinella sparsicostata Cushman & Laiming 1931, J. Paleontol., 5, p. 112, pl. 12, fig. 12a, b. – Renz 1948, p. 175, pl. 12, fig. 20.

For annotations see *U. capayana*.

UVIGERINA SPINICOSTATA Cushman & Jarvis
Figure **54.25–26**; 50. Figure **87.22**; 86

Uvigerina spinicostata Cushman & Jarvis 1929, p. 12, pl. 3, figs. 9, 10. – Cushman & Stainforth 1945, p. 48, pl. 7, fig. 16.- Beckmann 1953, p. 368, pl. 21, fig. 19. – Boersma 1984, p. 156, pl. 1, figs. 1–2.

For annotations see *U. seriata*.

UVIGERINA TUMEYENSIS Lamb
Figure **81.18**; 75

Uvigerina tumeyensis Lamb 1964*b*, p. 463, pl. 1, fig. 10. – Diaz de Gamero 1977*a*, p. 43, pl. 4, fig. 6, 7.

For annotations see *U. israelskyi*.

UVIGERINA YAZOOENSIS Cushman
Figure **54.22**; 50

Uvigerina yazooensis Cushman 1933, Contrib.
   Cushman Lab. foramin. Res., 9, p. 13, pl. 1,
   fig. 29. – Boersma 1984, p. 189, pl. 1, figs.
   1–3.

For annotations see *U. seriata*.

ANGULOGERINA ILLINGI Cushman &
Renz
Figure **80.31a–b**; 74

Angulogerina illingi Cushman & Renz 1941,
   p. 21, pl. 3, figs. 19–20. – Renz 1948, p. 114,
   pl. 7, figs. 32a–b.

TRIFARINA BRADYI Cushman
Figure **80.32**; 74

Trifarina bradyi Cushman 1923, Bull. U.S.
   natl. Mus., 104, pt. 4, p. 99, pl. 22, figs. 3–
   9. – Renz 1948, p. 172, pl. 7, fig. 33.

REUSSELLA SPINULOSA (Reuss)
Figure **80.25**; 74

Verneuilina spinulosa Reuss 1859, Denkschr.
   Akad. Wiss. Wien, 1, p. 374, fig. 12.
Reussella spinulosa (Reuss), Renz 1948, p. 156,
   pl. 7, figs. 16, 17.

Family Pleurostomellidae

ELLIPSODIMORPHINA
SUBCOMPACTA Liebus
Figure **87.32–34**; 86

Ellipsodimorphina subcompacta Liebus 1922,
   Lotos (Prag), 70, p. 57, pl. 2, fig. 13. –
   Beckmann 1953, p. 378, pl. 23, figs. 1–4. –
   Saunders *et al.* 1984, p. 407, pl. 3, fig. 8.

ELLIPSOGLANDULINA
MULTICOSTATA (Galloway & Morrey)
Figure **62.19**; 52

Daucina multicostata Galloway & Morrey 1929,
   Bull. Am. Paleontol., 15, no. 55, p. 42, pl. 6,
   fig. 13.

Ellipsonodosaria multicostata (Galloway &
   Morrey), Cushman & Stainforth 1945, p. 58,
   pl. 10, figs. 6, 7.

NODOSARELLA MAPPA (Cushman &
Jarvis)
Figure **63.15, 15a**; 52. Figure **87.37, 37a**; 86

Ellipsonodosaria mappa Cushman & Jarvis
   1934, p. 73, pl. 10, fig. 8. – Cushman &
   Stainforth 1945, p. 56, pl. 9, fig. 8.
Nodosarella mappa (Cushman & Jarvis),
   Beckmann 1953, p. 376, pl. 22, figs. 22, 23.

Family Stilostomellidae

ORTHOMORPHINA HAVANENSIS
(Cushman & Bermudez)
Figure **87.36**; 86

Nodogenerina havanensis Cushman &
   Bermudez 1937, Contrib. Cushman Lab.
   foramin. Res., 13, p. 14, pl. 1, figs. 47, 48.
Orthomorphina havanensis (Cushman &
   Bermudez), Beckmann 1953, p. 365, pl. 21,
   fig. 7.

ORTHOMORPHINA ROHRI (Cushman &
Stainforth)
Figure **62.17**; 52

Nodogenerina rohri Cushman & Stainforth
   1945, p. 39, pl. 5, fig. 26. – Cushman &
   Renz 1948, p. 24, pl. 5, figs. 9–11.

SIPHONODOSARIA ANNULIFERA
(Cushman & Bermudez)
Figure **87.40, 40a**; 86

Ellipsonodosaria annulifera Cushman &
   Bermudez 1936, Contrib. Cushman Lab.
   foramin. Res., 12, p. 28, pl. 5, figs. 8, 9.
Stilostomella annulifera (Cushman &
   Bermudez), Beckmann 1953, p. 370, pl. 21,
   fig. 23.

SIPHONODOSARIA CONSOBRINA
(d'Orbigny)
Figure **87.41;** 86

Dentalina consobrina d'Orbigny 1846, p. 46,
pl. 2, figs. 1–3.
Stilostomella consobrina (d'Orbigny),
Beckmann 1953, p. 370, pl. 21, figs. 24, 25.

SIPHONODOSARIA CURVATURA
(Cushman)
Figure **87.42;** 86

Ellipsonodosaria curvatura Cushman 1939,
Contrib. Cushman Lab. foramin. Res., 15,
p. 71, pl. 12, fig. 6.
Stilostomella curvatura (Cushman), Beckmann
1953, p. 370, pl. 21, figs. 26, 27.

SIPHONODOSARIA CURVATURA
SPINEA (Cushman)
Figure **87.43;** 86

Ellipsonodosaria curvatura Cushman var.
spinea Cushman 1939, Contrib. Cushman
Lab. foramin. Res., 15, p. 71, pl. 12, figs. 9–
11.
Stilostomella curvatura (Cushman) var. *spinea*
(Cushman), Beckmann 1953, p. 370, pl. 21,
fig. 28.

SIPHONODOSARIA NUTTALLI (Cushman
& Jarvis)
Figure **63.28;** 52

Ellipsonodosaria nuttalli Cushman & Jarvis
1934, p. 72, pl. 10, fig. 6. – Cushman &
Stainforth 1945, p. 55, pl. 9, fig. 13.

SIPHONODOSARIA NUTTALLI
ACULEATA (Cushman & Renz)
Figure **63.22;** 52

Ellipsonodosaria nuttalli var. *aculeata* Cushman
& Renz 1948, p. 32, pl. 6, fig. 10.

SIPHONODOSARIA NUTTALLI
GRACILLIMA (Cushman & Jarvis)
Figure **63.20;** 52

Ellipsonodosaria nuttalli var. *gracillima*
Cushman & Jarvis 1934, p. 72, pl. 10, fig. 7.
– Cushman & Stainforth 1945, p. 56, pl. 9,
figs. 14, 15.

SIPHONODOSARIA RECTA (Palmer &
Bermudez)
Figure **63.21;** 52

Ellipsonodosaria recta Palmer & Bermudez
1936, Mem. Soc. Cubana Hist. Nat., 10,
p. 297, pl. 18, figs. 6, 7. – Cushman &
Stainforth 1945, p. 57, pl. 10, figs. 4, 5.

SIPHONODOSARIA VERNEUILI
PAUCISTRIATA (Galloway & Morrey)
Figure **63.29;** 52

Nodosarella paucistriata Galloway & Morrey
1929, Bull. Am. Paleontol., 15, no. 55, p. 42,
pl. 6, fig. 12.
Ellipsonodosaria verneuili (d'Orbigny) var.
paucistriata (Galloway & Morrey), Cushman
& Stainforth 1945, p. 55, pl. 9, fig. 12.

STILOSTOMELLA MATANZANA (Palmer
& Bermudez)
Figure **87.39;** 86

Ellipsonodosaria? matanzana Palmer &
Bermudez 1936, Mem. Soc. Cub. Hist. Nat.,
10, p. 298, pl. 18, fig, 12.
Stilostomella matanzana (Palmer &
Bermudez), Beckmann 1953, p. 371, pl. 21,
fig. 31.

STILOSTOMELLA MODESTA (Bermudez)
Figure **87.38, 38a;** 86

Ellipsonodosaria modesta Bermudez 1937,
Mem. Soc. Cub. Hist. Nat., 11, p. 238,
pl. 20, fig. 3.
Stilostomella modesta (Bermudez), Beckmann
1953, p. 371, pl. 21, fig. 32.

STILOSTOMELLA SUBSPINOSA
(Cushman)
Figure **63.30, 30a;** 52

Ellipsonodosaria subspinosa Cushman 1943,
Contrib. Cushman Lab. foramin. Res., 19,
p. 92, pl. 16, figs. 6, 7. – Cushman &
Stainforth 1945, p. 56, pl. 9, figs. 9, 10.

Family Elphidiidae

CRIBROELPHIDIUM POEYANUM
(d'Orbigny)
Figure **80.11a–b;** 74

Polystomella poeyana d'Orbigny 1839, p. 55,
pl. 6, figs. 25–26.
Elphidium poeyanum (d'Orbigny), Renz 1948,
p. 132, pl. 6, figs. 6a–b.

ELPHIDIUM SAGRUM (d'Orbigny)
Figure **80.10a–b;** 74

Polystomella sagra d'Orbigny 1839, p. 74, pl. 6,
figs. 19–20.
Elphidium sagrum (d'Orbigny), Renz 1948,
p. 132, pl. 6, figs. 7a–b.

Calcareous trochospiral taxa

The calcareous trochospiral benthic foraminifera constitute a significant segment amongst the benthic foraminiferal associations in the Eocene to Miocene sediments of Trinidad, Venezuela and Barbados. Representatives of this group occur in both the deeper-water deposits such as the Eocene Navet and Oligocene-Miocene Cipero formations of Trinidad, or in shallower-water equivalents as the Agua Salada Formation of southeastern Falcon in Venezuela or the Ste. Croix/Brasso formations of Trinidad.

All calcareous trochospiral benthic taxa described and figured from Trinidad by Cushman & Renz (1947*b*, 1948), Cushman & Stainforth (1945), from southeastern Falcon by Cushman & Renz (1941), Renz (1948) and from Barbados by Beckmann (1953), Saunders *et al.* (1984) and Wood *et al.* (1985) are here incorporated as a coherent group in the systematic part, with some of them discussed in more detail in annotations. The illustrations of all taxa figured in

these publications are here reproduced, supplemented in numerous instances also by illustrations of the holotype. Furthermore, ranges of all taxa are shown for each of the studied areas, Trinidad, Falcon and Barbados.

The calcareous trochospiral benthic foraminifera are systematically subdivided into numerous families and many genera but individual taxa are on the species level mostly quite readily identifiable. Many of the taxa are rich in individuals and also have restricted ranges; they therefore are good stratigraphic markers. These are the principal reasons why this characteristic group has been included in this volume as completely as possible.

As is the case with other benthic forms the distribution of the calcareous trochospiral taxa is controlled by environment. This is shown for instance in their distribution pattern for Trinidad, plotted on Fig. 51. Here, numerous taxa appear first for instance in the shallow-water Upper Eocene San Fernando Formation (*Turborotalia cerroazulensis* s. l. Zone) which succeeds the deep-water Navet Formation. Noteworthy are also the numerous species extinctions in the Late Eocene and the several new appearances in the Oligocene *Globigerina ampliapertura* Zone. A strong turnover of calcareous trochospiral benthic taxa thus took place in Trinidad between Late Eocene and Early Oligocene. As is also evident from Fig. 51 numerous calcareous trochospiral benthic taxa make their first appearance in the *Catapsydrax stainforthi* and *Globigerinatella insueta* zones of the Miocene Ste. Croix and Brasso formations, which are the shallower-water equivalents of the deeper-water Cipero Formation.

The distributional pattern of the calcareous trochospiral benthic taxa through the Agua Salada Formation of southeastern Falcon is, as shown on Fig. 73, more continuous compared with Trinidad. The reason for this is that here we have almost throughout the formation an outer shelf-type facies that changed little in time, except in its uppermost part where due to further shallowing the calcareous trochospiral benthic taxa become very rare. But Fig. 73 also shows that many of the species have their restricted ranges also within the environmentally little changing part of the Agua Salada Formation.

Diaz de Gamero (1977*a*) listed a large number of calcareous trochospiral benthic species from the Oligocene-Miocene of central Falcon and plotted their ranges for the sections studied. However, these taxa were not discussed or illustrated by her. With only names and ranges available the central Falcon calcareous trochospiral benthic taxa could therefore not be treated in the same way as those from the other areas. Ranges of only a few characteristic and short-ranged forms are plotted on Figure 81.

The 93 calcareous trochospiral taxa included in the following systematic part fall into 23 different genera. 57 species or subspecies have been illustrated from Trinidad on Fig. 55–61 and their stratigraphic distribution plotted on Fig. 51. From SE-Falcon 32 taxa are illustrated on Fig. 79 and 80, with their ranges shown on Fig. 73. The 29 taxa from Barbados are illustrated on Figs. 88 and 89 and their ranges plotted on Fig. 86.

Family Epistominidae

HOEGLUNDINA ELEGANS (d'Orbigny)
Figure **55.6–8, 9a–e;** 51

Rotaliia (Turbinulina) elegans d'Orbigny 1826, Ann. Sci. Nat., 7, p. 276, No. 54.
Epistomina elegans (d'Orbigny), Cushman & Renz 1947*b*, p. 36, pl. 7, fig. 20.
Pulvinulina elegans (d'Orbigny), Nuttall 1928, p. 84, p. 101, pl. 7, figs. 9, 10.

D'Orbigny (1826) published a *Rotalia (Turbinulina) elegans* without providing a description, figure, type level and type locality. Parker, Jones & Brady (1871) illustrated as their fig. 142 the spiral and umbilical views of a specimen they assigned to *Rotalia elegans*, based on a figure R after Soldani (Fig. 55.9d–e), apparently from the Pliocene of Coroncina near Siena, Italy. Fornasini (1906) also illustrated as *Rotalia elegans* three views of a specimen, apparently from unpublished drawings by d'Orbigny (his figs. 10, 10a, 10b, here re-illustrated as Fig. 55.9a–c).

Both these interpretations are illustrated in the Catalogue of Foraminifera under *Rotalia (Rotalina) elegans* d'Orbigny 1826. A comparison of Parker, Jones & Brady's illustration of *Rotalia elegans* with that by Fornasini shows that we are dealing with two entirely different taxa belonging to two different genera.

When Brotzen (1948) established the genus *Hoeglundina* he selected *Rotalia elegans* d'Orbigny as genotype without however making reference to an illustration. In all likelihood he must have had in mind the specimen illustrated by Parker, Jones & Brady (1871) which seems to fit the definition for his new genus *Hoeglundina*.

Nuttall (1928) and Cushman & Renz (1947*b*) placed their *elegans* specimens from Trinidad in the genera *Pulvinulina* and *Epistomina* respectively, apparently also having in mind the illustrations by Parker, Jones & Brady which, though somewhat stylized, seem to fit their illustrated specimens. They are here placed in the genus *Hoeglundina*.

The Trinidad specimens have been recorded from Late Eocene to late Middle Miocene.

Family Bagginidae

BAGGINA COJIMARENSIS Palmer
Figure **55.1a–b, 2;** 51

Baggina cojimarensis Palmer 1941, Mem. Soc. Cubana Hist. Nat., 15, p. 198, pl. 16, figs. 13, 14. – Cushman & Renz 1947*b*, p. 37, pl. 7, fig. 22.

The spiral view of a specimen given by Cushman & Renz (1947*b*) from the Early Miocene Ste. Croix Formation (Fig. 55.2) displays the number of chambers forming the last whorl but does not show the strongly inflated nature of the chambers typical for the genus. The rather simplified illustration of the spiral and umbilical sides of the holotype by Palmer shows this feature better (Fig. 55.1a–b). In Trinidad the species has also been recorded from the Late Eocene San Fernando Formation.

CANCRIS CUBENSIS Cushman & Bermudez
Figure **55.4, 5a–c;** 51

Cancris cubensis Cushman & Bermudez 1937, p. 25, pl. 2, figs. 48–50. – Cushman & Renz 1947*b*, p. 37, pl. 7, fig. 21.

The umbilical view of a specimen given by Cushman & Renz (1947*b*) from the Ste. Croix Formation (Fig. 55.4) compares well with the corresponding view of the holotype (Fig. 55.5c). Characteristic is the large size of the final chamber in comparison with the earlier ones.

In addition to the Ste. Croix Formation (*Globigerinatella insueta* Zone) the species has in Trinidad also been recorded from the *Globigerina ampliapertura* Zone, Cipero Formation. Originally the species was described by Cushman & Bermudez from the Eocene of Cuba.

CANCRIS MAURYAE Cushman & Renz
Figure **55.3a–c**; 51

Cancris mauryae Cushman & Renz 1942, p. 11, pl. 2, figs. 17a–c.

Typical for the species is its size, strongly elongate test shape, large number of chambers forming the last whorl, rapidly increasing in size, and the presence of a distinct peripheral keel. Compared with *C. cubensis* the species is larger and has more chambers in the last whorl.

The species has been recorded in Trinidad only from the Paleocene to Early Eocene Soldado Formation.

CANCRIS PANAMENSIS Natland
Figure **79.1a–b**; 73

Cancris panamensis Natland 1938, Bull. Scripps Inst. Oceanogr., Tech. Ser., 4, no. 5, p. 148, pl. 6, figs. 1a–c. – Renz 1948, p. 123, pl. 12, figs. 22a–b.

The species has a similar though generally smaller test shape compared with *C. mauryae* but possesses thicker, occasionally slightly elevated sutures between the earlier chambers as seen from the spiral side.

The species occurs throughout the Middle Miocene of the Agua Salada Formation of SE-Falcon.

CANCRIS SAGRA (d'Orbigny)
Figure **79.2a–c**; 73

Rotalia sagra d'Orbigny 1839, p. 77, pl. 5, figs. 13–15. – Renz 1948, p. 123, pl. 9, figs. 3a–c.

In comparison with *C. mauryae* the species is smaller, has a distinctly elongate test, more inflated chambers and does not have a peripheral keel.

The species is irregularly present in the Middle to basal Late Miocene of the Agua Salada Formation, southeastern Falcon.

CIBICORBIS CONCENTRICUS (Cushman)
Figure **79.3a–c**; 73

Truncatulina concentrica Cushman 1918, Bull. U.S. geol. Surv., 676, p. 64, pl. 21, fig. 3.
Cibicides concentricus (Cushman), Renz 1948, p. 127, pl. 10, figs. 8a–c.

The species is characterized in its spiral side being plane to slightly concave, the umbilical one strongly inflated. Nine to 10 chambers, the final ones rapidly increasing in size, form the last whorl. Earlier whorls are not visible. Intercameral sutures are prominent on the spiral side, with a tendency to become limbate. On the umbilical side they remain delicate.

The species ranges in the Agua Salada Formation of southeastern Falcon from the Early Miocene *Globigerinatella insueta* Zone to the late Middle Miocene *Sphaeroidinellopsis seminulina* Zone.

CIBICORBIS HERRICKI Hadley
Figure **79.5a–c**; 73

Cibicorbis herricki Hadley 1934, Bull. Am. Paleontol., 20, no. 70A, p. 26, pl. 5, figs. 1–3.
Valvulineria herricki (Hadley), Renz 1948, p. 177, pl. 8, figs. 10a–c.

Instead of a plane to slightly concave spiral surface characteristic for *C. concentricus* (Fig. 79.3a–c) the final 3–4 chambers increase rapidly in size and remain distinctly inflated. The umbilical side is strongly inflated. Compared with the usually smaller *C. concentricus* the species has only about 8 chambers forming the last whorl, instead of 9–10. Sutures on the spiral side are prominent between the early not-inflated chambers, to become delicate and depressed between the final ones.

The species has a comparatively short range in the Agua Salada Formation of southeastern Falcon where it has been recorded in the Middle Miocene from the *Globorotalia fohsi fohsi* Zone to the *Globorotalia mayeri/Globorotalia lenguaensis* Zone.

CIBICORBIS ISIDROENSIS (Cushman & Renz)
Figure **79.4a–c**; 73

Cibicides isidroensis Cushman & Renz 1941, p. 26, pl. 4, fig. 10. – Renz 1948, p. 128, pl. 10, figs. 10a–c.

The species with about 9 chambers forming the last whorl differs from *C. concentricus* and *C. herricki* in a nearly biconvex test shape, with the umbilical side being slighly higher. The sutures are prominent between early chambers to become delicate between the later more inflated chambers.

The species occurs in the Agua Salada Formation of southeastern Falcon in the Middle Miocene *Globorotalia fohsi robusta* Zone to the *Sphaeroidinellopsis seminulina* Zone.

VALVULINERIA GEORGIANA Cushman
Figure **55.12a–c, 13**; 51

Valvulineria georgiana Cushman 1935, Contrib. Cushman Lab. foramin. Res., 11, p. 82, pl. 12, fig. 17. – Cushman & Renz 1947*b*, p. 34, pl. 7, fig. 15.

The species is, as the holotype (Fig. 55.12a–c) shows, characterized by the rapid size increase of the elongate chambers in the last whorl, resulting in a distinctly elongate test. The spiral view of a specimen illustrated from Trinidad (Fig. 55.13) seems to compare with the species concept.

The species has been recorded in Trinidad from the late Early Miocene Ste. Croix Formation.

VALVULINERIA INAEQUALIS LOBATA Cushman & Renz
Figure **79.6a–c**; 73

Valvulineria inaequalis (d'Orbigny) var. *lobata* Cushman & Renz 1941, p. 23, pl. 3, fig. 24. – Renz 1948, p. 178, pl. 8, figs. 8a–c.

The subspecies has a similar chamber arrangement and elongate test shape as *V. georgiana* (Fig. 55.12a–c) from which it differs primarily in the characteristic strongly lobate, cockscomb-like periphery.

The subspecies ranges in the Agua Salada Formation of southeastern Falcon throughout most of the Early and Middle Miocene.

VALVULINERIA PALMERAE Cushman & Todd
Figure **55.14, 15a–c**; 51

Valvulineria palmerae Cushman & Todd 1941, Mem. Soc. Cubana Hist. Nat.,15, p. 191, pl. 15, figs. 18a–c. – Cushman & Renz 1947*b*, p. 34, pl.7, fig.13.

Cushman & Renz (1947*b*) illustrate the spiral view of a specimen from the Ste. Croix Forma-

tion of Trinidad (Fig. 55.14) regarding it as conspecific with the holotype published from Jamaica (Fig. 55.15a–c). The general test shape is comparable to that of *V. venezuelana* with the same number of chambers forming the last whorl. Tests of *V. palmerae* are, however, distinctly smaller.

In Trinidad the species occurs from the Early Eocene *Catapsydrax stainforthi* Zone to the Middle Miocene *Globorotalia fohsi lobata* Zone.

VALVULINERIA VENEZUELANA Hedberg
Figure **55.10, 11a–c**; 51. Figure **79.7a–c**; 73

Valvulineria venezuelana Hedberg 1937*a*, p. 678, pl. 91, figs. 21a–c. – Cushman & Renz 1947*b*, p. 34, pl. 7, fig. 16.

The illustrated umbilical view of a Trinidad specimen (Fig. 55.10) and the specimen from Falcon (Fig. 79.7a–c) compare well with the holotype (Fig. 55.11a–c). The species differs from *V. georgiana* and *V. inaequalis lobata* in the last whorl consisting of slightly more chambers, about 7 against 6, and in the last chambers being less elongate, giving the test a more circular shape in equatorial view.

The range of the species is restricted in Trinidad from the Late Eocene *Globigerinatheka semiinvoluta* to the Middle Oligocene *Globigerina ampliapertura* Zone. In the Agua Salada Formation of southeastern Falcon the species has been recorded from the Early Miocene *Catapsydrax stainforthi* Zone to the Middle Miocene *Globorotalia fohsi fohsi* Zone.

Family Eponididae

EPONIDES BYRAMENSIS (Cushman)
Figure **55.19a–b**; 51

Pulvinulina byramensis Cushman 1922, p. 99, pl. 22, figs. 4, 5.
Eponides byramensis (Cushman), Hedberg 1937*a*, p. 679, pl. 92, figs. 2a–b. – Cushman & Renz 1947*b*, p. 36.

The species was only listed but not illustrated by Cushman & Renz (1947*b*). Specimens examined compare well with the holotype figures and also with the specimen illustrated by Hedberg

(1937*a*) from the Carapita Formation of eastern Venezuela, here reproduced on Fig. 55.19a–b.

The species occurs in Trinidad with interruptions from the Late Eocene *Turborotalia cerroazulensis* s.l. Zone to the Middle Miocene *Globorotalia fohsi robusta* Zone.

EPONIDES CREBBSI Hedberg
Figure **55.16, 17a–c**; 51. Figure **79.8a–b**; 73

Eponides crebbsi Hedberg 1937*a*, p. 679, pl. 92, figs. 1a–b. – Cushman & Renz 1947*b*, p. 36, pl. 7, fig. 19.

The umbilical view by Cushman & Renz (1947*b*) of a specimen from the Ste. Croix Formation of Trinidad (Fig. 55.16) showing the characteristic sigmoidally shaped intercameral sutures and distinct pores compares closely with the holotype (Fig. 55.17a–c) published by Hedberg (1937*a*) from the Carapita Formation of eastern Venezuela.

The species ranges in Trinidad throughout most of the Middle Miocene.

EPONIDES ELEVATUS (Plummer)
Figure **56.2a–c**; 51

Truncatulina elevata Plummer 1926, p. 142, pl. 11, figs. 1a–c.

The species differs from the also small *Eponides vicksburgensis* (Fig. 56.1a–c) whose test is nearly biconvex, in the distinctly higher convex spiral side and in the intercameral sutures of the spiral side being curved, not as straight and less strongly oblique.

In Trinidad the species has been recorded from the Early Eocene *Acarinina pentacamerata* Zone.

EPONIDES PARANTILLANUM Galloway & Heminway
Figure **55.18a–c**; 51

Eponides parantillanum Galloway & Heminway 1941, p. 374, pl. 18, figs. 1a–c.
Eponides parantillanum (Fig. 55.18a–c) has, compared with *E. byramensis* (Fig. 55.19a–c), a distinctly higher spiral and lower umbilical side.

The species is stratigraphically more restricted and is reported in Trinidad from the *Globigerina*

ampliapertura, the *Catapsydrax stainforthi* and the *Globigerinatella insueta* zones. It was not previously published from Trinidad.

EPONIDES VICKSBURGENSIS Cushman & Ellisor
Figure **56.1a–c**; 51

Eponides vicksburgensis Cushman & Ellisor 1931, Contrib. Cushman Lab. foramin. Res., 7, p. 56, pl. 7, figs. 8a–c.

The Trinidad specimens assigned as *Eponides* cf. *vicksburgensis* differ from the holotype (Fig. 56.1a–c) in possessing more chambers in the last whorl (9–10 against 7 in the holotype). The intercameral sutures on the spiral side appear slightly more curved in the Trinidad specimens compared with those of the holotype.

The species has a long range in Trinidad, from Early Eocene to late Middle Miocene.

Family Discorbididae

'DISCORBIS' CIPERENSIS Cushman & Stainforth
Figure **56.3a–c**; 51

Discorbis ciperensis Cushman & Stainforth 1945, p. 60, pl. 10, figs. 17a–c.

'*Discorbis*' *ciperensis* with its low biconvex text, broadly rounded periphery and about 7 chambers forming the last whorl has been found in Trinidad to be restricted to the *Globorotalia opima opima* Zone of the Cipero Formation.

Family Siphoninidae

SIPHONINA POZONENSIS Cushman & Renz
Figure **80.8a–b**; 73

Siphonina pozonensis Cushman & Renz 1941, p. 24, pl. 4, fig. 3.

For annotations see *S. pulchra*.

The species ranges in southeastern Falcon throughout most of the Early and Middle Miocene parts of the Agua Salada Formation.

SIPHONINA PULCHRA Cushman
Figure **56.4a–b**; 51

Siphonina pulchra Cushman 1919, Carnegie
  Inst.Washington, 291, p. 42, pl. 14, fig. 7. –
  Cushman & Stainforth 1945, p. 62, pl. 11,
  figs. 6a–8b.

The two species *S. pulchra* and *S. pozonensis*
are here regarded as being virtually synonymous.

The species has been recorded in Trinidad
from Late Eocene to Middle Miocene.

Family Parrelloididae

CIBICIDOIDES cf. ATRATIENSIS
(Tolmachoff)
Figure **56.12a–b, 13a–c**; 51

Cibicides atratiensis Tolmachoff 1934, Carnegie
  Mus. Ann. Pittsburgh, 23, p. 338, pl. 42,
  figs. 16–18. – Cushman & Stainforth 1945,
  p. 74, pl. 15, figs. 3a–b.

The Trinidad specimen illustrated by Cushman
& Stainforth (1945) from the Cipero Formation
(Fig.56.12a–b) as cf. *atratiensis* compares well
with the original species description given by
Tolmachoff. In contrast the illustrations by this
author of the type specimens (Fig. 56.13a–c)
do not show sufficient details for a close comparison. Characteristic for the species are the
number of chambers forming the last whorl (10–
13), an acute, partly keeled periphery and the
central portion of the spiral side slightly raised
on account of irregularly shaped shell material
covering the inner whorls. The spiral side is flat,
the umbilical one raised. Different is the size;
while it is given as 0.4 mm for the holotype the
illustrated Trinidad specimen measures 1.2 mm.

CIBICIDOIDES COOKEI (Cushman &
Garrett)
Figure **56.14a–c, 15a–b**; 51. Figure **88.9a–c**; 86

Cibicides cookei Cushman & Garrett 1938,
  Contrib. Cushman Lab. foramin. Res.,14,
  p. 65, pl. 11, fig. 3. – Cushman & Stainforth
  1945, p. 73, pl. 15, figs. 4a–b. – Beckmann
  1953, p. 402, pl. 27, fig. 6.

The specimen illustrated as *C. cookei* by Cushman & Stainforth (1945) from the Cipero Formation (Fig. 56.15a–b) does not compare with
the holotype figures of this species (Fig.56.14a–
c). Instead it appears to be much closer to the
holotype of *C. pippeni* Cushman & Garrett
(1938) (Fig.56.16a–c) published from the same
level. Other Cipero specimens examined are,
however, closely comparable to the holotype of
C. cookei in the general biconvex test shape,
rounded to subacute periphery, number of
chambers in last whorl (7–8), curvature of the
intercameral sutures, strong on spiral, only slight
on umbilical side, and coarsely perforate wall.
The Trinidad specimens are also characterized
by a distinct glassy transparent layer covering
the inner whorls and giving the spiral side a
smooth and rounded appearance.

The specimen illustrated by Cushman & Stainforth (1945) from Trinidad displays in contrast
a distinctly keeled periphery, has 12 chambers
forming the last whorl and only slightly curved
intercameral sutures on the spiral side, all features seen in *C. pippeni*. On the other hand, the
specimen illustrated by Beckmann from Barbados as *C. cookei* (Fig. 88.9a–c) compares
closely with the holotype.

CIBICIDOIDES FALCONENSIS (Renz)
Figure **79.10a–c, 11**; 73

Cibicides falconensis Renz 1948, p. 128, pl. 11,
  figs. 6a–c, 7.

C. falconensis and *C. matanzasensis* have certain
features in common such as overall test shape
and number of whorls visible on the spiral side.
According to Renz (1948), author of *C. falconensis, C. matanzasensis* differs from his
species in its larger size, being more biconvex,
having more chambers forming the last whorl,
14–15 against 10–11 in the respective holotypes,
and more strongly raised intercameral sutures.

Both species have been described and illustrated from the Agua Salada Formation, southeastern Falcon, where *C. falconensis* appears in
the Early Miocene *Catapsydrax stainforthi*
Zone, *C. matanzasensis* already in the *Catapsydrax dissimilis* Zone. Both taxa disappear in
the *Globorotalia mayeri/Globigerina nepenthes*
Zone.

CIBICIDOIDES GRIMSDALEI (Nuttall)
Figure **56**.17a–c, 18a–c; 51. Figure **88**.4a–c; 86

Cibicides grimsdalei Nuttall 1930, p. 291, pl. 25,
    figs. 7, 8, 11. – Cushman & Renz 1948, p. 41,
    pl. 8, figs. 17–19. – Beckmann 1953, p. 402,
    pl. 27, fig. 7. – Saunders *et al.* 1984, 30,
    p. 407, pl. 4, fig. 5. – Wood *et al.* 1985, pl. 2,
    figs. 7–9.

With its strongly inflated umbilical side, semi-circular in side view, plane spiral side and last whorl formed by 10–12 chambers, the species is clearly distinct from the other *Cibicidoides* taxa discussed here. The illustrated Trinidad specimen (Fig. 56.17a–c) seems to compare well with the type (Fig. 56.18a–b). The same is true for the specimen illustrated from Barbados (Fig. 88.4a–c).

The taxon originally described from the Eocene Aragon Formation of Mexico is restricted in Trinidad to the Middle and Late Eocene, but in Barbados extends into the Oligocene.

CIBICIDOIDES HAVANENSIS (Cushman & Bermudez)
Figure **88**.1a–c; 86

Cibicides havanensis Cushman & Bermudez
    1937, Contrib. Cushman Lab. foramin.
    Res., 13, p. 28, pl. 3, figs. 1–3. – Beckmann
    1953, p. 402, pl. 27, fig. 8.
Cibicidoides cf. *C. havanensis* (Cushman &
    Bermudez), Saunders *et al.* 1984,
    Micropaleontology, 30, p. 108, pl. 4, fig. 9.

The Barbados specimens are strongly calcified and therefore more rounded in appearance than the types from the Eocene of Cuba.

CIBICIDOIDES MARTINEZENSIS
(Cushman & Barksdale)
Figure **88**.5a–c; 86

Cibicides martinezensis Cushman & Barksdale
    1930, Contrib. Stanford Univ. Dept. Geol.,
    1, p. 68, pl. 12, fig. 9. – Beckmann 1953,
    p. 402, pl. 27, fig. 9.

This species is represented in the Eocene of Barbados by specimens which are often distinctly convex on the ventral side with a clear umbilical plug and almost radial sutures; they appear to be transitional to the variety *malloryi* (Smith 1957).

CIBICIDOIDES MATANZASENSIS
(Hadley)
Figure **79**.9a–b; 73

Planulina matanzasensis Hadley 1934, Bull.
    Am. Paleontol., 20, p. 27, pl. 4, figs. 1–3. –
    Renz 1948, p. 129, pl. 11, figs. 12a–b.

The most distinctive character of this species is a series of randomly arranged beads restricted to the early chambers on the spiral side. In larger specimens the beads may fuse to form a lattice-like mass over the early chambers.

For further annotations see *C. falconensis*.

CIBICIDOIDES PARIANUS (Hedberg)
Figure **56**.5, 6a–c; 51

Anomalina pariana Hedberg 1937*a*, p. 681,
    pl. 92, figs. 8a–c.
Cibicides parianus (Hedberg), Cushman &
    Renz 1947*b*, p. 45, pl. 8, fig. 9.

The early whorls are not recognizable on the spiral side of the specimen from the Ste. Croix Formation placed by Cushman & Renz (1947*b*) in *C. parianus* (Fig. 56.5). The number of chambers forming the final whorl and the strong, raised intercameral sutures compare well with the holotype (Fig. 56.6a–b) from the Carapita Formation of eastern Venezuela.

CIBICIDOIDES PERLUCIDUS (Nuttall)
Figure **79**.21a–b; 73. Figure **88**.3a–c; 86

Cibicides perlucida Nuttall 1932, p. 33, pl. 8,
    figs. 10–12. – Renz 1948, p. 129, pl. 11, figs.
    12a–b. – Beckmann 1953, p. 403, pl. 28,
    fig. 2.
Cibicidoides perlucidus (Nuttall), Saunders *et
    al.* 1984, p. 408, pl. 4, fig. 6.

The species differs from *Heterolepa mexicana* (Nuttall) in the spiral side being slightly convex against a plane to slightly concave one in *H. mexicana;* furthermore, in a slightly less inflated umbilical side and a more acute peripheral margin. In Barbados, the conspicuous, relatively large, planoconvex specimens of this species are particularly common in the Oligocene samples. They form part of a group which ranges throughout the Eocene and Oligocene and which may include *C. tuxpamensis* (Cole) and possibly *'Rotalia' eocaena* Guembel (Morkhoven *et al.* 1986).

The species occurs in the Agua Salada Formation of southeastern Falcon in the Early and in the lower part of the Middle Miocene.

CIBICIDOIDES PIPPENI (Cushman & Garrett)
Figure **56.16a–c**

Cibicides pippeni Cushman & Garrett 1938, p. 64, pl. 11, fig. 2.

For annotations see *Cibicidoides cookei* Cushman & Garrett.

CIBICIDOIDES PSEUDOUNGERIANUS (Cushman)
Figure **57.1**; 51. Figure **88.8a–c**; 86

Truncatulina pseudoungeriana Cushman 1922, Prof. Pap. U. S. Geol. Survey, 129-E, p. 97, pl. 20, fig. 9.
Cibicides pseudoungerianus (Cushman), Beckmann 1953, p. 403, pl. 28, figs. 3, 4.
Cibicidoides pseudoungerianus (Cushman), Saunders *et al.* 1984, p. 408, pl. 4, fig. 7.

Cushman illustrates only the umbilical side of the type specimen of his species (Fig. 57.1). The umbilical view of the specimen illustrated by Beckmann from Barbados (Fig. 88.8c) shows the same chamber arrangement but is without the slightly lobate periphery of the holotype figure.

CIBICIDOIDES cf. PSEUDOUNGERIANUS (Cushman)
Figure **88.2a–c**, Figure **89.14a–c**; 86

Cibicides cf. *trinitatensis* (Nuttall), Beckmann 1953, p. 404, pl. 28, figs. 9, 10.
Cibicidoides cf. *C. pseudoungerianus* (Cushman), Saunders *et al.* 1984, p. 408, pl. 4, fig. 8.

Beckmann (1953) illustrates from Barbados two specimens as *C.* cf. *trinitatensis* which in contrast to typical *C. pseudoungerianus* possess 13–15 chambers forming the last whorl, instead of only 8 in the species. Saunders *et al.* (1984) place one of them (Fig. 89.14) in *C.* cf. *pseudoungerianus*.

The range in Barbados of *C.* cf. *pseudoungerianus* is given as occurring throughout the Eocene, that of *C. pseudoungerianus* as Late Eocene to Early Miocene.

CIBICIDOIDES ROBERTSONIANUS HAITENSIS (Coryell & Rivero)
Figure **88.7a–c**; 86

Cibicides robertsonianus (Brady) var. *haitensis* Coryell & Rivero 1940, J. Paleontol., 14, p. 335, pl. 44, figs. 4–6. – Beckmann 1953, p. 404, pl. 28, fig. 5.

The species was originally described from the Middle Miocene but according to Tjalsma & Lohmann (1983) it is recorded from the Early Eocene upwards.

CIBICIDOIDES SUBSPIRATUS (Nuttall)
Figure **56.7a–c, 8**; 51

Cibicides subspiratus Nuttall 1930, p. 292, pl. 25, figs. 9, 10, 11. – Cushman & Renz 1948, p. 42, pl. 8, figs. 24, 25.

The species originally described from the Eocene of the Tampico Embayment, Mexico, is characterized by a spirally only slightly, umbilically more convex test with a peripheral keel, the last whorl formed by about 10 chambers, with intercameral sutures on both the spiral and umbilical side distinctly curved (Fig. 56.7a–c). The spiral view of a specimen illustrated by Cushman & Renz (1948) from the Eocene of Trinidad as *C. subspiratus* (Fig. 56.8) is atypical compared with the holotype illustration in that the intercameral sutures are radial to only slightly curved.

Specimens that conform with the illustrations of the type specimens (Fig. 56.7a–c) occur in Trinidad from Middle Oligocene to basal Middle Miocene.

CIBICIDOIDES SUBSPIRATUS LIMBATUS (Cita)
Figure **88.10a–c**; 86

Cibicides subspiratus Nuttall var. *limbatus* Cita 1950, Riv. Ital. Pal. Strat., 56, p. 102, pl. 9, fig. 7. – Beckmann 1953, p. 404, pl. 28, fig. 7.

The Barbados specimens are very similar to the lectotype of *C. subspiratus* (Nuttall) drawn by M. A. Kaminski and designated in Morkhoven *et al.* (1986, p. 314). The spiral thickenings are a rather constant feature in Barbados, and if the variability of the group is the same in the type area, the subspecies *limbatus* would become a synonym of *C. subspiratus*.

CIBICIDOIDES TRINCHERASENSIS (Bermudez)
Figure **88.6a–c**, 86

Cibicides trincherasensis Bermudez 1949,
  p. 305, pl. 25, figs. 1–3. – Beckmann 1953,
  p. 404, pl. 28, fig. 8.

The specimen figured by Beckmann (1953) from Barbados (Fig. 88.6a–c) compares well with the holotype illustration where the planoconvex test shape, number of chambers and surface ornamentation are concerned.

CIBICIDOIDES UNGERIANUS (d'Orbigny)
Figure **56.9a–c, 10, 11a–b**; 51

Rotalina ungeriana d'Orbigny 1846, p. 157,
  pl. 8, figs. 16–18.
Cibicides ungeriana (d'Orbigny), Nuttall 1932,
  p. 34, pl. 9, figs. 4–6.

As regards overall test shape, number of chambers forming the last whorl (10–11), nearly planoconvex test and peripheral keel, the specimens illustrated by Nuttall from the Oligocene of Mexico (Fig. 56.10,11a–b) compare well with the type (Fig. 56.9a–c) described by d'Orbigny from the Vienna Basin. A difference lies in the intercameral sutures on the spiral side which in the holotype are shown as nearly radial, in the Trinidad specimens and in the illustrated Mexican specimen as distinctly curved backwards.

The species has not previously been published from Trinidad where it is restricted to the Middle Oligocene *Globigerina ampliapertura* Zone of the Cipero Formation.

Family Discorbinellidae

LATICARININA BULLBROOKI Cushman & Todd
Figure **57.2a–b**; 51. Figure **88.11a–c**; 86

Laticarinina bullbrooki Cushman & Todd 1942,
  Contrib. Cushman Lab. foramin. Res., 18,
  p. 19, pl. 4, figs. 8, 9. – Cushman &
  Stainforth 1945, p. 73, pl. 15, figs. 2a–b. –
  Beckmann 1953, p. 401, pl. 27, fig. 5.
Laticarinina pauperata (Parker & Jones),
  Wood *et al.* 1985, pl. 4, figs. 7–9.

The specimen here illustrated from the Cipero Formation by Cushman & Stainforth (1945) (Fig. 57.2a–b) compares well with the holotype of the species described by Cushman & Todd (1942) from the Miocene of Trinidad where it occurs from the *Catapsydrax dissimilis* Zone, Cipero Formation, to the *Globorotalia menardii* Zone, Lengua Formation.

Family Planulinellidae

PLANULINA DOHERTYI (Galloway & Morrey)
Figure **79.12a–b**; 73

Cibicides dohertyi Galloway & Morrey 1929,
  Bull. Am. Paleontol., 15, no. 55, p. 30, pl. 4,
  figs. 7a–c.
Planulina dohertyi (Galloway & Morrey), Renz
  1948, p. 150, pl. 10, figs. 6a–b.

The specimen illustrated from southeastern Falcon (Fig. 79.12a–b) differs from the figures of the type specimen in the last chambers of the final whorl being more drawn backwards. Compared with *P. wuellerstorfi* (Fig. 57.4a–c) the species has only about 8 chambers forming the last whorl, instead of 10–11.

The species has been reported from the Agua Salada Formation, SE-Falcon as being restricted to the *Globorotalia fohsi peripheroronda* Zone.

PLANULINA MARIALANA Hadley
Figure **57.10a–c, 11a–b**; 51. Figure **79.15a–b**; 73

Planulina marialana Hadley 1934, Bull. Am.
  Paleontol., 20, no. 70A, p. 27, pl.4, figs. 4–
  6. – Cushman & Stainforth 1945, p. 72,
  pl. 14, figs. 3a–b.

The test of the holotype (Fig. 57.10a–c) is delicate and thin, with 8 chambers forming the last whorl. The specimen figured by Renz (1948) from the Agua Salada Formation (Fig. 79.15a–b) compares well with the holotype illustration of Hadley. The specimen illustrated by Cushman & Stainforth (Fig. 57.11a–b) on the other hand is in contrast more robust and the last whorl made up of 12–13 chambers which compares better with the specimen illustrated by Schwager as *P. wuellerstorfi* (Fig. 79.15a–b).

The species has been recorded in Trinidad

from Late Eocene to top Middle Miocene whereas in southeastern Falcon it is restricted from uppermost Late to lower part Middle Miocene.

PLANULINA cf. MEXICANA Cushman
Figure **79.14a–b**; 73

Planulina mexicana Cushman 1927, Contrib. Cushman Lab. foramin. Res., 3, p. 113, pl. 23, figs. 5a–b.
Planulina cf. *mexicana* Cushman, Renz 1948, p. 151, pl. 11, figs. 5a–c.

The specimen identified by Renz (1948) as *P.* cf. *mexicana* from southeastern Falcon (Fig. 79.14a–b) compares quite well with the illustration of the type specimen but also with that of *P. subtenuissima* (Fig. 79.13a–b). See also annotations on *P. subtenuissima*.

Specimens assigned by Renz (1948) to *P.* cf. *mexicana* are known in the Agua Salada Formation of southeastern Falcon from the upper part of the Middle Miocene.

PLANULINA RENZI Cushman & Stainforth
Figure **57.3a–c**; 51. Figure **88.12a–c**; 86

Planulina renzi Cushman & Stainforth 1945, p. 72, pl. 15, figs. 1a–c. – Beckmann 1953, p. 401, pl. 27, fig. 4. – Wood *et al.* 1985, pl. 3, figs. 7–9.

According to its authors the species differs from *P. marialana* (Fig. 57.10a–c) in having a peripheral keel, raised and thickened sutures and a greater number of chambers, about 18 forming the last whorl, against only about 8 in *P. marialana*. Based on these specific characteristics the species is also readily distinguished from the other *Planulina* species discussed here.
The species ranges in Trinidad from Middle Oligocene through Middle Miocene.

PLANULINA SUBTENUISSIMA (Nuttall)
Figure **57.9**. Figure **79.13a–b**; 73

Anomalina subtenuissima Nuttall 1928, p. 100, pl. 7, figs. 13, 15, textfig. 6.
Planulina subtenuissima (Nuttall), Renz 1948, p. 151, pl. 11, figs. 4a–b.

Compared with the other species discussed here, *P. subtenuissima* is characterized by the central

parts of both the spiral and umbilical side being covered by a small plug of clear shell material. The specimen illustrated by Renz (1948) from the Agua Salada Formation (Fig. 79.13a–b) is large, with the last whorl formed by 12 chambers and the intercameral sutures strongly curved. It has no peripheral keel. The specimen differs from Nuttall's type specimen (Fig. 57.9) in its large size and the absence of the central plug characteristic for the species.

The species is restricted in southeastern Falcon from lower Early to upper Late Miocene.

PLANULINA WUELLERSTORFI
(Schwager)
Figure **57.4a–c, 5a–c, 6a–b, 7–9**; 51

Anomalina wuellerstorfi Schwager 1866, Novara Exped. Geol. Theil, p. 258, pl. 7, figs. 105, 107.
Planulina cf. *wuellerstorfi* (Schwager), Cushman & Stainforth 1945, p. 71, pl. 14, figs. 6a–b.
Planulina wuellerstorfi (Schwager), Nuttall 1932, p. 31, pl. 4, figs. 14, 15.
Truncatulina wuellerstorfi (Schwager), Nuttall 1928, p. 98, pl. 7, fig. 12.

Schwager figured from the Pliocene of Kar Nikobar two different specimens as *P. wuellerstorfi*. One (Fig. 57.4a–c) is very flat and displays on both sides about 12 chambers forming the last whorl. Intercameral sutures are strongly drawn back. The other specimen (Fig. 57.5a–c) in comparison has a distinctly higher umbilical side and only 8–9 chambers visible on both sides, with intercameral sutures much less drawn backwards. The two specimens illustrated by Schwager also differ in the spiral being distinctly more open in the specimen on Fig. 57.5a compared with that on Fig. 57.4a.

The specimen figured by Cushman & Renz from the Cipero Formation of Trinidad (Fig. 57.6a–b) compares better with the second illustration of Schwager (Fig. 57.5a–c). The specimens illustrated by Nuttall from Mexico (Fig. 57.7) and from Trinidad (Fig. 57.8) as *P. wuellerstorfi* are poor photographic reproductions difficult to assign with certainty to this species. It would appear that they are closer to *P. renzi* (Fig. 57.3a–c).

Specimens identified in Trinidad as *P.* cf. *wuellerstorfi* occur there sporadically in the Late Eocene and again in the Middle Miocene.

Family Cibicididae

CIBICIDINA ANTIQUA (Cushman & Applin)
Figure **57.12–14**; 51

Truncatulina americana Cushman var. *antiqua* Cushman & Applin 1926, Bull. Am. Assoc. Petr. Geol., 10, p. 179, pl. 9, figs. 12, 13.
Cibicides americanus (Cushman) var. *antiquus* (Cushman & Applin), Cushman & Renz 1948, p. 42, pl. 8, fig. 26.

Cushman & Applin illustrated the spiral and umbilical views of their new taxon by apparently two different specimens (their figs. 12 and 13), without designating one of them as the holotype. For additional annotations see *C. cushmani*.

CIBICIDINA CUSHMANI (Nuttall)
Figure **57.15a–c, 16**; 51

Cibicides cushmani Nuttall 1930, p. 291, pl. 25, figs. 3, 5, 6. – Cushman & Renz 1948, p. 41, pl. 8, figs. 22, 23.

The type specimens of *C. cushmani* (Fig. 57.15a–c) differ from *C. americana antiqua* (Fig. 57.12–14) in the greater number of chambers forming the last whorl (10–11 against 8) and in the intercameral sutures on both the spiral and umbilical side being more strongly curved. *C. cushmani* was originally described from the Aragon Formation of Mexico. The umbilical view of a specimen illustrated by Cushman & Renz (1948) from the Eocene of Trinidad as *C. cushmani* (Fig. 57.16) displays 8 chambers in the last whorl with only moderately curved intercameral sutures, against 11 chambers and strongly curved sutures in umbilical view of the type specimen (Fig. 57.15c). The specimen illustrated by Cushman & Renz as *C. cushmani* therefore should rather be placed in *C. americana antiqua*.

Ranges given for the two taxa in Trinidad are parts of the Middle Eocene for *C. cushmani* and Late Eocene to Middle Oligocene for *C. americana antiqua*.

Family Acervulinidae

GYPSINA aff. VESICULARIS (Parker & Jones)
Figure **80.30a–b**; 74

Orbitolina vesicularis Parker & Jones 1860, Ann. & Mag. Nat. Hist., ser. 3, 6, p. 31, no. 5.
Gypsina? sp. aff. *vesicularis* (Parker & Jones), Renz 1948, p. 139, pl. 10, figs. 12a–b.

Family Epistomariidae

NUTTALLIDES CRASSAFORMIS (Cushman & Siegfus)
Figure **58.1a–c**; 51

Asterigerina crassaformis Cushman & Siegfus 1935, Contrib. Cushman Lab. foramin. Res., 11, p. 94, pl. 14, figs. 10a–c.

For annotations see *N. truempyi*.

NUTTALLIDES TRUEMPYI (Nuttall)
Figure **58.2a–c, 3a–b**; 51. Figure **88.13a–c**; 86

Eponides truempyi Nuttall 1930, pp. 274, 287, pl. 24, figs. 9, 13, 14. – Cushman & Renz 1948, p. 35, pl. 7, figs. 7, 8.
Nuttallides truempyi (Nuttall), Beckmann 1953, p. 384, pl. 24, figs. 2, 3. – Saunders *et al.* 1984, p. 408, pl. 4, fig. 11. – Wood *et al.* 1985, pl. 5, figs. 1–3.

The type specimens of *N. truempyi* (Fig. 58.2a–c) and *N. crassaformis* (Fig. 58.1a–c) are similar in spiral view. Both possess 6–7 chambers forming the last whorl, on the spiral side with strongly backwards bent but nearly straight intercameral sutures. The spiral side of *N. truempyi* is, however, slightly convex compared with a plane to slightly concave one in *N. crassaformis*. The type specimens differ in that *N. crassaformis* possesses a higher, more inflated umbilical side with nearly radial intercameral sutures which in *N. truempyi* are curved. The Trinidad specimens of *N. truempyi* compare well with those from Mexico. The same is true for the specimens illustrated by Beckmann (1953) from the Eocene part of the Oceanic Formation of Barbados (Fig. 88.13a–c). A 'lenticular' variety also illustrated by him (Beckmann, 1953, pl. 24, fig. 3) differs from the typical forms in possessing more chambers in the last whorl (10–11) and in being more equally biconvex.

N. truempyi was originally described by Nuttall (1930) from the Eocene of Mexico. In Trinidad the species ranges from Maastrichtian through Late Eocene; *N. crassaformis* with interruptions from Early Paleocene to Middle Oligocene.

Family Asterigerinidae

AMPHISTEGINA LESSONII d'Orbigny
Figure **80.23**; 74

Amphistegina lessonii d'Orbigny 1826, p. 304, modèle no. 98. – Renz 1948, p. 113, pl. 9, fig. 4.

Family Alabaminidae

ALABAMINA ATLANTISAE DISSONATA (Cushman & Renz)
Figure **58.4a–b**; 51. Figure **88.14a–c**; 86

Pulvinulinella atlantisae Cushman var. *dissonata* Cushman & Renz 1948, p. 35, pl. 7, figs. 11, 12.
Alabamina dissonata (Cushman & Renz), Beckmann 1953, p. 386, pl. 24, fig. 4. – Saunders *et al*. 1984, p. 407, pl. 4, fig, 1.

According to its authors the subspecies differs from the species in possessing an acute, slightly keeled periphery with sutures more curved on the umbilical side. The illustrated specimen from Trinidad (Fig. 58.4a–b) compares well with that from Barbados (Fig. 88.14a–c), except that the last whorl of the Trinidad specimens is formed by 5, that from Barbados by 6 chambers.

Range in Trinidad: late Early to Middle Eocene.

Family Osangulariidae

OSANGULARIA CULTER (Parker & Jones)
Figure **79.16a–c**; 73

Planorbulina fareta (Fichtel & Moll) var. *ungeriana* (d'Orbigny) subvar. *culter* Parker & Jones 1865, Trans. Royal Soc. London Philos., 155, pp. 382, 421, pl. 19, figs. 1a–b.
Pulvinulinella culter (Parker & Jones), Renz 1948, p. 155, pl. 9, figs. 6a–c.

The specimens illustrated from the Agua Salada Formation of southeastern Falcon as *O. culter* (Fig. 79.16a–c) and *O. jarvisi* (Fig. 79.17a–c) are both small. In spiral view their last whorl is formed by 6 respectively 5 chambers. The peripheral margin of *O. culter* is characterized by a wide and delicate keel, that of *O. jarvisi* by distinct angular projections at the base of each chamber of the last whorl, occasionally appearing almost spinose. The *O. culter* specimen figured from the Agua Salada Formation, southeastern Falcon (Fig. 79.16a–c) compares well with the Parker & Jones illustration of the type specimen.

O. culter ranges in the Agua Salada Formation through the Early and the basal part of the Middle Miocene, whereas *O. jarvisi* is restricted there to the late Early and early Middle Miocene.

OSANGULARIA JARVISI (Cushman & Renz)
Figure **79.17a–c**; 73

Pulvinulinella jarvisi Cushman & Renz 1941, p. 24, pl. 4, fig. 4.

For annotations see *O. culter*.

OSANGULARIA MEXICANA (Cole)
Figure **58.5, 6, 7a–b, 8, 9**; 51. Figure **88.15a–c**; 86

Pulvinulinella culter (Parker & Jones) var. *mexicana* Cole 1927, Bull. Am. Paleontol., 14, no. 51, p. 31, pl. 1, figs. 15, 17.
Pulvinulinella mexicana Cole, Cushman & Stainforth 1945, p. 63, pl. 11, figs. 5a–b.
Osangularia (Cribroparella) mexicana (Cole), Beckmann 1953, p. 384, pl. 24, fig. 1.
Osangularia mexicana (Cole), Saunders *et al*. 1984, p. 408, pl. 4, fig. 12.

Cole illustrates two specimens from the Eocene Guayabal Formation of Mexico, a spiral view (Fig. 58.6) with 9 chambers forming the last whorl and an umbilical one (Fig. 58.5) with 14 chambers. Tests are usually large, the periphery characterized by a wide, fairly delicate keel. The specimens figured from Trinidad (58.7a–b, 8, 9) and Barbados (Fig. 88.15a–c) compare well with Cole's type figures.

The species appears in Trinidad in the Early Eocene Upper Lizard Springs Formation to continue into the Middle Miocene Lengua Formation.

Family Oridorsalidae

ORIDORSALIS UMBONATUS (Reuss)
Figure **58.10a–c, 11a–b, 12, 13a–b; 51**

Rotalina umbonata Reuss 1851, Zeitschr. Deutsch. Geol. Ges., 3, p. 75, pl. 5, fig. 35.
Eponides umbonatus (Reuss), Nuttall 1932, p. 26, pl. 6, figs. 4, 5. – Cushman & Stainforth 1945, p. 62, pl. 11, figs. 4a–b. – Cushman & Renz 1948, p. 35, pl. 7, fig. 6.
Oridorsalis umbonatus (Reuss), Saunders _et al._ 1984, p. 408, pl. 4, fig. 10.
Oridorsalis sp., Wood _et al._ 1985, pl. 3, figs. 1–3.

The specimen illustrated by Cushman & Stainforth (1945) from the Cipero Formation (Fig. 58.13a–b) and that by Nuttall from the Early Oligocene of Mexico (Fig. 58.11a–b) compare reasonably well with Reuss' original illustrations (Fig.58.10a–c) where general test shape and number of chambers are concerned. Different are, in particular in the figured Trinidad specimen, the distinctly obliquely arranged spiral intercameral sutures, and on the umbilical side the sigmoidally shaped ones compared with the strictly radial ones on both sides in the type figures. The umbilical view of the specimen figured by Cushman & Renz (1948) from the Eocene Navet Formation (Fig. 58.12) possesses 6 chambers against only 5–6 in the other discussed specimens.

The Trinidad specimens with the sigmoidally shaped sutures on the umbilical side (Fig. 58.11a–b, 12, 13a–b) in fact fit the description for _O. umbonatus ecuadorensis_ better. Conse-

quently these specimens identified as _O. umbonatus_ should be placed in that subspecies.

The species ranges in Trinidad from Early Eocene to top Middle Miocene.

ORIDORSALIS UMBONATUS ECUADORENSIS (Galloway & Morrey)
Figure **79.18a–b**; 73. Figure **88.16a–c**; 86

Rotalia ecuadorensis Galloway & Morrey 1929, Bull. Am. Paleontol., 15, no. 55, p. 26, pl. 3, figs. 13a–c.
Eponides umbonatus (Reuss) var. _ecuadorensis_ (Galloway & Morrey), Hedberg 1937_a_, 11, p. 679, pl. 91, fig. 22. – Renz 1948, p. 133, pl. 12, figs. 25a–b. – Beckmann 1953, p. 383, pl. 23, fig. 27.

According to its authors the subspecies resembles _O. umbonatus_ but differs in the curved sutures on the umbilical side which according to them is a constant character of the subspecies.

The subspecies ranges in southeastern Falcon throughout most of the Miocene Agua Salada Formation.

For annotations see also _O. umbonatus_.

Family Heterolepidae

ANOMALINOIDES ABUILLOTENSIS (Bermudez)
Figure **58.20–22**; 51

Anomalina abuillotensis Bermudez 1949, p. 288, pl. 22, figs. 47–49.

Typical Trinidad specimens (Fig. 58.20–22) compare well with the illustrations of the holotype from the Early Eocene of Haiti. The species was not previously published from Trinidad.

Typical specimens occur in Trinidad from Late Eocene to Middle Oligocene. Specimens recorded as cf. _abuillotensis_ however are already recorded from Middle Eocene to Late Oligocene.

ANOMALINOIDES AFFINIS (Hantken)
Figure **59.13–14, 15a–c**; 51

Pulvinulina affinis Hantken 1875, A. magy. kir.
    földt. int. évkönyve, 4, p. 68, pl. 10, fig. 6.
Anomalina affinis (Hantken), Cushman &
    Renz 1948, p. 40, pl. 8, figs. 9, 10.

In comparing illustrations *A. affinis* (Fig. 59.13,
14, 15a–c) appears to be very close to identical
with *A. alazanensis spissiformis* (Fig. 59.10, 11,
12a–c).

The species was originally described from the
Late Eocene of Hungary. In Trinidad is has been
recorded from within the Middle Eocene part of
the Navet Formation.

ANOMALINOIDES ALAZANENSIS
(Nuttall)
Figure **59.5a–c, 6–9**; 51. Figure **79.20a–b**; 73

Anomalina alazanensis Nuttall 1932, p. 31,
    pl. 8, figs. 5–7. – Cushman & Renz 1947*b*,
    p. 42, pl. 8, fig. 5. – Galloway & Heminway
    1941, p. 387, pl. 22, figs. 1a–c.

For annotations see *Anomalinoides alazanensis
spissiformis*.

Fig. 59.5a–c shows a typical *A. alazanensis*
specimen from Galloway & Heminway's (1941)
study on Puerto Rico and Virgin Islands foraminifera.

ANOMALINOIDES ALAZANENSIS
SPISSIFORMIS (Cushman & Stainforth)
Figure **59.10–11, 12a–c**; 51. Figure **89.5a–c**; 86

Anomalina alazanensis Nuttall var. *spissiformis*
    Cushman & Stainforth 1945, p. 71, pl. 14, figs.
    5a–c. – Cushman & Renz 1948, p. 41, pl. 8,
    figs. 15, 16. – Beckmann 1953, p. 399, pl. 26,
    fig. 14.
Anomalinoides spissiformis (Cushman &
    Stainforth), Saunders *et al.* 1984, p. 407,
    pl. 3, fig. 21.

The subspecies of *A. alazanensis* described by
Cushman & Stainforth (1945) from the Cipero
Formation differs from the species according to
its authors 'in its larger size and much thicker
and more closely coiled test with a rounded periphery'. Comparing the holotype illustrations of
the subspecies *spissiformis* (Fig.59.12a–c) with
those of the figures of the primary species *alazanensis* (Fig.59.5a–c, 7–9) these distinctions

seem justified. The umbilical view of an *A. alazanensis* specimen from the Ste. Croix Formation
of Trinidad in Cushman & Renz (1947*b*) here
shown on Fig.59.6, also compares well with the
corresponding view of the holotype. The specimens of *A. alazanensis spissiformis* illustrated by
Cushman & Renz (1948) from the Navet Formation (Fig.59.10–11) differ from the holotype in
being more involute and in possessing only 11
chambers in the last whorl against 13 in the
holotype.

The subspecies has been recorded in the Oligocene-Miocene Cipero Formation from the *Globigerina ciperoensis ciperoensis* and the
Globigerinatella insueta zones, in the Eocene
throughout the Navet and San Fernando formations.

Specimens regarded as belonging to *A. alazanensis* s.s. have been recorded only from the
Globigerinatella insueta Zone of the Cipero Formation. The *spissiformis* specimen illustrated by
Beckmann (1953) from the Oceanic Formation
of Barbados (Fig.89.5a–c) compares closely
with the Navet Formation specimens (Fig.
59.10–11) illustrated by Cushman & Renz
(1948).

ANOMALINOIDES BADENENSIS
(d'Orbigny)
Figure **89.2a–c, 3a–c**; 86

Anomalina badenensis d'Orbigny 1846, p. 171,
    pl. 9, figs. 1–3. – Beckmann 1953, p. 399,
    pl. 26, figs. 16, 17.

The more or less planispiral specimens in the
Oligocene part of the Oceanic Formation (Fig.
89.2a–c) are fairly typical. In the Eocene
samples, these are associated with a more trochospiral variety (Fig. 89.3a–c). The general
aspect of d'Orbigny's type (revised in Papp &
Schmid 1985) is similar but there is little information on the variability of the species in the
type area.

ANOMALINOIDES BILATERALIS
(Cushman)
Figure **59.16–17, 18a–b**; 51. Figure **89.4a–c**; 86

Anomalina bilateralis Cushman 1922, p. 97,
    pl. 21, figs. 1, 2. – Cushman & Renz 1948,
    p. 41, pl. 8, figs. 11, 12. – Beckmann 1953,
    p. 399, pl. 26, fig. 15.

Anomalinoides cf. *A. bilateralis* (Cushman), Saunders *et al.* 1984, p. 407, pl. 3, fig. 22.

Cushman figured only the spiral view of his type specimen (Fig. 59.18a). In addition he illustrated an edge view of another specimen (Fig. 59.18b). Cushman & Renz (1948) figured under this name the spiral and umbilical views of two specimens from Trinidad. They apparently fall within the species concept of *A. bilateralis* as does the specimen illustrated from Barbados (Fig. 89.4a–c).

The species was originally described and illustrated from the Oligocene Byram Marl of Mississippi. In Trinidad it is recorded throughout the Eocene, but not from the Oligocene. In Barbados it occurs in both the Eocene and Oligocene parts of the Oceanic Formation.

ANOMALINOIDES CICATRICOSUS
(Schwager)
Figure **59.3a–c, 4a–b**; 51

Anomalina cicatricosa Schwager 1866, Novara-Exped. Geol. Theil, 2, p. 260, pl. 7, fig. 108, textfig. 4.
Cibicides cicatricosa (Schwager), Cushman & Stainforth 1945, p. 73, pl. 15, figs. 6a–b.

The specimen figured by Cushman & Stainforth (1945) from the Cipero Formation (Fig.59.4a–b) as *Cibicides cicatricosa* compares with the specimen originally figured by Schwager from the younger Tertiary of Kar Nikobar (Fig. 59.3a–c) as regards general morphology, number of chambers and size. The peripheral outline of the last 3–4 chambers in the original specimen, however, appears more lobate compared with the illustrated Trinidad specimen. Other Trinidad specimens examined display these more globular chambers as seen in the type specimen. Characteristic for the species and clearly seen in the illustration of the type species are the distinct pores and the thickened and raised sutures in the inner whorls as seen in spiral view.

The species has been reported in Trinidad from the Middle Eocene and again from the Early and Middle Miocene.

ANOMALINOIDES DORRI
ARAGONENSIS (Nuttall)
Figure **59.1–2**; 51. Figure **89.1a–c**; 86

Anomalina dorri Cole var. *aragonensis* Nuttall 1930, p. 291, pl. 24, fig. 18; pl. 25, fig. 1. – Beckmann 1953, p. 399, pl. 27, fig. 1.
Anomalina dorri aragonensis Nuttall, Cushman & Renz 1948, p. 41, pl. 8, figs. 13, 14.

According to its author the test of the subspecies is smaller, with sutures on the last whorl more sharply defined and also more depressed compared with the species. In comparing illustrations of the type specimens of *A. dorri aragonensis* (Fig. 59.1–2) and *A. cicatricosa* (Fig. 59.3a–c) the two taxa agree closely in general test shape, number of chambers and the distinct large pores on both sides. The specimen illustrated from Barbados (Fig. 89.1a–c) differs from the type in being more involute and having less chambers forming the last whorl (6–7 against 8–9). However, it displays the same coarse pores on both sides. In side view (Fig. 89.1b) the Barbados specimen appears more inflated compared with the type specimen of which such a view is lacking.

The species has been recorded in Trinidad from the Middle Paleocene part of the Lizard Springs Formation to the late Middle Eocene part of the Navet Formation; in Barbados from the late Middle Eocene Mount Hillaby member of the Oceanic Formation.

ANOMALINOIDES MECATEPECENSIS
(Nuttall)
Figure **59.19–20, 21a–b**; 51

Anomalina mecatepecensis Nuttall 1932, p. 30, pl. 7, figs. 4, 6. – Hedberg 1937*a*, p. 682, pl. 92, figs. 10a–b.

The characteristic species with its inner whorls as seen from the spiral side obscured by a knob-like shell structure has not previously been published from Trinidad. It was originally described from the Lower Oligocene Alazan Shale of the Tampico embayment, Mexico. Hedberg (1937*a*) illustrated the species from the Carapita Formation, eastern Venezuela. In Trinidad identical forms occur with some interruptions throughout the Cipero and Lengua formations.

ANOMALINOIDES POMPILIOIDES
(Galloway & Heminway)
Figure **58.14a–b, 15a–c**; 51

Anomalina pompilioides Galloway &
　Heminway 1941, p. 389, pl. 22, fig. 3. –
　Cushman & Stainforth 1945, p. 71, pl. 14,
　fig. 1a–b.
Anomalina grosserugosa Guembel, Nuttall
　1928, p. 99, pl. 7, figs. 18, 19.

The Trinidad specimens (Fig. 58.14a–b) compare morphologically well with the holotype
(Fig. 58.15a–b) except for size. In Trinidad they
measure up to 1.4 mm (Nuttall, 1928, p. 100),
compared to only 0.4 mm of the holotype. Specimens of this small size which also occur in the
Cipero Formation of Trinidad are regarded by
Cushman & Stainforth (1945) as juvenile forms.
The specimens described and illustrated by Nuttall (1928) from Trinidad as *Anomalina grosserugosa* Guembel are placed by Cushman &
Stainforth (1945) in *Anomalina pompilioides*.
The species appears in the literature sometimes
also as *Anomalinoides semicribrata* Beckmann,
which is the type species of *Linaresia*.

The species ranges in Trinidad from the Late
Cretaceous to the Middle Miocene *Globorotalia
menardii* Zone.

ANOMALINOIDES SUBBADENENSIS
(Pijpers)
Figure **58.16a–b, 17–19**; 51

Anomalina subbadenensis Pijpers 1933, Geol.
　Pal. Bonaire, p. 72, textfigs. 116–120. –
　Cushman & Stainforth 1945, p. 71, pl. 14,
　fig. 2a–b.

The specimen illustrated by Cushman & Stainforth (1945) as *A. subbadenensis* (Fig. 58.16a–
b) shows in shape and number of chambers considerable affinities to *A. pompilioides*
(Fig.58.14a–b, 15a–b), except for its size and
slightly more curved intercameral sutures, particularly so on the umbilical side. Compared with
A. pompilioides the Trinidad specimens of *A.
subbadenensis* are, however, considerably
smaller, measuring about 0.5 mm. In comparison the original specimens from Bonaire (Fig.
58.17–19) measure only between 0.17 and
0.29 mm. Pijpers did not designate a holotype
from his six illustrated specimens, in which the
number of chambers in the last whorl varies

between 6 and 8. His spiral view 116 (Fig. 58.17)
and umbilical view 119 (Fig. 58.19) compare
closest with the Trinidad specimen illustrated by
Cushman & Stainforth (1945), reproduced here
on Fig. 58.16a–b. Later, Cushman & Stainforth
(1951) believe that their Trinidad specimen is a
Globorotalia mayeri.

Pijpers described the species from the Late
Eocene of Bonaire. In Trinidad it was recorded
by Cushman & Stainforth from the Early to
Middle Miocene part of the Cipero Formation.
For reasons of size and stratigraphic occurrence
we place the specimen illustrated in Cushman &
Stainforth only with reservation in *A. subbad-
enensis*.

ANOMALINOIDES TRINITATENSIS
(Nuttall)
Figure **59.22–27**; 51. Figure **79.19a–c**; 73

Truncatulina trinitatensis Nuttall 1928, p. 97,
　pl. 7, figs. 3, 5, 6. – Nuttall 1932, p. 33,
　pl. 7, fig. 9.
Cibicides trinitatensis Cushman 1946*a*, p. 40,
　pl. 8, figs. 9–11.
Anomalinoides trinitatensis (Nuttall), Renz
　1948, p. 115, pl. 10, figs. 11a–c.

The species with 12–14 chambers forming the
last whorl, a fairly low spiral and higher umbilical side, is characterized by an irregularly
shaped, partly pitted glassy plug covering the
inner whorls on the spiral side. A similar smaller
structure, more regularly rounded and with a
smooth, glassy surface covers the narrow umbilical area. The periphery is rounded, the intercameral sutures on the spiral side slightly curved.
The holotype illustration by Nuttall (Fig. 59.26)
is a poor photograph which does not satisfac-
torily show the characteristic features. They are
better visible on the specimens figured by Cushman (1946*a*) from the Cocoa Sand of Alabama
(Fig. 59.22–24), the spiral view (Fig. 59.27) in
Nuttall (1932) and in the specimen illustrated by
Renz (1948) from the Agua Salada Formation,
southeastern Falcon (Fig. 79.19a–c).

HETEROLEPA COMPRESSA (Cushman & Renz)
Figure **79.22a–c**;73

Cibicides floridanus (Cushman) var. *compressa*
Cushman & Renz 1941, p. 26, pl. 4, fig. 9.
Cibicides compressus Cushman & Renz, Renz
1948, p. 127, pl. 10, figs. 9a–c.

According to its authors the species differs from *H. floridana,* of which it was originally regarded as a variety, in the much compressed test, somewhat convex ventral side and more strongly curved and narrower chambers.

The species occurs virtually throughout the Agua Salada Formation of southeastern Falcon.

HETEROLEPA MEXICANA (Nuttall)
Figure **60.1a–c, 2a–b, 3, 4, 5a–c**; 51. Figure **89.6a–c**; 86

Cibicides mexicanus Nuttall 1932, p. 33, pl. 9,
figs. 7–9. – Galloway & Heminway 1941,
p. 394, pl. 22, figs. 5a–c. – Cushman &
Stainforth 1945, p. 73, pl. 15, figs. 5a–b. –
Cushman & Renz 1948, p. 42, pl. 8, figs. 20,
21. – Beckmann 1953, p. 402, pl. 8, fig.1.

The illustrated Trinidad specimens from the Cipero (Fig.60. 2a–b) and Navet (Fig. 60.3–4) formations and the specimen from the Oceanic Formation of Barbados (Fig. 89.6a–c) compare well with the type figures of the specimen described by Nuttall (1932) from the Early Oligocene of Mexico, except for the 15 chambers visible on the umbilical side in the Cipero specimen (Fig. 60.2b) against only 9 in the spiral view of the type specimen (Fig. 60.1a).

In the Cipero Formation of Trinidad the species has been recorded from the Middle and Late Oligocene, as well as from the Middle to Late Eocene Navet Formation which, however, is not shown on the distribution chart Fig. 51. In Barbados the species occurs in the Oligocene Codrington College Shales.

Family Gavelinellidae

GYROIDINOIDES ALTIFORMIS (R. E. & K. C. Stewart)
Figure **79.24a–c**; 73. Figure **89.11a–c**; 86

Gyroidina soldanii d'Orbigny, var. *altiformis*
R. E. & K. C. Stewart 1930, J. Paleontol., 4,
p. 67, pl. 9, fig. 2. – Renz 1948, p. 140, pl. 8,
figs, 13a–b.
Gyroidinoides altiformis (R. E. & K. C.
Stewart), Beckmann 1953, p. 381, pl. 23,
fig. 22.

The species is closely comparable to *G. dissimilis* and *G. girardanus* in general test shape and number of chambers forming the last whorl and in the acute periphery.

The species ranges in the Agua Salada Formation of southeastern Falcon throughout the Early and Middle Miocene.

GYROIDINOIDES ALTISPIRUS (Cushman & Stainforth)
Figure **60.16a–b**; 51

Gyroidinoides altispira Cushman & Stainforth
1945, p. 61, pl. 11, figs. 1a–b.

With its distinctly convex spiral side displaying about 7 closely coiled whorls with 8–10 chambers forming the last whorl, the species is a characteristic form.

In Trinidad it occurs from the Late Oligocene to the late Middle Miocene.

GYROIDINOIDES BYRAMENSIS CAMPESTER (Palmer & Bermudez)
Figure **80.5a–c**; 73

Eponides byramensis (Cushman) var. *cubensis*
Palmer & Bermudez 1936, Mem. Soc.
Cubana Hist. Nat., 10, p. 302, pl. 20, figs.
4–6 (not *Eponides cubensis* Palmer &
Bermudez 1936, l. c., 9, p. 252, pl. 21, figs.
10–12).
Eponides byramensis (Cushman) var.
campester Palmer & Bermudez 1941, Mem.
Soc. Cubana Hist. Nat., 15, 1941, p. 192.
Gyroidinoides byramensis (Cushman) var.
campester (Palmer & Bermudez), Renz 1948,
p. 139, pl. 8, figs. 15a–b, pl. 9, fig. 1.

The large species differs from *G.* cf. *soldanii* (Fig. 80.4a–c) in its low conical spiral side displaying 4–5 whorls, the last formed by about 10 chambers. The umbilical side is slightly less inflated and the periphery more distinctly angular.

The species occurs throughout the Early and Middle Miocene of the Agua Salada Formation of southeastern Falcon.

GYROIDINOIDES COMPLANATUS
(Cushman & Stainforth)
Figure **60.6a–c**; 51

Gyroidina complanata Cushman & Stainforth 1945, p. 61, pl. 11, figs. 2a–c.

For annotations see *G. planulatus.*

The species ranges in Trinidad throughout the Cipero and Lengua formations.

GYROIDINOIDES DISSIMILIS (Cushman & Renz)
Figure **60.15a–b**; 51

Gyroidina dissimilis Cushman & Renz 1947*b*, p. 35, pl. 7, figs. 18a–b.

The species is characterized by the slightly concave spiral and highly convex umbilical side, an angled to slightly keeled periphery, and about two whorls visible on the spiral side. The last whorl consists of 8–9 chambers. *G. dissimilis* differs from *G. girardanus* in its more concave spiral side and the angular to slightly keeled periphery.

The species was described from the late Lower Miocene Ste. Croix Formation of Trinidad.

GYROIDINOIDES GIRARDANUS (Reuss)
Figure **60.13a–c, 14a–b**; 51. Figure **89.8a–c**; 86

Rotalina girardana Reuss 1851, Zeitschr. deutsch. geol. Ges., 3, p. 73, pl. 5, fig. 34.
Gyroidina girardana (Reuss), Cushman & Stainforth 1945, p. 60, pl. 10, figs. 18a–b.
Gyroidina girardana (Reuss), Beckmann, 1953, p. 382, pl. 23, fig. 23.
Gyroidinoides ssp. (part), Saunders *et al.* 1984, p. 408, pl. 4, fig. 13.
Gyroidinoides sp., Wood *et al.* 1985, pl. 3, figs. 4–6.

The specimens from Trinidad and Barbados (Fig. 60.14a–b; Fig. 89.8a–c) compare well with the type specimen (Fig. 60.13a–c) except that the illustrated specimens have only 8 or sometimes 9 chambers forming the last whorl, against 10 in the type specimen. Further, in side view the illustrated Trinidad and Barbados specimens appear slightly more inflated compared with the type species. For comparison of *G. girardanus* with *G. dissimilis* see annotations on that species. *G. subangulatus* (Plummer), common in the Lower Lizard Springs Formation, is probably the ancestor of *G. girardanus*.

The species originally described by Reuss from the Eocene ranges in Trinidad from the Lower Eocene Navet Formation to the late Middle Miocene Lengua Formation. In Barbados the species has been recorded throughout the Eocene to Late Oligocene Oceanic Formation.

GYROIDINOIDES GIRARDANUS PERAMPLUS (Cushman & Stainforth)
Figure **60.11a–b, 12**; 51. Figure **89.7a–c**; 86

Gyroidina girardana (Reuss) var. *perampla* Cushman & Stainforth 1945, p. 61, pl. 10, figs. 19a–b.
Gyroidinoides girardana (Reuss) var. *perampla* (Cushman & Stainforth), Beckmann 1953, p. 382, pl. 23, fig. 24.

The subspecies differs from the species in its distinctly larger size, the holotype measuring 1.2 mm against about 0.6 mm for the species. It has one or two chambers more forming the last whorl compared with the Trinidad and Barbados specimens.

The range of the subspecies is more restricted compared with that of the species. In Trinidad it occurs from the Late Eocene *Turborotalia cerroazulensis* Zone to the late Middle Miocene *Globorotalia menardii* Zone. In Barbados it was noted throughout the Eocene to Late Oligocene Oceanic Formation.

GYROIDINOIDES PARVUS (Cushman & Renz)
Figure **79.23a–c**; 73

Gyroidina parva Cushman & Renz 1941, p. 23, pl. 4, fig. 2. – Renz 1948, p. 139, pl. 8, figs. 12a–c.

The species is very small, has a high umbilical side, with 6 slightly inflated chambers forming the last whorl.

The species occurs throughout the Early and Middle Miocene of the Agua Salada Formation of southeastern Falcon.

GYROIDINOIDES PLANULATUS
(Cushman & Renz)
Figure **60.7a–c, 8**; 51. Figure **80.2a–c**; 73. Figure **89.10a–c**; 86

Gyroidina planulata Cushman & Renz 1941, p. 23, pl. 4, fig. 1. – Cushman & Renz 1947*b*, p. 35, pl. 7, fig. 17.
Gyroidinoides planulata (Cushman & Renz), Renz 1948, p. 140, pl. 8, figs. 11a–c. – Beckmann 1953, p. 383, pl. 23, fig. 25.

The species differs from *G. complanatus* (Fig. 60.6a–c) in its smaller size, a less inflated test and in spiral view displaying about three whorls against only two in *G. complanatus*.

The species was originally described from southeastern Falcon and subsequently reported also from Trinidad and Barbados.

GYROIDINOIDES cf. SOLDANII
(d'Orbigny)
Figure **80.4a–c**; 73

Gyroidina soldanii d'Orbigny 1826, p. 278, modèle 36.
Gyroidinoides cf. *soldanii* (d'Orbigny), Renz, 1948, p. 140, pl. 8, figs. 14a–c.

The species is large and has a similar overall test shape to *G. girardanus peramplus*. The two taxa differ slightly in the spiral intercameral sutures of the last whorl being straight and strongly drawn back in *G.* cf. *soldanii*, while those of *G. girardanus peramplus* run more radially.

The taxon occurs from the *Catapsydrax stainforthi* Zone to the basal part of the *Globorotalia fohsi fohsi* Zone of the Agua Salada Formation, southeastern Falcon.

GYROIDINOIDES SOLDANII OCTOCAMERATUS (Cushman & Hanna)
Figure **60.9a–b, 10a–c**; 51. Figure **89.9a–c**; 86

Gyroidina soldanii d'Orbigny var. *octocamerata* Cushman & D. G. Hanna, Proc. Calif. Acad. Sci. ser. 4,16, p. 223, pl. 14, figs. 16–18. – Cushman & Renz 1948, p. 34, pl. 7, figs. 2, 3.
Gyroidinoides soldanii (d'Orbigny) var. *octocamerata* (Cushman & Hanna), Beckmann 1953, p. 383, pl. 23, fig. 26.

The overall test shape of the illustrated Trinidad (Fig. 60.9a–b) and Barbados (Fig. 89.9a–c) specimens compare well with the holotype (Fig. 60.10a–c). As the name indicates the species has 8 chambers forming the last whorl. However, in both the illustrated Trinidad and Barbados specimens it consists of only 6 chambers.

In Trinidad the species ranges from late Early to Late Eocene.

GYROIDINOIDES VENEZUELANUS
Renz
Figure **80.1a–c**; 73

Gyroidinoides venezuelana Renz 1948, p. 141, pl. 12, figs. 21a–c.

The species resembles *G. planulatus* (Fig. 80.2a–c) but differs in the test being nearly planoconvex with a sharply edged periphery, against being biconvex and with a rounded periphery in *G. planulatus*.

The species ranges throughout the Early and Middle Miocene of the Agua Salada Formation of southeastern Falcon.

GYROIDINA JARVISI Cushman & Stainforth
Figure **60.17a–b, 18a–c**; 51

Gyroidina jarvisi Cushman & Stainforth 1945, p. 62, pl. 11, figs. 3a–c. – Cushman & Renz 1948, p. 34, pl. 7, figs. 4, 5.

The species originally established in Trinidad is comparatively small, with about three whorls clearly visible on the moderately convex spiral side. Characteristic are the large number of chambers, 11 forming the last whorl of the holotype.

The species ranges in Trinidad from Middle Eocene to top Middle Miocene. The illustrated

Trinidad specimens are from the Navet Formation (Fig. 60.17a–b) and the Cipero Formation (Fig. 60.18a–c, holotype).

HANZAWAIA AMERICANA (Cushman)
Figure **80.3a–c**; 73

Truncatulina americana Cushman 1918, Bull. U. S. Nat. Mus., 103, p. 68, pl. 23, figs. 2a–c.
Cibicides americanus (Cushman), Renz 1948, p. 126, pl. 11, figs. 10a–c.

The specimen illustrated from the Agua Salada Formation, southeastern Falcon compares in all respects with the figures of the type specimen.

The species occurs virtually throughout the Agua Salada Formation.

HANZAWAIA CARSTENSI (Cushman & Ellisor)
Figure **61.9a–c, 10, 11**; 51. Figure **80.6a–c**; 73

Cibicides carstensi Cushman & Ellisor 1939, Contrib. Cushman Lab. foramin. Res.,15, p. 13, pl. 2, fig. 8. – Cushman & Renz 1947*b*, p. 45, pl. 8, figs. 10, 11.

The spiral and umbilical views of the specimens from the Lower Miocene Ste. Croix Formation of Trinidad (Fig.61.10–11) compare well with the holotype illustrations (Fig.61.9a–c).

HANZAWAIA ILLINGI (Nuttall)
Figure **61.4a–b, 5, 6**; 51. Figure **89.12a–c**; 86

Truncatulina illingi Nuttall 1928, p. 99, pl. 7, figs. 11, 17, textfig. 5.
Planulina illingi (Nuttall), Cushman & Stainforth 1945, p. 72, pl. 14, fig. 4a–b.
Anomalina illingi (Nuttall), Beckmann 1953, p. 400, pl. 27, fig. 2.

Both the type specimen (Fig. 61.4a–b) and the specimen published later from the Cipero Formation of Trinidad (Fig. 61.5, 6) compare well. The species is characterized by its large size in comparison to other *Hanzawaia* species.

The species has been recorded in Trinidad from the upper part of the Middle Eocene Navet Formation to the early Middle Miocene part of the Cipero Formation. In Barbados it occurs in the Oligocene Bath and Codrington College beds of the Oceanic Formation.

HANZAWAIA MANTAENSIS (Galloway & Morrey)
Figure **61.1a–c, 2, 3a–c**; 51. Figure **80.7a–b**; 73

Anomalina mantaensis Galloway & Morrey 1929, Bull. Am. Paleontol., 15, no. 55, p. 28, pl.4, fig. 5.
Cibicides mantaensis (Galloway & Morrey), Hedberg 1937*a*, p. 683, pl. 92, figs. 12a–c. – Cushman & Renz 1947*b*, p. 44, pl. 8, fig. 7.

The spiral view of *H. mantaensis* given by Cushman & Renz (1947*b*) from the Ste. Croix Formation (Fig. 61.2) compares well with that of the type specimen (Fig. 61.1a–c) and also with that by Hedberg from the Carapita Formation, eastern Venezuela (Fig.61.3a–c).

HANZAWAIA MISSISSIPPIENSIS (Cushman)
Figure **61.7a–b, 8**; 51

Anomalina mississippiensis Cushman 1922, p. 98, pl. 21, figs. 6–8.
Cibicides mississippiensis (Cushman), Cushman & Renz 1947*b*, p. 44, pl. 8, fig. 8. – Cushman 1946*a*, p. 39, pl. 8, figs. 5, 6.

Cushman & Renz (1947*b*) only illustrate an umbilical view of a specimen (Fig. 61.8) they place in *C. mississippiensis* indicating that the specimens are very typical in the Ste. Croix beds. Cushman characterizes the species as being small with a flattened to slightly concave spiral and a very convex umbilical side, the last whorl made up of 6–8 chambers (Fig. 61.7a–b).

The Trinidad specimens are according to later investigations restricted to the Late Eocene-Early Oligocene. They fit well into the original species description and its occurrence in the Lower Oligocene Byram Marl of Vicksburg.

LINARESIA POMPILIOIDES SEMICRIBRATA Beckmann
Figure **89.13a–c**; 86

Anomalina pompilioides Galloway & Heminway var. *semicribrata* Beckmann 1953, p. 400, pl. 27, fig. 3.
Gavelinella semicribrata (Beckmann), Wood *et al.* 1985, pl. 4, figs. 1–3.

Typical specimens are found in all parts of the Oceanic Formation. The holotype is from the

Oligocene part of the formation (*Globorotalia opima opima* Zone). The subspecies differs from the species in the distinctly coarser chamber perforation on the umbilical side.

Family Rotaliidae

AMMONIA BECCARII (Linné)
Figure **80.9a–c;** 73

Nautilus beccarii Linné 1758, Syst. Nat. ed., 10, tomus 1, p. 710.
Streblus beccarii (Linné), Renz 1948, p. 167, pl. 9, figs. 2a–c.

References

Albers, J. 1952. Taxonomie und Entwicklung einiger Arten von *Vaginulina* d'Orb. aus dem Barrême bei Hannover (Foram.). *Mitt. geol. Staatsinst. Hamburg*, 21, 75–112.

Alth, A. 1850. Geognostisch-paläontologische Beschreibung der nächsten Umgebung von Lemberg. *Haidinger's naturwiss. Abhandl., Vienna*, 3, 171–284.

Aubert, J. & Bartenstein, H. 1976. *Lenticulina (L.) nodosa*. Additional observations in the worldwide Lower Cretaceous. *Bull. Centre Rech. Pau-SNPA*, 10, 1–33.

Aubert, J. & Berggren, W. A. 1976. Paleocene benthic foraminiferal biostratigraphy and paleoecology of Tunisia. *Bull. Centre Rech. Pau-SNPA*, 10, 379–469.

Bailey, H. W., Gale, A. S., Mortimore, R. N., Swiecicki, A. & Wood, Ch. J. 1983. The Coniacian-Maastrichtian stages of the United Kingdom, with particular reference to southern England. *Newsletter Stratigr.*, 12, 29–42.

Bandy, O. L. 1970. Upper Cretaceous-Cenozoic bathymetric cycles, eastern Panama and northern Colombia. *Trans. Gulf Coast Assoc. geol. Soc.*, 20, 181–193.

Banner, F. T. & Blow, W. H. 1965. Progress in the planktonic foraminiferal biostratigraphy of the Neogene. *Nature*, 208/5016, 1164–1166.

Barbeito, P. J. R., Pittelli, R. & Evans, A. M. 1985. Estudios estratigraficos del Eoceno en el area de Mara-Maracaibo, Venezuela occidental, basado en interpretaciones paleontologicas y palinologicas. *Mem. VI Congr. geol. Venez.*, 1, 109–139.

Barr, K. W. 1963. The geology of the Toco district, Trinidad, West Indies. *Overseas Geol. Min. Res.*, 8, no. 4, 379–415; 9, no. 1, 1–29.

Bartenstein, H. 1974. *Lenticulina (Lenticulina) nodosa* (Reuss 1863) and its subspecies – worldwide index foraminifera in the Lower Cretaceous. *Eclog. geol. Helv.*, 67, 539–562.

Bartenstein, H. 1976. Foraminiferal zonation of the Lower Cretaceous in North West Germany and Trinidad, West Indies. An attempt. *Neues Jahrb. Geol. Paläontol., Monatsh.*, 1976/3, 187–192.

Bartenstein, H. 1985. Stratigraphic pattern of index foraminifera in the Lower Cretaceous of Trinidad. *Newsletter Stratigr.*, 14, 110–117.

Bartenstein, H. 1987. Micropaleontological synopsis of the Lower Cretaceous in Trinidad, West Indies. Remarks on the Aptian/Albian boundary. *Newsletter Stratigr.*, 17, 143–152.

Bartenstein, H., Bettenstaedt, F. & Bolli, H. M. 1957. Die Foraminiferen der Unterkreide von Trinidad, B. W. I. Erster Teil: Cuche- und Toco-Formation. *Eclog. geol. Helv.*, 50, 5–67.

Bartenstein, H., Bettenstaedt, F. & Bolli, H. M. 1966. Die Foraminiferen der Unterkreide von Trinidad, W. I. Zweiter Teil: Maridale-Formation (Typlokalität). *Eclog. geol. Helv.*, 59, 129–177.

Bartenstein, H., Bettenstaedt, F. & Kovatcheva, T. 1971. Foraminiferen des bulgarischen Barrême. Ein Beitrag zur weltweiten Unterkreide-Stratigraphie. *Neues Jahrb. Geol. Paläontol., Abh.*, 139, 125–162.

Bartenstein, H. & Bolli, H. M. 1973. Die Foraminiferen der Unterkreide von Trinidad. Dritter Teil: Maridaleformation (Co-Typlokalität). *Eclog. geol. Helv.*, 66, 389–418.

Bartenstein, H. & Bolli, H. M. 1977. The Foraminifera in the Lower Cretaceous of Trinidad, W. I. Part 4: Cuche Formation, upper part; *Leupoldina protuberans* Zone. *Eclog. geol. Helv.*, 70, 543–573.

Bartenstein, H. & Bolli, H. M. 1986. The foraminifera in the Lower Cretaceous of Trinidad, W. I. Part 5: Maridale Formation, upper part: *Hedbergella rohri* Zone. *Eclog. geol. Helv.*, 79, 945–999.

Bartenstein, H. & Brand, E. 1949. New genera of foraminifera from the Lower Cretaceous of Germany and England. *J. Paleontol.*, 23, 669–672.

Bartenstein, H. & Brand, E. 1951. Mikropaläontologische Untersuchungen zur Stratigraphie des nordwestdeutschen Valendis. *Abhandl. Senckenberg. naturf. Ges.*, 485, 239–336.

Basov, I. A. & Krasheninnikov, V. A. 1983. Benthic foraminifers in Mesozoic and Cenozoic sediments of the southwestern Atlantic as an indicator of paleoenvironment, Deep Sea Drilling Project Leg 71. *Init. Rep. Deep Sea drill. Proj.*, 71, 739–787.

Batjes, D. A. J. 1968. Palaeoecology of foraminiferal assemblages in the Late Miocene Cruse and Forest Formations of Trinidad, Antilles. *Trans. 4th Carib. Geol. Conf.*, Trinidad, 141–156.

Beck, C. & Furrer, M. A. 1977. Sobre la existencia de sedimentos marinos no metamorfizados del Neocomiense en el noreste del Estado Guarico, Venezuela septentrional. Implicaciones paleogeograficas. *Mem. V Congr. Geol. Venez., Caracas*, 135–147.

Beckmann, J. P. 1953. Die Foraminiferen der Oceanic Formation (Eocaen-Oligocaen) von Barbados, Kl. Antillen. *Eclog. geol. Helv.*, 46, 301–412.

Beckmann, J. P. 1960. Distribution of benthonic foraminifera at the Cretaceous-Tertiary boundary of Trinidad (West Indies). *Rep. Int. Geol. Congr., XXI Session, Norden*, V, 57–69.

Beckmann, J. P. 1974. The new genera and species of benthonic foraminifera described from Trinidad. *Verhandl. Naturf. Ges. Basel*, 84, 234–244.

Beckmann, J. P. 1978. Late Cretaceous smaller benthic

foraminifers from Sites 363 and 364, DSDP Leg 40, Southeast Atlantic Ocean. *Init. Rep. Deep Sea drill. Proj.*, 40, 759–781.

Beckmann, J. P. 1991. New taxa of foraminifera from the Cretaceous and basal Tertiary of Trinidad, West Indies. *Eclog. geol. Helv.*, 84, 819–835.

Beckmann, J. P., Bolli, H. M., Kleboth, P. & Proto Decima, F. 1982. Micropaleontology and biostratigraphy of the Campanian to Paleocene of the Monte Giglio, Bergamo Province, Italy. *Mem. Ist. Geol. Min. Univ. Padova*, 35, 91–172.

Beckmann, J. P., Bolli, H. M., Perch-Nielsen, K., Proto Decima, F., Saunders, J. B. & Toumarkine, M. 1981. Major calcareous nannofossil and foraminiferal events between the Middle Eocene and Early Miocene. *Palaeogeograph., Palaeoclimatol., Palaeoecol.*, 36, 155–190.

Beckmann, J. P. & Koch, W. 1964. Vergleiche von *Bolivinoides*, *Aragonia* und *Tappanina* (Foraminifera) aus Trinidad (Westindien) und Mitteleuropa. *Geol. Jahrb., Hannover*, 83, 31–64.

Belford, D. J. 1960. Upper Cretaceous foraminifera from the Toolonga calcilutite and Gingin Chalk, Western Australia. *Bull. Bureau Min. Res., Geol. and Geoph., Canberra*, 57, 198 pp.

Berggren, W. A. 1972. Cenozoic biostratigraphy and paleobiogeography of the North Atlantic. *Init. Rep. Deep Sea drill. Proj.*, 12, 965–1001.

Berggren, W. A. 1977. North Atlantic Cenozoic foraminifera. In: Swain, F. M., ed. 1977. *Stratigraphic micropaleontology of Atlantic Basin and borderlands. Devel. Palaeontol. Stratigr.*, 6, 389–410.

Berggren, W. A. & Aubert, J. 1975. Paleocene benthonic foraminiferal biostratigraphy, paleobiogeography and paleoecology of Atlantic-Tethyan regions: Midway-type fauna. *Palaeogeogr., Palaeoclimatol., Palaeoecol.*, 18, 73–192

Berggren, W. A. & Miller, K. G. 1989. Cenozoic bathyal and abyssal calcareous benthic foraminiferal zonation. *Micropaleontology*, 35, 308–320.

Bermudez, P. J. 1949. Tertiary smaller foraminifera of the Dominican Republic. *Cushman Lab. foramin. Res., Spec. Publ.*, 18, 322 pp.

Bermudez, P. J. 1962. Foraminiferos de las lutitas de Punta Tolete. *Geos*, 8, 11–123.

Bermudez, P. J. 1963. Foraminiferos del Paleoceno del Departamento de El Peten, Guatemala. *Bol. Soc. geol. Mexicana*, 26, 1–56.

Bermudez, P. J. 1966. Consideraciones sobre los sedimentos del Mioceno medio al Reciente de las costas central y oriental de Venezuela. Primera parte. *Boletin de Geologia, Minist. de Min. e Hidrocarb.*, 7/14, 333–411.

Bermudez, P. J. & Bolli, H. M. 1969. Consideraciones sobre los sedimentos del Mioceno medio al Reciente de las costas central y oriental de Venezuela. Tercera parte: Los foraminiferos planctonicos. *Boletin de Geologia, Minist. de Min. e Hidrocarb.*, 10/20, 137–223.

Bermudez, P. J. & Farias, J. R. 1977. Bioestratigrafia Venezolana. Zonacion del Cenozoico al Reciente basada en el estudio de los foraminiferos planctonicos. *Rev. Esp. Micropaleontol.*, 9, 159–189.

Bermudez, P. J. & Fuenmayor, A. N. 1968. Consideraciones sobre los sedimentos del Mioceno medio al Reciente de las costas central y oriental de Venezuela. Segunda parte: Los foraminiferos bentonicos. *Boletin de Geologia, Minist. de Min. e Hidrocarb.*, 7/14, 413–611.

Bermudez, P. J. & Stainforth, R. M. 1975. Aplicaciones de foraminiferos planctonicos a la bioestratigrafia del Terciario en Venezuela. *Rev. Esp. Micropaleontol.*, 7, 373–389.

Berthelin, G. 1880. Mémoire sur les foraminifères fossiles de l'étage Albien de Moncley (Doubs). *Mém. Soc. géol. France*, ser. 3, 1, 1–84.

Bettenstaedt, F. 1952. Stratigraphisch wichtige Foraminiferen-Arten aus dem Barrême vorwiegend Nordwest-Deutschlands. *Senckenbergiana*, 33, 263–295.

Bettenstaedt, F. & Wicher, C. A. 1955. Stratigraphic correlation of Upper Cretaceous and Lower Cretaceous in the Tethys and Boreal by aid of microfossils (Germany). *Proc. Fourth World. Petr. Congr.*, sect. 1, 493–516.

Bignot, G. 1984. Les foraminifères benthiques n'ont par subi de crise majeure à l'extrème fin du Crétacé. *Bull. Soc. Sci.*, 6, 27–55.

Biju-Duval, B., Moore, J. C., et al. 1984. *Init. Rep. Deep Sea drill. Proj.*, 78A, 1–621.

Blow, W. H. 1959. Age, correlation, and biostratigraphy of the Upper Tocuyo (San Lorenzo) and Pozon formations, Eastern Falcon, Venezuela. *Bull. Am. Paleontol.*, 39 (178), 67–251.

Blow, W. H. & Banner, F. T. 1966. The morphology, taxonomy and biostratigraphy of *Globorotalia barisanensis* LeRoy, *Globorotalia fohsi* Cushman & Ellisor, and related taxa. *Micropaleontology*, 12, 286–303.

Boersma, A. 1977. Eocene to Early Miocene benthic foraminifera, DSDP Leg 39, South Atlantic. *Init. Rep. Deep Sea drill. Proj.*, 39, 643–656.

Boersma, A. 1984. Handbook of common Tertiary *Uvigerina*. *Microclimates Press, Stony Point, New York*, 207 pp.

Bolli, H. M. 1950. The direction of coiling in the evolution of some Globorotaliidae. *Contrib. Cushman Found. foramin. Res.*, 1, 82–89.

Bolli, H. M. 1951. The genus *Globotruncana* in Trinidad, B. W. I. *J. Paleontol.*, 25, 187–199.

Bolli, H. M. 1952. Note on the Cretaceous-Tertiary boundary in Trinidad. *J. Paleontol.*, 26, 669–675.

Bolli, H. M. 1957a. The genera *Praeglobotruncana, Rotalipora, Globotruncana*, and *Abathomphalus* in the Upper Cretaceous of Trinidad, B. W. I. *Bull. U. S. natl. Mus.*, 215, 51–60.

Bolli, H. M. 1957b. The genera *Globigerina* and *Globorotalia* in the Paleocene-Lower Eocene Lizard Springs Formation of Trinidad, B. W. I. *Bull. U. S. natl. Mus.*, 215, 61–81.

Bolli, H. M. 1957c. Planktonic foraminifera from the Oligocene-Miocene Cipero and Lengua formations of Trinidad, B. W. I. *Bull. U. S. natl. Mus.*, 215, 97–123.

Bolli, H. M. 1957d. Planktonic foraminifera from the Eocene Navet and San Fernando formations of Trinidad, B. W. I. *Bull. U.S. natl. Mus.*, 215, 155–172.

Bolli, H. M. 1957e. The foraminiferal genera *Schackoina* Thalmann, emended and *Leupoldina*, n. gen. in the Cretaceous of Trinidad, B. W. I. *Eclog. geol. Helv.*, 50, 271–278.

Bolli, H. M. 1959. Planktonic foraminifera from the Cretaceous of Trinidad, B. W. I. *Bull. Am. Paleontol.*, 39/179, 257–277.

Bolli, H. M. 1966. Zonation of Cretaceous to Pliocene marine sediments based on planktonic foraminifera. *Bol. Informativo, Asoc. Venez. Geol., Min. y Petroleo*, 9, 3–32.

Bolli, H. M. ed. 1975. Monografia micropaleontologica sul Paleocene e l'Eocene di Possagno, Provincia di Treviso, Italia. *Mem. svizzere Paleontol.*, 97, 1–223.

Bolli, H. M. & Bermudez, P. J. 1965. Zonation based on planktonic foraminifera of Middle Miocene to Pliocene warm water sediments. *Bol. Informativo, Asoc. Ven. Geol. Min. y Petroleo*, 8, 121–150.

Bolli, H. M., Loeblich, A. R. jr. & Tappan, H. 1957. Planktonic foraminiferal families Hantkeninidae, Orbulinidae, Globorotaliidae and Globotruncanidae. *Bull. U.S. natl. Mus.*, 215, 3–50.

Bolli, H. M., Saunders, J. B. & Perch-Nielsen, K. eds. 1985. Plankton stratigraphy. *Cambridge University Press*, 1032 pp.

Boltovskoy, E., 1965. Twilight of foraminiferology. *J. Paleontol.*, 39, 383–390.

Boltovskoy, E. 1978. Late Cenozoic benthonic foraminifera of the Ninetyeast Ridge (Indian Ocean). *Mar. Geology*, 26, 139–175.

Brabb, E. E., ed. 1983. Studies in Tertiary stratigraphy of the California Coast Ranges. *Prof. Pap. U. S. geol. Surv.*, 1213, 1–93.

Brady, H. B. 1884. Report on the foraminifera dredged by H.M.S. Challenger during the years 1873–1876. *Rep. Voy. Challenger, Zoology*, 9, 814 pp.

Broennimann, P. 1951. *Guppyella, Alveovalvulina*, and *Discamminoides*, new genera of arenaceous foraminifera from the Miocene of Trinidad, B.W.I. *Contrib. Cushman Found. foramin. Res.*, 2, 97–105.

Broennimann, P. 1953. Arenaceous foraminifera from the Oligo-Miocene of Trinidad. *Contrib. Cushman Found. foramin. Res.*, 4, 87–100.

Brotzen, F. 1936. Foraminiferen aus dem schwedischen untersten Senon von Eriksdal in Schonen. *Sver. geol. Unders.*, C 396, 206 pp.

Brotzen, F. 1948. The Swedish Paleocene and its foraminiferal fauna. *Sver. geol. Unders.*, C 493, 140 pp.

Caldwell, W. G. E., North, B. R., Stelck, C. R. & Wall, J. H. 1978. A foraminiferal zonal scheme for the Cretaceous system in the interior plains of Canada. *Spec. Pap. geol. Assoc. Canada*, 18, 495–575.

Caron, M. 1985. Cretaceous planktic foraminifera. In Bolli *et al.*, ed.: Plankton stratigraphy, *Cambridge University Press*, 17–86.

Carsey, D. O. 1926. Foraminifera of the Cretaceous of Central Texas. *Univ. Texas Bull.*, 2612, 56 pp.

Cati, F. 1964. Una microfauna campaniana dei Monti Berici (Vicenza). *Giorn. Geol., Bologna*, 2, 32, 199–271.

Chamney, T. P. 1978. Albian foraminifera of the Yukon Territory. *Bull. geol. Surv. Canada*, 253, 62 pp.

Charnock, M. A. & Jones R. W. 1990. Agglutinated foraminifera from the Palaeogene of the North Sea. In: Hemleben *et al.*, ed. 1990, *Paleoecology, biostratigraphy, paleoceanography and taxonomy of agglutinated foraminifera. NATO ASI Series*, ser. C, 327, 139–244.

Clark, M. W. & Wright, R. C. 1984. Paleogene abyssal foraminifers from the Cape and Angola Basins, South Atlantic Ocean. *Init. Rep. Deep Sea drill. Proj.*, 73, 459–480.

Colom, G. 1947. Foraminiferos del Cretáceo superior del Sáhara español, recogidos for el profesor M. Alía. *Bol. R. Soc. Esp. Hist. Nat.*, 45, 659–672.

Crittenden, S. 1983. *Osangularia schloenbachi* (Reuss, 1863): an index foraminiferid species from the Middle Albian to Late Aptian of the southern North Sea. *Neues Jahrb. Geol. Paläontol., Abhandl.*, 167, 40–64.

Crittenden, S. & Price, R. J. 1991. The foraminiferid *Osangularia schloenbachi* (Reuss, 1863); the erection of a neotype. *J. Micropalaeontol.*, 9, 253–256.

Cushman, J. A. 1922. The Byram calcareous marl of Mississippi and its Foraminifera. *Prof. Pap. U.S. geol. Surv.*, 129-E, 87–105.

Cushman, J. A. 1926. The foraminifera of the Velasco shale of the Tampico embayment. *Bull. Am. Assoc. Petroleum Geol.*, 10, p. 581–612.

Cushman, J. A. 1929*a*. Some species of *Siphogenerinoides* from the Cretaceous of Venezuela. *Contrib. Cushman Lab. foramin. Res.*, 5, 55–59.

Cushman, J. A. 1929*b*. A late Tertiary fauna of Venezuela and other related regions. *Contrib. Cushman Lab. foramin. Res.*, 5, 77–101.

Cushman, J. A. 1931. Foraminifera of Tennessee. *State of Tennessee, Div. Geol., Bull.*, 41, 114 pp.

Cushman, J. A. 1937*a*. A monograph of the foraminiferal family Verneuilinidae. *Cushman Lab. foramin. Res., Spec. Publ.*, 7, 157 pp.

Cushman, J. A. 1937*b*. A monograph of the foraminiferal family Valvulinidae. *Cushman Lab. foramin. Res., Spec. Publ.*, 8, 210 pp.

Cushman, J. A. 1937*c*. A monograph of the foraminiferal subfamily Virgulininae. *Cushman Lab. foramin. Res., Spec. Publ.*, 9, 228 pp.

Cushman, J. A. 1940. American Upper Cretaceous foraminifera of the genera *Dentalina* and *Nodosaria*. *Contrib. Cushman Lab. foramin. Res.*, 16, 75–96.

Cushman, J. A. 1946*a*. A rich foraminiferal fauna from the Cocoa Sand of Alabama. *Contrib. Cushman Lab. foramin. Res., Spec. Publ.*, 16, 1–40.

Cushman, J. A. 1946*b*. Upper Cretaceous foraminifera of the Gulf Coastal region of the United States and adjacent areas. *Prof. Pap. U. S. geol. Surv.*, 206, 241 pp.

Cushman, J. A. 1947. A foraminiferal fauna from the Santa

Anita Formation of Venezuela. *Contrib. Cushman Lab. foramin. Res.*, 23, 1–18.

Cushman, J. A. 1949. The foraminiferal fauna of the Upper Cretaceous Arkadelphia Marl of Arkansas. *Prof. Pap. U. S. geol. Surv.*, 221-A, 1–10.

Cushman, J. A. 1951. Paleocene foraminifera of the Gulf Coastal region of the United States and adjacent areas. *Prof. Pap. U. S. geol. Surv.*, 232, 75 pp.

Cushman, J. A. & Bermudez, P. J. 1937. Further new species of foraminifera from the Eocene of Cuba. *Contrib. Cushman Lab. foramin. Res.*, 13, 1–29.

Cushman, J. A. & Church, C. C. 1929. Some Upper Cretaceous foraminifera from near Coalinga, California. *Proc. California Acad. Sci.*, ser. 4, 18, 497–530.

Cushman, J. A. & Garrett, J. B. 1938. Three new rotaliform foraminifera from the Lower Oligocene and Upper Eocene of Alabama. *Contrib. Cushman Lab. foramin. Res.*, 14, 62–66.

Cushman, J. A. & Hedberg, H. D. 1930. Notes on some foraminifera from Venezuela and Colombia. *Contrib. Cushman Lab. foramin. Res.*, 6, 64–69.

Cushman, J. A. & Hedberg, H. D. 1941. Upper Cretaceous foraminifera from Santander del Norte, Colombia, S. A. *Contrib. Cushman Lab. foramin. Res.*, 17, 79–100.

Cushman, J. A. & Jarvis, P. W. 1928. Cretaceous foraminifera from Trinidad. *Contrib. Cushman Lab. foramin. Res.*, 4, 85–103.

Cushman, J. A. & Jarvis, P. W. 1929. New foraminifera from Trinidad. *Contrib. Cushman Lab. foramin. Res.*, 5, 6–17.

Cushman, J. A. & Jarvis, P. W. 1930. Miocene foraminifera from Buff Bay, Jamaica. *J. Paleontol.*, 4, 353–368.

Cushman, J. A. & Jarvis, P. M. 1932. Upper Cretaceous foraminifera from Trinidad. *Proc. U.S. natl. Mus.*, 80/14, 1–60.

Cushman, J. A. & Jarvis, P. M. 1934. Some interesting new uniserial foraminifera from Trinidad. *Contrib. Cushman Lab. foramin. Res.*, 10, 71–75.

Cushman, J. A. & Ozawa, Y. 1930. A monograph of the foraminiferal family Polymorphinidae, Recent and fossil. *Proc. U. S. natl. Mus.*, 77/6, 185 pp.

Cushman, J. A. & Parker, F. L. 1935. Some American Cretaceous Buliminas. *Contrib. Cushman Lab. foramin. Res.*, 11, 96–101.

Cushman J. A. & Renz, H. H. 1941. New Oligocene-Miocene foraminifera from Venezuela. *Contrib. Cushman Lab. foramin. Res.*, 17, 1–27.

Cushman, J. A. & Renz, H. H. 1942. Eocene, Midway, foraminifera from Soldado Rock, Trinidad. *Contrib. Cushman Lab. foramin. Res.*, 18, 1–14.

Cushman, J. A. & Renz, H. H. 1946. The foraminiferal fauna of the Lizard Springs Formation of Trinidad, British West Indies. *Cushman Lab. foramin. Res., Spec. Publ.*, 18, 1–48.

Cushman, J. A. & Renz, H. H. 1947*a*. Further notes on the Cretaceous foraminifera of Trinidad. *Contrib. Cushman Lab. foramin. Res.*, 23, 31–51.

Cushman, J. A. & Renz, H. H. 1947*b*. The foraminiferal

fauna of the Oligocene Ste. Croix Formation of Trinidad, B. W. I. *Cushman Lab. foramin Res., Spec. Publ.*, 22, 1–46.

Cushman, J. A. & Renz, H. H. 1948. Eocene foraminifera of the Navet and Hospital Hill formations of Trinidad, B. W. I. *Cushman Lab. foramin. Res.*, Spec. Publ., 24, 1–42.

Cushman, J. A. & Stainforth, R. M. 1945. The foraminifera of the Cipero Marl Formation of Trinidad, British West Indies. *Cushman Lab. foramin Res., Spec. Publ.*, 14, 1–75.

Cushman, J. A. & Stainforth, R. M. 1951. Tertiary foraminifera of coastal Ecuador: part I, Eocene. *J. Paleontol.*, 25, 129–164.

Cushman, J. A. & Todd, R. 1945. Miocene foraminifera from Buff Bay, Jamaica. *Cushman Lab. foramin. Res., Spec. Publ.*, 15, 1–73.

Cushman, J. A. & Todd, R. 1949. Species of the genera *Allomorphina* and *Quadrimorphina*. *Contrib. Cushman Lab. foramin. Res.*, 25, 59–72.

Dailey, D. H. 1970. Some new Cretaceous foraminifera from the Budden Canyon Formation, northwestern Sacramento Valley, California. *Contrib. Cushman Found. foramin. Res.*, 21, 100–111.

Dailey, D. H. 1973. Early Cretaceous foraminifera from the Budden Canyon Formation, northwestern Sacramento Valley, California. *University California Publ. geol. Sci.*, 106, 1–111.

Dailey, D. H. 1983. Late Cretaceous and Paleocene benthic foraminifers from Deep Sea Drilling Project Site 516, Rio Grande Rise, western South Atlantic Ocean. *Init. Rep. Deep Sea drill. Proj.*, 72, 757–782.

Dam, A. ten 1946. Arenaceous foraminifera and Lagenidae from the Neocomian of the Netherlands. *J. Paleontol.*, 20, 570–577.

Dam, A. ten & Magné, J. 1948. Les espèces du genre de foraminifères *Globorotalites* Brotzen. *Rev. Inst. Français Pétrole*, 1948/8, 222–228.

Dam, A. ten & Sigal, J. 1950. Some new species of foraminifera from the Dano-Montian of Algeria. *Contrib. Cushman Found. foramin. Res.*, 1, 31–37.

Diaz de Gamero, M. L. 1977a. Estratigrafia y micropaleontologia del Oligoceno y Mioceno inferior del centro de la Cuenca de Falcon, Venezuela. *Geos*, 22, 2–54.

Diaz de Gamero, M. L. 1977b. Revision de los edades de las unidades lithostratigraficas en Falcon Central en base a su contenido de foraminiferos planctonicos. *Mem. V Congr. Geol. Venez.*, 1, 81–86.

Diaz de Gamero, M. L. 1985a. Micropaleontologia de la Formacion Agua Salada, Falcon nororiental. *Mem. VI Congr. Geol. Venez.*, 1, 384–453.

Diaz de Gamero, M. L. 1985b. Estratigrafia de Falcon nororiental. *Mem. VI Congr. Geol. Venez.*, 1, 464–502.

Diaz de Gamero, M. L. 1986. Geological evolution of the northern Agua Salada Subbasin, East Falcon, Venezuela. *Trans. 1st. Geol. Conf. Geol. Soc. Trinidad & Tobago*, 1985, 288–301.

Diaz de Gamero, M. L. 1989. El Mioceno temprano y medio de Falcon septentrional (The Early and Middle Miocene of northern Falcon). *Geos*, 29, 25–35.

D'Orbigny, A. 1826. Tableau méthodique de la classe des Céphalopodes. *Ann. Sci. Nat., Paris*, 7, 245–314.

D'Orbigny, A. 1839. Foraminifères. In: Ramon de la Sagra, Histoire physique, politique et naturelle de l'île de Cuba. *Paris, Arthus Bertrand*, 224 pp.,

D'Orbigny, A. 1840. Mémoire sur les foraminifères de la Craie Blanche du Bassin de Paris. *Mém. Soc. géol. France*, 4, 1–51.

D'Orbigny, A. 1846. Foraminifères fossiles du Bassin Tertiaire de Vienne. *Paris:Gide et Comp., Libraires-Editeurs*, 312 pp.

Douglas, R. G. 1973. Benthonic foraminiferal biostratigraphy in the central North Pacific, Leg 17, Deep Sea Drilling Project. *Init. Rep. Deep Sea drill. Proj.*, 17, 607–671.

Douglas, R. G. & Woodruff, F. 1981. Deep-sea benthic foraminifera. In: Emiliani, C., ed., The oceanic lithosphere. *The Sea (New York: John Wiley & Sons)*, 7, 1233–1327.

Earland, A. 1933. Foraminifera. Part II. South Georgia. *Discovery Rep.*, 7, 27–138.

Falcon, R. A. 1989a. Estudio bioestratigrafico preliminar mediante foraminiferas en el Miembro Garcia de la Formacion El Cantil, Cretacico Inferior de Venezuela Oriental. *Geos*, 29, 36–47.

Falcon, R. A. 1989b. Revision y redefinicion del termino "Miembro Garcia" como "Formacion Garcia", Cretacico Inferior de Venezuela Oriental. *Geos*, 29, 48–58.

Finlay, H. J. 1940. New Zealand foraminifera: key species in stratigraphy no. 4. *Trans. Royal Soc. New Zealand*, 69, 448–472.

Fornasini, C. 1906. Illustrazione di specie orbignyane di "Rotalidi" istitute nel 1826. *Mem. R. Accad. Sci. Ist. Bologna, Sci. Nat.*, ser. 6, 3, 61–70.

Franke, A. 1928. Die Foraminiferen der oberen Kreide Nord- und Mitteldeutschlands. *Abhandl. Preuss. geol. Landesanst.*, n. Folge, 111, 207 pp.

Franklin, E. S. 1944. Microfauna from the Carapita Formation of Venezuela. *J. Paleontol.*, 18, 301–319.

Frizzell, D. L. 1943. Upper Cretaceous foraminifera from northwestern Peru. *J. Paleontol.*, 17, 331–353.

Frizzell, D. L. 1954. Handbook of Cretaceous foraminifera of Texas. *Rep. Investig. Bureau Econ. Geol. University Texas*, 22, 232 pp.

Fuenmayor, A. N. 1989. Manual de foraminiferos de la cuenca de Maracaibo. Maracaibo: *Maraven*, 191 pp.

Furrer, M. A. 1972. Fossil Tintinnids in Venezuela. *Trans. 6th Carib. Geol. Conf., Isla de Margarita, Venezuela*, 1971, 451–454.

Galea-Alvarez, F. A. 1985. Biostratigraphy and depositional environment of the Upper Cretaceous-Eocene Santa Anita Group (North Eastern Venezuela). *Thesis, Amsterdam*, 115 pp.

Galloway, J. J. & Heminway, C. E. 1941. The Tertiary foraminifera of Puerto Rico. *New York Acad. Sci., Surv. of Puerto Rico and the Virgin Islands*, 3/4, 275–491.

Gandolfi, R. 1942. Ricerche micropaleontologiche e stratigrafiche sulla Scaglia e sul Flysch cretacici dei dintorni di Balerna (Canton Ticino). *Rivista Ital. Paleontol.*, 48, mem. 4, 160 pp.

Gawor-Biedowa, E. 1972. The Albian, Cenomanian and Turonian foraminifers of Poland and their stratigraphic importance. *Acta palaeontol. Pol.*, 17, 1–155.

Gawor-Biedowa, E. 1980. Turonian and Coniacian foraminifera from the Nysa Trough, Sudetes, Poland. *Acta paleontol. Pol.*, 25, 3–54.

Gawor-Biedowa, E. 1987. New benthic foraminfers from the Late Cretaceous of Poland. *Acta palaeontol. Pol.*, 33, 49–71.

Glaessner, M. F. 1937. Studien ueber Foraminiferen aus der Kreide und dem Tertiaer des Kaukasus. 1. Die Foraminiferen der aeltesten Tertiaerschichten des Nordwestkaukasus. *Probl. Paleontol., Moscow*, 2–3, 349–410.

Glaessner, M. F. 1945. Principles of micropalaeontology. *Melbourne University Press*, 296 pp.

Goel, R. K. 1962. Contribution à l'étude des foraminifères du Crétacé supérieur de la Basse-Seine. *Thesis, University Bordeaux*, 257 pp.

Gonzales de Juana, C., Iturralde de Arozena, J. A. & Picard Chatillat, X. 1980. Geologia de Venezuela y sus Cuencas Petroliferas. *Ediciones Foninves.* 2 vols., 1031 pp.

Gradstein, F. M. 1978. Biostratigraphy of the Lower Cretaceous Blake Nose and Blake-Bahama basin foraminifers. DSDP Leg 44, Western North Atlantic Ocean. *Init. Rep. Deep Sea drill. Proj.*, 44, 663–701.

Gradstein F. M. & Berggren, W. A. 1981. Flysch-type agglutinated foraminifera and the Maastrichtian to Paleogene history of the Labrador and North Seas. *Mar. Micropaleontol.*, 6, 211–268.

Gradstein, F. M. & Kaminski, M. A. 1989. Taxonomy and biostratigraphy of new and emended species of Cenozoic deep-water agglutinated foraminifera from the Labrador and North Seas. *Micropaleontology*, 35, 72–92.

Graham, J. J. & Church, C. C. 1963. Campanian foraminifera from the Stanford University campus, California. *Stanford University Publ., Geol. Sci.*, 8, 1–90.

Graham, J. J. & Classen, W. J. 1955. A Lower Eocene foraminiferal faunule from the Woodside area, San Mateo County, California. *Contrib. Cushman Found. foramin. Res.*, 6, 1–38.

Gravell, D. W. 1933. Tertiary larger foraminifera of Venezuela. *Smithsonian Misc. Coll.*, 89/11, 1–44.

Grimsdale, T. F. 1947. Upper Cretaceous foraminifera: a criticism. *J. Paleontol.*, 21, 586–587.

Grün, W., Lauer, G., Niedermayr, G. & Schnabel, W. 1964. Die Kreide-Tertiär-Grenze im Wienerwaldflysch bei Hochstrass (Niederösterreich). *Vienna:Verhandl. geol. Bundesanst.*, 1964, 226–283.

Grzybowski, J. 1896. Otwornice czerwonych ilow z Wadowice. *Rozpr. Akad. Um. Krakow*, 10, 261–308.

Grzybowski, J. 1898. Otwornice pokladow naftonosnych okolicy Krosna. *Rozpr. Akad. Um. Krakow*, 33, 257–305.

Grzybowski, J. 1901. Otwornice warstw inoceramowych okolicy Gorlic. *Rozpr. Akad. Um. Krakow*, 41, 201–286.

Guillaume, H. A. 1961. Lower Cretaceous stratigraphy of the Serrania del Interior. *Compania Shell de Venezuela*, Report No. 1680, 72 pp.

Guillaume, H. A., Bolli, H. M. & Beckmann, J. P. 1972. Estratigrafia del Cretaceo Inferior en la Serrania del Interior, Oriente de Venezuela. *Mem. IV Congr. Geol. Venez., Bol. Geol., Publ. Especial*, 5, 1619–1658.

Guppy, R. J. L. 1863. On the occurrence of foraminifera in the Tertiary beds at San Fernando, Trinidad. *The Geologist*, 159.

Guppy, R. J. L. 1873. On foraminifera from the Tertiaries of San Fernando, Trinidad. *Geol. Mag.*, 10, 362–363.

Guppy, R. J. L. 1894. On some forminifera from the microzoic deposits of Trinidad, West Indies. *Proc. Zool. Soc. London*, 647–653.

Haas, M. W. & Hubman, R. G. 1937. Notes on the stratigraphy of the Bolivar coastal fields, Maracaibo Basin, Venezuela. *Bol. Geol.*, 1, 115–155.

Hagn, H. 1953. Die Foraminiferen der Pinswanger Schichten (Unteres Obercampan). *Palaeontographica*, A 104, 1–119.

Haig, D. W. 1979. Global distribution patterns for mid-Cretaceous foraminiferids. *J. foramin. Res.*, 9, 29–40.

Haig, D. W. 1980. Early Cretaceous textulariine foraminiferids from Queensland. *Palaeontographica*, A 170, 87–138.

Haig, D. W. 1982. Early Cretaceous Milioline and Rotaliine benthic foraminiferids from Queensland. *Palaeontographica*, A 177, 1–88.

Hanzlikova, E. 1972. Carpathian Upper Cretaceous Foraminiferida of Moravia (Turonian-Maastrichtian). *Rozpr. Ustredn. Ustavu Geol.*, 39, 160 pp.

Haque, A. F. M. M. 1956. The smaller foraminifera of the Ranikot and the Laki of the Nammal Gorge, Salt Range. *Mem. geol. Surv. Pakistan, Palaeontol. Pakist.*, 1, 300 pp.

Hay, W. W. 1959. A study of the Velasco Formation of northeastern Mexico. *Unpublished thesis, Stanford University, California*, 388 pp.

Haynes, J. 1954. Taxonomic position of some British Paleocene Buliminidae. *Contrib. Cushman Found. foramin. Res.*, 5, 185–191.

Hecht, F. E. 1938. Standard-Gliederung der nordwestdeutschen Unterkreide nach Foraminiferen. *Abh. Senckenberg. naturf. Ges.*, 443, 42 pp.

Hedberg, H. D. 1937a. Foraminifera of the Middle Tertiary Carapita Formation of northeastern Venezeula. *J. Paleontol.*, 8, 661–697.

Hedberg, H. D. 1937b. Stratigraphy of the Rio Querecual section of northeastern Venezuela. *Bull. Geol. Soc. Am.*, 48, 1971–2024.

Hedberg, H. D. 1950. Geology of the eastern Venezuela Basin (Anzoategui-Monagas-Sucre-eastern Guarico portion). *Bull. Geol. Soc. Am.*, 61, 1173–1216.

Hedberg, H. D. & Pyre, A. 1944. Stratigraphy of northeastern Anzoategui, Venezuela. *Bull. Am. Assoc. Petrol. Geol.*, 20, 1–28.

Hemleben, Ch. & Troester, J. 1984. Campanian-Maastrichtian deep-water foraminifers from Hole 543A, Deep Sea Drilling Project. *Init. Rep. Deep Sea drill. Proj.*, 78, 509–532.

Hercogova, J. 1987. New findings of arenaceous foraminifera in the Cenomanian of the Bohemian Massif. *Sbor. geol. Ved, Paleontol.*, 28, 179–227.

Hillebrandt, A. von 1962. Das Paleozän und seine Foraminiferenfauna im Becken von Reichenhall und Salzburg. *Abh. Bayer. Akad. Wiss., math.-naturwiss. Kl.*, n. Folge, 108, 1–182.

Hiltermann, H. 1974. *Rzehakina epigona* und Unterarten dieser Foraminifere. *Paläontol. Zeitschr.*, 48, 36–56.

Hoffmeister, W. S. 1938. Aspecto e division en zonas de la fauna de molluscos en las formaciones La Rosa y Lagunillas, campos costaneros de Bolivar, Venezuela. *Bol. Geol. y Min., Caracas*, 2, 103–122.

Huber, B. T. 1988. Upper Campanian-Paleocene foraminifera from the James Ross Island region, Antarctic Peninsula. *Mem. geol. Soc. Am.*, 169, 163–252.

Hulsbos, R. E. 1987. Eocene benthic foraminifers from the upper continental rise off New Jersey, Deep Sea Drilling Project Site 605. *Init. Rep. Deep Sea drill. Proj.*, 93, 525–538.

Hutchison, A. G. 1938. A note upon the Cretaceous of Trinidad and Venezuela. *Bol. Geol. y Min.*, 2, 226–235.

Imlay, R. W. 1954. Barremian ammonites from Trinidad, British West Indies. *J. Paleontol.*, 28, 662–667.

Israelsky, M. C. 1951. Foraminifera of the Lodo Formation, Central California. General introduction and Part 1, Arenaceous Foraminifera. *Prof. Pap. U. S. geol. Surv.*, 240-A, 1–29.

Israelsky, M. C. 1955. Foraminifera of the Lodo Formation, Central California. Part 2, Calcareous Foraminifera (Miliolidae and Lagenidae, part). *Prof. Pap. U. S. geol. Surv.*, 240-B, 31–79.

Jarvis, P. W. 1929. Some notes on Cretaceous occurrences at Lizard Springs, Trinidad. *J. Inst. Petrol. Technol. London*, 15, 440–442.

Jenkins, D. G. & Murray, J. W. 1989. Stratigraphical atlas of fossil foraminifera. 2nd edition. *New York: John Wiley & Sons*, 593 pp.

Jennings, Ph. H. 1936. A microfauna from the Monmouth and basal Rancocas Groups of New Jersey. *Bull. Am. Paleontol.*, 23 (78), 3–77.

Jukes-Browne, A. J. & Harrison, J. B. 1892. The geology of Barbados. Part. 2. The Oceanic deposits. *Quart. J. geol. Soc. London*, 48, 170–226.

Kaever, M. 1961. Morphologie, Taxionomie und Biostratigraphie von *Globorotalites* und *Conorotalites* (Kreide-Foram.). *Geol. Jahrb., Hannover*, 78, 387–438.

Kaiho, K. 1988. Uppermost Cretaceous to Paleogene bathyal benthic foraminiferal biostratigraphy of Japan and New Zealand: Latest Paleocene-Middle Eocene benthic foraminiferal species turnover. *Rev. Paléobiol., Vol. Spéc.*, 2 (Benthos '86), 553–559.

Kaminski, M. A., Gradstein, F. M., Berggren, W. A., Geroch, S. & Beckmann, J. P. 1988. Flysch-type agglutinated foraminferal assemblages from Trinidad: Taxonomy, stratigraphy and paleobathymetry. *Abhandl. geol. Bundesanst., Vienna*, 41, 155–227.

Kellough, G. R. 1965. Paleoecology of the Foraminiferida of the Wills Point Formation (Midway Group) in Northeast Texas. *Trans. Gulf Coast Assoc. geol. Soc.*, 15, 73–153.

Key, C. E. 1960. Estratigrafia del subsuelo de Alturitas. *Mem. III Congr. Geol. Venez., Caracas 1959*, 3, 511–545.

Klasz, I. de & Hinte, J. E. van 1977. Remarques sur le genre *Gabonella* (Foraminifères) et description de deux nouvelles espèces du Crétacé Supérieur du Gabon. *Actes 6e Coll. Afr. Micropaléontol., Ann. Mines Géol., Tunis*, 28, 481–497.

Klasz, I. de, Klasz, S. de, Colin, J.-P., Du Chène, R. J., Ausseil-Badie, J., Bellion, Y. & Peypouquet, J.-P. 1987a. Apports de la micropaléontologie (foraminifères, ostracodes, dinoflagellés) à la connaissance stratigraphique et paléoécologique de la Formation des Madeleines (Danien du Sénégal). *Cahiers Micropaléontol.*, n. sér., 2, 5–27.

Klasz, I. de, Klasz, S. de & Ausseil-Badie, J. 1987b. Etude systématique des foraminifères du Danien de la Formation des Madeleines de Dakar (Sénégal). *Cahiers Micropaléontol.*, n. sér., 2, 29–76.

Kleinpell, R. M. 1938. Miocene stratigraphy of California. *Am. Assoc. Petrol. Geol.*, 450 pp.

Koch, W. 1968. Zur Mikropaläontologie und Biostratigraphie der Oberkreide und des Alttertiärs von Jordanien. I. Oberkreide. *Geol. Jahrb., Hannover*, 85, 627–668.

Koutsoukos, E. A. & Merrick, K. A. 1986. Foraminiferal paleoenvironments from the Barremian to Maastrichtian of Trinidad, West Indies. *Trans. First Conf. geol. Soc. Trinidad & Tobago (1985)*, 85–101.

Kristan-Tollmann, E. & Tollmann, A. 1976. Neue Neoflabellinen (Foraminifera) aus dem Senon der Gamser Gosau, Oesterreich. *Sitzungsber. Oesterr. Akad. Wiss, math.-naturwiss. Kl., Abt. I*, 185, 307–321.

Kugler, H. G. 1936. Summary digest of geology of Trinidad. *Bull. Am. Assoc. Petrol. Geol.*, 20, 1439–1453.

Kugler, H. G. 1950. Resumen de la historia geologica de Trinidad. *Bol. Asoc. Venez. Geol. Min. y Petrol.*, 2, 48–78.

Kugler, H. G. 1953. Jurassic to Recent sedimentary environments in Trinidad. *Bull. Verein. Schweiz. Petrol. Geol. und Ing.*, 20(59), 27–60.

Kugler, H. G. 1956. Trinidad. *Paris:Lexique Strat. Internat., Centre National Rech. Scientifique*, V, fasc 2b, 39–116.

Kugler, H. G. & Bolli, H. M. 1967. Cretaceous biostratigraphy in Trinidad. *Bol. Informativo, Asoc. Venez. Geol., Min. y Petrol.*, 10, 209–236.

Kugler, H. G. & Caudri, C. M. B. 1975. Geology and paleontology of Soldado Rock, Trinidad (West Indies). Part 1: geology and biostratigraphy. *Eclog. geol. Helv.*, 68, 365–430.

Kuhnt, W. 1990. Agglutinated foraminifera of western Mediterranean Upper Cretaceous pelagic limestones

(Umbrian Apennines, Italy, and Betic Cordillera, southern Spain). *Micropaleontology*, 36, 297–330.

Kuhnt, W. & Moullade, M. 1991. Quantitative analysis of Upper Cretaceous abyssal agglutinated foraminiferal distribution in the North Atlantic – paleoceanographic implications. *Rev. Micropaléontol.*, 34, 313–349.

Kuyl, O. S., Muller, J. & Waterbolk, H. T. 1955. The application of palynology to oil geology with reference to western Venezuela. *Geol. en Mijnb.*, n. ser., 17/3, 49–76.

Lalicker, C. G. 1935. New Cretaceous Textulariidae. *Contrib. Cushman Lab. foramin. Res.*, 11, 1–13.

Lamb, J. L. 1964a. The geology and paleontology of the Rio Aragua surface section, Serrania del Interior, Estado Monagas, Venezuela. *Bol. Informativo, Asoc. Venez, Geol. Min. y Petrol.*, 7/4, 111–123.

Lamb, J. L. 1964b. The stratigraphic occurrences and relationship of some mid-Tertiary Uvigerinas and Siphogenerinas. *Micropaleontology*, 10, 457–476.

Lamb, J. L. & Sulek, J. A. 1965. Definition of the Cachipo Member of the Carapita Formation. *Bol. Informativo, Asoc. Venez. Geol. Min. y Petrol.*, 8/4, 111–114.

Lamb, J. L. & Sulek, J. A. 1968. Miocene turbidites in the Carapita Formation of eastern Venezuela. *Trans. 4th Caribbean Geol. Conf., Trinidad 1965*, 111–119.

Lexico Estratigrafico de Venezuela. 1970. 2nd edition. *Bul. Geol., Publ. Especial*, 4, 756 pp.

Liddle, R. A. 1928. The geology of Venezuela and Trinidad. *Fort Worth (Texas)*, 552 pp.

Liddle, R. A. 1946. The geology of Venezuela and Trinidad. 2nd. edition, *Ithaca (New York)*, 890 pp.

Liszka, S. & Liszkova J. 1981. Revision of J. Grzybowski's paper (1896) "Foraminifera of the Red Clays from Wadowice". *Rocznik Polsk. Towar. Geol.*, 51, 153–208.

Loeblich, A. R. jr. & Tappan, H. 1952. The foraminiferal genus *Triplasia* Reuss, 1854. *Smithsonian misc. Coll.*, 117, 1–61.

Loeblich, A. R. jr. & Tappan, H. 1964. Protista 2. Sarcodina; chiefly "Thecamoebians" and Foraminiferida. *Treatise Invertebr. Paleontol.*, Part C, 900 pp.

Loeblich, A. R. jr. & Tappan, H. 1982. A revision of mid-Cretaceous textularian foraminifers from Texas. *J. Micropalaeontol.*, 1, 55–69.

Loeblich, A. R. jr. & Tappan, H. 1987. Foraminiferal genera and their classification. *New York: Van Nostrand Reinhold*, 2 vols., 970 & 212 pp., 847 pls.

Macbeth, J. I. & Schmidt, R. A. M. 1973. Upper Cretaceous foraminifera from Ocean Point, northern Alaska. *J. Paleontol.*, 47, 1047–1061.

Mallory, V. S. 1959. Lower Tertiary biostratigraphy of the California Coast Ranges. *Am. Assoc. Petrol. Geol.*, 416 pp.

Marie, P. 1941. Les foraminifères de la Craie à Belemnitella mucronata du Bassin de Paris. *Mém. Mus. National Hist. Nat.*, n. sér, 12, 296 pp.

Marsson, Th. 1878. Die Foraminiferen der Weissen Schreibkreide der Insel Rügen. *Mitt. naturwiss. Verein Neu-Vorpommern u. Rügen*, 10, 115–196.

Martin, L. 1964. Upper Cretaceous and Lower Tertiary foraminifera from Fresno County, California. *Jahrb. geol. Bundesanst., Vienna, Sonderbd.* 9, 128 pp.

Martinez, J. I. R. 1989. Foraminiferal biostratigraphy and paleoenvironments of the Maastrichtian Colon mudstones of northern South America. *Micropaleontology*, 35, 97–113.

Martini, E. 1971. Standard Tertiary and Quaternary calcareous nannoplankton zonation. In: Farinacci ed., *Proc. IId Planctonic Conf., Rome 1970*, 2, 739–785.

Maync, W. 1950. The foraminiferal genus *Choffatella* Schlumberger in the Lower Cretaceous (Urgonian) of the Caribbean Region (Venezuela, Cuba, Mexico, and Florida). *Eclog. geol. Helv.*, 42, 529–547.

Maync, W. 1953. *Pseudocyclammina hedbergi* n. sp. from the Urgo-Aptian of Venezuela. *Contrib. Cushman Found. foramin. Res.*, 4, 101–103.

Maync, W. 1955a. *Dictyoconus walnutensis* (Carsey) in the Middle Albian Guacharo limestone of eastern Venezuela. *Contrib. Cushman Found. foramin. Res.*, 6, 85–93.

Maync, W. 1955b. *Coscinolina sunnilandensis* n. sp., a Lower Cretaceous (Urgo-Aptian) species. *Contrib. Cushman Found. foramin. Res.*, 6, 105–111.

Maync, W. 1956. On the age of *Choffatella*-bearing beds in Venezuela. *Micropaleontology*, 2, 92.

McGugan, A. 1964. Upper Cretaceous zone foraminifera, Vancouver Island, British Columbia, Canada. *J. Paleontol.*, 38, 933–951.

McNeil, D. H. & Caldwell, W. G. E. 1981. Cretaceous rocks and their foraminifera in the Manitoba Escarpment. *Spec. Publ. geol. Assoc. Canada*, 21, 439 pp.

McNulty, C. L. 1984. Cretaceous foraminifers of Hole 530A, Leg 75, Deep Sea Drilling Project. *Init. Rep. Deep Sea drill. Proj.*, 75, 547–564.

Medizza, F. 1967. I generi *Bolivinoides*, *Aragonia* e *Neoflabellina* (Foraminifera) nelle formazioni Cretaceo-Eoceniche del Veneto. *Mem. Ist. Geol. Min. Univ. Padova*, 26, 1–44.

Mello, J. F. 1969. Foraminifera and stratigraphy of the upper part of the Pierre Shale and lower part of the Fox Hills Sandstone (Cretaceous), North-Central South Dakota. *Prof. Pap. U. S. geol. Surv.*, 611, 121 pp.

Miller, K. G. & Katz, M. E. 1987. Eocene benthic foraminiferal biofacies of the New Jersey transect. *Init. Rep. Deep Sea drill. Proj.*, 95, 267–298.

Moore, M. C., Mascle, A., *et al.* 1990. *Proc. Ocean. drill. Program, Sci. Res.*, 110, 1–448.

Morkhoven, F. P. C. M. van 1981. Cosmopolitan Tertiary bathyal benthic foraminifera. *Trans. Gulf Coast Assoc. geol. Soc.*, Suppl. 31, 445.

Morkhoven, F. P. C. M. van, Berggren, W. A. & Edwards, A. S. 1986. Cenozoic cosmopolitan deep-water benthic foraminifera. *Bull. Centre Rech. Expl.-Prod. Elf-Aquitaine*, Mem. 11, 421 pp.

Morrow, A. L. 1934. Foraminifera and ostracoda from the Upper Cretaceous of Kansas. *J. Paleontol.*, 8, 186–205.

Moullade, M. 1984. Intérêt des petits foraminifères benthiques «profonds» pour la biostratigraphie et l'analyse des paléoenvironnements océaniques mésozoïques. In: Oertli, H. J., ed., Benthos '83. *2nd Int. Symp. Benthic Foramin., Pau and Bordeaux*, 429–464.

Moullade, M., Kuhnt, W. & Thurow, J. 1988. Agglutinated benthic foraminifers from Upper Cretaceous variegated clays of the North Atlantic Ocean (DSDP Leg 93 and ODP Leg 103). *Proc. Ocean drill. Prog., Sci. Results*, 103, 349–377.

Myatlyuk, E. V. 1953. Iskopaemye Foraminifery SSSR. Spirillinidy, Rotaliidy, Epistominidy i Asterigerinidy. *Trudy VNIGRI, Leningrad*, 71, 274 pp.

Myatlyuk, E. V. 1970. Foraminifery flischovych otlozhenii Vostochnych Karpat. *Trudy VNIGRI, Leningrad*, 282, 360 pp.

Natland, M. L., Gonzalez, E. P., Cañon, A. & Ernst, M. 1974. A system of stages for correlation of Magallanes Basin sediments. *Mem. Geol. Soc. Am.*, 139, 126 pp.

Nogan, D. S. 1964. Foraminifera, stratigraphy, and paleoecology of the Aquia formation of Maryland and Virginia. *Cushman Found. foramin. Res., Spec. Publ.*, 7, 1–50.

Noth, R. 1951. Foraminiferen aus Unter- und Oberkreide des Oesterreichischen Anteils an Flysch, Helvetikum und Vorlandvorkommen. *Jahrb. geol. Bundesanst., Vienna, Sonderbd.* 3, 91 pp.

Nuttall, W. L. F. 1928. Tertiary foraminifera from the Naparima region of Trinidad (British West Indies). *Quart. J. Geol. Soc.*, 84, 57–116.

Nuttall, W. L. F. 1930. Eocene foraminifera from Mexico. *J. Paleontol.*, 4, 271–293.

Nuttall, W. L. F. 1932. Lower Oligocene Foraminifera from Mexico. *J. Paleontol.*, 6, 3–35.

Nuttall, W. L. F. 1935. Upper Eocene foraminifera from Venezuela. *J. Paleontol.*, 9, 121–131.

Nyong, E. E. & Olsson R. K. 1983. A paleoslope model of Campanian to Lower Maastrichtian foraminifera in the North American Basin and adjacent continental margin. *Mar. Micropaleontol.*, 8, 437–477.

Odehnal, M. A. & Falcon, R. A. 1989. La Zona de *Schackoina cabri* en Venezuela. *Mem. VII Congr. Geol. Venez., Barquisimeto*, 2, 524–549.

Olsson, R. K. 1960. Foraminifera of latest Cretaceous and earliest Tertiary age in the New Jersey coastal plain. *J. Paleontol.*, 34, 1–58.

Olsson, R. K. 1977. Mesozoic foraminifera – Western Atlantic. In: Swain, F. M., ed. 1977. *Stratigraphic micropaleontology of Atlantic Basin and borderlands. Devel. Palaeontol. Stratigr.*, 6, 205–230.

Papp, A. & Schmid, M. E. 1985. Die fossilen Foraminiferen des tertiären Beckens von Wien. Revision der Monographie von Alcide d'Orbigny (1846). *Abhandl. geol. Bundesanst., Vienna*, 37, 1–311.

Parker, W. K., Jones, T. R. & Brady, H. B. 1871. The species founded upon the figures in Soldani's "Testaceographia ac Zoophytographia". *Ann. Mag. Nat. Hist.*, ser. 4, 8, 145–149, 238–266.

Petri, S. 1962. Foraminiferos Cretaceos de Sergipe. *Bol. Fac. Filos. Ciencias Letras Univ. Sao Paulo*, 265 (*Geologia*, 20), 1–140.

Pitelli Viapiana, R. 1989. Eocene stratigraphical studies, Maracaibo Basin, northwestern Venezuela. *Trans. 12th Caribbean Geol. Conf. St. Croix, Virgin Islands*, 485–494.

Plummer, H. J. 1926. Foraminifera of the Midway Formation in Texas. *University of Texas Bull.*, 2644, 1–206.

Plummer, H. J. 1931. Some Cretaceous foraminifera in Texas. *University of Texas Bull.*, 3101, 109–239.

Pozaryska, K. 1957. Lagenidae du Crétacé supérieur de Pologne. *Palaeontol. Polon.*, 8, 1–190.

Premoli Silva, I. & Bolli, H. M. 1973. Late Cretaceous to Eocene planktonic foraminifera and stratigraphy of Leg 15 sites in the Caribbean Sea. *Init. Rep. Deep Sea drill. Proj.*, 15, 499–547.

Proto Decima, F. & Bolli, H. M. 1978. Southeast Atlantic DSDP Leg 40 Paleogene benthic foraminifers. *Init. Rep. Deep Sea drill. Proj.*, 40, 783–809.

Raadshoven, van, B. 1951. On some Paleocene and Eocene larger foraminifera of western Venezuela. *Proc 3rd World Petrol. Congr., The Hague*, sec. 1, 476–498.

Renz, H. H. 1942. Stratigraphy of northern South America, Trinidad, and Barbados. *Proc. Eighth Amer. Sci. Congr., Washington D. C., IV, Geol. Sci.*, 513–571.

Renz, H. H., 1948. Stratigraphy and fauna of the Agua Salada Group, State of Falcon, Venezuela. *Geol. Soc. Am. Mem.*, 32, 219 pp.

Renz, H. H. 1950. A new name for an Eocene foraminifer from Trinidad, British West Indies. *Contrib. Cushman Found. foramin. Res.*, 1, 45–46.

Renz, O. 1959. Estratigrafia del Cretaceo en Venezuela occidental. *Bol. Geol., Caracas*, 5/10, 3–48.

Renz, O. 1982. The Cretaceous ammonites of Venezuela. *Birkhäuser, Basel*: 132 pp.

Reuss, A. E. 1845. Die Versteinerungen der böhmischen Kreideformation, Erste Abtheilung. *Stuttgart: Schweizerbart*, 1–57.

Reuss, A. E. 1846. Die Versteinerungen der böhmischen Kreideformation, Zweite Abtheilung. *Stuttgart: Schweizerbart*, 1–148.

Reuss, A. E. 1851. Die Foraminiferen und Entomostraceen des Kreidemergels von Lemberg. *Haidinger's naturwiss. Abhandl., Vienna*, 4, 15–52.

Reuss, A. E. 1860. Die Foraminiferen der westphälischen Kreideformation. *Sitzungsber. K. Akad. Wiss, math.-naturwiss. Kl., Vienna*, 40, 147–238.

Reuss, A. E. 1860. Die Foraminiferen der westphälischen Kreideformation. *Sitzungsber. K. Akad. Wiss, math.-naturwiss. Kl., Vienna*, 40, 147–238.

Reuss, A. E. 1863. Die Foraminiferen des norddeutschen Hils und Gault. *Sitzungsber. K. Akad. Wiss, math.-naturwiss. Kl., Vienna*, 46 (1862), 5–100.

Roemer, F. A. 1841. Die Versteinerungen des norddeutschen Kreidegebirges. *Hahn'sche Hofbuchhandlung*, 1–45.

Rod, E. & Maync, W. 1954. Revision of Lower Cretaceous stratigraphy of Venezuela. *Bull. Am. Assoc. Petrol. Geol.*, 38, 123–211.

Saint-Marc, P. 1986. Qualitative and quantitative analysis

of benthic foraminifers in Paleocene deep-sea sediments of the Sierra Leone Rise, central Atlantic. *J. foramin. Res.*, 16, 244–253.

Saint-Marc, P. 1987. Biostratigraphic and paleoenvironmental study of Paleocene benthic and planktonic foraminifers, Site 605, Deep Sea Drilling Project Leg 93. *Init. Rep. Deep Sea drill. Proj.*, 93, 539–547.

Salvador, A. & Rosales, H. 1960. Guia de la Excursion A-3, Jusepin – Cumana. *Mem. III Congr. Geol. Venez.*, 1, 63–74.

Sandidge, J. R. 1932. Foraminifera from the Ripley formation of western Alabama. *J. Paleontol.*, 6, 265–278.

Sandulescu, J. 1973. Etude micropaléontologique et stratigraphique du flysch du Crétacé supérieur – Paleocène de la région de Bretcu – Comandau (secteur interne méridional de la nappe de Tarcau – Carpates orientales). *Mem. Inst. Géol., Bucarest*, 17, 1–52.

Sastry, M. V. A. & Sastri, V. V. 1966. Foraminifera from the Utatur stage of the Cretaceous formations of Trichinopoly district, Madras. *Rec. geol. Survey India*, 94, 277–296.

Saunders, J. B., Bernoulli, D., Müller-Merz, E., Oberhänsli, H., Perch-Nielsen, K., Riedel, W. R., Sanfilippo, A. & Torrini, R. jr. 1984. Stratigraphy of the late Middle Eocene to Early Oligocene in the Bath Cliff section, Barbados, West Indies. *Micropaleontology*, 30, 390–425.

Saunders, J. B. & Bolli, H. M. 1985. Trinidad's contribution to world biostratigraphy. *Trans. Fourth Latin Am. geol. Congr., Trinidad and Tobago (1979)*, 2, 781–795.

Saunders, J. B. & Cordey, W. G. 1968. The biostratigraphy of the Oceanic Formation in the Bath Cliff section, Barbados. *Trans. 4th Carib. Geol. Conf., Trinidad*, 179–181.

Scheibnerova, V. 1974. Aptian-Albian benthonic foraminifera from DSDP Leg 27, Sites 259, 260, and 263, Eastern Indian Ocean. *Init. Rep. Deep Sea drill. Proj.*, 27, 697–741.

Schnitker, D. 1979. Cenozoic deep water benthic foraminifers, Bay of Biscay. *Init. Rep. Deep Sea drill. Proj.*, 68, 377–413.

Schreiber, O. S. 1980. Benthonische Foraminiferen der Pemberger-Folge (Oberkreide) von Klein-Sankt Paul am Krappfeld (Kärnten). *Beitr. Paläontol. Oesterreich*, 7, 119–237.

Schubert, R. J. 1901. Neue und interessante Foraminiferen aus dem südtiroler Alttertiär. *Beitr. Paläontol. Geol. Oesterreich-Ungarn Orients*, 14, 9–26.

Sellier de Civrieux, J. M. 1952. Estudio de la microfauna de la seccion-tipo del miembro Socuy de la Formacion Colon, Distrito Mara, Estado Zulia. *Boletin de Geologia, Caracas*, 2, 231–310.

Senn, A. 1935. Die stratigraphische Verbreitung der tertiären Orbitoiden mit spezieller Berücksichtigung ihres Vorkommens in Nord-Venezuela und Marokko. *Eclog. geol. Helv.*, 28, 51–113, 369–373.

Senn, A. 1940. Paleogene of Barbados and its bearing on history and structure of Antillean-Caribbean region. *Bull. Am. Assoc. Petrol. Geol.*, 24, 1548–1610.

Senn, A. 1947. Die Geologie der Insel Barbados, B. W. I.

(Kleine Antillen) und die Morphogenese der umliegenden marinen Grossformen. *Eclog. geol. Helv.*, 40, 199–222.

Sigal, J. 1952. Aperçu stratigraphique sur la micropaléontologie du Crétacé. *XIXe Congr. Géol. Internat., Monogr. Rég., ser 1., Algérie*, 26, 1–45.

Skelton, R. H. 1929. Some notes on a portion of the Lizard Springs Anticline. *J. Inst. Petrol. Technol. London*, 15, 443–455.

Sliter, W. V. 1968. Upper Cretaceous foraminifera from southern California and northwestern Baja California, Mexico. *Univ. Kansas paleontol. Contrib.*, 49, 1–141.

Sliter, W. V. 1972. Cretaceous foraminifera – Depth habitats and their origin. *Nature*, 239/5374, 514–515.

Sliter, W. V. 1977. Cretaceous benthic foraminifers from the western South Atlantic, Leg 39, Deep Sea Drilling Project. *Init. Rep. Deep Sea drill. Proj.*, 39, 657–697.

Sliter, W. V. 1980. Mesozoic foraminifers and deep-sea benthic environments from Deep Sea Drilling Project Sites 415 and 416, eastern North Atlantic. *Init. Rep. Deep Sea drill. Proj.*, 50, 353–427.

Sliter, W. V. 1985. Cretaceous redeposited benthic foraminifers from Deep Sea Drilling Project Site 585 in the East Mariana Basin, western Equatorial Pacific, and implications for the geologic history of the region. *Init. Rep. Deep Sea drill. Proj.*, 89, 327–361.

Sliter, W. V. & Baker, R. A. 1972. Cretaceous bathymetric distribution of benthic foraminifers. *J. foramin. Res.*, 2, 167–183.

Smith, B. Y. 1957. Lower Tertiary foraminifera from Contra Costa County, California. *Univ. California Publ. geol. Sci.*, 32/3, p. 127–242.

Speed, R. C. 1981. Geology of Barbados: implications for an accretionary origin. *Proc. 26th Internat. Geol. Congr., Paris. Oceanol. Acta*, 1981, 259–265.

Speed, R. C. 1988. Geologic history of Barbados: a preliminary synthesis. *Trans. 11th Carib. Geol. Conf., Barbados*, 29, 1–11.

Stacy, H. E. 1966. The Lower Cretaceous microfauna from Trinidad and adjacent areas. *Unpublished doctoral thesis, University of Michigan*, 228 pp.

Stainforth, R. M. 1948. Description, correlation, and paleoecology of Tertiary Cipero Marl Formation, Trinidad, B. W. I. *Bull. Am. Assoc. Petrol. Geol.*, 32, 1292–1230.

Stanley, D. J. 1960. Stratigraphy and foraminifera of Lower Tertiary Vidoño shale, near Puerto La Cruz, Venezuela. *Bull. Am. Assoc. Petrol. Geol.*, 44, 616–627.

Subbotina, N. N. 1964. Foraminifery melovykh i paleogenovykh otlozhenii zapadno-sibirskoi nizmennosti. *Trudy VNIGRI, Leningrad*, 234, 456 pp.

Sulek, J. A. 1961. Miocene correlation in the Maturin Sub-Basin. *Bol. Informativo, Asoc. Venez. Geol. Min. y Petrol.*, 4/4, 131–139.

Sutton, F. A. 1946. Geology of Maracaibo Basin, Venezuela. *Bull. Am. Assoc. Petrol. Geol.*, 30, 1621–1741.

Takayanagi, Y. 1960. Cretaceous foraminifera from Hokkaido, Japan. *Sci. Rep. Tohoku University, Sendai, 2nd. ser. (Geology)*, 32, 1–154.

Tappan, H. 1940. Foraminifera from the Grayson Formation of Texas. *J. Paleontol.*, 14, 93–126.

Tappan, H. 1943. Foraminifera from the Duck Creek Formation of Oklahoma and Texas. *J. Paleontol.*, 17, 476–517.

Tappan, H. 1957. New Cretaceous index foraminifera from northern Alaska. *Bull. U. S. Natl. Mus.*, 215, 201–222.

Tappan, H. 1962. Foraminifera from the Arctic Slope of Alaska. Part 3, Cretaceous foraminifera. *Prof. Pap. U. S. geol. Surv.*, 236-C, 91–209.

Thomas, E. 1990. Late Cretaceous through Neogene deep-sea benthic foraminifers (Maud Rise, Weddell Sea, Antarctica). *Proc. Ocean drill. Prog., Sci. Results*, 113, 571–594.

Thomas, H. D. 1935. On some sponges and a coral of Upper Cretaceous from Toco Bay, Trinidad. *Geol. Mag.*, 72, no. 850, 175–179.

Thompson, L. B., Heine, C. J., Percival, S. F. jr. & Selznick, M. R. 1991. Stratigraphy and micropaleontology of the Campanian shelf in northeast Texas. *Spec. Publ. Micropaleontol. Press*, 5, 148 pp.

Titova, G. H. 1975. Foraminifery gruppy *Neoflabellina reticulata* maastrikhtskogo i datskogo yarusov yugo-vostoka russkoi plity. *Paleontol. Sbornik, Lvov*, 12, 27–34.

Tjalsma, R. C. & Lohmann, G. P. 1983. Paleocene – Eocene bathyal and abyssal benthic foraminifera from the Atlantic Ocean. *Micropaleontology, Spec. Publ.*, 4, 1–90.

Todd, R. 1970. Maastrichtian (Late Cretaceous) foraminifera from a deep-sea core off southwestern Africa. *Rev. Esp. Micropaleontol.*, 2, 135–154.

Torrini, R. jr. 1988. Structure and kinematics of the Oceanic nappes of Barbados. *Trans. 11th Carib. Geol. Conf., Barbados*, 15, 1–15.

Torrini, R. jr., Speed, R. C. & Mattioli, G. S. 1985. Tectonic relationships between forearc-basin strata and the accretionary complex at Bath, Barbados. *Bull. geol. Soc. Am.*, 96, 861–874.

Toulmin, L. D. 1941. Eocene smaller foraminifera from the Salt Mountain limestone of Alabama. *J. Paleontol.*, 15, 567–611.

Trechmann, C. T. 1935. Fossils from the Northern Range of Trinidad. *Geol. Mag.*, 72, no. 850, 166–175.

Trujillo, E. F. 1960. Upper Cretaceous foraminifera from near Redding, Shasta County, California. *J. Paleontol.*, 34, 290–346.

Turenko, T. V. 1983. Novye vidy foraminifer semeistva Rzehakinidae iz verchnego mela Sakhalina. *Paleontol. Zhurnal*, 1983/3, 13–21.

Vaughan, T. W. & Cole, W. S. 1941. A preliminary report on the Cretaceous and Tertiary larger foraminifera of Trinidad, B. W. I. *Spec. Pap. geol. Soc. Am.*, 30, 1–137.

Vincent, E., Gibson, J. M. & Brun, L. 1974. Paleocene and Early Eocene microfacies, benthonic foraminifera, and paleobathymetry of Deep Sea Drilling Project Sites 236 and 237, western Indian Ocean. *Init. Rep. Deep Sea drill. Proj.*, 24, 859–885.

Von der Osten, E. 1954. Geologia de la region de la Bahia de Santa Fe, Estado Sucre. *Bol. Geol., Caracas*, 3/8, 123–211.

Wall, G. P. & Sawkins, J. G. 1860. Report on the geology of Trinidad. *Geol. Surv. Mem., London*, 211 pp.

Walton, W. M. 1966. Contributions of the AVGMP Maracaibo Basin Eocene Nomenclature Committee. II. The Pauji and Mene Grande formations. *Bol. Informativo, Asoc. Venez. Geol. Min. y Petrol.*, 9/12, 325–337.

Walton, W. M. 1967. Contributions of the AVGMP Maracaibo Basin Eocene Nomenclature Committee. IV. The informal units of the subsurface Eocene. *Bol. Informativo, Asoc. Venez. Geol. y Petrol.*, 10/1, 21–30.

Waring, G. A. 1926. The geology of the island of Trinidad, B. W. I. *Johns Hopkins Univ. Studies Geol.*, 7, 1–180.

Wells, J. W. 1948. Lower Cretaceous corals from Trinidad, B. W. I. *J. Paleontol.*, 22, 608–616.

Wheeler, C. B. 1960. Estratigrafia del Oligoceno y Mioceno inferior de Falcon occidental y nororiental. *Mem. III Congr. Geol. Venez.*, 1, 407–465.

Wheeler, C. B. 1963. Oligocene and Lower Miocene stratigraphy of western and northeastern Falcon Basin, Venezuela. *Bull. Am. Assoc. Petrol. Geol.*, 47, 35–68.

White, M. P. 1928. Some index foraminifera of the Tampico Embayment area of Mexico. *J. Paleontol.*, 2, 177–215, 280–317.

White, M. P. 1929. Some index foraminifera of the Tampico Embayment area of Mexico. *J. Paleontol.*, 3, 30–58.

Whittaker, J. E., 1988. Benthic Cenozoic foraminifera from Ecuador. *London: British Museum (Natural History)*, 194 pp.

Wicher, C. A. 1949. On the age of the higher Upper Cretaceous of the Tampico Embayment area in Mexico, as an example of the worldwide existence of microfossils and the practical consequences arising from this. *Bull. Mus. Hist. nat. du Pays Serbe*, (A), 2, 76–105.

Wicher, C. A. 1956. Die Gosau-Schichten im Becken von Gams (Oesterreich) und die Foraminiferengliederung der höheren Oberkreide in der Tethys. *Paläontol. Zeitschr.*, 30, *Sonderheft*, 87–136.

Widmark, J. G. V. & Malmgren, B. 1992. Benthic foraminiferal changes across the Cretaceous-Tertiary boundary in the deep sea: DSDP Sites 525, 527, and 465. *J. foramin. Res.*, 22, 81–113.

Wood, K. Ch., Miller, K. G. & Lohmann, G. P. 1985. Middle Eocene to Oligocene benthic foraminifera from the Oceanic Formation, Barbados. *Micropaleontology*, 31, 181–197.

Yasuda, H. 1986. Cretaceous and Paleocene foraminifera from northern Hokkaido, Japan. *Sci. Rep. Tohoku University, Sendai, 2nd ser. (Geology)*, 57, 1–101.

Index

General index

Taxonomic index

The page references are given as follows:
Bold – description of taxon. This reference will also give access to figure numbers for illustrations and ranges
Roman – other references to taxa in text
Italic – additional references to charts and tables